APPLICATIONS OF MATHEMATICS IN ENGINEERING AND ECONOMICS '34

To learn more about AIP Conference Proceedings, including the Conference Proceedings Series, please visit the webpage **http://proceedings.aip.org/proceedings**

APPLICATIONS OF MATHEMATICS IN ENGINEERING AND ECONOMICS '34

Proceedings of the 34th International Conference (AMEE '08)

Sozopol, Bulgaria   8 – 14 June 2008

EDITOR

Michail D. Todorov
Technical University of Sofia
Sofia, Bulgaria

All papers have been peer reviewed

SPONSORING ORGANIZATION
Research and Development Sector of the Technical University of Sofia, Bulgaria

Melville, New York, 2008
AIP CONFERENCE PROCEEDINGS ■ VOLUME 1067

Editor:

Michail D. Todorov
Faculty of Applied Mathematics and Informatics
Technical University of Sofia
8, Kliment Ohridski Blvd.
1000 Sofia
Bulgaria

E-mail: mtod@tu-sofia.bg

L.C. Catalog Card No. 2008938271
ISBN 978-0-7354-0598-1
ISSN 0094-243X
Printed in the United States of America

CONTENTS

KEYNOTE LECTURES

SPECIAL SESSION

ANALYTICAL AND COMPUTATIONAL TECHNIQUES FOR NONLINEAR SYSTEMS AND PROCESSES

SPECIAL SESSION

TEACHING ENVIRONMENTS FOR MATHEMATICS COURSES

NUMERICAL METHODS IN MATHEMATICAL MODELING

OPERATIONS RESEARCH AND STATISTICS

SOFTWARE INNOVATIONS AND ALGEBRAIC METHODS

WORKSHOP ON GRID AND SCIENTIFIC ENGINEERING APPLICATION
(GRID AND SEA)

Preface

It is a good tradition of the Faculty of Applied Mathematics and Informatics by the Technical University of Sofia to organize an annual meeting at the Black Sea coast called International Conference "Applications of Mathematics in Engineering and Economics" (AMEE). This year it was the 34th one.

The Conference is originally aimed for the scientists from the Technical University of Sofia and prominent specialists in the field from Bulgaria and abroad. During that long period of time we preserve the main ideas of AMEE, that are:

- to make and develop contacts with colleagues from other countries;
- to exchange opinions and experiences related to the topics of the Conference;
- to get and discuss new ideas in the field of activity;
- to cooperate in writing joint papers, books, or in preparing and applying for common projects.

One can represent his results informally. After discussions supposed to be fruitfully he can prepare the manuscript and submit it to the Program Committee for consideration. Keeping the big success of previous AMEE'07 this year for the second time all contributed manuscripts as provided were reviewed from anonymous referees – prominent specialists of their field of competence and only improved after revision and accepted works in the volume are published.

The subject area of the event is rather large, so that everyone can find his favorite field. This year we had participants from 18 countries from all over the world. The main activities again were subject to the conference motto "Nonlinear Phenomena – Mathematical Theory and Environmental Reality" and were related to two special sessions, the topics: Numerical Methods in Mathematical Modeling, Analysis and Applications, Differential Equations, Operations Research and Statistics, Software Innovations and Algebraic Methods, Workshop on Grid and Scientific and Engineering Applications, Student Session, Poster Session, Conference Tutorial "Introduction to Software Agents and Their Applications" as well as a "Round Table - Presentations and Discussion - on Mathematics Education in Bachelor Degree Programs and in Master Degree Programs".

I would like to thank the organizers of the special and contributed sessions for the good arrangement and the high level of talks. We are highly keep in esteem the numerous anonymous referees who involved enthusiastically and very professionally and essentially helped us to improve the quality of publications.

An essential role for advertisement and attracting high qualified scientists played the members of International Program Committee and especially its chairman Prof. Christo I. Christov. His contribution is undoubted and highly appreciated.

We are grateful to our sponsor – Research and Development Sector by the Technical University of Sofia. Special thanks are due to the Governing Body of the Technical University of Sofia, who placed the Holiday House at the disposal of the Conference.

We are very grateful to our publisher International Conference Series by the American Institute of Physics, New York, USA for the fruitful and stimulating cooperation. It laid the conference on a qualitatively new foundation and stimulated its future development, improvement, and vogue.

Michail D. Todorov

General Chair of AMEE'08 and Volume Editor

International Program Committee

I. ANANIEVSKI
Russian Academy of Sciences, Moscow, Russia

C. I. CHRISTOV (Chairman)
University of Louisiana at Lafayette, Lafayette, USA

B. DIMITROV
Kettering University, Flint, MI, USA

A. DZHANOEV
Universidad Rey Juan Carlos, Madrid, Spain

M. FLIKOP (American Institute of Physics representative)
Melville, New York, USA

H. KOJOUHAROV
University of Texas at Arlington, Arlington, USA

R. LAZAROV
Texas A&M University, College Station, USA

A. LOSKUTOV
Moscow State University, Moscow, Russia

P. MINEV
University of Alberta, Edmonton, Canada

M. PAPRZYCKI
Polish Academy of Science, Warsaw, Poland

M. TODOROV
Technical University of Sofia, Sofia, Bulgaria

List of Participants

I. ANANIEVSKI
Institute for Problems in Mechanics
Russian Academy of Sciences
101-1 prosp. Vernadskogo
Moscow 117526
RUSSIA
Email: *anan@ipmnet.ru*

L. ANDREEV
Faculty of Applied Mathematics and Informatics
Technical University of Sofia
8 Kl.Ohridski Blvd.
1000 Sofia
BULGARIA
Email: *l_v_a@abv.bg*

I. ANGELOV
FMI at Paisii Hilendarski University of Plovdiv
Plovdiv
BULGARIA
Email: *ivan_ang@gmx.net*

A. ANTONOVA
National Aviation University/ Management and Economics Institute
Kyiv
UKRAINE
Email: *anna_antonova_08mail.ru*

E. ATANASSOV
Dept. of Parallel Algorithms
Institute for Parallel Processing
Bulgarian Academy of Sciences
Sofia
BULGARIA
Email: *emanouil@parallel.bas.bg*

S. AVRAMSKA
Faculty of Applied Mathematics and Informatics
Technical University of Sofia
8 Kliment Ohridski Blvd.
1000 Sofia
BULGARIA
Email: *s.avramska@gmail.com*

S. BAEVA
Faculty of Applied Mathematics and Informatics
Technical University of Sofia
8 Kl.Ohridski Blvd.
1000 Sofia
Bulgaria

G. BIJEV
Faculty of Applied Mathematics and Informatics
Technical University of Sofia
8 Kl.Ohridski Blvd.
1000 Sofia
BULGARIA
Email: *gbijev@tu-sofia.bg*

D. BOYADZHIEV
Dept. of Applied Mathematics and Modeling
Paisii Hilendarski University of Plovdiv
Plovdiv
BULGARIA
Email: *dtb@uni-plovdiv.bg*

B. CHESHANKOV
Faculty of Applied Mathematics and Informatics
Technical University of Sofia
8 Kl.Ohridski Blvd.
1000 Sofia
BULGARIA
Email: *chesh@tu-sofia.bg*

M. CHRISTOU
Dept. of Mathematics and Statistics
University of Cyprus
Nicosia
CYPRUS
Email: *mxc0927@yahoo.com*

C. I. CHRISTOV
Dept. of Mathematics

University of Louisiana at Lafayette
Lafayette, LA 70504-1010
USA
Email: *christov@louisiana.edu*

D. DANKOV
Institute of Mechanics, Bulgarian Academy of Sciences
Sofia
BULGARIA
Email: *dobri_dankov@abv.bg*

B. DELIJSKA
University of Forestry
Sofia
BULGARIA
Email: *delijska@mail.bg*

A. DIDENKO
Art & Science Program
The Petroleum Institute
P.O. Box 2533
Abu Dhabi
UNITED ARAB EMIRATES
Email:*adidenko@pi.ac.ae*

S. DIMITROV
Faculty of Applied Mathematics and Informatics
Technical University of Sofia
8 Kl.Ohridski Blvd.
1000 Sofia
BULGARIA
Email: *hdobl@abv.bg*

R. DJULGEROVA
Institute of Solid State Physics
Bulgarian Academy of Sciences
72 Tzarigradsko Chaussee
1784 Sofia
BULGARIA
Email: *renna@issp.bas.bg*

Tz. DONCHEV
Dept. of Mathematics
University of Architecture and Civil Engineering
Sofia
BULGARIA
Email: *tzankodd@gmail.com*

F. A. FERREIRA
Instituto Politecnico do Porto
ESEIG
Vila do Conde
PORTUGAL
Email: *fernandaamelia@eseig.ipp.pt*

Fl. FERREIRA
Instituto Politecnico do Porto
ESEIG
Vila do Conde
PORTUGAL
Email: *flavioferreira@eseig.ipp.pt*

F. FEUILLEBOIS
Laboratoire PMMH, CNRS UMR 7636
Paris
FRANCE
Email: *feuilleb@pmmh.espci.fr*

M. GANZHA
Systems Research Institute
Polish Academy of Science
Warsaw
POLAND
Email: *ganzha@ibspan.waw.pl*

G. GEORGIEV
Faculty of Mathematics and Informatics
Konstantin Preslavsky University of Shoumen
115 Universitetska str.
9712 Shoumen
BULGARIA
Email: *georgiev@fmi.shu-bg.net*

B. GILEV
Faculty of Applied Mathematics and Informatics
Technical University of Sofia
8 Kl.Ohridski Blvd.
1000 Sofia
BULGARIA
Email: *b_gilev@tu-sofia.bg*

S. GOCHEVA-ILIEVA
Dept. of Applied Mathematics and Modeling
Paisii Hilendarski University of Plovdiv
Plovdiv

BULGARIA
Email: *snegocheva@yahoo.com*

R. GORANOVA
St.Kliment Ohridski University of Sofia
Faculty of Mathematics and Informatics
Sofia
BULGARIA
Email: *radoslava@fmi.uni-sofia.bg*

G. GOTCHEV
Dept. of Computer Science
Technical University of Sofia
Sofia
BULGARIA
Email: *gotchev@tu-sofia.bg*

U. GURLER
Bilkent University
Industrial Engineering Dept.
Ankara
TURKEY
Email: *ulku@bilkent.edu.tr*

T. GUROV
Dept. of Parallel Algorithms
Institute for Parallel Processing
Bulgarian Academy of Sciences
Sofia
BULGARIA
Email: *gurov@parallel.bas.bg*

A. HACHIKYAN
Dept. of Computer Science
Technical University of Sofia
Sofia
BULGARIA
Email: *ahh@tu-sofia.bg*

S.-E. HAN
Honam University
Dept. of Computer & Applied Mathematics
59-1 Seobongdong Gwangsangu
Gwangju
REPUBLIC OF KOREA
Email: *sehan2020@yahoo.co.kr*

I. ILIEV
Dept. of Physics
Technical University of Plovdiv
25 Tzanko Djusstabanov str.
4000 Plovdiv
BULGARIA
Email: *iliev55@abv.bg*

J. KANDILAROV
University of Rousse
Dept. of Mathematics
7017 Rousse
BULGARIA
Email: *ukandilarov@ru.acad.bg*

H. KOJOUHAROV
Dept. of Mathematics
University of Texas at Arlington
Arlington
USA
Email: *hristo@uta.edu*

M. KOLEVA
Rousse University
Rousse
BULGARIA
Email: *mkoleva@ru.acad.bg*

M. KONSTANTINOV
Dept. of Mathematics
University of Architecture, Civil Engineering and Geodesy
1 Hr. Smirnenski Blvd.
1421 Sofia
BULGARIA
Email: *mmk_fte@uasg.bg*

S. MARGENOV
Institute for Parallel Processing
Bulgarian Academy of Sciences
Sofia
BULGARIA
Email: *margenov@parallel.bas.bg*

B. MAROVIC
University of Belgrade
SERBIA

M. MCCALL
Dept. of Physics
Imperial College of London
London
UNITED KINGDOM
Email: *m.mccall@imperial.ac.uk*

T. MELTON
Louisiana State University at Alexandria
Dept. of Mathematics and Physical Sciences
Alexandria, LA 71302-9121
USA
Email: *tmelton@lsua.edu*

L. MONIER
IRMAR, Rennes
FRANCE
Email: *laurent.monier@insa-rennes.fr*

C. NEDEVA
Faculty of Applied Mathematics and Informatics
Technical University of Sofia
8 Kl.Ohridski Blvd.
1000 Sofia
BULGARIA
Email: *cnedeva@tu-sofia.bg*

A.OVSEEVICH
Institute for Problems in Mechanics
Russian Academy of Sciences
101-1 prosp. Vernadskogo
Moscow 119526
RUSSIA
Email: *ovseev@ipmnet.ru*

N. PAPANICOLAOU
Dept. of Computer Science
University of Nicosia
46 Makedonitissas Ave.
P.O.Box 24005
1700 Nicosia
CYPRUS
Email: *papanicolaou.n@unic.ac.cy*

M. PAPRZYCKI
Computer Science Institute

at the Warsaw School of Social Psychology
Warsaw
POLAND
Email: *marcin.paprzycki@swps.edu.pl*

V. PASHEVA
Faculty of Applied Mathematics and Informatics
Technical University of Sofia
8 Kl.Ohridski Blvd.
1000 Sofia
BULGARIA
Email: *vvp@tu-sofia.bg*

K. PEEVA
Faculty of Applied Mathematics and Informatics
Technical University of Sofia
8 Kl.Ohridski Blvd.
1000 Sofia
BULGARIA
Email: *kgp@tu-sofia.bg*

Z. PETROVA
Faculty of Applied Mathematics and Informatics
Technical University of Sofia
8 Kl.Ohridski Blvd.
1000 Sofia
BULGARIA
Email: *zap@tu-sofia.bg*

J. PONTES
Metallurgy and Materials Engineering Department
Federal University of Rio de Janeiro
Rio de Janeiro
BRAZIL
Email: *jopontes@metalmat.ufrj.br*

A. PORUBOV
Institute for Problems in Mechanical Engineering
Russian Academy of Sciences
V.O. 61 Bolshoy av.
Saint-Petersburg 199178
RUSSIA
Email: *porubov.math@mail.ioffe.ru*

K. PRODANOVA
Faculty of Applied Mathematics and Informatics
Technical University of Sofia

8 Kl.Ohridski Blvd.
1000 Sofia
BULGARIA
Email: *kprod@tu-sofia.bg*

M. RAEVA
Faculty of Applied Mathematics and Informatics
Technical University of Sofia
8 Kl.Ohridski Blvd.
1000 Sofia
BULGARIA
Email: *mraeva@tu-sofia.bg*

M. RIEDEL
Institute for Advanced Simulation
Juelich Supercomputing Centre
Forschungszentrum Juelich
Juelich
GERMANY
Email: *m.riedel@fz-juelich.de*

A. SALUPERE
Institute of Cybernetics at Tallinn University of Technology
Tallinn
ESTONIA
Email: *salupere@ioc.ee*

M. SANJUAN
Grupo de Dinamica No Lineal y Teoria del Caos
Departamento de Fisica
Escuela Superior de Ciencias Experimentales y Tecnologia
Universidad Rey Juan Carlos
Tulipán s/n, 28933 Mostoles
Madrid
SPAIN
Email: *Miguel.sanjuan@urjc.es*

T. STANCHEVA
Faculty of Applied Mathematics and Informatics
Technical University of Sofia
8 Kl.Ohridski Blvd.
1000 Sofia
BULGARIA
Email: *tms@tu-sofia.bg*

S. STOICHEV
Computer Systems Dept.

Technical University of Sofia
8 Kl.Ohridski Blvd.
1000 Sofia
BULGARIA
Email: *stoi@tu-sofia.bg*

T. STOYANOV
Dept. of Mathematics
Economic University
77 Knyaz Boris I Blvd.
9002 Varna
BULGARIA
Email: *todorstoy@mail.bg*

S. TABAKOVA
Dept. of Mechanics
Technical University of Sofia, branch Plovdiv
Plovdiv
BULGARIA
Emails: *stabakova@hotmail.com, sonia@tu-plovdiv.bg*

M. TODOROV
Faculty of Applied Mathematics and Informatics
Technical University of Sofia
8 Kl.Ohridski Blvd.
1000 Sofia
BULGARIA
Email: *mtod@tu-sofia.bg*

L. TOPCHIJSKA
Faculty of Applied Mathematics and Informatics
Technical University of Sofia
8 Kl.Ohridski Blvd.
1000 Sofia
BULGARIA
Email: *lgt@tu-sofia.bg*

D. VAKOVSKY
Faculty of Applied Mathematics and Informatics
Technical University of Sofia
8 Kl.Ohridski Blvd.
1000 Sofia
BULGARIA
Email: *vakovsky@tu-sofia.bg*

E. VARBANOVA
Faculty of Applied Mathematics and Informatics

Technical University of Sofia
8 Kl.Ohridski Blvd.
1000 Sofia
BULGARIA
Email: *elvar@tu-sofia.bg*

A. VATSALA
Dept. of Mathematics
University of Louisiana at Lafayette
P.O. Box 41010
Lafayette, LA 70504-10101
USA
Email: *asv5357@louisiana.edu*

P. VEL'MISOV
Ulyanovsk State Technical University
Ulyanovsk
RUSSIA
Email: *velmisov@ulstu.ru*

S. VEL'MISOVA
Ulyanovsk State University
Ulyanovsk
RUSSIA
Email: *velmisova@mail.ru*

G. VENKOV
Faculty of Applied Mathematics and Informatics
Technical University of Sofia
8 Kliment Ohridski Blvd.
1000 Sofia
BULGARIA
Email: *gvenkov@tu-sofia.bg*

P. VOLKOV
Ugra State University
Faculty of Geology, Oil and Gas
Khanty-Mansiysk
RUSSIA
Email: *volkovpk@mail.ru*

L. VULKOV
Dept. of Mathematics, FNSE
University of Rousse
8 Studentska str.
7017 Rousse
BULGARIA

Email: *vulkov@ami.ru.acad.bg*

S. XIANG
Central South University
Applied Mathematics and Software
Changsha
CHINA
Email: *xiangsh@mail.csu.edu.cn*

Z. ZAHARIEV
Faculty of Applied Mathematics and Informatics
Technical University of Sofia
8 Kl.Ohridski Blvd.
1000 Sofia
BULGARIA
Email: *zlatko@tu-sofia.bg*

Y. ZHOU
Dept. of Mathematics
Guangdong Ocean University
Zhanjiang
CHINA
Email: *zhouyongxiong@126.com*

A. ZHUROV
Institute for Problems in Mechanics
Russian Academy of Sciences
101-1 prosp. Vernadskogo
Moscow 1195268
RUSSIA
Email: *zhurov@ipmnet.ru*

xxix

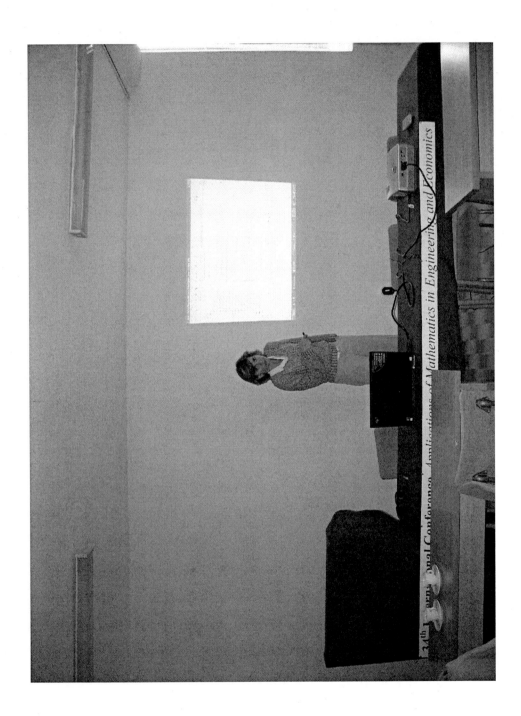

Problem Formulation: Initial Conditions

If we assume that, for each of the functions ϕ, w the initial condition is of the form of a simple propagating soliton, then we consider linearly polarized solitons. In this case

$$\begin{Bmatrix} \phi(x,t) \\ w(x,t) \end{Bmatrix} = \begin{Bmatrix} A^\phi \\ A^w \end{Bmatrix} \operatorname{sech}[b(x-X)-vt] \exp\left\{ i\left[\frac{c}{2\beta}(x-X)-nt \right] \right\}$$ (2)

$$\beta = \frac{1}{\delta}\left(n + \frac{c^2}{4\beta}\right), \quad A = b\sqrt{\frac{2\beta}{\alpha_1}}, \quad v_s = \frac{2n\beta}{c}.$$

where X is the spatial position (center of soliton), c is the phase speed, n is the carrier frequency, and b^{-1} – a measure of the support of the localized wave.

Let now assume that for each of the functions ϕ, w the initial

KEYNOTE LECTURES

Space as a Material Continuum and the Cosmological Redshift of Spectral Lines

C. I. Christov

Dept. of Mathematics, University of Louisiana at Lafayette, LA 70504-1010, USA

Abstract. The idea that space is a material continuum with standard mechanical properties such as elasticity and viscosity is applied to the propagation of shear waves (electromagnetic waves). The viscosity is shown to introduce dispersive dissipation in the governing system which affects differently the propagation of waves of different frequencies. This results in redshifting of a spectral line without the need to assume Doppler dilation of each of the frequencies constituting the packet. It is shown that the wider spectral lines experience larger redshift than the thinner one which is in very good agreement with the observation of the high-redshift objects, such as Quasi-Stellar Objects (QSOs or Quasars). The data for the spectral lines of the QSO 3C273 is analyzed in detail from the point of view of the proposed model. The conclusion is the the mere dissipation is not enough for good quantitative explanation of the redshift, and that dispersive effects should also be taken into account.

Keywords: Cosmological Redshift, metacontinuum, dissipation, dispersion, Gaussian apodization function.

PACS: 42.25.Bs, 95.75.Pq, 98.54.Aj

INTRODUCTION

Modern astrophysics is based on the interpretation of spectroscopy experiments in which different spectral lines are identified. A spectral line is actually a wave packet containing waves with close frequencies to the central frequency. The distribution of the power as function of the frequency is called the apodization function.

One of the most important discoveries made originally by Hubble shows that the spectral lines of distant light sources (especially the emission lines) are predominantly shifted to the lower frequencies. This effect is called the *cosmological redshift*. Doppler effect is the most natural explanation for the red or blue shifts of a monochromatic wave. In Doppler effect, shifting to the red means that the sources move away from the observer. This gave rise to the theory of inflating universe, that culminated in the so-called Big-Bang model.

Yet, until now nobody has actually measured a single frequency: the very experimental concept deals *by definition* with spectral lines because atoms never emit light as a single frequency. The distribution of energy as function of frequency (the apodization function) is usually sharply localized, but is never a Dirac delta function.

However, the theorists treated the measurements as if the apodization functions of the spectral lines are Dirac deltas. Then, the only explanation of the redshift is the "dilation" of the carrier wave. The reality, however, is rather different and the spectral lines are *always* of finite width. It is well established (see [1]) that the spectral lines of sources with larger redshifts are several times wider than the spectral lines of some

CP1067, *Applications of Mathematics in Engineering and Economics '34—AMEE '08,* edited by M. D. Todorov
© 2008 American Institute of Physics 978-0-7354-0598-1/08/$23.00

nearby benchmark stars like the Sun, Vega and others. This suggests that the width of the spectral line can have an impact on its shifting to the red. The width of a spectral line is the result of different *thickening* mechanisms. If the cause is the Doppler shift due to the relative motion of the emitting atoms, then the distribution of the wave length around the main frequency is given by the Gaussian distribution function.

If one is to consider spectral lines of finite width, one must investigate the propagation of a wave packet, rather than the Doppler effect on a wave of a single monochromatic frequency. There can be different causes for shifting of a packet. In [2], we have showed that when dissipation is present, a wave packet of finite width behaves rather differently than a single monochromatic frequency. During a long evolution, the wave number (or frequency) of the maximal amplitude of the packet ('central wave number') can change its position making the packet appear shifted in the frequency space in comparison of its profile in the initial instant of time.

Lossless propagation of light is a useful abstraction but even the smallest amount of interstellar dust will cause dissipation and/or dispersion. A term with a small parameter in the governing equations can safely be neglected locally but it may have a cumulative effect at large distances and long times. Dispersion due to dissipation means that different wave modes are damped differently and after some time (or distance) the constitution of the wave packet can change drastically in comparison with the initial one in the context of spectroscopy measurements.

Present work deals with the effect of dissipation on the evolution of wave packets.

MATERIAL SPACE AS A VISCOELASTIC CONTINUUM

In recent works [3, 4, 5], the present author proposed that space is an absolute material continuum (called the *metacontinumm*). He showed that Maxwell equations and the Lorentz force can be derived as corollaries of the governing equation of an elastic continuum.

In the limit of small displacements/velocities of the points of the metacontinuum, we can assume the Kelvin-Voigt model of elastic medium, which stipulates that the stress tensor is a linear combination of the strain tensor and the rate of deformation tensor. For this kind of viscoelastic constitutive relation, the so-called Navier equations (see, e.g., [6, p.117] for the displacement vector $u(x,t)$ can be generalized to incorporate the viscosity, namely

$$\mu \frac{\partial^2 u}{\partial t^2} = (\lambda + \eta)\nabla(\nabla \cdot u) + \eta \nabla^2 u + (\zeta + \mu\delta)\nabla(\nabla \cdot v) + \mu\delta\nabla^2 v$$

$$= -\nabla\phi - \eta\nabla \times \nabla \times u - \mu\delta\nabla \times \nabla \times \frac{\partial u}{\partial t}, \quad (1)$$

where λ and η are the Lamé coefficients (see, e.g., [7, 6]). Respectively, ζ is the bulk coefficient of viscosity. The shear viscosity coefficient is written as $\mu\delta$, where δ is the kinematic shear viscosity coefficient, and μ is the density of the metacontinuum identified as magnetic permeability in [3, 4, 5] . Thus we avoid the abuse of the notation μ usually used for the dynamic shear coefficient of viscosity. The notation ϕ stands for

4

the pressure that must arise in a continuum to enforce the incompressibility condition. Note that the second line of equation (1) is valid for incompressible metacontinuum for which $\nabla \cdot \mathbf{v} = 0$ and $\nabla \cdot \mathbf{u} = const$.

One can use 'nabla' operator ∇ because the linearized equations are the same as if they are written in the current description. For a judicious distinction between the referential and current descriptions of the metacontinuum see [5].

Since the light is a shear wave, we can consider the case when we have only one component $u(x,t) = u_y(x,t)$ of the displacement vector (say y or z- transverse component) which is a function of the longitudinal coordinate x and time t. For propagation from distant sources, the fact that the one is faced actually with spherical waves can be easily accounted for by a scaling factor proportional to the inverse distance to the emitting source. Then equations (1) reduce to the following scalar equation for $u(x,t)$:

$$\frac{\partial^2 u}{\partial t^2} - \delta \frac{\partial}{\partial t} \frac{\partial^2 u}{\partial x^2} = c^2 \frac{\partial^2 u}{\partial x^2}, \tag{2}$$

where $c^2 = \eta \mu^{-1}$ is the speed of the shear waves (called 'light' when they are in the visible spectrum), and δ is the above introduced kinematic coefficient of shear viscosity.

The last equation is often called Jeffrey's equation. Equation (2) was derived by Alfven [8] for the magnetic field in the interstellar plasma when a rarefied gas of moving electrons was stipulated to exist in space. This means that even if one can neglect the viscosity of the metacontinuum, a dissipation can be present in the interstellar space because of the small impurities.

PROPAGATION OF WAVE PACKETS IN PRESENCE OF DISSIPATION

The viscosity of the metacontinuum (parameterized by δ) introduces a dissipation of dispersive type. It does not just bring about the decay of a wave in time or space (attenuation), but also acts to change the dispersion relation and causes the original wave packet to disperse and eventually dissipate (disappear as an entity).

Following [2], we consider here the propagation of harmonic waves $e^{ikx+i\hat{\omega}t}$. We will focus on the case when a wave of given spatial wave number k is excited in the initial moment of time, and then left to evolve in accordance with equation (2). The case when the wave is excited at one of the spatial boundaries as function of time can be treated in a similar fashion and is shown in the end of this section. The harmonic waves satisfy the following dispersion relation

$$-\hat{\omega}^2 + i\delta\hat{\omega}k^2 = -c^2 k^2. \tag{3}$$

For a given real wave number k, the complex frequency $\hat{\omega}$ is given by

$$\hat{\omega} = \omega + is, \qquad s = \frac{\delta k^2}{2}, \qquad \omega = \sqrt{c^2 k^2 - \frac{\delta^2 k^4}{4}}, \tag{4}$$

where ω is the actual (real-valued) frequency of the wave, and s gives the decay of the wave with time. It is interesting to note that the traveling waves exist only for $k < 2c/\delta$, which condition is easily satisfied since the dissipation is supposed to be very small, while the speed of light is a large parameter.

As already mentioned, the Doppler widening of the spectral lines usually results in a Gaussian apodization function, described by

$$E(k) = e^{-\frac{\beta}{2}(k/\tilde{k}-1)^2},$$ (5)

where $\tilde{k}$ is the central wave number of the packet.

A larger β means a sharper spectral line in terms of k. Let $k = q$ be the wave number k for which the amplitude is exactly $e^{-\pi/4}$ times smaller than the maximal amplitude. Then

$$d = \sqrt{\frac{\pi}{2\beta}} \quad \text{or} \quad \beta = \frac{\pi}{2d^2}, \quad \text{where} \quad d = \frac{q-\tilde{k}}{\tilde{k}},$$ (6)

and $d\tilde{k}$ gives the effective half-width of the packet.

The actual wave as function of the spatial variable in the initial moment of time is given by the following Fourier integral

$$u_0(x) = \int\limits_{-\infty}^{\infty} e^{-\frac{\beta}{2}(k/\tilde{k}-1)^2} e^{ikx} dk.$$ (7)

Then, the evolution of Fourier density of the wave packet according to equation (2) is given by the following simple relation

$$e^{i\omega t + ikx} e^{-\frac{1}{2}\beta(k-\tilde{k})^2} e^{-\frac{1}{2}\delta k^2 t},$$ (8)

for the amplitude of a component with specific wave number k. The first term in equation (8) corresponds to a the traveling wave with phase speed

$$\frac{\omega}{k} = c\sqrt{1 - \frac{\delta^2 k^2}{4c^2}},$$ (9)

which means that the shorter waves travel slower than the longer waves, and are damped faster. After a time $t = T$ has passed, the amplitude in Fourier space of the wave of wave number k (the spectrum) is given by

$$E^T(k) = e^{-\frac{1}{2}\beta(k-\tilde{k})^2} e^{-\frac{1}{2}\delta k^2 T}.$$ (10)

SHIFTING OF CENTRAL WAVE NUMBER OF A LOCALIZED PACKET

According to equation (10), after time $t = T$ has passed, the maximum of $E^T(k)$ is not longer at $k = \tilde{k}$. Its position is defined by

$$\frac{dE^T(k)}{dk} = \left[-\frac{\beta}{\tilde{k}}\left(\frac{k}{\tilde{k}}-1\right) - \delta T k \right] e^{-\frac{\beta}{2}(k/\tilde{k}-1)^2} e^{-\frac{\delta k^2}{2}T} = 0,$$

6

which is a linear equation in k and its root is

$$\bar{k} = \frac{\beta \tilde{k}}{\beta + \delta \tilde{k}^2 T} = \frac{\tilde{k}}{1 + \delta \beta^{-1} \tilde{k}^2 T}, \tag{11}$$

For the shift of the central wavenumber one gets

$$\Delta k \overset{\text{def}}{=} \bar{k} - \tilde{k} = -\frac{\tilde{k} \delta \beta^{-1} \tilde{k}^2 T}{1 + \delta \beta^{-1} \tilde{k}^2 T} \quad \text{or} \quad \frac{\Delta k}{\tilde{k}} = -\frac{\delta \beta^{-1} \tilde{k}^2 T}{1 + \delta \beta^{-1} \tilde{k}^2 T} \tag{12}$$

which means that at time T, the new central wave number $\bar{k}$ of a packet is shifted to the smaller wavenumbers.

In terms of the wavelength we have

$$\lambda = \frac{2\pi}{k}, \quad k = \frac{2c\pi}{\lambda}, \quad \Delta k = -2c\pi \frac{\Delta \lambda}{\bar{\lambda} \tilde{\lambda}} \quad \Rightarrow \quad \frac{\Delta k}{\tilde{k}} = -\frac{\Delta \lambda}{\tilde{\lambda}}.$$

Then equation (12) can be recast as the following "Time-Hubble Law"

$$\frac{\Delta \lambda}{\tilde{\lambda}} = \delta \beta^{-1} \frac{4c^2 \pi^2}{\tilde{\lambda}^2} T \overset{\text{def}}{=} z, \quad \text{where} \quad \Delta \lambda = \lambda - \tilde{\lambda}. \tag{13}$$

The central wave length of the packet is shifted to longer waves (*redshifted*).

Equation (9) gives that $c_g = \omega(\tilde{k})/\tilde{k}$. Then, the propagation time is $T \approx c_g^{-1} L$, where L is the length traveled by the packet. Thus,

$$\frac{\Delta \lambda}{\tilde{\lambda}} = \frac{4c^2 \pi^2 \delta}{c_g \beta \tilde{\lambda}^2} L = HL, \quad \text{where} \quad H = \frac{4c^2 \pi^2 \delta}{\beta c_g \tilde{\lambda}^2}, \tag{14}$$

and can be called "Distance-Hubble law."

The most important difference between the evolution of a wave packet (spectral line) and the redshift of a single monochromatic wave, is the fact that the redshift depends not merely on the distance travelled (time in transit), but also on the width of the initial spectral line according to equation (13). It depends also on the central wave length (central wave number) of the packet. This calls for reinterpretation of the spectral-lines observations, and reassessing the the actual redshifts.

In the view of this theory, the distance is not in one-to-one correspondence with the redshift, because of the dependence on the width of the packet, even if we choose the dissipation coefficient δ to be a universal constant. This means that two sources that are in a close proximity in cosmological sense, can have different redshifts depending on the parameter β for the spectral line under consideration. A more active (hotter) source (smaller β) will appear to the observer as much more redshifted than a quieter (cooler) source (larger β). This effect is observed experimentally, and is recently the point of a raging controversy (see, among others, [9, 10] and the literature cited therein).

Note that the Hubble constant is also a function of the central wave number $\tilde{k}$ through the possible dependence $\beta = \beta(\tilde{k})$. The latter means that parameter β can have a different value depending on the central wave number despite that in absolute units two widths of different spectral lines may appear equal. Yet, this is a smaller effect, and should be taken into account only if the respective evidence is present.

Finally, we mention that equation (10) can be rewritten as follows

$$E^T(k) = E^T_{max} e^{-\frac{(\beta+\delta T)}{2}(k/\bar{k}-1)^2}, \quad E^T_{max} = e^{-\frac{\delta T \bar{k}^2}{2(1+\delta\beta^{-1}\bar{k}^2 T)}}, \quad (15)$$

which is once again a Gaussian distribution with somewhat larger effective β, namely

$$\beta^T = \beta + \delta T = \beta(1 + \delta\beta^{-1}T) = \beta(1+z), \quad (16)$$

where z is the redshift.

Before proceeding with the application to Cosmology, we reformulate the above theory for the case of spatial propagation. Consider a wave that is excited at distance L from the observer, i.e. when the wave is created from a boundary condition at $x = 0$ and the observer is situated at $x = L$. Once again we are interested in superposition of harmonic waves of type $e^{i\omega t}$ with amplitudes with a Gaussian apodization function

$$E(\omega) = e^{-\frac{\beta}{2}(\omega/\bar{\omega}-1)^2}. \quad (17)$$

Note that in [2] a factor c^2 was introduced which results in insignificant differences of the actual outlook of the formulas. We can rewrite the dispersion equation as

$$k = \frac{\omega}{\sqrt{c^2 + i\delta\omega}}. \quad (18)$$

The sign is selected to have waves that decay at $x \to \infty$. The dispersion relation can be simplified if we also assume $\delta\omega \ll c^2$ and then

$$\Im[k] \approx -\frac{1}{2}\frac{\delta\omega^2}{c^3}, \quad \Re[k] = \frac{\omega}{c}\left[1 - \frac{3}{8}\frac{\delta^2\omega^2}{c^4}\right]. \quad (19)$$

A packet produced at $x = 0$ with a Gaussian distribution will arrive at $x = L$ as follows

$$e^{-\frac{1}{2}\beta\bar{\omega}^{-2}(\omega/\bar{\omega}-1)^2}e^{-\frac{1}{2}(\delta\omega^2)c^{-3}L} = e^{-\delta\bar{\omega}\bar{\omega}c^{-3}L}e^{-\frac{1}{2}\bar{\beta}(\omega/\bar{\omega}-1)^2}, \quad (20)$$

which gives for a new central frequency of the packet:

$$\bar{\omega} = \frac{\tilde{\omega}}{1+\delta\beta^{-1}c^{-3}\tilde{\omega}^2 L}, \quad \Rightarrow \quad \frac{\Delta\omega}{\bar{\omega}} = \frac{\tilde{\omega}-\bar{\omega}}{\bar{\omega}} = \frac{\delta\tilde{\omega}^2 L}{\beta c^3} = HL = z. \quad (21)$$

Respectively, the inverse thickness of the spectral line at the site of measurement is

$$\bar{\beta} = \beta\left(1 + \frac{\delta\tilde{\omega}^2 L}{c^3}\right) = \beta(1+z). \quad (22)$$

Now the Distance-Hubble law is linear with Hubble constant $H = \delta\beta^{-1}c^{-3}\tilde{\omega}^2$. The most important feature of the above expression is the explicit dependence on the central frequency.

Apart from the shifting of the central frequency to the red, the term $e^{-\delta\bar{\omega}\bar{\omega}c^{-3}L}$ in equation (20) gives that the amplitude of the wave will also diminish, i.e., there is an attenuation that depends on the frequency. The presence of an attenuation makes the stars and galaxies appear dimmer and thus the estimated distance to be exaggerated. The attenuation can provide an alternative approach to the apparent missing mass in the Universe, without assuming the presence of dark matter (see also [4]).

Since the attenuation is frequency dependent, it will result into a different distribution of the intensity of the different colors (the so-called continuum of the spectrum). This means that the spectrum at the site of the light source will be different from the observed if the above described attenuation is taken into account.

APPLICATION TO COSMOLOGY

Many stars and galaxies emit visible electromagnetic waves predominantly in the hydrogen spectrum, called H_α, H_β, and H_γ lines (or Balmer series), because the abundance of hydrogen in the universe. Only first three Balmer lines are in the visible spectrum. Many high-temperature objects emit also in so-called Lyman series in the ultraviolet spectrum. Usually, the L_α line is the only ultraviolet hydrogen spectral line that is reliably measured. When the redshift is very large, the original L_α line can even appear in the visible spectrum.

The hydrogen lines can be broadened by the Doppler effect connected with the thermal motion of the atoms during the emission of the photons. According to the approach proposed in the present work, the magnitude of the redshift depends on the effective thickness of the respective line. For this reason, we outline in this section the way the experimental data for the spectral lines has to be treated in order to estimate the thickness of the line at the source.

Characterization of a Spectral Line

In most of the cases, the spectral line is presented as an experimentally measured apodization function of the wave length λ. After fitting, in the least-square sense, a Gaussian function, one can have the following reasonable approximation for the spectral line:

$$e^{-\frac{\beta}{2}(\lambda/\tilde{\lambda}-1)^2},\qquad(23)$$

where $\tilde{\lambda}$ is called 'central wave length' of the wave packet that represents the particular spectral line. If the process of widening of the lines is similar, the constant β is expected to be almost the same for that set of spectral lines. For instance, it is quite reasonable to expect that all of the hydrogen lines will have similar β's, because of the Doppler broadening due to the mean-square velocity of the hydrogen atoms. It is interesting to investigate this conjecture on larger selection of stars and QSO's, but it goes beyond the scope of the present paper.

The Gaussian apodization function of λ, equation (23) does not necessarily entail that the distribution for $k = 2\pi/\lambda$ will also be Gaussian, but for $|\lambda - \tilde{\lambda}| \ll \tilde{\lambda}$ both

9

representations of a spectral line are virtually Gaussian. Thus, for $\tilde{\lambda} \ll 1$ ($\tilde{k} \gg 1$), and for very narrow packet $\beta \gg 1$, the Gaussian shape, equation (5), is virtually equivalent to equation (23), namely for the spectral line under consideration

$$\frac{\beta}{2}\left(\frac{\lambda}{\tilde{\lambda}}-1\right)^2 = \frac{\beta}{2}\left(\frac{\tilde{\omega}}{\omega}-1\right)^2 = \frac{\beta}{2}\left(\frac{\omega}{\tilde{\omega}}-1\right)^2 + \beta \cdot O\left(\frac{\omega}{\tilde{\omega}}-1\right)^3$$

when $(\omega/\tilde{\omega}-1) \ll 1$. This means that if we consider the characterization of a spectral line as

$$e^{-\frac{\beta}{2}(\omega/\tilde{\omega}-1)^2}, \tag{24}$$

then β is the same number as in equation (23)

The above formulas give us the opportunity to either fit a best approximation to the experimental data (as measured in terms of wave length λ) or to first render the spectrum as a function of the frequency and then find the best-fit Gaussian approximation. Up to terms of third asymptotic order of smallness, the two approaches give the same results.

The dependence of the redshift on the central frequency of the spectral line has another important impact on the interpretation of the observations. The superposition of different spectral lines emitted by a source give the background (called 'continuum') against which the most prominent lines can be discerned. Very often, two central frequencies of two different lines can appear very enough in the measurements. In such a case, one tries to find the possible physical mechanism responsible for the unusually looking line. The hydrogen lines (Balmer or Lyman or even Paschen lines) are supposed to always be present because the fuel of most of the stars is predominantly hydrogen. If the redshift can be a function of the width of the spectral line, then upon arrival at the site of the detecting device, different lines can appear in a juxtaposition, while they are, actually, quite different at the source. This is illustrated in Figure 1. One of the lines is the H_β

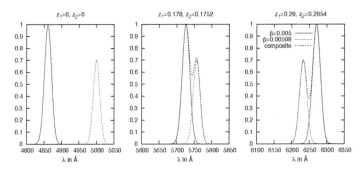

Figure 1. The comparison of two different emitted lines with their position at arrival

line which has been used by Schmidt for defining the redshift of 3C273. The other is a hypothetical line with larger initial central wavelength but with smaller initial width.

In the end of the section, we reiterate that the dependence of the redshift on the width of the spectral line, allows to have objects of different redshift situated in the vicinity of each other. The observational evidence for this is discussed in [9].

Featuring Example: QSO 3C273

The Quasi-Stellar Object (QSO or 'Quasar') #273 (called 3C273 for brevity) from the third Cambridge catalog was one of the first QSOs to be thoroughly investigated because it is the brightest QSO ever observed (see [1]). In fact, the peculiar positions of the brightest visible spectral lines of 3C273 were identified by M. Schmidt [11] as the hydrogen Balmer lines shifted with $z \approx 0.158$ and the first ever object with significant redsifit was discovered. With the advance of the spectrophotometers, much fainter objects with redshifts as large as $z > 5$ have recently been discovered. Yet, the luminosity of 3C237 and the intensity of its spectral lines makes it the most desirable benchmark for any new redshift theory. We have been able to find only three independent spectroscopy measurements of the visible and near infrared spectrum of 3C273 [12, 13, 14], and one detailed study of the ultraviolet spectrum [15]. Verifying the finding about the spectral lines is of importance of our work, because in the present approach we can make the subtle distinction between the redshifts of different spectral lines as functions of the width of the line.

Figure 2 shows the comparison between the three known experimental spectroscopic studies of 3C273. In order to make the comparison presented in Figure 2, we digitalized the respective graphs from the source papers. Despite the seeming disagreement (most

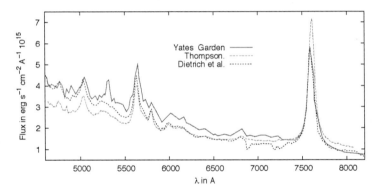

Figure 2. Comparison of three different experimental investigations of the line spectra of 3C273

probably due to some renormalizations), an important conclusion that can be reached from Figure 2, namely that the spectral lines are almost in the same positions (the redshifts of the same lines are the same) and that their widths are in good agreement with each other across the set of independent experiments. This means that the results are reliable enough for the purpose of putting the present theory to the test.

The result in Figure 2 give the first and foremost qualitative confirmation of the proposed here model: the spectral lines of the highly redshifted object, 3C273, are wider than the spectral lines of the ordinary stars. According to [1, Pg.67] this fact holds true for all observed QSOs.

Best-Fits for the major Spectral Lines of 3C273. Limitations of the Model with Dissipation only

Since the width of a spectral line plays a major role in the above presented model of the cosmological redshift, it is important to fit Gaussian function to the available major spectral lines of a QSO, in order to get an estimate for the parameter β. Figure 3 presents the best-fits of type of equation (24) for five major spectral lines of 3C273 which are believed to be (from left to right in the figure): Paschen P_α, Balmer H_α, H_β, H_γ, and Lyman L_α lines in the spectrum of 3C273. It is seen that Gaussian functions fits

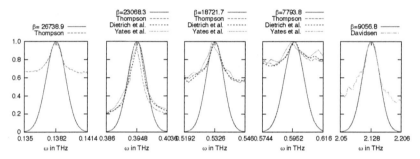

Figure 3. The best fits to the spectral lines as functions of the frequency $\exp[-\frac{B}{2}(\omega - \bar{\omega})^2]$. Left to right are the supposed P_α, H_α, H_β, H_γ, and L_α lines

reasonably well the measurements, but the three experimental results do not agree very well between each other. The disagreement is not consistent, and different combinations of measurements agree for different spectral lines.

Here it should be noted that the results presented in Figure 3 are obtained after the graphs from the respective articles are digitalized. The eventual distortion of the scales during the reproduction of the graphs in the original appears, or during the digitalizing process, could also be a cause for disagreement.

The differences in estimating β from wave-length data and from frequency data show that the procedure is very sensitive. This calls for more experimental data with denser set of frequencies. Yet, the best-fit approximations clearly demonstrate that coefficient β is roughly of the same order for the different spectral line under consideration. This means that the redshift is proportional to the square of the of the original central frequency as shown in equation (21). This prediction is not supported by the observation, since all of the hydrogen lines appear with virtually the same redshift.

There are two possible explanation for this disagreement. The first is that what we identify as redshifted Balmer lines are actually some other lines that are in quite different positions originally, and only appear as a set of equally shifted Balmer lines. This explanation does not seem very likely, because involves events with very low probability: to have some disorganized array of spectral lines, each shifted with very special amount to appear as Balmer lines. The second explanation is that the model that involves solely dissipation cannot quantitatively predict the redshift. This means that along with the dissipation, another mechanism is at play in the shifting of the central frequency of a packet. This additional mechanism can be dispersion.

The Role of Dispersion

As it can be seen from equation (19), the dissipation has a dispersive effect because it changes the dependence of the spatial wave number on the frequency, i.e., the phase speed of waves with different frequencies is different. Unfortunately the dissipative dispersion does not change the redshift, but rather acts to simply increase the width of the spectral line. The only way to have a sufficiently strong dispersion is to consider dispersive terms in the equation, i.e., instead of the original equation, one has to consider the following, more general, equation

$$u_{tt} = c^2 u_{xx} - u_{txx} + \chi_1 u_{ttxx} - \chi_2 u_{xxxx}, \tag{25}$$

where $\chi_1, \chi_2 > 0$ are the two dispersion coefficients. The fourth-order spatial derivative was shown (see [5, 16] for the details) to be important for eliminating singularities when considering localized torsional waves (identified as the charges) in metacontinuum. The mixed fourth derivative is called rotational inertia and it stems from the same micro-polar models, so-called Cosserat continua, (see [17]) as the spatial fourth derivatives. The dispersion relation for he model described by equation (25) reads

$$c^2 k^2 + i\delta\omega k^2 - \chi_1 \omega^2 k^2 + \chi_2 k^4 - \omega^2 = 0. \tag{26}$$

The last equation presents a radically different relationship between ω and k which has to be investigated in depth in order to understand its quantitative effects on the redshift. This will be done in a consecutive work.

CONCLUSIONS

The present paper is devoted to the possible cosmological application of a model of space as a dissipative material continuum. The spectral lines in the spectra of distant sources (Quasi-Stellar Objects or QSOs) are considered as wave packets, and the evolution of the packets under the influence of dissipation is investigated. Only wave packets that are well-approximated by Gaussian apodization function are considered. It is shown that the redshift is given by a Hubble-type law in which the change of the frequency/wave number is proportional to the distance traveled by the wave (or the time in transit).

Physically speaking, the effect is related to the fact that dissipation damps the higher frequencies stronger and causes the maximal frequency of the packet to shift to lower frequencies (longer wave lengths). The redshifting is a property of the packet. No actual *dilation* of the different harmonics is needed as in Doppler effect. This means that even a slightest dissipation in the interstellar medium will result in a persistent (cosmological) redshift of the light propagating throughout the Universe.

The important trait of the new model is that the Hubble constant depends on the initial width of the spectral line investigated. Then, two sources, that are in a close proximity in cosmological sense, can have different redshifts depending on the width, d, of the spectral line (as represented by the parameter $\beta \propto d^{-2}$). A more active (hotter) source (smaller β or wider spectral line) will appear to the observer as much more redshifted than a more quieter (cooler) source (larger β or thinner spectral line). This conclusion is in very good qualitative agreement with the actual experimental observations.

The examination, from the point of view of the proposed theory of, the experimental data on hydrogen emission lines in the QSO 3C273, shows that dissipation alone cannot provide a satisfactory quantitative prediction and some higher-order dispersion should be taken into account in order to improve the proposed model.

ACKNOWLEDGMENTS

This work is sponsored in part by the Communitas Foundation, Bulgaria.

The author is indebted to Dr. N.P. Moshkin for help with digitalizing the spectroscopy data from the literature.

REFERENCES

1. G. Burbidge, and M. Burbidge, *Quasi-Stellar Objects*, W.H. Freedman, San Francisco, 1967.
2. C. I. Christov, *Wave Motion* **45**, 154–161 (2008).
3. C. I. Christov, *Foundations of Physics* **36**, 1701–1717 (2006).
4. C. I. Christov, *Math. Comput. Simul.* **74**, 93–103 (2007).
5. C. I. Christov, *http://arxiv.org/pdf/0804.4253* (2008).
6. L. A. Segel, *Mathematics Applied to Continuum Mechanics*, Dover, New York, 1987.
7. L. D. Landau, and M. Lifschitz, *The Classical Theory of Fields*, Reed Educational and Professional Publishing, New York, 2000, 4 edn.
8. H. Alfvén, *Cosmic Plasma*, Holland, 1981.
9. H. Arp, *Seeing Red: Redshifts, Cosmology and Academic Science*, Apeiron, 1998.
10. G. Burbidge, *Nature* **282**, 451 – 455 (1979).
11. M. Schmidt, *Nature* **197**, 1040 (1963).
12. M. Dietrich, S. J. Wagner, T.-L. Courvoisier, H. Bock, and P. North, *Astron. Astrophys.* **351**, 31–42 (1999).
13. van den Berk, and et al., *Astrophys. J.* **122**, 549– (2001).
14. M. Yates, and R. Garden, *Mon. Not. R. ast. Soc.* **241**, 167–194 (1989).
15. A. Davidsen, G. F. Hartig, and W. Fastie, *Nature* **269**, 302–206 (1977).
16. C. I. Christov, "Dynamics of Patterns on Elastic Hypersurfaces. Part I. Shear Waves in the Middle Surface," in *"ISIS International Symposium on Interdisciplinary Science" Proceedings, Natchitoches, October 6-8, 2004*, AIP Conference Proceedings 755, Washington D.C., 2005, pp. 46–52.
17. M. Ciarletta, and D. Iesan, *Non-classical Elastic Solids*, CRC Press, 1993.

Dilute Suspensions near Walls

F. Feuillebois and S. Yahiaoui

PMMH, UMR 7636 CNRS, ESPCI, 10 rue Vauquelin, 75231 Paris Cedex 05, France

Abstract. A precise account of particle-wall hydrodynamic interactions is important for modelling the motion of dilute suspensions near walls. Particles considered here are solid, spherical and small enough for the Reynolds number of the flow around each particle to be small compared with unity. At first order in Reynolds number, Stokes equations apply. Ambient flow fields are expressed as polynomials in terms of the coordinates. Various flow fields perturbed by a spherical particle near a wall may be superimposed by linearity of Stokes equations. Their solutions are recalled; they were calculated with the bispherical coordinates technique, providing precise results for the drag and torque on a particle. For particles in a gas, particle inertia may be important even though for low Reynolds number fluid inertia is negligible. It is shown that collision of a particle on a wall is not possible in creeping flow even if particle inertia is not negligible. At second order in the low Reynolds number, a small fluid inertia appears. For a sphere moving in a linear shear flow and a quadratic shear flow along a wall, there is then a lift force. The various pair couplings between sphere translation, rotation and the shear flows provide contributions to the lift. For a particle moving normal to a wall in a quiescent fluid at constant velocity, fluid inertia provides steady and unsteady contributions to the drag force.

Keywords: Creeping flow, collision, lift force, suspension.
PACS: 47.15.G, 47.57.Ef, 47.61.Jd

INTRODUCTION

Microscale flow fields raise a recent interest, with applications like microfluidics, biological flow fields, separation techniques in analytical chemistry and in general small scale transport phenomena. Flows of suspensions at microscale are considered here. For dilute suspensions, that if for a low volume fraction of particles, particles may be considered as independent. The Reynolds number for the flow relative to a particle is:

$$\text{Re} = \frac{aU}{\nu}$$

where a is the particle characteristic dimension (e.g., the radius for a sphere), U a characteristic velocity and ν the fluid kinematic viscosity. At microscales the Reynolds number is usually small: $\text{Re} \ll 1$.

We start from Navier-Stokes equations for the flow field, written here in a dimensionless form:

$$\text{Re}\frac{\partial \mathbf{v}}{\partial t} + \text{Re}\, \mathbf{v} \cdot \nabla \mathbf{v} = -\nabla p + \nabla^2 \mathbf{v}, \quad \nabla \cdot \mathbf{v} = 0 \tag{1}$$

Here $\mathbf{v}$ stands for the fluid velocity, p for its pressure. Expanding for $\text{Re} \ll 1$, the Navier-Stokes equations reduce at first order to the linear Stokes equations:

$$-\nabla p + \nabla^2 \mathbf{v} = 0 \tag{2a}$$

$$\nabla \cdot \mathbf{v} = 0 \tag{2b}$$

CP1067, *Applications of Mathematics in Engineering and Economics '34—AMEE '08*, edited by M. D. Todorov
© 2008 American Institute of Physics 978-0-7354-0598-1/08/$23.00

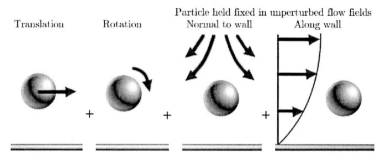

FIGURE 1. Superposition of flow fields

The Green function of Stokes equations in infinite space gives a fluid velocity which decays like the inverse of the distance to the singularity, thus very slowly. As a consequence of these long range hydrodynamic interactions, wall effects are important. Moreover, for decaying scales, walls effects become in general important relative to the bulk. A recurrent question when modeling the motion of suspensions of particles is then to ascertain boundary conditions. These are motivations for studying in detail particle-wall interactions.

The outline of the paper is as follows. Various solutions for a sphere in creeping flow will be presented in the Section . Then the problem of collision of a particle with a wall will be discussed. At order $O(\text{Re})$ in the low Reynolds number, some fluid inertia is taken into account (the terms in the left-hand-side of Navier-Stokes equation (1)). Section is concerned with these terms for a steady sphere motion along the wall. There is then a lift force. Section is concerned also with $O(\text{Re})$ but for a sphere motion normal to the wall. In that case, there is an unsteady contribution to the drag force.

SPHERE IN CREEPING FLOW NEAR A WALL

Calculation of Flow Field and Friction Factors on the Sphere

Consider Stokes equations (2). By linearity, the flow field due to a sphere translating and rotating in an ambient flow may be obtained as the sum of various flow fields (Figure 1). Consider the cases of a sphere held fixed in an ambient flow. The perturbed flow field is sought as the sum of this ambient (unperturbed) flow field and a perturbation: $\mathbf{u}_\infty + \mathbf{v}$. Then Stokes equations (2) apply for the perturbation. The boundary conditions are: $\mathbf{v} = -u^\infty \mathbf{e}_x (\text{or } -u^\infty \mathbf{e}_z)$ on the sphere and $\mathbf{v} = 0$ on the wall and at infinity.

As an example, consider an axisymmetric unperturbed flow normal to a wall [1]. Assuming that the stream function of the unperturbed flow field is expressed as a polynomial in the cylindrical coordinates $\tilde{\rho}, \tilde{z}$, its form should be:

$$\psi = \sum_M \tilde{\rho}^2 C_M \tilde{z}^M$$

TABLE 1. Unperturbed axisymmetric fluid velocity

degree of polynomial in $\tilde{z}$	$\tilde{u}_\rho^\infty$	$\tilde{u}_z^\infty$	$\tilde{p}^\infty$
$M = 0$	0	$-S_0$	0
$M = 2$	$S_2 \tilde{\rho} \tilde{z}$	$-S_2 \tilde{z}^2$	$-2S_2 \mu_f \tilde{z}$
$M = 3$	$3S_3 \tilde{\rho} \tilde{z}^2$	$-2S_3 \tilde{z}^3$	$-3S_3 \mu_f (2\tilde{z}^2 - \tilde{\rho}^2)$

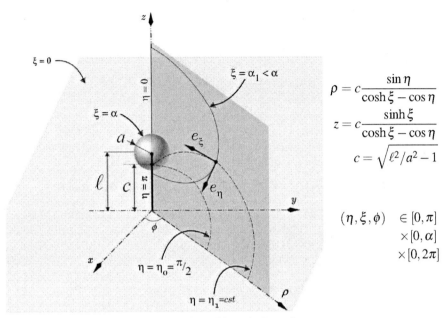

$$\rho = c \frac{\sin \eta}{\cosh \xi - \cos \eta}$$

$$z = c \frac{\sinh \xi}{\cosh \xi - \cos \eta}$$

$$c = \sqrt{\ell^2 / a^2 - 1}$$

$$(\eta, \xi, \phi) \in [0, \pi] \times [0, \alpha] \times [0, 2\pi]$$

FIGURE 2. Bispherical coordinates

where the C_M's are constants. All quantities here are dimensional. Our solutions are limited to $M \leq 3$. The case $M = 0$ gives a uniform flow. For $M = 1$, this is the classical stagnation point flow, an exact solution of Navier-Stokes equations, but applying the no-slip condition on wall gives trivially a fluid at rest. The cases $M = 2, 3$ are given in Table 1. Here the ambient fluid velocity $\tilde{\mathbf{u}}^\infty$ is in dimensional form, μ_f is the fluid viscosity. S_2 and S_3 are constants.

For this geometry involving a sphere and a plane wall, bispherical coordinates (η, ξ, ϕ) have been used for various problems [2]. They are defined from the cylindrical coordinates $(\rho = \tilde{\rho}/a, z = \tilde{z}/a, \phi)$ as shown in Figure 2. The plane is represented by $\xi = 0$ and the sphere by $\xi = \alpha$ with $\cosh \alpha = \ell/a$, where ℓ is the distance from the sphere centre to the wall.

The solution of the Stokes momentum equation (2a) for the pressure p and velocity (v_ρ, v_z) is sought in the following form:

$$p = \frac{Q_0}{c}, \quad v_\rho = \frac{\rho Q_0}{2c} + V_1, \quad v_z = \frac{z Q_0}{2c} + W_0, \quad \text{where:} \tag{3a}$$

17

$$Q_0 = c^M (\cosh \xi - \mu)^{1/2} \sum_{n=0}^{\infty} [A_n, B_n] P_n(\mu) \tag{3b}$$

$$V_1 = c^M (\cosh \xi - \mu)^{1/2} \sin \eta \sum_{n=1}^{\infty} [C_n, D_n] P_n'(\mu) \tag{3c}$$

$$W_0 = c^M (\cosh \xi - \mu)^{1/2} \sum_{n=0}^{\infty} [0, F_n] P_n(\mu) \tag{3d}$$

with $\mu = \cos \eta$ and with the shorthand notation:

$$[A_n, B_n] = A_n \cosh \left(n + \frac{1}{2} \right) \xi + B_n \sinh \left(n + \frac{1}{2} \right) \xi$$

The velocity satisfies the no-slip condition on the wall. It also vanishes at infinity since $\xi = \eta = 0$ there. Applying then the no-slip boundary condition on the sphere surface, the coefficients A_n, B_n, C_n, D_n are expressed in term of the F_n's [1]. Applying finally the continuity equation (2b) gives an iteration relationship between the F_n's.

$$f_{n,n-1}(\alpha) F_{n-1} + f_{n,n}(\alpha) F_n + f_{n,n+1}(\alpha) F_{n+1} = b_n^M(\alpha), \quad n \geq 0 \quad (M = 0, 2, 3)$$

where $f_{0,-1} = 0$ and the other f's and the b's are known in term of the dimensionless sphere to wall distance $\ell/a = \cosh \alpha$. The first coefficient F_0 is unknown. But we know that since the series in (3) should converge, $F_n \to 0$ for $n \to \infty$. Thus the infinite linear system may be solved by truncating for some large n. An appropriate solution uses an iterative technique [3, 4, 1]. The coefficients F_n are written in the form $F_n = t_n + F_0 u_n$ for $n \geq 1$ in which the t_n's and u_n's are defined by iteration:

$$t_1 = 0 \quad ; \quad f_{n,n-1} t_{n-1} + f_{n,n} t_n + f_{n,n+1} t_{n+1} = b_n \quad (n \geq 1)$$
$$u_1 = 1 \quad ; \quad f_{n,n-1} u_{n-1} + f_{n,n} u_n + f_{n,n+1} u_{n+1} = 0 \quad (n \geq 1)$$

$F_n \to 0$ for $n \to \infty$ gives the result: $F_0 = \lim_{n \to \infty} (-t_n/u_n)$. Other F_n's are calculated by iteration. Since this solution is explicit, numerical calculations have to be accurate. We used Maple software for its "infinite" precision, allowing the number of digits to grow upon request. The number of terms in the series increases when the gap between the sphere and wall decreases. Results for small gaps, viz. in lubrication region required typically thousands of terms. Since the required computer memory is small, this calculation may be performed on a standard PC. For example for the case $M = 0$, in order to obtain a 10^{-16} precision at a gap $\ell/a - 1 = 10^{-5}$ required typically 35 digits and 8188 terms. This calculation was possible here thanks to the present iteration technique.

The drag force acting on the particle may by linearity be written as:

$$\mathbf{F}^{(M)} = 6 \pi \mu_f a \left[\tilde{u}_z^{\infty} \right]_C \left[f_{zz}^{(M)} \right]_C \mathbf{e}_z$$

with the unperturbed fluid velocity $\left[\tilde{u}_z^{\infty} \right]_C$ at the sphere centre and the friction factor:

$$\left[f_{zz}^{(M)} \right]_C = -\frac{2\sqrt{2}}{3(M-1)} \tanh^M \alpha \sinh \alpha \sum_{n=0}^{\infty} F_n, \quad (M = 2, 3).$$

18

The solution (3) provides the fluid velocity and pressure. The perturbed flow field around the sphere in a stagnation point ambient flow shows a set of toroidal vortices between the sphere and wall [1]. An alternative solution for this axisymmetric problem was obtained using the stream function. The velocity components are obtained as its derivatives. But calculating the pressure then requires some developments; the solution in [1] uses a technique derived first in [5] and presented in [6].

Problems of a particle translating and rotating along a wall and held fixed in an unperturbed flow along the wall were solved by a similar technique in terms of velocity and pressure using bispherical coordinates [4] [7]. The linear shear flow $\tilde{u}_x^\infty = k^s \tilde{z}$ was considered in [4] and the quadratic shear flow $\tilde{u}_x^\infty = k^q \tilde{z}$ in [7]. Results for the hydrodynamic force and torque on the particle may be written in terms of friction factors:

$$F_x^t = -6\pi a \mu_f f_{xx}^t \tilde{V}_x \quad C_y^t = 8\pi a^2 \mu_f c_{yx}^t \tilde{V}_x \quad \text{translation}$$
$$F_x^r = 6\pi a^2 \mu_f f_{xy}^r \tilde{\Omega}_y \quad C_y^r = -8\pi a^3 \mu_f c_{yy}^r \tilde{\Omega}_y \quad \text{rotation}$$
$$F_x^s = 6\pi a \mu_f f_{xx}^s k^s \ell \quad C_y^s = 4\pi a^3 \mu_f c_{yx}^s k^s \quad \text{linear shear flow}$$
$$F_x^q = 6\pi a \mu_f f_{xx}^q k^q \ell^2 \quad C_y^q = 8\pi a^3 \mu_f c_{yx}^q k^q \ell \quad \text{quadratic shear flow}$$

All friction factors for the force f and for the torque c are scaled so that their limit is unity for $\ell/a \to \infty$. Results are presented elsewhere [4] [7]. The case of a Poiseuille flow, taking only the nearest wall into account, may be obtained by superposition of a linear shear flow and a quadratic shear flow, Figure 3: $\tilde{u}(z) = k^s \tilde{z} + k^q \tilde{z}^2$, where:

$$k^s = \frac{h}{2\mu_f}\left(-\frac{d\tilde{p}}{d\tilde{x}}\right), \quad k^q = \frac{1}{2\mu_f}\left(\frac{d\tilde{p}}{d\tilde{x}}\right) \tag{4}$$

and $d\tilde{p}/d\tilde{x}$ is the pressure gradient along the Poiseuille flow.

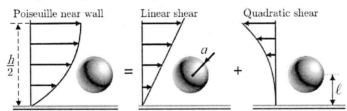

FIGURE 3.   Poiseuille flow near wall as a superposition of linear and quadratic shear flows

Ambient flows normal and parallel to the wall may be superposed, again by linearity of the Stokes equations.

Example of a Sphere in Axisymmetric Flow. Entrainment and the Collision Problem

Consider for example a spherical particle moving in an ambient axisymmetric flow (see Table 1):
$$\tilde{\mathbf{u}}_\infty = (S_2 \tilde{\rho} \tilde{z} + 3S_3 \tilde{\rho} \tilde{z}^2)\mathbf{e}_\rho - (S_2 \tilde{z}^2 + 2S_3 \tilde{z}^3)\mathbf{e}_z$$

Let the particle be centered at $(\tilde{X}, \tilde{Y}, \tilde{Z})$ off the $\tilde{z}$ axis. In a fixed frame $(\tilde{x}', \tilde{y}', \tilde{z}')$ with origin at $(\tilde{X}, \tilde{Y}, 0)$ at the considered time, the local ambient flow is composed of these axisymmetric flows plus linear and quadratic shear flows:

$$\tilde{\mathbf{u}}'_\infty = (S_2\tilde{z}' + 3S_3\tilde{z}'^2)(\tilde{x}'\mathbf{e}_x + \tilde{y}'\mathbf{e}_y) - (S_2\tilde{z}'^2 + 2S_3\tilde{z}'^3)\mathbf{e}_z$$
$$+ \tilde{X}S_2\tilde{z}'\mathbf{e}_x + \tilde{Y}S_2\tilde{z}'\mathbf{e}_y + 3\tilde{X}S_3\tilde{z}'^2\mathbf{e}_x + 3\tilde{Y}S_3\tilde{z}'^2\mathbf{e}_y$$

To construct dimensionless quantities, take now $\tau_e = (S_2a)^{-1}$ as characteristic time for the flow field. Let also $\lambda_3 = aS_3/S_2$. The characteristic time for accelerating a particle is from the equation of motion of a single particle: $\tau_p = m_p/(6\pi a\mu_f)$, where m_p is the mass of the particle. The *Stokes number* is defined traditionally as the ratio:

$$\mathscr{S} = \frac{\tau_p}{\tau_e} = \frac{2}{9}\frac{S_2a^3}{v_f}\frac{\rho_p}{\rho_f}$$

where ρ_p, ρ_f are the particle and fluid density and $v_f = \mu_f/\rho_f$. Note that $S_2a^3/v_f = \mathrm{Re} \ll 1$ is a small Reynolds number relative to the particle. Thus, $\mathscr{S}$ may be not small for $\rho_p/\rho_f \gg 1$ that is for a particle in a gas. In that case particle inertia may be non-negligible even though fluid inertia is neglected. The normalised weight plus buoyancy of the particle are introduced by defining V_S, ratio of the Stokes velocity to S_2a^2. Introducing all forces and torques due to the preceding ambient flows in the equations of motion of the particle, we obtain in dimensionless form:

$$\mathscr{S}\frac{dV_x}{dt} = -f_{xx}^t V_x + f_{xy}^r \Omega_y - f_{xx}^s XZ - 3f_{xx}^q \lambda_3 XZ^2$$
$$\mathscr{S}\frac{dV_y}{dt} = -f_{xx}^t V_y - f_{xy}^r \Omega_x - f_{xx}^s YZ - 3f_{xx}^q \lambda_3 YZ^2$$
$$\mathscr{S}\frac{dV_z}{dt} = -f_{zz}^t V_z + f_{zz}^{(2)} + f_{zz}^{(3)} \lambda_3 - V_S$$
$$\mathscr{S}\frac{d\Omega_y}{dt} = \frac{10}{3}c_{yx}^t V_x - \frac{10}{3}c_{yy}^r \Omega_y - \frac{5}{3}c_{yx}^s X - 10c_{yx}^q \lambda_3 XZ$$
$$\mathscr{S}\frac{d\Omega_x}{dt} = -\frac{10}{3}c_{yx}^t V_y - \frac{10}{3}c_{yy}^r \Omega_x + \frac{5}{3}c_{yx}^s Y + 10c_{yx}^q \lambda_3 YZ.$$

This system together with the definitions of velocities $(dX/dt = V_x, dY/dt = V_y, dZ/dt = V_z)$ is solved in order to obtain the particle trajectories. A difficulty is that f_{zz}^t, the normalised force for the motion normal to the wall, varies for small normalised gap $\zeta = Z - 1$ like $1/\zeta$. Thus the system is stiff. It is then solved numerically with a predictor-corrector method.

Trajectories of particles entrained from a wall for a Stokes number $\mathscr{S} = 1$ are plotted in Figure 4. The departure points are at $X = -1000, -800, -600, -400, -200$ and $Z = 1.01$. It is observed that the particle spends a long time moving along a wall before being lifted away by the axisymmetric flow. There is then an overshoot at values $X > 0$ because of particle inertia. When the particle reaches the position $Z = 3.1$, we reverse the ambient flow so that it goes down. Then the particle trajectories go directly towards the wall. This non-reversibility of particle trajectories is due to $\mathscr{S} = 1$. For very small

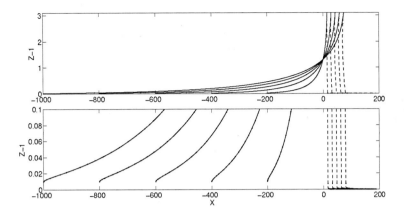

FIGURE 4. Trajectories of a solid particle in a gas in stagnation point flow. Solid lines: going up. Dashed lines: going down from the end point of the preceding trajectory. Bottom figure: zoom of top one

$\mathcal{S}$, without particle inertia, trajectories would be practically reversible. In the present case $\mathcal{S} = 1$, particles come very close to the wall but do no collide on it.

This subtle point can be seen more clearly by studying analytically a simplified typical equation of motion. For a particle on the axis, when the gap becomes small $\zeta \ll 1$, the equation of motion reads:

$$\mathcal{S}\ddot{\zeta} = -\frac{\dot{\zeta}}{\zeta} - 1.$$

The unity on the rhs comes from the first order term of the force due to the axisymmetric flow field, after normalising. Integrating from some $\zeta_0 \ll 1$ with the initial velocity $\dot{\zeta}_0$:

$$\mathcal{S}(\dot{\zeta} - \dot{\zeta}_0) + \log\frac{\zeta}{\zeta_0} + \tau = 0.$$

The 1st term is bounded since the velocity is physically bounded. There would be a possibility of collision if $\zeta \to 0$ in the 2nd term. This is only possible if the normalised time $\tau \to \infty$ in the 3rd term. Thus there is *no collision in a finite time, for any value of the Stokes number*. This result is in contradiction with earlier approaches [8] which used a simplified drag force ignoring lubrication effects.

Then how can small particles in a liquid or a gas collide with a wall ? Actually, particles may come so close to the wall that other physical effects become more important, e.g., the roughness of the particle and wall, short range forces like van der Waals. Thus, we now calculate the time τ^* for ζ to reach some small value ζ^*, e.g., due to roughness. As we will see, τ^* gets smaller for larger $\mathcal{S}$. Consider, again, the model equation for the gap variation:

$$\mathcal{S}\zeta\ddot{\zeta} + \dot{\zeta} + \zeta = 0$$

with initial conditions $\zeta = \zeta_0 \ll 1, \dot{\zeta} = \dot{\zeta}_0$. For $\zeta_0 \to 0$, the 1st term is negligible and the equation looses one order; then we cannot apply both initial conditions. That is, the perturbation problem is singular. We use the method of matched asymptotic expansions for small ζ_0. Let $\zeta = \zeta_0 \tilde{\zeta}$ with $\tilde{\zeta} = O(1)$.

Outer equation: $\dot{\tilde{\zeta}} + \tilde{\zeta} = 0$ has the solution $\tilde{\zeta} = Ce^{-\tau}$.

Inner equation: Let the inner variable $\hat{\tau} = \tau/\zeta_0$. Rename the unknown $\tilde{\zeta}$ as $\hat{\zeta}$. The inner equation is: $\mathscr{S}\hat{\zeta}\ddot{\hat{\zeta}} + \dot{\hat{\zeta}} = 0$. The solution using the initial conditions is:

$$\hat{\tau} = \mathscr{S}e^{\mathscr{S}\dot{\zeta}_0} \int\limits_{-\mathscr{S}\dot{\zeta}_0 + \log \hat{\zeta}}^{-\mathscr{S}\dot{\zeta}_0} \frac{e^\theta}{\theta} d\theta.$$

Matching: $\lim_{\hat{\tau} \to \infty} \hat{\zeta} = \lim_{\tau \to 0} \tilde{\zeta}$. The inner limit $\hat{\tau} \to \infty$ is obtained as: $-\mathscr{S}\dot{\zeta}_0 + \log \hat{\zeta} \to 0$ because then the integral diverges in $\theta = 0$. The outer limit is $\lim_{\tau \to 0} \tilde{\zeta} = C$. Thus the matching gives the constant $C = \exp \mathscr{S}\dot{\zeta}_0$.

The result for the outer solution is in the original variable:

$$\zeta = \zeta_0 e^{\mathscr{S}\dot{\zeta}_0 - \tau}.$$

A small value ζ^* is thus attained after a time $\tau^* = \log(\zeta_0/\zeta^*) + \mathscr{S}\dot{\zeta}_0$. For larger Stokes number $\mathscr{S}$, since $\dot{\zeta}_0 < 0$, a small value ζ^* is attained more rapidly. But in any case, the value $\zeta = 0$ would be attained only after an infinite time.

SMALL EFFECTS OF FLUID INERTIA ON SPHERE MOTION

Order $O(\text{Re})$ Steady Terms in Motion along Wall. Lift Force

Consider now the second order in the expansion for low Reynolds number, that is the terms of order $O(\text{Re})$. A small fluid inertia is thus taken into account.

In this section we consider the motion of a sphere along a wall with a constant velocity $\mathbf{V} = V_x\mathbf{e}_x$. The sphere may also be submitted to an ambient flow field in the same direction along the wall, $\mathbf{u}_\infty = u_\infty\mathbf{e}_x$. It is also rotating along the wall in the perpendicular direction with a velocity $\mathbf{\Omega} = \Omega_y\mathbf{e}_y$.

As before, let us search the unknown perturbed velocity as: $\mathbf{u}_\infty + \mathbf{v}$. In a reference frame attached to the sphere centre, the perturbed flow field is steady. The steady Navier-Stokes equations in this frame are:

$$\nabla.\mathbf{v} = 0, \qquad \Delta\mathbf{v} - \nabla p = \text{Re}\,(\mathbf{v}.\nabla\mathbf{v} + \mathbf{v}.\nabla\mathbf{u}_\infty + (\mathbf{u}_\infty - \mathbf{V}).\nabla\mathbf{v})$$

with boundary conditions:

$$\mathbf{v} = \mathbf{V} + \mathbf{\Omega} \wedge \mathbf{r} - \mathbf{u}_\infty \quad \text{on sphere surface}$$
$$\mathbf{v} = 0 \quad \text{on plane}, \quad \mathbf{v} \to 0 \quad \text{at infinity}.$$

Consider here an ambient flow field which is a combination of a linear shear flow and a quadratic shear flow along the wall: $\mathbf{u}_\infty = \mathbf{u}_\infty^s + \mathbf{u}_\infty^q$. An example is a Poiseuille flow in the near wall region for which k^s, k^q are given in (4). The fluid velocity and pressure are expanded as:

$$\mathbf{v} = \mathbf{v}^0 + \mathrm{Re}\,\mathbf{v}^1 + O(\mathrm{Re}^2), \qquad p = p^0 + \mathrm{Re}\,p^1 + O(\mathrm{Re}^2).$$

Here, the particle is assumed to be close enough to the wall, or the Reynolds number small enough so that the distance of the particle to wall $\ell \ll a/\mathrm{Re}$. Then the wall is in the inner region of expansion [9] and the perturbation problem at order $O(\mathrm{Re})$ is regular. This is because at order $O(1)$ perturbations due to the sphere together with their images in the wall decay fast enough at infinity so that the problem at next order $O(\mathrm{Re})$ is regular. Note that the problem without wall would be singular and a second, Oseen type, expansion would then be necessary.

The force on the sphere is obtained as the expansion: $\mathbf{F} = \mathbf{F}^0 + \mathrm{Re}\,\mathbf{F}^1 + O(\mathrm{Re}^2)$. To calculate $\mathbf{F}^1$, calculating the whole flow field $\mathbf{v}^1, p^1$ is unnecessary. It is sufficient to use a variation of the Lorentz reciprocity theorem proposed by [9]. The force $\mathbf{F}^0$ is a drag force along the wall. The force $\mathbf{F}^1$ contains a correction to the drag plus a lift force which is perpendicular to the ambient flow relative to the sphere, thus normal to the wall. We will limit ourselves to the calculation of this lift force which has various applications for separation techniques, etc. Using the reciprocity theorem as in [10], the following integral expression is obtained for the z component of $\mathbf{F}^1$:

$$F_z^1 = -L = -\int_{V_f} \left(\mathbf{v}^0 \cdot \nabla \mathbf{v}^0 + \mathbf{v}^0 \cdot \nabla \mathbf{u}_\infty + (\mathbf{u}_\infty - \mathbf{V}) \cdot \nabla \mathbf{v}^0\right) . \mathbf{w}\, dV$$

where $\mathbf{v}^0 = T_t \mathbf{v}^t + T_r \mathbf{v}^r + T_s \mathbf{v}^s + T_q \mathbf{v}^q$, the T's are constants and $\mathbf{w}$ is the velocity field in Stokes flow for a sphere moving normal to the wall with a unit velocity. Since $\mathbf{v}^0$ is a linear combination of four terms, the expression of the integral L in the expression for the lift is composed of ten coupling terms:

$$
\begin{aligned}
L = {}& T_s^2 L_s + T_q^2 L_q + T_r^2 L_r + T_t^2 L_t \\
& + T_s T_q L_{sq} + T_s T_r L_{sr} + T_s T_t L_{st} \\
& + T_q T_r L_{qr} + T_q T_t L_{qt} + T_r T_t L_{rt}.
\end{aligned}
$$

with:

$$L_s = \int_{V_f} (\mathbf{v}_s . \nabla \mathbf{v}_s^\infty + (\mathbf{v}_s + \mathbf{v}_s^\infty) . \nabla \mathbf{v}_s) . \mathbf{w}\, dV$$

$$L_t = \int_{V_f} ((\mathbf{v}_t - \mathbf{V}) . \nabla \mathbf{v}_t) . \mathbf{w}\, dV$$

$$L_{sr} = \int_{V_f} (\mathbf{v}_r . \nabla (\mathbf{v}_s + \mathbf{v}_s^\infty) + (\mathbf{v}_s + \mathbf{v}_s^\infty) . \nabla \mathbf{v}_r) . \mathbf{w}\, dV$$

$$L_r = \int_{V_f} (\mathbf{v}_r . \nabla \mathbf{v}_r) . \mathbf{w}\, dV$$

$$L_{st} = \int_{V_f} (\mathbf{v}_t . \nabla (\mathbf{v}_s + \mathbf{v}_s^\infty) + (\mathbf{v}_s + \mathbf{v}_s^\infty) . \nabla \mathbf{v}_t - \mathbf{V} . \nabla \mathbf{v}_s) . \mathbf{w}\, dV$$

$$L_{rt} = \int_{V_f} (\mathbf{v}_r . \nabla \mathbf{v}_t + (\mathbf{v}_t - \mathbf{V}) . \nabla \mathbf{v}_r) . \mathbf{w}\, dV$$

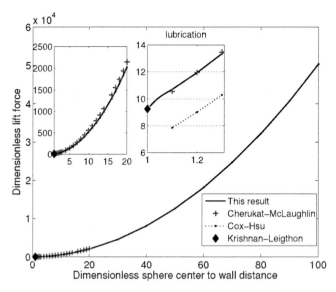

FIGURE 5. Dimensionless lift force $F_z^1 = F_z/\text{Re}$ on a fixed sphere in a linear shear flow. Our result bridges gaps between earlier ones

$$L_{sq} = \int\limits_{V_f} \left((\mathbf{v}_s + \mathbf{v}_s^\infty) . \nabla \mathbf{v}_q + (\mathbf{v}_q + \mathbf{v}_q^\infty) . \nabla \mathbf{v}_s + \mathbf{v}_q . \nabla \mathbf{v}_s^\infty + \mathbf{v}_s . \nabla \mathbf{v}_q^\infty \right) . \mathbf{w} \; dV.$$

The expressions for L_q, L_{qr}, L_{qt} are formally like L_s, L_{sr}, L_{st}. All these integrals were calculated exploiting the precision of our results for Stokes flow in bispherical coordinates. Consider, e.g., a fixed sphere in an ambient linear shear flow $\mathbf{u}_\infty = \mathbf{u}_\infty^s$. Our result for F_z^1 shown in Figure 5 is in excellent agreement with those of Krishnan & Leighton [11] for a sphere touching the plane and of Cherukat & Mc Laughlin [12] at intermediate distances. It converges to that of Cox & Hsu [13] for large distances. It bridges various gaps.

Order $O(\text{Re})$ Unsteady Terms in Motion Normal to Wall. Correction to the Drag

Consider now the motion of a sphere normal to a wall. The boundary conditions are: $\mathbf{v} = \mathbf{V} = V_z \mathbf{e}_z$ on the sphere surface, $\mathbf{v} = 0$ on the plane and at infinity. The Reynolds number is here based on the dimensional sphere velocity $\bar{V}_z$; it is negative for a sphere moving towards the wall. Navier-Stokes equations (1) are expanded for small $|Re| \ll 1$. At order $O(1)$, Stokes equations apply with these conditions. At $O(\text{Re})$:

$$\nabla . \mathbf{v}^{(1)} = 0 \quad , \quad \Delta \mathbf{v}^{(1)} - \nabla p^{(1)} = \frac{\partial \mathbf{v}^{(0)}}{\partial t} + \mathbf{v}^{(0)} . \nabla \mathbf{v}^{(0)} \tag{5}$$

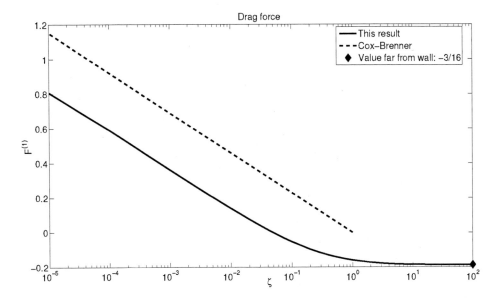

FIGURE 6. $O(\text{Re})$ term in the drag force on a sphere moving towards a plane with unit velocity, versus the normalised gap $\zeta = \delta - 1$ between the sphere and wall. Comparison with the calculation of Cox & Brenner [14] for small gaps

with the boundary conditions: $\mathbf{v}^{(1)} = 0$ on the sphere surface, on the plane and at infinity. Like in §3.1, the problem is regular because of the wall. Note that here the unsteady term $\partial \mathbf{v}^{(0)}/\partial t$ remains in any reference frame (wall or sphere).

The drag force on the sphere, after normalizing by $\mu_f a |\tilde{V}_z|$, is expanded as: $F = F^{(0)} + \text{Re}\, F^{(1)}$. Generalizing the reciprocity theorem of §3.1 to include $\partial \mathbf{v}^{(0)}/\partial t$ gives:

$$F^{(1)} = - \left[\int_{V_f} \frac{\partial \mathbf{v}^{(0)}}{\partial \delta} \frac{\partial \delta}{\partial t} \cdot \mathbf{w}'\, dV + \int_{V_f} \left(\mathbf{v}^{(0)} \cdot \nabla \mathbf{v}^{(0)} \right) \cdot \mathbf{w}'\, dV \right] \qquad (6)$$

where $\delta = \ell/a$ is the dimensionless sphere centre to wall distance and $\mathbf{w}'$ is the creeping flow velocity for a sphere moving *towards* the wall with unit velocity. The last integral in (6) vanishes by symmetry. In the first integral in (6), note that $\partial \delta/\partial t = \text{sgn}(V_z)$; this shows that the drag force depends on the direction of the motion.

Results for $F^{(1)}$ for a sphere moving towards the wall with unit velocity, $V_z = -1$, are displayed in Figure 6. They are compared with the calculation of [14] for small gaps. It is observed that the results match approximatively for very small gaps. We have no explanation for this small discrepancy. For a given distance from the wall, the drag is different for a sphere moving towards it (like in Figure 6) and away from it (for then the wake interacts with wall). Close to the wall ($\delta < 1.05$), $F^{(1)}$ is increased for a motion towards the wall and decreased for a motion away from it. For distances $\delta > 1.05$, this is the reverse. Far from the wall, $\delta \gg 1$, $F^{(1)} \rightarrow -\frac{3}{16}$ for a motion towards the wall and $\frac{3}{16}$ for a motion away from it (the value 3/16 was obtained from an approximation of numerical results but suggests that an analytical calculation might be performed). Far from the wall, the net dimensional drag force may be written as $\mathbf{F}_t = -6\pi a \mu_f \, \tilde{\mathbf{V}} f_t$ with:

$$f_t = 1 + \frac{9}{8\delta} - \mathrm{sgn}(V_z)\frac{3}{16}|\mathrm{Re}| + o\left(\mathrm{Re}, \frac{1}{\delta}\right).$$

The second term is the classical $O(1)$ term due to Lorentz. The third one is new.

Note that the sphere motion being steady, there is neither added-mass term nor historical (Boussinesq-Basset) term in the force.

Returning now to the collision problem, §2.2, no possibility of collision is brought by the $O(\mathrm{Re})$ term calculated here since it is small compared with the quasi-steady lubrication term. Thus the only possibility to have a collision based solely on hydrodynamics is that the Reynolds number be not small compared with unity.

CONCLUSION

Comprehensive results have been presented for a sphere in creeping flow : translating, rotating, held fixed in unperturbed flows that are axisymmetric (polynomial up to degree 3) and along the wall (polynomial up to degree 2). All these creeping flows may be superimposed by linearity of Stokes equations. Solutions for all these flow fields were calculated in bispherical coordinates using an iteration technique which provides very precise results; it makes it possible to obtain details of the slow flow field and obtain results for the drag force for small gaps, in the lubrication region. An example of trajectories of a sphere moving in an axisymmetric flow field was given. A discussion shows that there is in principle no collision in creeping flow, even for an $O(1)$ Stokes number; a "collision" can thus only occur because of various other physical effects.

$O(\mathrm{Re})$ terms for a sphere moving along a wall in an ambient shear flow give a lift force. This force involves all couplings between translation, rotation, linear shear flow, quadratic shear flow. The $O(\mathrm{Re})$ unsteady drag force on a sphere moving normal to the wall was calculated. Values of the drag for motions towards the wall and away from it are different. This new term does not create any possibility of collision with the wall. Thus a non-small Re is essential for a possibility of collision due to hydrodynamics.

REFERENCES

1. L. Pasol, M. Chaoui, S. Yahiaoui, and F. Feuillebois, *Phys. Fluids* **17**(7), 073602 (2005).
2. J. Happel and H. Brenner, *Low Reynolds Number Hydrodynamics*, Martinus Nijhoff, 1973.
3. M. E. O'Neill and B. S. Bhatt, *Q. J. Mech. Appl. Math.* **44**, 91–104 (1991).
4. M. Chaoui and F. Feuillebois, *Quart. J. Mech. Applied Math.* **56**(3), 381–410 (2003).
5. E. Chervenivanova and Z. Zapryanov, *Int. J. Multiphase flow* **11**(5), 721–738 (1985).
6. Z. Zapryanov and S. Tabakova, *Dynamics of Bubbles, Drops and Rigid Particles*, Kluwer, 1998.
7. L. Pasol, A. Sellier, and F. Feuillebois, *Quart. J. Mech. Applied Math.* **59**,587–614 (2006).
8. D. H. Michael, *J. Fluid Mech.* **31**, 175–192 (1968).

9. R. G. Cox and H. Brenner, *Chem. Eng. Sci.* **23**, 147–173 (1968).
10. B. P. Ho and L. G. Leal, *J. Fluid Mech.* **65**, 365–400 (1974).
11. G.P. Krishnan and D.T. Jr Leighton, *Phys. Fluids* **7**(11), 2538–2545 (1995).
12. P. Cherukat and J. B. Mc Laughlin, *Int. J. Multiph. Flow* **20**(5), 339–353 (1994).
13. R. G. Cox and S. K. Hsu, *Int. J. Multiph. Flow* **3**, 201–222 (1977).
14. R. G. Cox and H. Brenner, *Chem. Eng. Sci.* **22**, 1753–1777 (1967).

Compatible Discretizations for Continuous Dynamical Systems

H. V. Kojouharov* and D. T. Dimitrov†

*Dept. of Mathematics, The University of Texas at Arlington
P.O. Box 19408, Arlington, TX 76019-0408, USA
†Statistical Center for HIV/AIDS Research and Prevention (SCHARP)
Vaccine & Infectious Disease Institute, Fred Hutchinson Cancer Research Center
Seattle, WA 98109-1024, USA

Abstract. In this paper, we briefly review the history of the nonstandard finite difference (NSFD) methods and their application to a wide range of physical and biological problems. Emphasis is given on the use of the nonstandard discretization ideas to the numerical solution of continuous dynamical systems. In that context, specific examples of positive and elementary stable nonstandard (PESN) finite difference schemes are presented. We also outline possible future research directions for the construction of structural property-preserving numerical methods and discuss some open problems in the general area of nonstandard discretizations of differential equations.

Keywords: Compatible discretizations, nonstandard, finite difference, dynamical systems.
PACS: 82.39.-k; 87.10.Ed; 87.23-n

INTRODUCTION

Numerical simulations are important in order to explore the implications of assumptions and sensitivity of model behavior to parameter changes; to predict the behavior of the systems; and to draw conclusions about the possible dynamical scenarios. For complicated systems which try to include all relevant physical effects, numerical simulations may be the only way through which to view the phenomena. One major effort in most of the publications in the natural and social sciences is to present arguments and data, that justify the choice of the mathematical model and to define the space in which the model represents the analyzed system realistically. However, it is also crucial to understand the behavior of numerical simulations of nonlinear dynamical systems in order (1) to interpret the data obtained from such simulations; and (2) to facilitate the design of algorithms, which provide the correct qualitative information without being excessively expensive.

Standard numerical methods, such as Euler, Runge-Kutta and Adams methods, often impose significant requirements on the initial conditions and parameter values as well as on the time step-size to guarantee consistency or convergence. Their use raises questions about truncation errors, stability regions and, from a dynamical point of view, the accuracy at which the dynamics of the continuous system are represented by the numerical approximation. Dynamical systems very often express special structural (physical or behavioral) properties, which are not transformed properly by standard numerical methods. Therefore, the goal is to develop nonstandard numerical methods that, under mild or no restriction on the time-step, preserve the long-time properties of

CP1067, *Applications of Mathematics in Engineering and Economics '34—AMEE '08*, edited by M. D. Todorov
© 2008 American Institute of Physics 978-0-7354-0598-1/08/$23.00

the underlying nonlinear dynamical systems.

In this review paper we consider systems of the following form:

$$D(\bar{x}) = f(\bar{x}, t); \quad \bar{x}(t_0) = \bar{x}_0 \in \mathscr{R}^n, \ n \geq 2 \tag{1}$$

where $\bar{x} = [x^1, x^2, \ldots, x^n]^T : [t_0, T) \to \mathscr{R}^n$, D is a differential operator in $\mathscr{R}^n$, and the function $f = [f^1, f^2, \ldots, f^n]^T : \mathscr{R}^n \mapsto \mathscr{R}^n$ is differentiable. Since, we regard the dimension of the problem as being arbitrary and since it is always possible, at the cost of raising the dimension by 1, to write a non-autonomous problem in autonomous form, there is clearly no loss of generality in assuming that the general n-dimensional system (1) is autonomous.

A finite-difference scheme with a step size h, that approximates the solution $\bar{x}(t_k)$ of System (1) can be written in the form:

$$D_h(\bar{x}_k) = F_h(f; \bar{x}_k, h), \tag{2}$$

where $D_h(\bar{x}_k) \approx D(\bar{x}(t_k))$, $\bar{x}_k \approx \bar{x}(t_k)$, $F_h(f; \bar{x}_k, h) \approx f(\bar{x}(t_k), t_k)$, and $t_k = t_0 + kh$. There is a wide variety of methods available to numerically integrate dynamical systems yet the choice of a suitable numerical method is not obvious. Methods are usually chosen for their order of accuracy, stability properties, or ease of implementation. Finding numerical methods which replicate the global properties of dynamical systems under mild or no restrictions on the time-step is of great importance in many areas of science and engineering and has been an integrable part of the numerical analysis of differential equations for the last several decades. There are considerable number of results in the literature where conditions have been investigated for numerical methods for ordinary differential equations to generate discrete dynamical systems, which are similar to those generated by the differential equation. The books [16, 39, 8], among others, and the references therein give a comprehensive summary of the work done on the subject.

NONSTANDARD FINITE DIFFERENCE METHODS

The numerical methods discussed in this paper are most closely related to the class of methods called nonstandard finite difference (NSFD) methods, in the sense of the following definition [3]:

Definition 1 *The numerical method (2) is called nonstandard if at least one of the following conditions is true:*

- *Derivatives of order r in $D(\bar{x})$ at time step k are replaced by expressions $\dfrac{d_r(\bar{x}_k)}{\varphi(h)}$, where $d_r(\bar{x}_k)$ is a linear combination of r consecutive approximations of $\bar{x}$ in time and $\varphi(h) = h^r + \mathscr{O}(h^{r+1})$ is a differentiable function;*
- *The right-hand side function $f(\bar{x})$ is approximated at time t_k by a function $g(\ldots, \bar{x}_{k-1}, \bar{x}_k, \bar{x}_{k+1}, \ldots, h)$, obtained by a non-local treatment of each copy of $\bar{x}$ in $f(\bar{x})$. For instance, a term like $\bar{x}^2$ can be replaced by $\bar{x}_k \bar{x}_{k-1}$, $\bar{x}_k \bar{x}_{k+1}$, $2\bar{x}_k^2 - \bar{x}_k \bar{x}_{k-1}$, etc.*

Each item of the definition has been proved to affect different aspects of the dynamical behavior of the numerical methods, such as accuracy, existence and number of fixed points, local and global stability of those fixed points, conservativity of the system, positivity of numerical solutions, and other [9, 24].

The idea of NSFD methods was introduced by Mickens in a 1989 article [29]. The discretization rules were developed empirically for solving practical problems in science and engineering. In most of the cases, the underlying differential equations were analytically solvable and the exact solutions were used to construct the nonstandard numerical schemes. The books [1, 2, 30] and the references therein give a complete listing and a comprehensive summary of the work done up to date on the subject. During the last decade, Mickens has published a number of papers in which he developed nonstandard schemes for specific differential equations, where a nontraditional discrete derivative has naturally arisen [28, 31, 32]. Anguelov and Lubuma [3] were able to address questions about equilibrium stability of some classes of autonomous ODE systems using "elementary stable" nonstandard numerical methods in the sense of the following definition:

Definition 2 *The finite-difference method (2) is called elementary stable if, for any value of the step size h, its only fixed points $\bar{x}^*$ are those of the dynamical system (1), the linear stability properties of each $\bar{x}^*$ being the same for both the dynamical system and the discrete method.*

Some of the initial progress into establishing a theoretical basis for the NSFD methods was also made in [3]. In [19, 35] the nonstandard technique has been applied to some epidemic models for the design of positive and "elementary stable" methods, however with no specific implication made to more general models. The NSFD methods to date have been of rather low accuracy in the time-step, although Twizell and collaborators have made some efforts toward second-order nonstandard schemes for a particular reaction-diffusion system [40]. An essentially complete listing and summary of publications using NSFD methods, up to 2004, is presented in the paper by Patidar [34].

Most of our work in the area of NSFD methods has been concentrated on developing different types of numerical methods of the form:

$$\frac{\bar{x}_{k+1} - \bar{x}_k}{\varphi(h)} = g(\bar{x}_k, \bar{x}_{k+1}, h), \tag{3}$$

for continuous dynamical systems (1). We have applied different denominator functions $\varphi(h)$ in the derivative approximation and different non-local representations $g(\bar{x}_k, \bar{x}_{k+1}, h)$ to numerically preserve specific structural properties of System (1).

In [20, 21, 23, 25, 26], we experimented with nonstandard methods using the classical nonstandard denominator function:

$$\varphi(h) = \frac{e^{ah} - 1}{a},$$

where a is a constant. In [7], we showed that this class of functions can be used to design numerical methods with higher than first-order accuracy and with better dynamical performance for differential equations with polynomial right-hand sides. Our discretization techniques were based on a linear combination of exact nonstandard discretizations for each nonlinear term [20], which makes the corresponding numerical method "locally" exact. An additional advantage of those nonstandard numerical schemes is their improved stability properties, which was demonstrated through a set of numerical simulation (see Figure 2-left). In [7], we also formulated a set of necessary and sufficient conditions for second- and third-order accuracy of the general one-step nonstandard finite-difference methods. Subsequently, we used those results to develop a set of conditions for the discretization $g(\bar{x}_k, \bar{x}_{k+1}, h)$ of the right-hand side function of a differential equation, which is equivalent to the second-order accuracy of the numerical scheme. We proved that the accuracy of nonstandard methods (3) can be regulated by appropriate selection of non-local approximations $g(\bar{x}_k, \bar{x}_{k+1}, h)$.

Another class of nonstandard denominator functions that improves the stability properties of the numerical method was analyzed in [13, 14]. There we showed that nonstandard numerical methods, using the more general denominator function:

$$\varphi(h) = \phi(hq)/q,$$

where

$$\phi(h) = h + \mathcal{O}(h^2) \text{ and } 0 < \phi(h) < 1 \text{ for all } h > 0 \tag{4}$$

with the constant q depending on the corresponding dynamical system, are "elementary stable." This property is especially important in ecology where one would like to predict the long term behavior of a given ecosystem, without being concerned about possible stability problems due to the embedded numerical scheme. In [12, 14] we presented theoretical results for autonomous systems with hyperbolic equilibria by designing "elementary stable" versions of the standard θ- and Runge-Kutta methods. We have successfully applied those methods to several predator-prey, phytoplankton-nutrient, and epidemic models (see Figure 1-left). As an example, the elementary-stable version of the second-order Runge-Kutta method is presented below:

Theorem 1 *Let ϕ be a real-valued function of the class (4) and*

$$q > \max_{\lambda \in \Omega} \left(\frac{|\lambda|^2}{2|Re(\lambda)|} \right), \tag{5}$$

$$\Omega = \bigcup_{\bar{x}^* \in \Gamma} \sigma(J(\bar{x}^*)),$$

where Γ represents the set of all equilibria of System (1) and $\sigma(J(\bar{x}^*))$ denotes the spectrum of the Jacobian $J(\bar{x}^*)$ of System (1) at $\bar{x}^*$. Then the following numerical scheme, based on the standard second-order Runge-Kutta method, represents an elementary stable method:

$$\frac{\bar{x}_{k+1} - \bar{x}_k}{\phi(hq)/q} = \frac{f(\bar{x}_k) + f(\bar{x}_k + (\phi(hq)/q)f(\bar{x}_k))}{2}. \tag{6}$$

In order to also preserve the positivity of numerical solutions, in addition to the local and global stability properties of the corresponding continuous dynamical system, we used non-local approximations of the right-hand side, based on ideas presented by

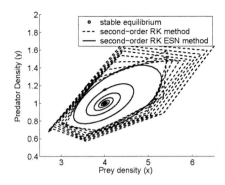 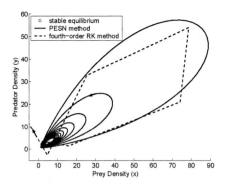

FIGURE 1. Numerical approximations of the solutions of System (9) in the case of a stable interior equilibrium. Comparison of the performance of the standard second-order Runge-Kutta method to its corresponding elementary-stable nonstandard (ESN) version (6) for a large time-step size (left); and a comparison of the performance of the positive and elementary-stable nonstandard (PESN) method (10) to the standard fourth-order Runge-Kutta method for a large time-step size (right)

Patankar in [33]. In general, the procedure can be formulated as follows: *If System (1) is a production-destruction system, i.e.,*

$$f^i(\bar{x}) = P^i(\bar{x}) - N^i(\bar{x}), \quad i = 1, 2, \dots, n, \tag{7}$$

where $P^i(\bar{x}) \geq 0$ and $N^i(\bar{x}) \geq 0$ for all $\bar{x} \in \mathscr{R}_+^n$, then the ith-component of the approximation function $g(\bar{x}_k, \bar{x}_{k+1}, h)$ in the numerical scheme (3) should be selected as follows:

$$g^i(\bar{x}_k, \bar{x}_{k+1}, h) = P^i(\bar{x}_k) - N^i(\bar{x}_k)\frac{x_{k+1}^i}{x_k^i}. \tag{8}$$

In [12] we designed new positive and "elementary stable" numerical methods for a class of phytoplankton-nutrient systems with nutrient loss. In [15] we applied similar techniques to Rosenzweig-MacArthur predator-prey models, which are systems characterized by prey-dependent functional response and include the most popular functions of Holling type I, II and III [18]. We have also extended those results to systems with predator-dependent functional response [10]. As an illustration of our work, let us consider the following Beddington-DeAngelis system:

$$\frac{dx}{dt} = x - \frac{axy}{1 + x + y}; \quad x(0) = x_0 \geq 0,$$

$$\frac{dy}{dt} = \frac{exy}{1 + x + y} - dy; \quad y(0) = y_0 \geq 0, \tag{9}$$

where x and y represent the prey and predator population sizes and the positive constants a, e and d represent the generalized feeding rate, generalized conversion efficiency and generalized mortality rate of the predator, respectively. Our numerical efforts in simulating System (9) are summarized in the following theorem:

Theorem 2 *Let ϕ be a real-valued function on $\mathscr{R}$ that satisfies the property (4). Then the following scheme for solving System (9) represents a positive and "elementary stable" method:*

$$\frac{x_{k+1}-x_k}{\varphi(h)} = x_k - \frac{ax_{k+1}y_k}{1+x_k+y_k};$$

$$\frac{y_{k+1}-y_k}{\varphi(h)} = \frac{ex_k y_k}{1+x_k+y_k} - dy_{k+1};$$

(10)

where $\varphi(h)$ belongs to the following class of functions:

1. *If $ae - e - ad \leq 0$ then $\varphi(h) = \phi(hq)/q$ for all $q \geq 0$, with $\varphi(h) = h$ for $q = 0$.*

2. *If $ae - e - ad > 0$ and $a \neq e$ then $\varphi(h) = \phi(hq)/q$ for $q > \dfrac{e|a-2|}{|e-a|}$.*

The discrete system (10) can be solved explicitly for x_{k+1} and y_{k+1} which decreases the introduced numerical errors and increases the computational speed of the method (see Figure 1-right).

STRUCTURAL PROPERTY-PRESERVING NUMERICAL METHODS

Our current goal in this research area is to develop nonstandard numerical methods for autonomous dynamical systems (1) that preserve:

(a) fixed points and their stability properties;

(b) positivity;

(c) conservativity, limit cycles and other periodic solutions; and

(d) time/trajectory speed.

Although the research work described above has answered some basic questions in this area, still many fundamental questions remain unanswered.

Stability-preserving Numerical Methods

The existence and stability of the equilibria are the most widely examined properties addressed by nonstandard numerical methods for dynamical systems [3, 4, 5, 17, 19, 27]. The numerical schemes, based on explicit exact solutions [20] of the differential systems are effective, but unfortunately practically applicable in only a very few cases. The "elementary stability" is a realistic and a necessary improvement of the standard techniques, which makes the modeling process independent of the limitations of the further simulation process.

Currently we are working in this research direction by:

- developing effective computational tools for calculating/estimating the threshold values of the denominator constant q; and also

- extending the validity of our results to systems with non-hyperbolic equilibria.

Positivity-preserving Elementary-stable Numerical Methods

The question about positivity of numerical solutions arises naturally when solving problems in many areas of the natural and social sciences. The mathematical models expressed by differential equations have solutions with only positive components while the same is not always true for the corresponding difference systems. Most of the published papers, which discuss the problem of numerical positivity, deal with a single system or a specific class of systems [17, 28, 35].

We are extending our numerical approach [10, 12, 15] to multi-dimensional production-destruction systems. In a preliminary result, we have formulated [9] a universal procedure for designing positive and "elementary stable" methods, of the form (3), for a class of production-destruction systems, which includes the major models used in ecology. It contains the following two components:

- Select an appropriate non-local approximation $g(\bar{x}_k, \bar{x}_{k+1}, h)$ of the right-hand side function, as shown in (8), which guarantees an unconditional positivity of the discrete solutions of Scheme (3);
- Choose a denominator function $\varphi(h)$, which also makes the scheme elementary stable. The appropriate selection requires dynamical correspondence between the behavior of the differential and numerical systems around each equilibrium point. This correspondence can be established by matching the stability conditions of both systems, using different tools of analysis of characteristic polynomials of the Jacobians, such as Routh-Hurwitz conditions, and Jury conditions.

Symplectic and Conservative Numerical Methods

Limit cycles and other periodic solutions are dynamical objects of particular importance in physics, chemistry and the life sciences. Mickens [28] and Roeger [36] have discussed the classical Lotka-Volterra system in the context of NSFD methods, which preserve its periodic solutions. Mounim and Dormale [38] have developed a modification of Mickens' nonstandard scheme, which is not only symplectic, but also preserves some symmetrical properties of the solutions. Many other biological systems express similar periodic behavior, such as Beddington-DeAngelis models with non-hyperbolic interior equilibrium [11].

We are currently developing symplectic numerical methods for such biological systems, based on the nonstandard discretization techniques and Kahan's methods [16, 37], and plan to analyze how well they preserve the other characteristics of the periodic solutions, such as size, shape and symmetries.

We are also analyzing the effects of a more general class of non-local representations, in which the terms of the right-hand side function $f(\bar{x})$ of System (1) are replaced with

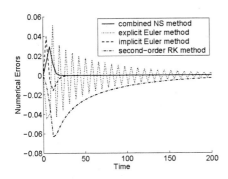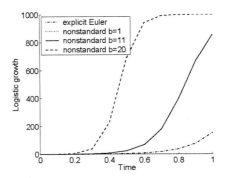

FIGURE 2. Comparison of numerical errors from the combined nonstandard scheme [7], the explicit and implicit Euler schemes, and the second-order Runge-Kutta method when solving the modified logistic-growth equation [41, p. 115] for a large time-step size (left). Comparison of numerical solutions for the logistic equation using the explicit Euler scheme and the nonstandard scheme (11) for different values of b in the denominator function (right)

the following θ-type discretizations:

$$x^i \quad \rightarrow \quad \theta x_n^i - (1-\theta)x_{n+1}^i,$$

$$x^i x^j \quad \rightarrow \quad (\theta_1 x_n^i - (1-\theta_1)x_{n+1}^i)(\theta_2 x_n^j - (1-\theta_2)x_{n+1}^j),$$

etc. The above class of θ-transformations includes the standard implicit ($\theta = 0$) and explicit ($\theta = 1$) forms. Representations, used in [19, 35], are also covered when $\theta = 2$, $\theta_1 = 1$ and $\theta_2 = 0$. This class of non-local approximations promises to be able to not only improve the stability of the equilibria, but to also establish consistency between approximations and dynamical systems regarding limit cycles and other periodic solutions.

Time-preserving Numerical Methods

Approximations of the differential systems with numerical schemes introduce numerical errors in solutions and also in the pace of time. Changes in the nonstandard denominator function $\varphi(h)$, can be used to affect the time pace, slowing down or accelerating the motion of the numerical solutions along the trajectories.

As an illustration of the idea, let us consider the logistic equation $dx/dt = ax(1 - x/K)$, where a and K are given constants, and its numerical approximations by the standard explicit Euler scheme and the following nonstandard scheme:

$$\frac{x_{k+1} - x_k}{\dfrac{e^{bh} - 1}{b}} = ax_k(1 - \frac{x_{k+1}}{K}), \tag{11}$$

for different values of the constant b. In the case when $b = a$ the above nonstandard scheme is exact, i.e., $x(t_k) = x_k$. Numerical simulations for $a = 11$, $K = 1000$, and

$h = 0.1$ (see Figure 2-right) show that an increase of the value of b speeds up the solutions, while a decrease in b slows them down. The difference justifies the use of the denominator function $\dfrac{e^{ah} - 1}{a}$, which corrects the speed of the solutions. This example demonstrates that nonstandard schemes can be used to regulate the pace of time, which makes them important tool not only for long-term modeling purposes but also for short-term predictions. The correct time scaling is very important when modelers are interpreting results of numerical simulations, especially in cases when periodic or oscillatory behavior is observed and an accurate estimate of the period is needed.

We are currently examining the time-scaling effects of the constant q in the nonstandard denominator functions $\varphi(h) = \phi(hq)/q$, as PESN methods are being applied to a variety of dynamical systems in ecology.

CONCLUDING REMARKS

Compatible finite difference methods improve our ability to accurately model more complex processes as well as the ability to understand complex systems. A variety of such numerical methods have been developed and explored over the last several decades. These types of methods enable researchers to make testable predictions concerning the dynamics of numerical simulations of mathematical models. Development of successful property-preserving numerical techniques can be expected to carry over to many applications in ecology, chemistry, physics, economics, and engineering.

The nonstandard discretization approach promises to be a powerful tool, presenting opportunities to design numerical methods for each specific continuous dynamical system and to address any specific point of analytical interest.

REFERENCES

1. R.E. Mickens (Ed), *Advances in the Applications of the Nonstandard Finite Difference Schemes*, World Scientific Publishing, River Edge, NJ, 2005.
2. R.E. Mickens (Ed), *Applications of Nonstandard Finite Difference Schemes*, World Scientific Publishing, River Edge, NJ, 2000.
3. R. Anguelov and J.M.-S. Lubuma, *Numer. Methods Partial Diff. Equations* **17**(5), 518–543 (2001).
4. R. Anguelov and J.M.-S. Lubuma, *Math. Comput. Simulation* **61**(3-6), 465–475 (2003).
5. R. Anguelov, P. Kama, and J.M.-S. Lubuma, *J. Comput. Appl. Math.* **175**(1), 11–29 (2005).
6. H. Burchard, E. Deleersnijder, and A. Meister, *Appl. Numer. Math.* **47**(1), 1–30 (2003).
7. B.M., Chen-Charpentier, D.T. Dimitrov, and H.V. Kojouharov, *Math. Comput. Simulation* **73**(1-4), 105–113 (2006).
8. D. Estep and S. Tavener (Eds), *Collected Lectures on the Preservation of Stability under Discretization (Fort Collins, CO, 2001)*, SIAM, Philadelphia, PA, 2002.
9. D.T. Dimitrov, H.V. Kojouharov, and B.M. Chen, "Reliable Finite Difference Schemes with Applications in Mathematical Ecology", in *Advances in the Applications of the Nonstandard Finite Difference Schemes*, edited by R.E. Mickens, World Scientific Publishing, River Edge, NJ, 2005, pp. 249–286.
10. D.T. Dimitrov and H.V. Kojouharov, *Electron. J. Differential Equations* **15**, 67–75 (2007).
11. D.T. Dimitrov and H.V. Kojouharov, *Appl. Math. Comput.* **162**(2), 523–538 (2005).
12. D.T. Dimitrov and H.V. Kojouharov, *Math. Comput. Simulation* **70**(1), 33–43 (2005).
13. D.T. Dimitrov and H.V. Kojouharov, *Appl. Math. Lett.* **18**(7), 769–774 (2005).
14. D.T. Dimitrov. and H.V. Kojouharov, *Int. J. Numer. Anal. Model.* **4**(2), 282–292 (2007).

15. D.T. Dimitrov and H.V. Kojouharov, *J. Comput. Appl. Math.* **189**(1-2), 98–108 (2006).
16. M.J. Gander and R. Meyer-Spasche, "An Introduction to Numerical Integrators Preserving Physical Properties, Chapter 5," in *Applications of Nonstandard Finite Difference Schemes*, edited by R.E. Mickens, World Scientific Publishing, River Edge, NJ, 2000.
17. A.B. Gumel, R.E. Mickens, and B.D. Corbett, *J. Comput. Meth. Sci. Engin.* **3**(1), 91–98 (2003).
18. C.S. Holling, *Mem. Entomol. Soc. Canada* **45**, 1–60 (1965).
19. H. Jansen and E.H. Twizell, *Math. Comput. Simulation* **58**, 147–158 (2002).
20. H. V. Kojouharov and B.M. Chen, *Appl. Numer. Math.* **49**(2), 225–243 (2004).
21. H.V. Kojouharov and B.D. Welfert, *Lecture Notes in Comput. Sci.* **290**7), 465–472 (2004).
22. H.V. Kojouharov and B.D. Welfert, *J. Comput. Appl. Math.* **151**(2), 335–353 (2003).
23. H.V. Kojouharov and B.D. Welfert, *Internat. J. Appl. Sci. Comput.* **8**(2), 119–126 (2001).
24. H.V. Kojouharov and B.M. Chen, "Nonstandard Eulerian-Lagrangian Methods for Advection-Diffusion-Reaction Equations," in *Applications of Nonstandard Finite Difference Schemes*, edited by R.E. Mickens, World Scientific Publishing, River Edge, NJ, 2000, pp. 55–108.
25. H.V. Kojouharov and B.M. Chen, *Numer. Methods Partial Differential Equations* **15**(6), 617–624 (1999).
26. H.V. Kojouharov and B.M. Chen, *Numer. Methods Partial Differential Equations* **14**(4), 467–485 (1998).
27. J.M.-S. Lubuma and A. Roux, *J. Differ. Equations Appl.* **9**(11), 1023–1035 (2003).
28. A.S. de Markus and R.E. Mickens, *J. Comput. Appl. Math.* **106**(2), 317–324 (1999).
29. R.E. Mickens, *Numer. Methods Partial Diff. Eq.* **5**, 313–3125 (1989).
30. R.E. Mickens, *Nonstandard finite difference model of differential equations*, World Scientific, Singapore, 1994.
31. R.E. Mickens, *Numer. Methods Partial Differential Equations* **13**(1), 51–55 (1997).
32. R.E. Mickens, *J. Differ. Equations Appl.* **8**(9), 823–847 (2002).
33. S.V. Patankar, *Numerical Heat Transfer and Fluid Flow*, McGraw-Hill, New York, 1980.
34. K.C. Patidar, *J. Difference Equ. Appl.* **11**(8), 735–758 (2005).
35. W. Piyawong, E.H. Twizell, and A.B. Gumel, *Appl. Math. Comput.* **146**, 611–625 (2003).
36. Lih-Ing W. Roeger, *J. Difference Equ. Appl.* **11**(8), 721–733 (2005).
37. J.M. Sanz-Serna, *Appl. Numer. Math.* **16**(1-2), 245–250 (1994).
38. Abdellatif Serghini Mounim and Bernard M. de Dormale, *Appl. Numer. Math.* **51**(2-3), 341–344 (2004).
39. A.M. Stuart and A.R. Humphries, *Dynamical Systems and Numerical Analysis*, Cambridge University Press, 1996.
40. E.H. Twizell, A.B. Gumel, and Q. Rao, *Journal of Mathematical Chemistry* **26**, 297–316 (1999).
41. E.K. Yeargers, R. W. Shonkwiler, and J. V. Herod, *An Introduction to the Mathematics of Biology: Computer Algebra Models*, Birkhäuser, Boston, 1996.

Condition and Error Estimates in Numerical Matrix Computations

M.M. Konstantinov* and P.H. Petkov†

*University of Architecture, Civil Engineering and Geodesy, 1046 Sofia, Bulgaria
†Technical University of Sofia, 1000 Sofia, Bulgaria

Abstract. This tutorial paper deals with sensitivity and error estimates in matrix computational processes. The main factors determining the accuracy of the result computed in floating–point machine arithmetics are considered. Special attention is paid to the perturbation analysis of matrix algebraic equations and unitary matrix decompositions.

Keywords: Matrix computations, sensitivity analysis, condition estimates, error estimates.
PACS: 02.60.Dc, 02.60.Cb

INTRODUCTION

When solving matrix computational problems (MCP) in a computing environment there are several sources of errors in the computed solution. Among them the following three are worth mentioning: 1) the *machine arithmetic* and in particular its rounding unit **u** which is of order 10^{-16} for the double precision arithmetics according to the IEEE Standard for floating point calculations [4]; 2) the computational problem and in particular its *sensitivity* to perturbations in the initial data; 3) the computational algorithm and in particular its *numerical stability*. The proper accounting of these factors leads to the derivation of *error bounds* on the computed solution. A numerical solution of MCP may not be recognized as reliable without taking into account the such bounds.

We recall that in floating–point machine arithmetics a real number x in the standard range of the arithmetics, i.e., $10^{-308} < |x| < 10^{308}$, is rounded to the nearest machine number x^*. The rounding and the performance of arithmetic operations in machine arithmetic are governed by the following assumptions.

Assumption 1 *The relative error of rounding is bounded by* **u**, *i.e.,* $|x^* - x|/|x| \leq$ **u**.

Assumption 2 *Let* • *be a binary arithmetic operation and let the numbers* x, y *and* $x \bullet y$ *lie in the standard range of the machine arithmetic. Then the machine result* $(x \bullet y)^*$ *satisfies* $(x \bullet y)^* = (x \bullet y)(1 + \varepsilon)$, *where* ε *is a small multiple of* **u**.

The sensitivity of computational problems may be revealed and taken into account by the *methods and techniques of perturbation analysis*, see, e.g., [9, 6]. In turn, a detailed analysis of computational algorithms in linear algebra and control may be found in [1, 8, 3].

In what follows we use the following notation: $R^{m \times n}$ and $C^{m \times n}$ – the spaces of $m \times n$ matrices over the field of real R and complex C numbers; $R^n = R^{n \times 1}$, I_n – the identity $n \times n$ matrix; $A^\top$ and A^H – the transpose and the complex conjugate transpose of the

CP1067, *Applications of Mathematics in Engineering and Economics '34—AMEE '08*, edited by M. D. Todorov
© 2008 American Institute of Physics 978-0-7354-0598-1/08/$23.00

matrix A; $\text{vec}(A)$ – the column–wise vectorization of the matrix A; $A \otimes B$ – the Kronecker product of the matrices A and B; $\|\cdot\|$ – a vector or a matrix norm; $\|\cdot\|_F$ and $\|\cdot\|_2$ – the Frobenius norm and the 2–norm of a matrix or a vector, respectively; Low, Diag, and Up – the projectors of $C^{n\times n}$ onto the subspaces of lower triangular, diagonal and upper triangular matrices, respectively; $\mathbf{u}$ – the rounding unit of the finite machine arithmetic. The notation ':=' means 'equal by definition'.

PERTURBATION ANALYSIS PROBLEMS

Most computational problems in science and engineering may be formulated in one of the following two ways: problems with *explicit solution*, e.g., evaluating functions defined by explicit computable expressions, and problems with *implicit solution*, e.g., solving equations. A complicated problem may include a chain of explicit and implicit subproblems.

Consider a function $\Phi : \mathscr{A} \to \mathscr{X}$, where the set $\mathscr{A}$ of data and the set $\mathscr{X}$ or results are (subsets) of normed linear spaces which are usually finite-dimensional. The function Φ is at least continuous and often is Lipschitzian or differentiable. Thus for every data $A \in \mathscr{A}$ we have the result $X = \Phi(A) \in \mathscr{X}$.

Example 3 The computation of matrix valued functions, e.g., the matrix exponential $X = \exp(A)$ is a common computational problem in linear system theory.

The dependence of X on A may not be functional. In this case we consider a set-valued function $\widehat{\Phi} : \mathscr{A} \to 2^{\mathscr{X}}$ which assigns a set $\widehat{\Phi}(A)$ of solutions to each data $A \in \mathscr{A}$. The computational problem is identified with the pair (Φ, A) or $(\Phi, \mathscr{A})$.

The function Φ may also be defined implicitly via an equation $F(A, X) = 0$, where $F : \mathscr{A} \times \mathscr{X} \to \mathscr{X}$ is a given continuous function. Here the function Φ satisfies $F(B, \Phi(B)) = 0$ for all $B \in \mathscr{A}$ from a neighborhood of A.

Example 4 The solution of the Lyapunov $Q + A^H X + XA = 0$, $Q + A^H XA - X = 0$ and Riccati $Q + A^H X + XA - XMX = 0$, $Q - X + A^H X (I + MX)^{-1} A = 0$ equations is an implicit computational problem, where the matrix triple (Q, A, M) is the data and X is the solution.

Modifications of implicit problems are the matrix decompositions and the canonical, or condensed forms of matrices and control systems. They are formulated by the equation

$$\mathscr{P}(C(A, U)) = 0, \tag{1}$$

where U is a transformation matrix from a certain multiplicative group Γ, $C(A, U)$ is the condensed form of A relative to Γ and $\mathscr{P}$ is a projector. The problem here is to compute both U and C, where $U = U(A)$ is implicitly defined by A via the projection equation (1).

Example 5 Consider the QR decomposition $A = QR$ of the matrix $A \in C^{n\times m}$, where the matrix Q is unitary and the matrix R is upper triangular. The matrix $R = C(A, Q) = Q^H A$ is the condensed form of A relative to the left multiplicative action of the unitary matrix group $\mathscr{U}(n) \subset C^{n\times n}$ and we have $\mathscr{P} = \text{Low}$ and $\text{Low}(R) = 0$.

Let now the data in the computational problem (Φ, A) be perturbed from A to $A + \delta A$. Then the result X is also perturbed from X to $X + \delta X$, where

$$\delta X = \Psi(\delta A, A) := \Phi(A + \delta A) - \Phi(A).$$

Here Ψ is the *perturbation operator* of the corresponding computational problem. It is important to note that Ψ depends not only on δA but also on the data A.

We stress that *studying the properties of the perturbation operator is the aim of perturbation analysis.*

Suppose that the function Φ is locally Lipschitzian in a neighborhood of the data A.

Definition 6 *The quantity*

$$K(A) := \lim_{\alpha \to 0} \sup \left\{ \frac{\|\Psi(\delta A, A)\|}{\|\delta A\|} : 0 < \|\delta A\| \le \alpha \right\}$$

is the absolute condition number *of the problem $X = \Phi(A)$.*

For problems with explicit solution $X = \Phi(A)$ the absolute condition number is the Lipschitz constant $K(A)$ of Φ at the point A. In particular if the function Φ is differentiable we have $K(A) = \|\Phi'(A)\|$, where the linear operator $\Phi'(A)$ is the Fréchet derivative of Φ computed at the point A. When A and X are vectors then $\Phi'(A)$ is the Jacobi matrix of the vector function Φ.

For problems with implicit solution $F(A, X) = 0$ things are more complicated. Suppose that the partial Frécher derivatives F_X and F_A of F in X and A respectively exist and that the linear operator $F_X : \mathscr{X} \to \mathscr{X}$ is invertible. Within terms of first order in δA, δX we have $\delta X \approx -F_X^{-1}(A, X) \circ F_A(A, X)(\delta A)$ and hence

$$K(A) = \left\| F_X^{-1}(A, X) \circ F_A(A, X) \right\|.$$

When $A \ne 0$ and $X \ne 0$ we may define the *relative condition number* from

$$k(A) := K(A) \frac{\|A\|}{\|X\|}.$$

For small δA we have

$$\delta_X \le k(A)\delta_A + O(\delta_A^2), \ \delta_X := \frac{\|\delta X\|}{\|X\|}, \ \delta_A := \frac{\|\delta A\|}{\|A\|}.$$

So the main problem of *local perturbation analysis* is to find estimates for the condition number $k(A)$. More sophisticated perturbation schemes are aimed at obtaining improved local perturbation bounds which may not be based on condition numbers.

A more detailed perturbation analysis leads to estimates of the form

$$|\delta X| \preceq L(A)|\delta A|,$$

where $|\cdot| : \mathscr{X} \to \mathscr{K}$ is a generalized norm and $\mathscr{K}$ is a non-negative cone, say R_+^m. In this case L is a non–negative matrix. In particular, if A and X are vectors or matrices, then $|A|$ and $|X|$ may be vector or matrix absolute values.

Sometimes it is hard to distinguish explicit and implicit problems.

Example 7 The solution of the general Sylvester equation $A_1XB_1 + A_2XB_2 + \cdots + A_rXB_r = C$, where the coefficients A_k, B_k, C and the solution X are matrices of compatible size, is an implicit problem. But we may also write the formal solution as

$$X = \text{vec}^{-1}(M^{-1}\text{vec}(C)), \; M := \sum_{k=1}^{r} B_k^{\top} \otimes A_k,$$

thus obtaining a problem with explicit solution.

Let a numerically stable algorithm be applied to solve the problem $X = \Phi(A)$ in a computing environment with rounding unit $\mathbf{u}$. Then, within first order terms in $\mathbf{u}$, the computed solution X^* shall be close to the exact solution of a near problem in the sense that $\|X^* - \Phi(A^*)\| \leq a\mathbf{u}\|X\|$ and $\|A^* - A\| \leq b\mathbf{u}\|A\|$, where the constants a and b depend on the computational algorithm.

Note that for $a = 0$ the computed solution X^* is the exact solution $\Phi(A^*)$ of a near problem since A^* and A are $\mathbf{u}$–close. Thus the algorithm is *backwardly stable* in the sense of Wilkinson. When $b = 0$ we have $A^* = A$ and the algorithm is *forwardly stable*.

Now we may estimate the actual error in the computed solution as

$$
\begin{aligned}
\|X^* - X\| &= \|X^* - \Phi(A)\| = \|X^* - \Phi(A^*) + \Phi(A^*) - \Phi(A)\| \\
&\leq \|X^* - \Phi(A^*)\| + \|\Phi(A^*) - \Phi(A)\| \\
&\leq a\mathbf{u}\|X\| + K(A)\|A^* - A\| \leq a\mathbf{u}\|X\| + Kb\mathbf{u}\|A\|.
\end{aligned}
$$

Dividing both sides of the last inequality by $\|X\|$ (when $X \neq 0$) we get the important accuracy estimate

$$\frac{\|X^* - X\|}{\|X\|} \leq \mathbf{u}(a + bk(A)). \tag{2}$$

The inequality (2) clearly reveals the three main factors determining the relative error: the parameters of the computing environment via the rounding unit $\mathbf{u}$; the sensitivity of the computational problem via the relative condition number $k(A)$, and the properties of the computational algorithm via the constants a and b.

In practice it is difficult to estimate the constants a and b. If we set $a = 0$ and $b = 1$ the following heuristic estimate is obtained

$$\frac{\|X^* - X\|}{\|X\|} \leq \mathbf{u}k(A). \tag{3}$$

Thus we have the following heuristic rule.

When $\mathbf{u}k(A) < 1$ we may expect about $-\log_{10}(\mathbf{u}k(A))$ true decimal digits in the computed solution.

Another error estimate concerns the equivalent data perturbation. Suppose that we have an approximate solution $X^* = X + \delta X$ which in particular may be any computed solution. Suppose that an equivalent perturbation δA exists such that $X^* = \Phi(A + \delta A)$. Then we may define the absolute backward error as

$$E(X^*) := \inf\{\|\delta A\| : X^* = \Phi(A + \delta A)\}.$$

In case of an implicit problem $F(A, X) = 0$ the backward error is given by

$$E(X^*) := \inf\{\|\delta A\| : F(A + \delta A, X^*) = 0\}.$$

In both cases $E(X) = 0$.

When X^* is close to X then δA will be small and we have $X + \Phi'(A)(\delta A) \approx X^*$. Hence $E(X^*)$ may be estimated from

$$E(X^*) \leq \|(\Phi'(A))^\dagger\| \|X^* - X\|,$$

where $L^\dagger$ is the normal pseudo-inverse of the linear operator L.

PERTURBATION BOUNDS

Perturbation bounds may be local and non–local, linear and non–linear, forward and backward, norm–wise and component–wise, and even sharp, non–improvable and so long. In this section we shall analyze local and non–local bounds.

Suppose that the data A is a collection of matrices $A_1, A_2, \ldots, A_m$ which are perturbed as $A_i \to A_i + \delta A_i$, and let $X + \delta X$ be the solution of the perturbed problem. Using the Fréchet derivatives or pseudo-derivatives of the functions Φ for explicit problems, or and F for implicit problems, it is possible to derive expressions as

$$x \approx \sum_{i=1}^m M_i a_i,$$

where $x := \text{vec}(\delta X)$, $a_i := \text{vec}(\delta A_i)$ and M_i are easily computable matrices. We note that $\|\delta X\|_F = \|x\|_2$. Let $\delta_i := \|\delta A_i\|_F$ and $\delta := [\delta_1, \delta_2, \ldots, \delta_m]^\top \in R_+^m$.

The first local bound is $\|x\|_2 \leq \text{est}_1(\delta) + O(\|\delta\|^2)$, where $\text{est}_1(\delta) := K\delta = K_1\delta_1 + K_2\delta_2 + \cdots + K_m\delta_m$, and $K := [K_1, K_2, \ldots, K_m]$, $K_i := \|M_i\|_2$. This is a condition number based estimate since K_i is the absolute condition number relative to A_i and K is the absolute condition vector of the problem.

Another local estimate is $\|x\|_2 \leq \text{est}_2(\delta) + O(\|\delta\|^2)$, where

$$\text{est}_2(\delta) := \|[M_1, M_2, \ldots, M_m]\|_2 \|\delta\|_2.$$

A third easily computable local bound is $\|x\|_2 \leq \text{est}_3(\delta) + O(\|\delta\|^2)$, where $\text{est}_3(\delta) := \sqrt{\delta^\top M \delta}$ and the non–negative matrix M is defined from $M := [M_{ij}] \in R_+^{m \times m}$, $M_{ij} := \|M_i^H M_j\|_2$.

It may be shown that $\text{est}_3(\delta) \leq \text{est}_1(\delta)$. Hence we have the following improved local perturbation estimate.

Theorem 8 *Within first order terms in* $\|\delta\|$ *the following estimate is valid*

$$\|\delta X\|_F \leq \text{est}(\delta) + O(\|\delta\|^2), \quad \text{est}(\delta) := \min\{\text{est}_2(\delta), \text{est}_3(\delta)\}.$$

Non–local *norm–wise* perturbation bounds are expressions of the form

$$\|\delta X\|_F \leq f(\delta), \ \delta \in \Omega \subset R_+^r,$$

where $f : \Omega \to R_+$ is a given continuous function, non–decreasing in each of its arguments and satisfying $f(0) = 0$.

There are many techniques to derive non–local bounds. One of the most effective ways is using the technique of Lyapunov majorants which is briefly considered below.

When generalized rather than usual norms are used we have *component–wise* perturbation bounds. In this case $f(\delta)$ may be a non–negative vector. Such bounds may be derived by the use of vector Lyapunov majorants.

PERTURBATION ANALYSIS OF MATRIX EQUATIONS

Matrix equations take a significant part in the mathematical modelling of systems and processes in science and engineering and in particular in control theory.

Consider the algebraic matrix equation $F(A,X) = 0$, $A = (A_1, A_2, \ldots, A_m)$, where A_i are matrix coefficients subject to perturbations $A_i \to A_i + \delta A_i$. Let $X + \delta X$ be the solution of the perturbed equation, i.e., $F(A + \delta A, X + \delta X) = 0$.

The aim of norm–wise perturbation analysis here is to estimate the norm $\|\delta X\|_F$ of δX as a function of the perturbation vector $\delta = [\delta_1, \delta_2, \ldots, \delta_m]^\top$, where $\delta_i := \|\delta A_i\|_F$.

Under some differentiability and regularity conditions the perturbed equation may be written as an equivalent operator equation $\delta X = \Pi(\delta A, \delta X)$, where

$$
\begin{aligned}
\Pi(E,Y) &:= -F_X^{-1}(A,X)(F_A(A,X)(E) + G(A,X,E,Y)), \\
G(A,X,E,Y) &:= F(A+E, X+Y) - F(A,X) \\
&\quad - F_A(A,X)(E) - F_X(A,X)(Y).
\end{aligned}
$$

It may be shown that the operator $\Pi(\delta A, \cdot)$ transforms into itself a 'small' set $\mathscr{B}_\rho$ of radius ρ vanishing with $\|\delta A\|$. Thus according to the Schauder fixed point principle there is a 'small' solution for δX with $\|\delta X\|_F \leq \rho$. The last inequality is the desired non–local perturbation estimate.

In this case the use of the Schauder principle is based on the technique of Lyapunov majorants which goes back to the monograph [2]. Further developments on this subject may be found in [6].

Consider again the operator equation $\delta X = \Pi(\delta A, \delta X)$ for the perturbation δX, where $A = (A_1, A_2, \ldots, A_m)$. Set accordingly $E = (E_1, E_2, \ldots, E_m)$ and let $\delta := [\delta_1, \delta_2, \ldots, \delta_m]^\top$ be a given non–negative vector.

Definition 9 *The function $h : R_+^r \times R_+ \to R_+$, defined by*

$$h(\delta, \rho) := \sup\{\|\Pi(E,Y)\|_F : \|E_i\|_F \leq \delta_i, \ \|Y\|_F \leq \rho\}$$

is called a Lyapunov majorant *for the operator Π.*

Since it is difficult to construct the exact Lyapunov majorant, we use an easily computable function $\widehat{h}$ such that $h(\delta, \rho) \leq \widehat{h}(\delta, \rho)$. The function h is non–decreasing and convex and satisfies $h(0,0) = 0$ and $h_\rho'(0,0) < 1$.

The technique of Lyapunov majorants is based on the *majorant equation* $\rho = h(\delta, \rho)$. Under the above conditions we have the following result.

Theorem 10 *There exists a domain $\Omega \subset R_+^m$ such that:*
(i) for $\delta \in \Omega$ the majorant equation has two roots $\rho_1(\delta) < \rho_2(\delta)$ which are differentiable in δ and $\rho_1(0) = 0$;
(ii) for some points δ on the boundary $\partial\Omega$ of Ω it is fulfilled $\rho_1(\delta) = \delta_2(\rho)$; at such points the function ρ_1 may not be differentiable but is still continuous.

Denote by $\overline{\Omega}$ the closure of Ω and set $f(\delta) := \rho_1(\delta)$ and $\mathscr{B}(\delta) := \{Y : \|Y\|_F \leq f(\delta)\}$. For $\delta \in \overline{\Omega}$ and $\|Y\|_F \leq f(\delta)$ we have $\|\Pi(E,Y)\|_F \leq f(\delta)$. Thus the operator $\Pi(E, \cdot)$ transforms the set $\mathscr{B}(\delta)$ into itself. Hence, according to the Schauder fixed point principle, there is a solution $Y \in \mathscr{B}(\delta)$ of the operator equation $Y = \Pi(E,Y)$. As a corollary we have the following estimate.

Theorem 11 *The non–local non–linear perturbation estimate holds*

$$\|\delta X\|_F \leq f(\delta), \ \delta \in \overline{\Omega}.$$

In practice the domain Ω is not constructed explicitly. Rather, the inclusion $\delta \in \overline{\Omega}$ is checked directly by a single inequality.

When component–wise estimates are derived then the technique of vector Lyapunov majorants may be used.

For linear matrix equations $Q + A_1 X B_1 + A_2 X B_2 + \cdots + A_r X B_r = 0$ the technique of Lyapunov majorants may also be used. However, in this case the component–wise bounds are more interesting. The corresponding results may be found in [6]. The most simple vector case is $Ax = c$. Let the matrix A and the vector c be perturbed to $A + E$ and $c + e$, where component–wise bounds for E and e are known. If the matrix A is invertible and the spectral radius of the matrix $|A^{-1}||E|$ is less than 1, then the perturbation δx in x satisfies the component–wise estimate

$$|\delta x| \preceq (I - |A^{-1}||E|)^{-1}(|e| + |E||x|).$$

For quadratic matrix equations

$$Q + \sum_{k=1}^{l} A_k X B_k + \sum_{i=1}^{m} C_i X D_i X E_i = 0$$

the Lyapunov majorant is quadratic,

$$h(\delta, \rho) = a_0(\delta) + a_1(\delta)\rho + a_2(\delta)\rho^2,$$

where $a_0(\delta)$ is a function of type est(δ) and $a_1(0) = 0$. The concrete expressions for the coefficients $a_k(\delta)$ for this and many other equations may be found in the monograph [6].

The majorant equation here is $a_0(\delta) + (1 - a_1(\delta))\rho + a_2(\delta)\rho^2 = 0$ and hence the domain $\overline{\Omega}$ (pretty complicated in the general case) is defined by the inequality $a_1(\delta) + 2\sqrt{a_0(\delta)a_2(\delta)} \leq 1$.

Finally, the non–local bound is

$$\|\delta X\|_F \le f(\delta) := \frac{2a_0(\delta)}{1 - a_1(\delta) + \sqrt{(1 - a_1(\delta))^2 - 4a_0(\delta)a_2(\delta)}}$$

provided that $\delta \in \overline{\Omega}$.

Matrix algebraic equations involving m–th degree expressions ($m \ge 3$) in X are treated similarly. For $m = 3$ a special technique is applied and the 3-rd degree Lyapunov majorant is replaced by a quadratic one as in the previous subsection. For $m \ge 4$ the technique is more involved but the m-th degree Lyapunov majorant is again replaced by a quadratic one.

Fractional affine matrix equations involve terms $G^{-1}(X)$, where $G(X)$ is an affine expression in X. Typical example here is the discrete-time matrix Riccati equation $X = Q + A^H X(I + MX)^{-1}A$. Using some matrix manipulations, the Lyapunov majorants $h(\rho, \delta)$ for such equations are again reduced to second degree polynomials in ρ.

THE METHOD OF SPLITTING OPERATORS

The method of splitting operators (MSO) had been proposed by the authors in 1993 and has been recently justified in the paper [7]. It is based on splitting of matrices and operators into lower, diagonal and upper triangular forms by the operators Low, Diag and Up, considered above. This method is very effective in the perturbation analysis of problems, involving unitary (or orthogonal) matrix factorizations.

Consider the QR decomposition $A = QR$ of the matrix A, where the matrix Q is unitary and the matrix R is upper triangular. Perturbing the corresponding matrices and setting $X := \delta Q^H Q$ and $E := Q^H \delta A$ we get $XR = \delta R - (I + X)E$. Since $I + X$ is unitary we have $X + X^H + X^H X = 0$.

If we split the matrix X as $X = X_1 + X_2 + X_3$, where $X_1 := \mathrm{Low}(X)$, $X_2 := \mathrm{Diag}(X)$, $X_3 := \mathrm{Up}(X)$, and use the equation for X above, we get

$$X_1 = \Pi_1(\delta A, X), \tag{4}$$

where $\Pi_1(B, X)$ is an easily computable expression.

Two more equations for X_2 and X_3 follow from the unitarity of X, namely

$$X_2 = \Pi_2(X) := -0.5\,\mathrm{Diag}(X^H X), \tag{5}$$
$$X_3 = \Pi_3(X) := -\mathrm{Up}(X^H) - \mathrm{Up}(X^H X). \tag{6}$$

Equations (4) and (5), (6) constitute and operator equation

$$X = \Pi(\delta A, X) \tag{7}$$

for X. Note that equations (5), (6) do not depend on the particular problem and are universal equations for MSO.

Furthermore a vector Lyapunov majorant is constructed for equation (7) which yields a non–local non–linear perturbation bound for $\|\delta X\|_F$ and hence for $\|\delta R\|_F$.

The MSO is a powerful tool to obtain local and non–local perturbation bounds for matrix factorizations involving unitary or orthogonal factors. Among the most important applications of MSO are the Schur, QR, Hamilton–Schur and block–Schur decompositions of a matrix. A brief description of this applications is given below.

The application of MSO to the perturbation analysis of the QR decomposition has already been outlined. Consider now the Schur decomposition $A = UTU^H$ of the matrix $A \in C^{n \times n}$, where the matrix U is unitary and the Schur form T of A is an upper triangular matrix. The perturbed decomposition is $A + \delta A = (U + \delta U)(T + \delta T)(U + \delta U)^H$. Here the application of MSO is more involved but it allows to find an equivalent operator equation for $X = U^H \delta U$ and the corresponding vector Lyapunov majorant [7].

Consider next the Hamiltonian $2n \times 2n$ matrix $H = \begin{bmatrix} H_1 & H_3 \\ H_2 & -H_1^H \end{bmatrix}$, where the matrices H_2 and H_3 are Hermitian. Then the Hamilton Schur decomposition of H is $H = U\Sigma U^H$, where $\Sigma = \begin{bmatrix} \Sigma_1 & \Sigma_2 \\ 0 & -\Sigma_1^H \end{bmatrix}$ and U is an unitary simplectic matrix. Here the MSO and the technique of vector Lyapunov majorants again allows to perform a complete non–local perturbation analysis of the problem.

In the block–Schur decomposition the Schur form of A is $T = \begin{bmatrix} T_{11} & T_{12} \\ 0 & T_{22} \end{bmatrix}$, where the matrices T_{11} and T_{22} are upper triangular and have disjoint spectra. Here MSO again solves the perturbation problem.

MSO finds surprisingly efficient applications in control theory. Two such applications are considered below.

The controllable pair $[A, B] \in C^{n \times n} \times C^{n \times m}$ may be reduced into a canonical form $[A^0, B^0] = [U^H AU, U^H B)$, where the matrix U is unitary and A^0 is an upper block-Hessenberg matrix. The perturbation analysis of such forms was done by MSO in a number of papers of the authors and G. Sun.

Consider now the controllable pair $[A, B]$ and let $\lambda \subset C$ be a prescribed collection of eigenvalues of the closed–loop system matrix. The *pole assignment problem* is to find a gain matrix K such that the spectrum of the matrix $A + BK$ to be equal to λ.

In a more general framework let M be an unitarily attainable form of the closed–loop system matrix. Then the *general problem of synthesis of linear static feedback* is to find K so that $A + BK$ to be unitarily similar to the matrix M. Perturbation analysis for both feedback synthesis problems have been done by using MSO.

CONCLUSIONS AND UNSOLVED PROBLEMS

In this paper we have considered a number of important computational problems in linear algebra and control together with the corresponding perturbation methods for the analysis of their sensitivity. There are many unsolved problems in this area and some of them are listed below.

- Perturbation analysis of the multi–block Schur form

$$T = U^{\mathrm{H}}AU = \begin{bmatrix} T_{11} & T_{12} & \cdots & T_{1m} \\ 0 & T_{22} & \cdots & T_{2m} \\ \vdots & \vdots & \ddots & \vdots \\ 0 & 0 & \cdots & T_{mm} \end{bmatrix}$$

of the $n \times n$ matrix A, where $2 < m < n$ and the matrices T_{kk} have disjoint spectra.

- Perturbation analysis of Jordan-like forms $J = P^{-1}AP$ of the square matrix A which are upper bi-diagonal matrices with arbitrary elements on the super-diagonal, e.g., $J = \mathrm{diag}(J_1, J_2, \ldots, J_m)$, where J_i have the form $J_i = \lambda$, or

$$J_i = \begin{bmatrix} \lambda & a_1 & \cdots & 0 & 0 \\ 0 & \lambda & \cdots & 0 & 0 \\ \vdots & \vdots & \ddots & \lambda & a_{k-1} \\ 0 & 0 & \cdots & 0 & \lambda \end{bmatrix} \in C^{k \times k}.$$

We recall that Jordan-like forms have better numerical behavior than the standard Jordan forms with $a_1 = a_2 = \cdots = a_{k-1} = 1$.

- Perturbation analysis of Hamiltonian matrix pencils $\{\lambda L - \mu M : \lambda, \mu \in C\} \subset C^{2n \times 2n}$ under the action of the group of symplectic $2n \times 2n$ matrices.
- Perturbation analysis of pole assignment problems via dynamic feedback.
- Perturbation analysis of Hamiltonian–Schur and block–Schur forms of linear operators in Hilbert spaces.
- Backward perturbation analysis of the above mentioned problems

REFERENCES

1. G. Golub and C. Van Loan, *Matrix Computations*, John Hopkins Univ. Press, Baltimore, 1996.
2. E. Grebenikov and Yu. Ryabov, *Constructive Methods for Analysis of Nonlinear Systems*. Nauka, Moscow, 1979. (in Russian)
3. N. Higham, *Accuracy and Stability of Numerical Algorithms*, SIAM, Philadelphia, 2006.
4. *IEEE Standard on Floating Point Computations*, IEEE, New York, 1985.
5. M. Konstantinov, *Foundations of Numerical Analysis*, Publ House UACEG, Sofia, 2005, PDF Text available at www.uacg.bg/books.
6. M. Konstantinov, D. Gu, V. Mehrmann, and P. Petkov, *Matrix Perturbation Theory*, Elsevier, Amsterdam, 2003.
7. M. Konstantinov and P. Petkov, *Numer. Func. Anal. Appl.* **23**, 529–572 (2002).
8. P. Petkov, N. Christov, and M. Konstantinov, *Computational Methods for Linear Control Problems*, Prentice Hall, Hemel Hempstead, 1991.
9. G. Stewart and J. Sun, *Matrix Perturbation Theory*, Academic Press, New York, 1990.
10. N. Vulchanov and M. Konstantinov, *Modern Mathematical Methods for Computer Calculations. Part 1: Foundations of Computer Calculations*, BIAR Studies in Mathematics, vol. 1, Demetra, Sofia, 1996, 2006.

Electromagnetics: from Covariance to Cloaking

M. W. McCall

Dept. of Physics, Imperial College London, SW72AZ, UK.

Abstract. An overview of some topical themes in electromagnetism is presented. Recent interest in metamaterials research has enabled earlier theoretical speculations concerning electromagnetic media displaying a negative refractive index to be experimentally realized. Such media can act as perfect lenses. The mathematical criterion of what signals such unusual electromagnetic behavior is discussed, showing that a covariant (or coordinate free) perspective is essential. Coordinate transformations have also become significant in the theme of transformation optics, where the interplay between a coordinate transformation and metamaterial behavior has led to the concept of an electromagnetic cloak.

Keywords: Electromagnetics, negative index of refraction, covariance, relativity, transformation optics, cloaking.
PACS: 03.50.De, 41.20.-q

INTRODUCTION

The aim of this review is to provide a bridge from the old to the new. All practising physicists and most practising mathematicians are familiar with the equations developed by James Clerk Maxwell in the nineteenth century [16] that provide the most complete descriptions we know of electromagnetic phenomena. Despite being studied for over a century, a quiet revolution is in progress over the last decade in which some hitherto cherished notions are now being challenged. Key to this revolution is that technology has now provided the means to manufacture materials that, from an electromagnetic perspective, are *radically* different from those encountered previously. Snell's law, for example, which we learned at high school, has undergone a reformulation, and the field of *negative refraction* has been born. As we move into this new electromagnetic era, new mathematical descriptions are found to be appropriate. Maxwell's equations are still Maxwell's equations, but it turns out that coordinate-free, or co-variant descriptions, are particularly clarifying, particularly in relation to some of the controversies that have emerged as the field has developed. The emergence of so-called *transformation optics* in which an imagined electromagnetic space becomes physically realized, is also naturally suited to a covariant description, where theory and experiment have merged to produce the electromagnetic cloak of Harry Potter fame.

CP1067, *Applications of Mathematics in Engineering and Economics '34—AMEE '08,* edited by M. D. Todorov
© 2008 American Institute of Physics 978-0-7354-0598-1/08/$23.00

CLASSICAL SOURCE-FREE ELECTROMAGNETICS

In the absence of free charges and currents Maxwell's equations are ([2] pp. 1–2.)

$$\nabla \times \mathbf{E} = -\frac{\partial \mathbf{B}}{\partial t}, \quad \nabla \times \mathbf{H} = \frac{\partial \mathbf{D}}{\partial t},$$

$$\nabla \cdot \mathbf{D} = 0, \quad \nabla \cdot \mathbf{B} = 0. \tag{1}$$

In fact the last two equations can be derived from the first two using the differential vector identity $\nabla \cdot (\nabla \times \mathbf{A}) = 0$ for any vector field $\mathbf{A}$. We thus have effectively two vector equations, or six scalar equations, for the unknown field vectors $\mathbf{E}, \mathbf{B}, \mathbf{D}$ and $\mathbf{H}$, a total of twelve unknown scalar components. The theory is therefore incomplete without further equations, and these are provided by the so-called *constitutive relations* that characterize the media within which the fields are propagating. They can take various forms; for instantaneously responding *linear* media we have

$$\mathbf{D} = = \underline{\varepsilon} \cdot \mathbf{E} + \underline{\alpha} \cdot \mathbf{B}$$

$$\mathbf{H} = = \underline{\beta} \cdot \mathbf{E} + \underline{\mu}^{-1} \cdot \mathbf{B} \tag{2}$$

where the tensors $\underline{\varepsilon}$, $\underline{\mu}^{-1}$, $\underline{\alpha}$ and $\underline{\beta}$ specify the nature of the electromagnetic medium. Equations (2) provide the six additional equations required to specify the fields. Equations (1) and (2) are sufficient, together with appropriate boundary conditions to solve problems in linear electromagnetics. One of the simplest situations to consider is where

(a) the medium is homogeneous (i.e., the constitutive parameters are not dependent on position),

(b) the so-called *magneto-electric* terms in (2), (i.e., $\underline{\alpha}$ and $\underline{\beta}$) are absent.

(c) the magnetic response is isotropic, i.e., $\underline{\mu}^{-1} = \mu^{-1}\mathrm{diag}(1,1,1)$ where μ^{-1} is a constant.

(d) the electric field is assumed to be a time harmonic plane wave of the form $\mathbf{E} = \mathrm{Re}\left[\mathbf{E}_0 e^{i(\mathbf{k} \cdot \mathbf{r} - \omega t)}\right]$.

Under these assumptions it is straightforward to show that, $\mathbf{E}_\perp$, the electric field perpendicular to the wave vector $\mathbf{k}$ must satisfy [17]

$$\left[\omega^2 \mu \left(\underline{\varepsilon}^{-1}\right)_\perp^{-1} - k^2\right] \cdot \mathbf{E}_\perp, \tag{3}$$

where $k^2 = \mathbf{k} \cdot \mathbf{k}$, and $\left(\underline{\varepsilon}^{-1}\right)_\perp$ is the restriction of the inverse of $\underline{\varepsilon}$ to the plane perpendicular to $\mathbf{k}$. Equation (3) is an eigen-relation for the field $\mathbf{E}_\perp$ and the wave number k. The latter is found from the *dispersion relation*

$$\det\left|\omega^2 \mu \left(\underline{\varepsilon}^{-1}\right)_\perp^{-1} - k^2\right| = 0, \tag{4}$$

which is a bi-quadratic equation yielding the solutions $\pm k_i$, $i = 1, 2$. For each of the k_i there is a corresponding polarization of the electric field $\mathbf{E}_i$, $i = 1, 2$. From the theory of eigen-systems, these polarizations are orthogonal. The above procedure is the classical algorithm for finding the plane wave electromagnetic modes for a birefringent medium such as calcite. When the constitutive parameters are more complicated than this simple case, the procedure for calculating the plane electromagnetic modes is harder, but basically similar. Once the electric field is determined, the remaining fields follow from equations (1).

The direction of flow of electromagnetic power is determined from the Poynting vector, given by

$$\mathbf{P} = \frac{1}{2}\text{Re}\left(\mathbf{E} \times \mathbf{H}^*\right) . \tag{5}$$

In the context of this general discussion, it is important to note that the direction of $\mathbf{P}$ need not be in the direction of the wave vector $\mathbf{k}$. In birefringent media, for example, there is a small angular separation. However, for isotropic media where all the tensors in equations (2) are all multiples of the identity, then $\mathbf{P}$ and $\mathbf{k}$ are parallel. For example, the electric field in an isotropic, non-magnetoelectric medium, satisfies

$$\nabla^2 \mathbf{E} - \varepsilon\mu \frac{\partial^2 \mathbf{E}}{\partial t^2} = 0 , \tag{6}$$

together with a similar equation for $\mathbf{H}$.

VESELAGO'S SPECULATION

In 1968 Victor Veselago made an interesting, but highly provocative theoretical speculation [32]. He said, let us imagine a medium in which the scalar parameters ε and μ are simultaneously *negative*. Now of course the wave equation, equation (6) is unaffected, so that the field $\mathbf{E}$ is the same. The difference comes in consideration of the curl relation, the first of equations (1). This dictates that the orientation of the field vectors must be as shown in Figure 1, wherein the Poynting vector $\mathbf{P}$ points oppositely to the wave vector $\mathbf{k}$. How strange! The wave's phase advances in one direction, but the power flow is in the opposite direction. This impacts many physical laws, but the most striking is that of refraction at an interface - Snell's law. Take a plane interface between a conventional medium characterized by $\varepsilon, \mu > 0$ (for example vacuum, where $\varepsilon = \varepsilon_0$ and $\mu = \mu_0$) and a Veselago medium for which $\varepsilon, \mu < 0$. Imagine light is incident in the conventional medium from the left. At the interface the light refracts, but not in the usual way that we learned at school. Rather it refracts on the *same* side of the normal to the interface as shown in Figure 2 for light passing through a slab of such a medium. As well, the light's energy travels away from the interface, but the wave vector points back towards the interface as discussed above. Snell's law, which for a vacuum-medium interface states that

$$\sin\theta_0 = n\sin\theta \tag{7}$$

is still satisfied perfectly, because the negative sign of n dictates that θ must also be negative (i.e., the same side of normal). How this arises from the wave equation (6) is

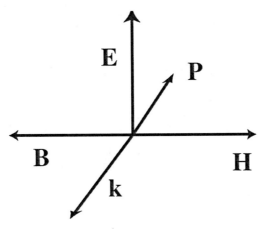

FIGURE 1. Field vectors $\mathbf{E}, \mathbf{B}, \mathbf{H}, \mathbf{k}$ and $\mathbf{P} = \frac{1}{2}(\mathbf{E} \times \mathbf{H}^*)$ in a medium for which ε and μ are simultaneously negative

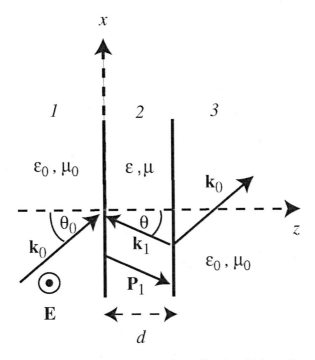

FIGURE 2. Light incident from vacuum to a slab of a medium for which ε and μ are both negative. Inside the medium the wave vector $\mathbf{k}_1$ points oppositely to the Poynting vector $\mathbf{P}_1$

that the latter dictates that $k^2 = \varepsilon\mu\omega^2$, but taking the negative square root yields the negative refractive index according to

$$n = c\frac{k}{\omega} = -\left(\frac{\varepsilon\mu}{\varepsilon_0\mu_0}\right)^{1/2}. \tag{8}$$

NOVEL ELECTROMAGNETIC MEDIA

Veselago's speculation remained just that - a speculation, since at the time there was no way to produce materials that simultaneously had negative ε and negative μ. Negative ε is easy. The electrons in metals behave like a plasma for which the dielectric response is given by[1]

$$\varepsilon(\omega) = 1 - \frac{\omega_p^2}{\omega^2}, \tag{9}$$

which is negative for frequencies above the plasma frequency ω_p. The problem is with the magnetic part. In a much quoted text from Landau and Lifschitz [4] it was said

> Unlike ε, μ ceases to have any physical meaning above a few GHz. To take account of $\mu(\omega)$ would be an unwarrantable refinement ... there is certainly no meaning in using the magnetic susceptibility from optical frequencies onwards and in discussion of such phenomena we must put $\mu = 1$.

Naturally occurring media are magnetically inert (i.e., $\mu = 1$) above a few GHz. The breakthrough came in 2001 when it was shown that magnetic activity leading to negative index could result from artificial structures called metamaterials [30]. The so-called split-ring structure consisted of a pair of concentric metallic rings as shown in Figure 3. This structure is capacitively resonant and behaves magnetically with the response function shown in Figure 4. As with all driven oscillator systems, the response is contra to the driving force (in this case the exciting magnetic field) above the resonance frequency, ω_{mp}. If the resonance is sufficiently sharp, then the μ for this system can be negative over a limited bandwidth above ω_{mp}. With high frequency magnetic response achieved, the way was open to produce a negative refractive index meta-material. The breakthrough experiments of Shelby et al [30] were based on a metamaterial consisting of an array of the split ring resonators discussed above, together with an array of wires that mimicked the purely metallic response, but with a plasma frequency close to the ω_{mp} of the split rings. With the dielectric and magnetic responses now in the same (GHz) spectral region, the composite metamaterial did indeed show negatively refracting behavior. A prism of the material refracted light the 'wrong' way.

PERFECT LENSES

We all learn that the resolution limit for lenses is limited by diffraction to be about one wavelength. Such has been known since the days of Abbé (see, e.g., [2], p. 461. However, transmission through a slab of negatively refracting material radically changes this conclusion because of the way the slab treats the higher spatial frequencies within

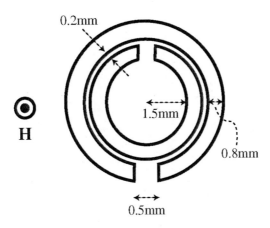

FIGURE 3. Split-ring resonator

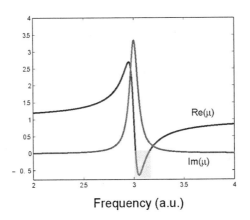

FIGURE 4. Resonant response of a split-ring resonator. In the shaded region μ is negative

the object to be imaged. Firstly, it is a simple exercise to show from Figure 5 that the unusual form of Snell's law operating at the interface of a slab of thickness d leads to all rays being imaged according to

$$d = d_1 + d_2 . \tag{10}$$

So unlike conventional lenses that are necessarily curved, a 'Veselago lens' images using a flat slab. However, Pendry [26] showed that, remarkably, an image formed by an ideal Veselago lens is 'perfect', i.e., its resolution is not limited by diffraction to $\sim \lambda$. A detailed proof of this remarkable feature can be found in [21], although a brief outline is sketched here. When light passes through a slab of material characterized by ε, μ that is

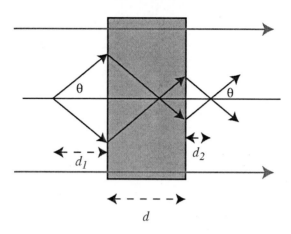

FIGURE 5. Imaging with a slab of negative index material

surrounded by vacuum the fraction, T, of light transmitted is given by

$$T = \left| (1 - \rho^2) \left(\frac{e^{-i(k_{0z} - k_z)d}}{1 - \rho^2 e^{-2ik_z d}} \right) \right|^2 , \tag{11}$$

where

$$\rho = \frac{k_z - \mu k_{0z}}{k_z + \mu k_{0z}} , \tag{12}$$

is the amplitude reflection coefficient at the vacuum-medium interface. Here the z-components of the wave vectors outside and inside the medium are determined from

$$k_{0z} = \left(\frac{\omega^2}{c^2} - k_x^2 \right)^{1/2} , \tag{13}$$

$$k_z = -\left(\varepsilon\mu \frac{\omega^2}{c^2} - k_x^2 \right)^{1/2} , \tag{14}$$

being the dispersion relations (4) for the two cases. The transverse wave number k_x, which is the same in both media on account of Snell's law, is related to the incidence angle via $k_x = \frac{\omega}{c} \sin\theta_0$. We do not exclude, however, the possibility that $k_x > \omega/c$ in which case k_{0z} is purely imaginary, corresponding to an imaginary incidence angle. The most interesting case is when the negatively refracting slab consists of 'anti-vacuum', where $\varepsilon = -\varepsilon_0$ and $\mu = -\mu_0$. For this case $\rho = 0$ and we have

$$T = \left| e^{-i(k_{0z} - k_z)d} \right|^2 . \tag{15}$$

Now we know on causality grounds that energy must flow away from the interface, and also that when ε and μ are negative, the power flow opposes the wave vector. Hence for this case we must take the negative root in equation (14), and for $\varepsilon = -\varepsilon_0$ and $\mu = -\mu_0$ (= 'anti-vacuum') we have $k_z = -k_{0z}$. What about k_{0z}? When k_x exceeds ω/c, $k_{0z} = i\kappa$, say, is purely imaginary, as seen from equation (13). This corresponds to a wave that decays exponentially from the source. But the fraction of transmitted light is found from (15) to be

$$T = e^{4\kappa d}. \tag{16}$$

Thus the transmission increases exponentially with slab thickness. After the slab, the evanescent wave starts to decay again as $e^{-\kappa z}$. For a unit amplitude object placed a distance d_1 from the slab, it is straightforward to show that unit amplitude is restored again at a distance $d_2 = d - d_1$ beyond the slab. This is indeed a remarkable result. It says that evanescent waves, associated with high spatial frequencies ($k_x > \omega/c$) that are beyond the diffraction limit of a conventional lens, are 'imaged' by the Veselago lens at exactly the point where the conventional rays converge! The image is thus a 'perfect' reconstruction of the original object, untethered by the constraints of diffraction.

Since Pendry's famous paper [26] expounding the theory of the perfect lens, there has been limited progress towards achieving this ideal. The technological difficulties in producing a meta-material that approaches the ideal 'perfect lens' condition with $\varepsilon = -\varepsilon_0$ and $\mu = -\mu_0$ are considerable. Nevertheless, we believe now that the resolution of Veselago slab lenses, are limited by technology, rather than diffraction physics.

HOW CAN BACKWARD WAVES BE IDENTIFIED MATHEMATICALLY?

As much as these developments are exciting, it is equally important to place the phenomenon of negative refraction on a sound mathematical basis. Above we have discussed examples of negative refraction that are signalled by waves whose phase advances in the opposite sense to which the power flows. So what is the appropriate mathematical characterization of a backward wave that leads to negative refraction? However we idealize, for example by assuming an infinite plane wave within an infinite homogeneous medium, the issue is fraught with pitfalls, some of which have led to considerable controversy in the literature. To start with there is no fixed nomenclature. The phrase 'negative refraction' that has been used widely, should perhaps be restricted to the unusual refraction that takes place between two media as in Figure 2. Other appellations such as double negative media [33] or left-handed media [9] have been used to characterize the media within which the disposition of the field vectors is as in Figure 1. As regards a name that actually characterizes the propagational phenomenon, the author and his co-workers have suggested the name Negative Phase Velocity Propagation (NPVP) [18], as capturing the essential feature of electromagnetic energy propagating contra to the direction of phase advance. But how should NPVP be characterized mathematically? Firstly, is it sufficient to simply require that a *component* k_j, of the wave vector, be oppositely signed to the corresponding component, $P_j = \frac{1}{2}\text{Re}\,(\mathbf{E} \times \mathbf{H}^*)_j$, of the Poynting vector? No. from Figure 6 it is clear that this requirement can be fulfilled for classical birefringence. Moreover, this criterion is actually dependent on the orientation of the cartesian axes as shown in Figure 6. Such a coordinate dependence is a sign that the

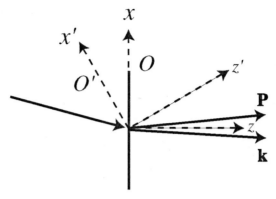

FIGURE 6. Light passing from vacuum into a conventional birefringent medium. In coordinate system O we have that $\mathrm{sgn}(k_x) \neq \mathrm{sgn}(P_x)$, whereas in the rotated system O', $\mathrm{sgn}(k_{x'}) = \mathrm{sgn}(P_{x'})$

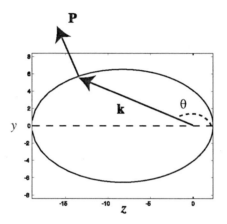

FIGURE 7. Phase (wave vector) surfaces for simple medium moving with speed $v < (\varepsilon\mu)^{-1/2}$

putative criterion $\mathrm{sgn}(k_j) \neq \mathrm{sgn}(P_j)$ for some $j = 1, 2, 3$ is unsuitable for identifying the phenomenon of NPVP. A much better possibility is the inequality

$$\mathbf{P} \cdot \mathbf{k} < 0 \,. \tag{17}$$

This inequality apparently identifies the 'true' backward wave phenomenon occurring in Veselago media, as opposed to classical birefringence. For the latter $\mathbf{P} \cdot \mathbf{k} > 0$ always. Moreover, the inequality (17) is invariant with respect to spatial transformations (e.g., rotations) and is therefore acceptable from the point of view of spatial covariance. However, it has recently been discovered that (17) is also insufficient in general [19]. Although (17) is invariant with respect to spatial transformations, it is *not* invariant under transformations that involve both space and time, such as Lorentz transformations. Thus we anticipate that for moving media (17) will not be invariant, and must be replaced by a covariant criterion, just as $\mathbf{P} \cdot \mathbf{k} < 0$ superseded the criterion that $\mathrm{sgn}(k_j) \neq \mathrm{sgn}(P_j)$ for some $j = 1,2,3$. The situation is best illustrated by the example of a medium for which ε and μ are both real and positive in the medium's rest frame. In this frame the k value that solves equation (4) is just $k = +(\varepsilon\mu)^{1/2}\omega$ where here the positive root is necessary so that $\mathbf{k}$ and $\mathbf{P}$ point in the same direction. How does this situation change if we observe from a frame moving to the left with velocity $\mathbf{v} = -v\hat{\mathbf{z}}$. In this frame the medium appears to move to the right with speed v? Now in the new frame the medium is no longer electromagnetically isotropic. In fact it is characterized by the constitutive parameters [19]

$$\underline{\underline{\varepsilon}} \;=\; \varepsilon \begin{pmatrix} a\gamma^2 & 0 & 0 \\ 0 & a\gamma^2 & 0 \\ 0 & 0 & 1 \end{pmatrix} \tag{18}$$

$$\underline{\underline{\mu}}^{-1} \;=\; \mu^{-1} \begin{pmatrix} b\gamma^2 & 0 & 0 \\ 0 & b\gamma^2 & 0 \\ 0 & 0 & 1 \end{pmatrix} \tag{19}$$

$$\underline{\underline{\alpha}} = \underline{\underline{\beta}} \;=\; \gamma^2 v\varepsilon d \begin{pmatrix} 0 & 1 & 0 \\ -1 & 0 & 0 \\ 0 & 0 & 0 \end{pmatrix} \tag{20}$$

where $a = [1 - v^2(\varepsilon\mu)^{-1}]$, $b = (1 - \varepsilon\mu v^2)$ and $d = [1 - (\varepsilon\mu)^{-1}]$.

The process of finding the eigenmodes is now more complicated that the prescription given earlier for non-magneto-electric media. For light propagating at an angle θ to the $+z$ axis, the value of k is given by [19, 20]

$$\frac{k}{\omega} = \frac{-v\xi \cos\theta \pm \left[1 + \xi\left(1 - v^2\cos^2\theta\right)\right]^{1/2}}{1 - v^2\xi\cos^2\theta} , \tag{21}$$

where $\xi = \frac{\varepsilon\mu - 1}{1 - v^2}$. The phase surfaces $k(\theta)$ are illustrated in Figures 7 and 8. For low medium speed $v < (\varepsilon\mu)^{-1/2}$ the phase surface is ellipsoidal (Figure 7), being a distortion of the spherical phase surface that prevails in the medium rest frame. For $v > (\varepsilon\mu)^{-1/2}$ the phase surface is hyperbolic (Figure 8). Now the direction of power flow, is normal to the phase surface [2]. Whilst for the elliptical surface the direction of power flow is obviously the *outward* normal, for the hyperbolic surface which direction is 'outward' is no longer clear. In fact, a careful calculation shows that the power flow directions are as shown in Figure 8[1]. The rightmost branch of the hyperbola in Figure 8 results from attempting to point light in the $-z$ direction, only for it to be blown back ('Fresnel dragged' [20]) back into the forward direction. Nevertheless, the power flows back against the medium as indicated. In fact for this branch we have that $\mathbf{P} \cdot \mathbf{k} < 0$. So we are back to the situation that our criterion that is supposed to identify NPVP can be induced

---

[1] Note, that the direction is given incorrectly in [3], p.310.

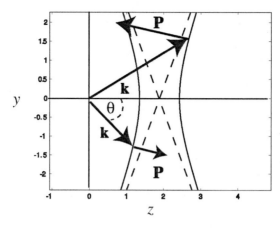

FIGURE 8. Phase (wave vector) surfaces for simple medium moving with speed $v > (\varepsilon\mu)^{-1/2}$

by a coordinate transformation - in the present case a Lorentz transformation that mixes space and time.

COVARIANT CONSIDERATIONS

Actually, if the full four-geometry of spacetime is embraced, the non-covariance of $\mathbf{P}\cdot\mathbf{k}$ is immediately clear, even for waves propagating in vacuum. Understanding how this arises assists in resolving a number of controversies that have appeared in recent literature. In vacuum, the four wave vector is given by $K^\mu = (\omega/c, \mathbf{k})$. If we assume the flat spacetime minkowski metric $\eta_{\alpha\beta} = \text{diag}(-1,1,1,1)$, then the (squared) length of the four wave vector is given by

$$\eta_{\alpha\beta}K^\alpha K^\beta = -\frac{\omega^2}{c^2} + \mathbf{k}\cdot\mathbf{k}, \tag{22}$$

where the summation convention is used. In fact this quantity is zero, since the dispersion relation in vacuum is simply $k^2 = \omega^2/c^2$. An alternative way of writing equation (22) is

$$K_\alpha K^\alpha = -\frac{\omega^2}{c^2} + \mathbf{k}\cdot\mathbf{k} = 0. \tag{23}$$

where the covariant components $K_\alpha = (-\omega, \mathbf{k})$ are obtained from the contravariant components K^α via $K_\alpha = \eta_{\alpha\beta}K^\beta$. Now multiplying K^μ by $\hbar$ generates a new four vector $P^\mu = (\mathscr{E}, \mathbf{P})$, the enegy-momentum four vector of a photon. For a collection of such photons the energy-momentum four vector is given by the same expression, except that now $\mathscr{E}$ is the total energy, and $\mathbf{P}$ is the Poynting vector associated with the electromagnetic wave. Analogously, therefore, we have that

$$\eta_{\alpha\beta}P^\alpha K^\beta = P_\alpha K^\alpha = -\mathscr{E}\omega + \mathbf{P}\cdot\mathbf{k} = 0. \tag{24}$$

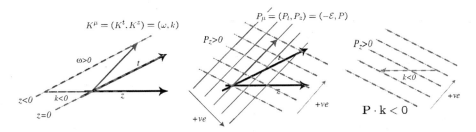

FIGURE 9. A linear transformation of spacetime coordinates that changes $\bar{\mathbf{P}} \cdot \bar{\mathbf{k}} > 0$ into $\mathbf{P} \cdot \mathbf{k} < 0$

Crucially, we see that the 3-scalar product $\mathbf{P} \cdot \mathbf{k}$ occurs as part of the expansion of a 4-scalar product. Its sign therefore has no intrinsic physical significance, and depends on the spacetime coordinates. In fact, as illustrated in Figure 9, the simple spacetime transformation given by

$$\begin{bmatrix} K^t \\ K^z \end{bmatrix} = \begin{bmatrix} \omega/c \\ k \end{bmatrix} = \begin{bmatrix} 1 & 0 \\ -a & 1 \end{bmatrix} \begin{bmatrix} K^{\bar{t}} \\ K^{\bar{z}} \end{bmatrix} = \begin{bmatrix} \bar{\omega}/c \\ -a\frac{\bar{\omega}}{c}+\bar{k} \end{bmatrix} = \begin{bmatrix} \bar{\omega}/c \\ \frac{\bar{\omega}}{c}(1-a) \end{bmatrix} ,\tag{25}$$

$$[P_t, P_z] = [-\bar{\mathscr{E}}/c, P_{\bar{z}}] \begin{bmatrix} 1 & 0 \\ a & 1 \end{bmatrix} = [(\bar{\mathscr{E}}+aP_{\bar{z}})/c, P_{\bar{z}}] ,\tag{26}$$

changes the sign of $\mathbf{P} \cdot \mathbf{k}$. In the original frame we have that $\bar{\mathbf{P}} \cdot \bar{\mathbf{k}}\cdot > 0$, whereas in the new frame we have

$$\mathbf{P} \cdot \mathbf{k} = P_{\bar{z}}\bar{\omega}(1-a) ,\tag{27}$$

which is negative for $a > 1$. When spacetime transformations are considered, it becomes necessary to supplant $\mathbf{P} \cdot \mathbf{k} < 0$ with one that is covariantly valid in spacetime. The above considerations have not been widely appreciated until quite recently. Indeed, for a time it was believed that by taking the metric of spacetime to be $g_{\alpha\beta}$, rather than the Minkowski form, it was possible for a gravitational field to induce NPVP [5, 6, 10, 11, 7, 12, 13, 31]. Such a discovery would have been very significant because it would mean that light reaching us from distant stars and galaxies might have been significantly deviated as a result of NPVP through the intervening region of spacetime. Star maps might be inaccurate. The calculations reaching this conclusion were non-covariant, being based on an effective medium approach [28] and the flaw therefore went undetected. From a covariant perspective it is easy to locate the geometrical mechanism that resulted in this conclusion. Any nonsingular region of spacetime is locally Lorentzian, i.e., the metric can be locally brought into the Minkowski form, $\eta_{\bar{\alpha}\bar{\beta}}$. However, one is not obliged to use such coordinates. A coordinate system that is isomorphic to Minkowski coordinates is equally valid, i.e., one in which $g_{\alpha\beta} = L^{\bar{\alpha}}_{\alpha}\eta_{\bar{\alpha}\bar{\beta}}L^{\bar{\beta}}_{\beta}$, where $L^{\bar{\alpha}}_{\alpha}$ is the isomorphism. Of course for the most general coordinate systems describing curved spacetime, the local isomorphism necessarily depends on the spacetime coordinates, X^{α}, so that $L^{\bar{\alpha}}_{\alpha} = L^{\bar{\alpha}}_{\alpha}(X^{\mu})$. Now calculating the invariant quantity of equation (24), with $P_{\mu}K^{\mu} = g_{\mu\nu}P^{\mu}K^{\nu} = -\mathscr{E}\omega + \mathbf{P} \cdot \mathbf{k}$, may well result in $\mathbf{P} \cdot \mathbf{k} < 0$, whereas $\bar{\mathbf{P}} \cdot \bar{\mathbf{k}} > 0$

in the original Minkowski coordinates. This was the situation, for example, for the transformation of (25) wherein the Minkowski metric in the slanted coordinate system becomes

$$g_{\alpha\beta} = \begin{pmatrix} -1+a^2 & 0 & 0 & a \\ 0 & 1 & 0 & 0 \\ 0 & 0 & 1 & 0 \\ a & 0 & 0 & 1 \end{pmatrix}. \tag{28}$$

As we showed above, this leads to $\mathbf{P} \cdot \mathbf{k} < 0$ for $a > 1$. Thus claims that gravity can induce negative refraction turn out be based on a geometrical artifact [22, 23, 24, 25].

A COVARIANT CRITERION FOR NPVP

As well as identifying the disease, covariant methods in spacetime also provide the cure - a fully covariant definition of NPVP that is robust to arbitrary spactime transformations [19]. In Minkowski coordinates the rest frame of a medium can be associated with a four velocity $U^{\bar{\mu}} = (1,0,0,0)$, the overbar on the index indicating rest-frame coordinates. The four momentum of an electromagnetic wave cannot be calculated as simply as for vacuum, but is rather given by $P_{\bar{\mu}} = -T_{\bar{\mu}\bar{\nu}}U^{\bar{\nu}}$, where $T_{\bar{\mu}\bar{\nu}}$ is the electromagnetic stress energy tensor for the medium. Finally, the key four vector turns out to be the projection of $K^{\bar{\mu}}$ orthogonal to the four velocity $U^{\bar{\mu}}$:

$$K_{\perp}^{\bar{\mu}} = K^{\bar{\mu}} - \left(\frac{U_{\bar{\nu}}K^{\bar{\nu}}}{U_{\bar{\alpha}}U^{\bar{\alpha}}} \right) U^{\bar{\mu}}. \tag{29}$$

In the Minkowski coordinates associated with the medium's rest frame $K_{\perp}^{\bar{\mu}} = (0,\mathbf{k})$. These quantities combine to provide the covariant inequality

$$P_{\mu}K_{\perp}^{\mu} < 0, \tag{30}$$

that is valid in arbitrary spacetime coordinates (and hence the overbars on the indices are dropped). The inequality of (30) reduces to $\bar{\mathbf{P}} \cdot \bar{\mathbf{k}} < 0$ in the medium's rest frame. Thus an equivalent, and equally precise formulation of the covariant criterion is the statement that NPVP occurs if and only if it occurs in the medium's rest frame. Application of (30) to the moving medium example considered earlier shows that $P_{\mu}K_{\perp}^{\mu} > 0$ in both the medium rest frame *and* in the moving frame. Studies that probe the sign of $\mathbf{P} \cdot \mathbf{k}$ under Lorentz transformations for various medium parameters [14, 15] are not physcially significant.

OPTICAL CLOAKING

It is fortunate that the theory of general coordinate transformations has been developed in relation to defining NPVP. It turns out that how the fields behave under transformation has quite another, very exciting application, namely the possibility of creating an electromagnetic 'invisibility cloak', of Harry Potter fame[27, 8].

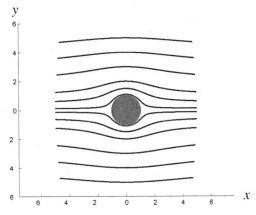

FIGURE 10. Illustrating the transformation (31)

The idea is based on coordinate transformations. Consider the transforming cylindrical polar coordinates (r, θ, z) according to

$$r' = \left(1 - \frac{a}{b}\right) r + a , \tag{31}$$

$$\theta' = \theta , \tag{32}$$

$$z' = z , \tag{33}$$

where a and b are constants with $b > a$. Since $r > 0$ then $r' > a$, and effectively in the transformed space a 'hole' of radius a has appeared. The straight lines for which $y = r \sin \theta$ is constant in the original coordinate system become the lines shown in Figure 10.

If the lines of constant $y = r \sin \theta$ represent straight ray paths going from left to right in the original system, the curved lines represent light rays bending round an obstacle in the new system. After curving round the obstacle the rays eventually become parallel again so that an observer looking towards the object from the right would not detect its presence and the object is hidden. This is quite different from technologies that seek to render an object undetectable by minimizing the backscatter when illuminated by radar, such as for the stealth bomber.

The transformation could be effected in physical (r, θ, z) space by replacing vacuum with a dielectric medium with appropriate properties. The coordinate transformation of equations (31)-(33) provides the required recipe. The vacuum electromagnetic quantities ε_0, μ_0 become, after the stretching and hole-punching of equation (31)-(33), anisotropic. The resultant medium then responds in the directions of increasing r, θ, z according to

$$(\varepsilon, \mu)_r = \left(1 - \frac{a}{r}\right) (\varepsilon, \mu)_0 , \tag{34}$$

$$(\varepsilon, \mu)_\theta = \left(1 - \frac{a}{r}\right)^{-1} (\varepsilon, \mu)_0 , \tag{35}$$

$$(\varepsilon,\mu)_z = \left(1 - \frac{a}{b}\right)^{-2}\left(1 - \frac{a}{r}\right)(\varepsilon,\mu)_0 . \tag{36}$$

The medium surrounds a cylindrical object of radius $r = a$. Details of how the electromagnetic constitutive parameters are transformed from one coordinate system to another may be found in the book by Post [29].

As currently conceived the electromagnetic cloak is highly restricted in its operation. It works only at one frequency, the cloak and the object it shrouds (which cannot be too large to preserve the effective medium approximation) must be stationary (since the constitutive parameters of a medium depend on its velocity relative to the observer's reference frame [29]), only guided waves polarized parallel and propagating normal to the axis of a given cylinder are allowed, and the observer must ignore any changes in overall intensity. Nevertheless, the first experimental demonstrations [27] are impressive, and it is likely that technological improvements and theoretical insights will advance the concept further. Although fully covariant techniques are not yet routinely applied, it is the author's belief that applying such techniques will yield further clarification and insight into transformation optics.

CONCLUSION

This review has attempted to capture some of the key modern aspects of electromagnetics that are currently exciting its practitioners. Perfect lenses and cloaks are certainly exciting goals, although as this paper has tried to emphasize, a precise mathematical description assists greatly in clarifying and understanding the underlying geometry.

ACKNOWLEDGMENTS

This work was supported by EPSRC Grant nos. EP/E031463/1 and EP/G000964/1.

REFERENCES

1. N W. Ashcroft and N. D Mermin, *Solid State Physics*, Holt, Rinehart and Winston, Philadelphia, 1976, pp.11–18.
2. M. Born, and E. Wolf, *Principles of Optics: Electromagnetic Theory of Propagation, Interference and Diffraction of Light,* Cambridge University Press, Cambridge, 1999, 7th ed.
3. H. C. Chen, *Theory of Electromagnetic Waves: A Coordinate-Free Approach*, McGraw-Hill, 1983.
4. L. D. Landau and E. M. Lifschitz, *Electrodynamics of Continuous Media*, Pergamon Press, Oxford, 1963.
5. A. Lakhtakia and T. G. Mackay, *J. Phys. A: Math. Gen.* **37**, L505 (2004).
6. A. Lakhtakia and T. G. Mackay, *Curr. Sci.* **90**, 640 (2006).
7. A. Lakhtakia, T. G. Mackay, and S. Setiawan, *Phys. Lett. A* **336**, 89 (2005).
8. U. Leonhardt and T. G. Philbin, *New Journal of Physics* **8**, 247 (2006).
9. I. V. Lindell, S. A. Tretyakov, K. I. Nikoskinen, and S. Ilvonen, *Microwave and Optical Tech. Lett.* **31**, 129–133 (2001).
10. T. G. Mackay, S. Setiawan, and A. Lakhtakia, *Eur. Phys. J.C.* **41**, 1 (2005).
11. T. G. Mackay, A. Lakhtakia, and S. Setiawan, *Europhysics Letters* **71**, 925 (2005).
12. T. G. Mackay, A. Lakhtakia, and S. Setiawan, *New Journal of Phys* **7**, 75 (2005).

13. T. G. Mackay, A. Lakhtakia, and S. Setiawan, *New Journal of Phys* **7**, 171 (2005).
14. T. G. Mackay and A. Lakhtakia, *J. Phys. A: Math. Gen.* **37**, 5697-Ũ5711 (2004).
15. T. G. Mackay, A. Lakhtakia, and S. Setiawan, *Optik* **118**, 195–202 (2007).
16. J. C. Maxwell, *A Treatise on Electricity and Magnetism, Vol.1 and Vol. 2*, Oxford at the Clarendon Press, Oxford, 1904, 3rd ed., See http://rack1.ul.cs.cmu.edu/is/maxwell1.
17. M. W. McCall and A. Lakhtakia, *J. Modern Optics* **51**, 111–127 (2004).
18. M. W. McCall, A. Lakhtakia, and W. S. Weiglhofer", *Eur. J. Phys.* **23**, 353–359 (2002).
19. M. W. McCall, *Metamaterials*, doi:10.1016/j.metmat.2008.05.001, (2008).
20. M. W. McCall and D. Censor *American Journal of Physics* **75**, 1134–1140 (2008).
21. M. W. McCall and G. Dewar, "Negative Refraction," in *Tutorials in Complex Photonic Media*, edited by M. Noginov, M. W. McCall, G. Dewar, and N. Zheludev, SPIE press, Bellingham, 2008.
22. M. W. McCall,*Journal of Modern Optics* **54**, 119–128 (2006).
23. M. W. McCall,*Journal of Modern Optics* **55**, 329–332 (2008).
24. M. W. McCall, *Phys. Rev. Lett.* **98**, 091102 (2007).
25. M. W. McCall, *Journal of Modern Optics* **55**, 333–340 (2008).
26. J. B. Pendry, *Phys. Rev. Lett.* **85**, 3966–3969 (2000).
27. , J. B. Pendry, D. Schurig, and D. R. Smith, *Science* **312**, 1780–1782 (2006).
28. J. Plebanski, *Phys. Rev.* **118**, 1396–1408 (1960).
29. E. J. Post, *Formal Structure of Electromagnetics*, Dover, Mineola, N.Y., 1997.
30. R. A. Shelby, D. R. Smith, and S. Schultz, *Science* **292**, 77–79 (2001).
31. S. Setiawan, T. G. Mackay and A. Lakhtakia, *Phys. Lett. A* **341**, 15 (2005).
32. V. G. Veselago, *Sov. Phys. USPEKHI* **10**, 517–526 (1968).
33. M. Y. Wang, J. Xu, J. Wu, Y. B. Yan, and H. L. Li, *Journal of Electromagnetic Waves and Applications* **21**, 1905–1914 (2007).

Exact Solutions to Nonlinear Equations and Systems of Equations of General Form in Mathematical Physics

A. D. Polyanin[*], A. I. Zhurov[†] and E. A. Vyazmina[**]

[*]*Institute for Problems in Mechanics, Russian Academy of Sciences*
101 Vernadsky Ave., bldg, 119526 Moscow, Russia
[†]*Cardiff University, School of Dentistry, Heath Park, Cardiff CF14 4XY, UK*
[**]*École Polytechnique, 91128 Palaiseau Cedex, France*

Abstract. The paper gives an overview of recent results in exact solutions to nonlinear equations and system of equations of general form with functional arbitrariness. Special attention is paid to equations that arise in heat and mass transfer, wave theory, and mathematical biology. The solutions considered below have been obtained, for the most part, with the method of generalized separation of variables and that of functional separation of variables. Some of the solutions presented are new.

Keywords: Exact solutions, generalized separable solutions, functional separable solutions, nonlinear equations, systems of equations, mathematical physics.
PACS: 02.30.Jr

STRUCTURE OF GENERALIZED AND FUNCTIONAL SEPARABLE SOLUTIONS

To simplify the presentation, we confine ourselves to the case of mathematical physics equations in two independent variables, x and t, and a dependent variable, w.

Definition 1. Exact solutions to nonlinear PDEs in the form

$$w(x,t) = \varphi_1(x)\psi_1(t) + \varphi_2(x)\psi_2(t) + \cdots + \varphi_n(x)\psi_n(t) \tag{1}$$

are called *generalized separable solutions*.

Definition 2. Exact solutions to nonlinear PDEs in the form

$$w(x,t) = F(z), \quad z = \varphi_1(x)\psi_1(t) + \varphi_2(x)\psi_2(t) + \cdots + \varphi_n(x)\psi_n(t) \tag{2}$$

are called *functional separable solutions*.

Substituting (1) (or (2)) into the equations under consideration leads to functional-differential equations for $\varphi_i(x)$ and $\psi_j(t)$ (and $F(z)$). These functional-differential equations may be solved using the method of differentiation and the splitting method, which are described in the books [1–3] (see also the Titov–Galaktionov method [4–6], based on using invariant subspaces).

Presented below are original nonlinear equations and respective final exact solutions, without intermediate results.

CP1067, *Applications of Mathematics in Engineering and Economics '34—AMEE '08*, edited by M. D. Todorov
© 2008 American Institute of Physics 978-0-7354-0598-1/08/$23.00

GENERALIZED POROUS MEDIUM EQUATION WITH A NONLINEAR SOURCE

Consider the following mass transfer equation for a medium with complex rheology and a power Fick's law for the concentration gradient:

$$\frac{\partial w}{\partial t} = \frac{\partial}{\partial x}\left[f(w)\left(\frac{\partial w}{\partial x}\right)^n\right] + g(w). \qquad (3)$$

This is a generalized porous medium equation with nonlinear source [7].

In the special case $n = 1$, equation (3) is the ordinary diffusion (heat) equation with volume reaction (heat source), whose invariant solutions are described in [8]. Generalized and functional separable exact solutions to equation (3) with can be found in [1, 3, 9].

In the general case, equation (3) admits an obvious traveling-wave solution $w = w(z)$, $z = x + \lambda t$, which will not be considered. Given below are more complicated exact solutions obtained in [10, 11].

$1°$. Suppose the function $f = f(w)$ is arbitrary and the function $g = g(w)$ is defined as

$$g(w) = A[f(w)]^{-1/n} - B,$$

where A and B are some numbers. In this case, equation (3) admits a functional separable solution, which is defined implicitly by

$$\int [f(w)]^{1/n}\,dw = At + \frac{n}{B(n+1)}(Bx + C_1)^{\frac{n+1}{n}} + C_2,$$

where C_1 and C_2 are arbitrary constants.

$2°$. If $g = g(w)$ is arbitrary and $f = f(w)$ is defined as

$$f(w) = \frac{A}{[g(w)]^n}\exp\left[Bn\int\frac{dw}{g(w)}\right],$$

where A and B are some numbers, then equation (3) admits a functional separable solution given implicitly by

$$\int\frac{dw}{g(w)} = t + \frac{1}{B}\ln|Bx + C_1| + C_2.$$

$3°$. If $g = g(w)$ is arbitrary and $f = f(w)$ is defined as

$$f(w) = \frac{A_1 A_2^n w + B}{[g(w)]^n} + \frac{A_2^n A_3}{[g(w)]^n}\int Z\,dw, \qquad (4)$$

$$Z = A_2\int\frac{dw}{g(w)}, \qquad (5)$$

65

where A_1, A_2, and A_3 are some numbers, then equation (3) admits generalized traveling-wave solutions of the form

$$w = w(Z), \quad Z = \varphi(t)x + \psi(t),$$

where the function $w(Z)$ is determined by the inversion of (5), while $\varphi(t)$ and $\psi(t)$ are expressed as

$$\varphi(t) = \left[\frac{1}{C_1 - A_3(n+1)t}\right]^{\frac{1}{n+1}},$$

$$\psi(t) = \varphi(t)\left[A_1 \int [\varphi(t)]^n \, dt + A_2 \int \frac{dt}{\varphi(t)} + C_2\right],$$

with C_1 and C_2 being arbitrary constants.

$4°$. If $g = g(w)$ is arbitrary and $f = f(w)$ is defined as

$$f(w) = \frac{1}{[(g(w)]^n}\left(A_1 w + A_3 \int Z \, dw\right) \exp\left[nA_4 \int \frac{dw}{g(w)}\right], \tag{6}$$

$$Z = \frac{1}{A_4}\exp\left[A_4 \int \frac{dw}{g(w)}\right] - \frac{A_2}{A_4}, \tag{7}$$

where A_1, A_2, A_3, and A_4 are some numbers ($A_4 \neq 0$), then equation (3) admits generalized traveling-wave solutions of the form

$$w = w(Z), \quad Z = \varphi(t)x + \psi(t),$$

where the function $w(Z)$ is determined by the inversion of (5), while $\varphi(t)$ and $\psi(t)$ are expressed as

$$\varphi(t) = \left[C_1 e^{-(n+1)A_4 t} - \frac{A_3}{A_4}\right]^{-\frac{1}{n+1}},$$

$$\psi(t) = \varphi(t)\left[A_1 \int [\varphi(t)]^n dt + A_2 \int \frac{dt}{\varphi(t)} + C_2\right],$$

with C_1 and C_2 being arbitrary constants.

$5°$. Suppose the functions $f(w)$ and $g(w)$ are defined as

$$f(w) = A\exp[-bn\varphi(w)][\varphi_w'(w)]^n, \quad g(w) = \frac{a}{\varphi_w'(w)} - Acn\exp[-bn\varphi(w)],$$

where $\varphi(w)$ is an arbitrary function and A, a, b, and c are arbitrary numbers. In this case, equation (6) admits a functional separable solution defined implicitly by

$$\varphi(w) = at + \theta(x),$$

where $\theta(x)$ is determined by the autonomous ordinary differential equation

$$(\theta'_x)^{n-1}\theta''_{xx} = b(\theta'_x)^{n+1} + c.$$

The general solution of this equation is obtained in parametric form and can be written as

$$\theta = \int \frac{u^n\,du}{bu^{n+1}+c} + C_1, \qquad x = \int \frac{u^{n-1}\,du}{bu^{n+1}+c} + C_2.$$

6°. If $f(w)$ and $g(w)$ are defined as

$$f(w) = \frac{1}{\lambda^n}\left(a - \frac{n\lambda}{n+1}w\right)\left[\varphi'_w(w)\right]^n, \qquad g(w) = \lambda\frac{\varphi(w)}{\varphi'_w(w)} + \frac{n\lambda}{n+1}w - a,$$

with arbitrary function $\varphi(w)$ and arbitrary numbers a and λ, equation (3) admits a functional separable solution written in implicit form as

$$\varphi(w) = C_1 e^{\lambda t} + \frac{n\lambda}{n+1}(x+C_2)^{\frac{n+1}{n}},$$

with arbitrary constants C_1 and C_2.

7°. Let $f(w)$ and $g(w)$ be defined as

$$f(w) = A\frac{\zeta(z)}{[\zeta'_z(z)]^n}z^{n-1}, \qquad g(w) = B(n+1)\zeta'_z(z)z^{-\frac{n}{n+1}} + B\zeta(z)z^{-\frac{2n+1}{n+1}},$$

where $\zeta(z)$ is an arbitrary function, A and B are arbitrary constants such that $AB \neq 0$, and $z = z(w)$ is implicitly given by

$$w = \int z^{-\frac{n}{n+1}}\zeta'_z(z)\,dz + C_1 \qquad (8)$$

with an arbitrary constant C_1. Then equation (3) has a functional separable solution of the form (8) with

$$z = \left[\frac{x^{n+1}}{C_2 - (n+1)k^n A t}\right]^{\frac{1}{n}} + (n+1)Bt - \frac{(n+1)BC_2}{nk^{n+1}A},$$

where C_2 is an arbitrary constant and $k = (n+1)/n$.

Remark 1. Some equations (3) that admit functional separable solutions of the form

$$w = F(\xi), \qquad \xi = \varphi(x) + \psi(t)$$

were treated in [12]. The equations considered there do not contain functional arbitrariness unlike the equations presented above in Items 1° to 7°.

TABLE 1. Functional separable solutions of the form $w = w(\zeta)$, $\zeta^2 = \varphi(x) + \psi(y)$, for heat equations in an anisotropic inhomogeneous medium with an arbitrary nonlinear source. Notation: C, a, b, m, n, μ, and ν are free parameters ($C \neq 0$, $m \neq 2$, $n \neq 2$, $\mu \neq 0$, and $\nu \neq 0$)

Heat equation	Functions $\varphi(x)$ and $\psi(y)$	Equation for $w = w(\zeta)$
$\dfrac{\partial}{\partial x}\left(ax^m \dfrac{\partial w}{\partial x}\right) + \dfrac{\partial}{\partial y}\left(by^n \dfrac{\partial w}{\partial y}\right) = f(w)$	$\varphi = \dfrac{Cx^{2-m}}{a(2-m)^2}$, $\psi = \dfrac{Cy^{2-n}}{b(2-n)^2}$	$w'' + \dfrac{4-mn}{(2-m)(2-n)\zeta}w' = \dfrac{4}{C}f(w)$
$\dfrac{\partial}{\partial x}\left(ae^{\mu x} \dfrac{\partial w}{\partial x}\right) + \dfrac{\partial}{\partial y}\left(be^{\nu y} \dfrac{\partial w}{\partial y}\right) = f(w)$	$\varphi = \dfrac{Ce^{-\mu x}}{a\mu^2}$, $\psi = \dfrac{Ce^{-\nu y}}{b\nu^2}$	$w'' - \dfrac{1}{\zeta}w' = \dfrac{4}{C}f(w)$
$\dfrac{\partial}{\partial x}\left(ae^{\mu x} \dfrac{\partial w}{\partial x}\right) + \dfrac{\partial}{\partial y}\left(by^n \dfrac{\partial w}{\partial y}\right) = f(w)$	$\varphi = \dfrac{Ce^{-\mu x}}{a\mu^2}$, $\psi = \dfrac{Cy^{2-n}}{b(2-n)^2}$	$w'' + \dfrac{n}{2-n}\dfrac{1}{\zeta}w' = \dfrac{4}{C}f(w)$

EQUATION OF HEAT TRANSFER IN AN ANISOTROPIC INHOMOGENEOUS MEDIUM WITH A NONLINEAR SOURCE

$1°$. Consider the following equation of steady-state heat transfer in an anisotropic inhomogeneous medium with a nonlinear source:

$$\frac{\partial}{\partial x}\left[\alpha(x)\frac{\partial w}{\partial x}\right] + \frac{\partial}{\partial y}\left[\beta(y)\frac{\partial w}{\partial y}\right] = f(w). \tag{9}$$

The right-hand side $f(w)$ is considered arbitrary.

Table 1 lists some functional separable solutions to the nonlinear equation (9); a more detailed list of solutions to this equation and its generalizations can be found in [1, 13].

$2°$. Consider the multidimensional nonlinear heat equation

$$\sum_{k=1}^{s}\frac{\partial}{\partial x_k}\left(a_k x_k^{m_k}\frac{\partial w}{\partial x_k}\right) + \sum_{k=s+1}^{n}\frac{\partial}{\partial x_k}\left(b_k e^{\lambda_k x_k}\frac{\partial w}{\partial x_k}\right) = f(w), \tag{10}$$

which is a natural generalization of the equations displayed in Table 1.

Functional separable solution [1]:

$$w = w(\zeta), \qquad \zeta^2 = C\sum_{k=1}^{s}\frac{x_k^{2-m_k}}{a_k(2-m_k)^2} + C\sum_{k=s+1}^{n}\frac{e^{-\lambda_k x_k}}{b_k \lambda_k^2},$$

where the function $w(\zeta)$ is determined by the ordinary differential equation

$$\frac{d^2 w}{d\zeta^2} + \frac{B}{\zeta}\frac{dw}{d\zeta} = \frac{4}{C}f(w), \qquad B = \sum_{k=1}^{s}\frac{2}{2-m_k} - 1.$$

Remark 2. The above results are also valid for hyperbolic equations of the form (10) where some of the coefficients a_k and b_k have opposite signs.

3°. The results presented in Table 1 are easily extensible to systems of two or more equations. For instance, consider the nonlinear system

$$\frac{\partial}{\partial x}\left(ax^m\frac{\partial u}{\partial x}\right) + \frac{\partial}{\partial y}\left(by^n\frac{\partial u}{\partial y}\right) = F(u,w),$$

$$\frac{\partial}{\partial x}\left(ax^m\frac{\partial w}{\partial x}\right) + \frac{\partial}{\partial y}\left(by^n\frac{\partial w}{\partial y}\right) = G(u,w),$$

which describes steady-state mass transfer with volume chemical reaction in an anisotropic inhomogeneous medium, with u and w being the concentrations of reacting components. The chemical reaction rates, $F(u,w)$ and $G(u,w)$, are considered to be arbitrary. The system admits an exact solution of the form

$$u = u(\zeta), \quad w = w(\zeta), \quad \zeta^2 = C\left[\frac{x^{2-m}}{a(2-m)^2} + \frac{y^{2-n}}{b(2-n)^2}\right],$$

where the functions $u(\zeta)$ and $w(\zeta)$ are determined by the system of ordinary differential equations

$$u''_{\zeta\zeta} + \frac{4-mn}{(2-m)(2-n)}\frac{1}{\zeta}u'_\zeta = \frac{4}{C}F(u,w),$$

$$w''_{\zeta\zeta} + \frac{4-mn}{(2-m)(2-n)}\frac{1}{\zeta}w'_\zeta = \frac{4}{C}G(u,w).$$

4°. Consider the more general nonlinear system of equations

$$L[u] = F(u,w), \quad L[w] = G(u,w),$$

$$L[u] \equiv \sum_{k=1}^{s}\frac{\partial}{\partial x_k}\left(a_kx_k^{m_k}\frac{\partial u}{\partial x_k}\right) + \sum_{k=s+1}^{n}\frac{\partial}{\partial x_k}\left(b_ke^{\lambda_kx_k}\frac{\partial u}{\partial x_k}\right).$$

It admits a functional separable solution

$$u = u(z), \quad w = w(z), \quad z^2 = C\sum_{k=1}^{s}\frac{x_k^{2-m_k}}{a_k(2-m_k)^2} + C\sum_{k=s+1}^{n}\frac{e^{-\lambda_kx_k}}{b_k\lambda_k^2},$$

where C is an arbitrary constant (other than zero) and the functions $u(z)$ and $w(z)$ are determined by the system of ordinary differential equations

$$\frac{d^2u}{dz^2} + \frac{B}{z}\frac{du}{dz} = \frac{4}{C}F(u,w), \quad \frac{d^2w}{dz^2} + \frac{B}{z}\frac{dw}{dz} = \frac{4}{C}G(u,w),$$

$$B = \sum_{k=1}^{s}\frac{2}{2-m_k} - 1, \quad m_k \neq 2.$$

69

EQUATIONS OF NONLINEAR WAVE THEORY

1°. Consider the nonlinear equation

$$\frac{\partial^2 w}{\partial t^2} = \frac{\partial}{\partial x}\left[f(w)\frac{\partial w}{\partial x}\right],$$

(11)

which arises in wave and gas dynamics.

It has exact solutions in implicit form:

$$
\begin{aligned}
x + t\sqrt{f(w)} &= \varphi(w), \\
x - t\sqrt{f(w)} &= \psi(w),
\end{aligned}
$$

where $\varphi(w)$ and $\psi(w)$ are arbitrary functions.

Remark 3. The papers [1, 14, 15] present other exact solutions to equation (11) and show that it can be linearized.

2°. The nonlinear two-dimensional wave equation

$$\frac{\partial^2 w}{\partial t^2} = \frac{\partial}{\partial x}\left[f(w)\frac{\partial w}{\partial x}\right] + \frac{\partial}{\partial y}\left[g(w)\frac{\partial w}{\partial y}\right]$$

(12)

has exact solutions in implicit form:

$$
\begin{aligned}
x\frac{\sin\varphi(w)}{\sqrt{f(w)}} + y\frac{\cos\varphi(w)}{\sqrt{g(w)}} + t &= \psi_1(w), \\
x\frac{\sin\varphi(w)}{\sqrt{f(w)}} + y\frac{\cos\varphi(w)}{\sqrt{g(w)}} - t &= \psi_2(w),
\end{aligned}
$$

where $\varphi(w)$, $\psi_1(w)$, and $\psi_2(w)$ are arbitrary functions.

3°. The nonlinear n-dimensional wave equation

$$\frac{\partial^2 w}{\partial t^2} = \sum_{k=1}^{n}\frac{\partial}{\partial x_k}\left[f_k(w)\frac{\partial w}{\partial x_k}\right]$$

(13)

admits similar exact solutions in implicit form:

$$
\begin{aligned}
\sum_{k=1}^{n} x_k\varphi_k(w) &= \psi_1(w) + t, \\
\sum_{k=1}^{n} x_k\varphi_k(w) &= \psi_2(w) - t,
\end{aligned}
$$

where $\varphi_1(w)$, ..., $\varphi_{n-1}(w)$, $\psi_1(w)$, and $\psi_2(w)$ are arbitrary functions; the function $\varphi_n(w)$ is determined by the normalization condition

$$\sum_{k=1}^{n} f_k(w)\varphi_k^2(w) = 1.$$

TABLE 2. Exact solutions to the nonhomogeneous Monge–Ampère equation (14); $f(x)$ and $g(x)$ are arbitrary functions; C_1, C_2, C_3, and β are arbitrary constants; a, b, k, and λ are some numbers ($k \neq -1, -2$)

No.	Function $F(x,y)$	Solution $w(x,y)$
1	$f(x)$	$C_1 y^2 + C_2 xy + \dfrac{C_2^2}{4C_1} x^2 - \dfrac{1}{2C_1} \displaystyle\int_a^x (x-t)f(t)\,dt$
2	$f(x)$	$\dfrac{1}{x+C_1}\left(C_2 y^2 + C_3 y + \dfrac{C_3^2}{4C_2} \right) - \dfrac{1}{2C_2} \displaystyle\int_a^x (x-t)(t+C_1)f(t)\,dt$
3	$f(x)y^k$	$\dfrac{C_1 y^{k+2}}{(k+1)(k+2)} - \dfrac{1}{C_1} \displaystyle\int_a^x (x-t)f(t)\,dt$
4	$f(x)y^k$	$\dfrac{y^{k+2}}{(x+C_1)^{k+1}} - \dfrac{1}{(k+1)(k+2)} \displaystyle\int_a^x (x-t)(t+C_1)^{k+1}f(t)\,dt$
5	$f(x)y^{2k+2} + g(x)y^k$	$\varphi(x)y^{k+2} - \dfrac{1}{(k+1)(k+2)} \displaystyle\int_a^x (x-t)\dfrac{g(t)}{\varphi(t)}\,dt$ (φ is determined by an ODE)
6	$f(x)e^{\lambda y}$	$C_1 e^{\beta x + \lambda y} - \dfrac{1}{C_1 \lambda^2} \displaystyle\int_a^x (x-t)e^{-\beta t}f(t)\,dt$
7	$f(x)e^{2\lambda y} + g(x)e^{\lambda y}$	$\varphi(x)e^{\lambda y} - \dfrac{1}{\lambda^2} \displaystyle\int_a^x (x-t)\dfrac{g(t)}{\varphi(t)}\,dt$ (φ is determined by an ODE)
8	$f(x)g(y) + \lambda^2$	$C_1 \displaystyle\int_a^x (x-t)f(t)\,dt - \dfrac{1}{C_1} \displaystyle\int_b^y (y-\xi)g(\xi)\,d\xi \pm \lambda xy$

Remark 4. For other exact solutions to equations (12) and (13), see [1, 14, 15].

NONHOMOGENEOUS MONGE–AMPÈRE EQUATION

Consider the nonhomogeneous Monge–Ampère equation

$$\left(\frac{\partial^2 w}{\partial x \partial y} \right)^2 - \frac{\partial^2 w}{\partial x^2}\frac{\partial^2 w}{\partial y^2} = F(x,y). \tag{14}$$

Equations of this form are encountered in differential geometry, gas dynamics, and meteorology.

Table 2 lists generalized separable solutions to the nonhomogeneous Monge–Ampère equation (14) (a more extensive list of exact solutions can be found in [1, 3, 14, 16]).

FIRST-ORDER NONLINEAR SYSTEMS OF REACTION-CONVECTIVE EQUATIONS

Consider nonlinear systems of the form

$$\frac{\partial u}{\partial x} = F(u,w), \qquad \frac{\partial w}{\partial t} = G(u,w). \tag{15}$$

The more general system

$$\frac{\partial u}{\partial \tau} + a_1 \frac{\partial u}{\partial \xi} = F(u,w), \qquad \frac{\partial w}{\partial \tau} + a_2 \frac{\partial w}{\partial \xi} = G(u,w),$$

which describes convective mass transfer in a two-component system with a volume chemical reaction in the case where the diffusion of both components is negligible, can be reduced to system (15). The reduction is performed by passing from ξ, τ to the characteristic coordinates

$$x = \frac{\xi - a_2 \tau}{a_1 - a_2}, \qquad t = \frac{\xi - a_1 \tau}{a_2 - a_1} \qquad (a_1 \neq a_2).$$

If the first (resp., second) medium is quiescent, then $a_1 = 0$ (resp., $a_2 = 0$).

Some exact solutions to system (15) are displayed in Table 3 (a more extensive list of exact solutions can be found in [3, 17]).

SECOND-ORDER NONLINEAR SYSTEMS OF REACTION-DIFFUSION EQUATIONS AND THEIR GENERALIZATIONS

Consider the nonlinear systems of reaction-diffusion equations

$$\frac{\partial u}{\partial t} = a_1 \frac{\partial^2 u}{\partial x^2} + F(u,w), \qquad \frac{\partial w}{\partial t} = a_2 \frac{\partial^2 w}{\partial x^2} + G(u,w). \tag{16}$$

Systems of this form often arise in the theory of heat and mass transfer in chemically reactive media, theory of chemical reactors, combustion theory, mathematical biology, and biophysics.

Such systems are invariant under translations in the independent variables (and under the change of x to $-x$) and admit traveling-wave solutions $u = u(kx - \lambda t)$, $w = w(kx - \lambda t)$. These solutions as well as those with one of the unknown functions being identically zero are not considered further in this section.

The papers [18–26] give a Lie group classification of and present exact solutions to system (16) and its modifications. Some exact solutions to system (16) are displayed in Table 4 (a more extensive list of exact solutions can be found in [3, 24, 26]).

Below are new exact solutions to system (16) and its generalizations.

TABLE 3. Generalized and functional separable solutions to some systems of the form (15); $f(\cdots)$, $g(\cdots)$, and $h(\cdots)$ are arbitrary functions

Right-hand sides of (15)	Solution structure	Determining equations
$F = u^m f(w)$, $G = u^k g(w)$	$u = t^{-1/k} v(z)$, $w = w(z)$, $z = x + \psi(t)$, $\psi(t)$ is an arbitrary function	$v'_z = f(w)v$, $w'_z = g(w)v^k$
$F = uf(w)$, $G = u^k g(w)$	$u = [\psi'(t)]^{1/k} v(z)$, $w = w(z)$, $z = x + \psi(t)$, $\psi(t)$ is an arbitrary function	$v'_z = f(w)v$, $w'_z = g(w)v^k$
$F = f(au+bw)$, $G = g(au+bw)$	$u = b(k_1 x - \lambda_1 t) + y(\xi)$, $w = -a(k_1 x - \lambda_1 t) + z(\xi)$, $\xi = k_2 x - \lambda_2 t$.	$k_2 y'_\xi + bk_1 = f(ay+bz)$, $-\lambda_2 z'_\xi + a\lambda_1 = g(ay+bz)$
$F = f(a_1 u + b_1 w)$, $G = g(a_2 u + b_2 w)$	$u = \frac{1}{\Delta}[b_2 \varphi(x) - b_1 \psi(t)]$, $w = \frac{1}{\Delta}[a_1 \psi(t) - a_2 \varphi(x)]$, $\Delta = a_1 b_2 - a_2 b_1 \neq 0$	$b_2 \varphi'_x = \Delta f(\varphi)$, $a_1 \psi'_t = \Delta g(\psi)$
$F = f(au-bw)$, $G = ug(au-bw) + h(au-bw)$	$u = \varphi(t) + b\theta(t)x$, $w = \psi(t) + a\theta(t)x$	$b\theta = f(a\varphi - b\psi)$, $a\theta'_t = b\theta g(a\varphi - b\psi) + h(a\varphi - b\psi)$, $\psi'_t = \varphi g(a\varphi - b\psi) + h(a\varphi - b\psi)$
$F = f(au-bw) + cw$, $G = ug(au-bw) + h(au-bw)$	$u = \varphi(t) + b\theta(t)e^{\lambda x}$, $w = \psi(t) + a\theta(t)e^{\lambda x}$, $\lambda = ac/b$	$f(a\varphi - b\psi) + c\psi = 0$, $\psi'_t = \varphi g(a\varphi - b\psi) + h(a\varphi - b\psi)$, $a\theta'_t = b\theta g(a\varphi - b\psi)$
$F = au\ln u + uf(u^n w^m)$, $G = wg(u^n w^m)$	$u = \exp(Cme^{ax})y(\xi)$, $w = \exp(-Cne^{ax})z(\xi)$, $\xi = kx - \lambda t$	$ky'_\xi = ay\ln y + yf(y^n z^m)$, $-\lambda z'_\xi = zg(y^n z^m)$
$F = uf(au^n + bw)$, $G = u^k g(au^n + bw)$	$u = (C_1 t + C_2)^{\frac{1}{n-k}} \theta(x)$, $w = \varphi(x) - \frac{a}{b}(C_1 t + C_2)^{\frac{n}{n-k}} [\theta(x)]^n$	$\theta'_x = \theta f(b\varphi)$, $\theta^{n-k} = \frac{b(k-n)}{aC_1 n} g(b\varphi)$

$1°$. The system

$$\frac{\partial u}{\partial t} = \frac{\partial}{\partial x}\left[f(au+bw)\frac{\partial u}{\partial x}\right] + g(au+bw),$$
$$\frac{\partial w}{\partial t} = \frac{\partial}{\partial x}\left[f(au+bw)\frac{\partial w}{\partial x}\right] + h(au+bw),$$

which involves three arbitrary functions with a composite argument, admits the exact solution

$$u = y(\xi) + b\theta(\xi,t), \quad w = z(\xi) - a\theta(\xi,t), \quad \xi = kx - \lambda t,$$

TABLE 4. Generalized and functional separable solutions to some systems of the form (16); $f(\cdots)$, $g(\cdots)$, and $h(\cdots)$ are arbitrary functions

Right-hand sides of (16)	Solution structure	Determining equations
$F = f(bu+cw)$, $G = g(bu+cw)$, $a_1 = a_2 = a$	$u = c\theta(x,t) + y(\xi)$, $w = -b\theta(x,t) + z(\xi)$, $\xi = kx - \lambda t$	$ak^2 y''_{\xi\xi} + \lambda y'_\xi + f(by+cz) = 0$, $ak^2 z''_{\xi\xi} + \lambda z'_\xi + g(by+cz) = 0$, $\dfrac{\partial\theta}{\partial t} = a\dfrac{\partial^2\theta}{\partial x^2}$
$F = f(bu+cw)$, $G = g(bu+cw)$, a_1, a_2 are any	$u = c(\alpha x^2 + \beta x + \gamma t) + y(\xi)$, $w = -b(\alpha x^2 + \beta x + \gamma t) + z(\xi)$, $\xi = kx - \lambda t$	$a_1 k^2 y''_{\xi\xi} + \lambda y'_\xi + 2a_1 c\alpha$ $\quad -c\gamma + f(by+cz) = 0$, $a_2 k^2 z''_{\xi\xi} + \lambda z'_\xi - 2a_2 b\alpha$ $\quad +by + g(by+cz) = 0$
$F = uf(bu-cw) + g(bu-cw)$, $G = wf(bu-cw) + h(bu-cw)$, $a_1 = a_2 = a$	$u = \varphi(t) + c\exp\!\left[\int f(b\varphi - c\psi)\,dt\right]\theta(x,t)$, $w = \psi(t) + b\exp\!\left[\int f(b\varphi - c\psi)\,dt\right]\theta(x,t)$	$\varphi'_t = \varphi f(b\varphi - c\psi) + g(b\varphi - c\psi)$, $\psi'_t = \psi f(b\varphi - c\psi) + h(b\varphi - c\psi)$, $\dfrac{\partial\theta}{\partial t} = a\dfrac{\partial^2\theta}{\partial x^2}$
$F = uf(u/w)$, $G = wg(u/w)$, $a_1 = a_2 = a$	$u = \left|C_1\sin(kx) + C_2\cos(kx)\right|\varphi(t)$, $w = \left|C_1\sin(kx) + C_2\cos(kx)\right|\psi(t)$	$\varphi'_t = -a_1 k^2\varphi + \varphi f(\varphi/\psi)$, $\psi'_t = -a_2 k^2\psi + \psi g(\varphi/\psi)$
$F = uf(u/w)$, $G = wg(u/w)$, $a_1 = a_2 = a$	$u = \varphi(t)\exp\!\left[\int g(\varphi(t))\,dt\right]\theta(x,t)$, $w = \exp\!\left[\int g(\varphi(t))\,dt\right]\theta(x,t)$	$\varphi'_t = [f(\varphi) - g(\varphi)]\varphi$, $\dfrac{\partial\theta}{\partial t} = a\dfrac{\partial^2\theta}{\partial x^2}$
$F = uf(u^2 + w^2) - wg(u^2 + w^2)$, $G = ug(u^2 + w^2) + wf(u^2 + w^2)$, $a_1 = a_2 = a$	$u = r(x)\cos\!\left[\theta(x) + C_1 t + C_2\right]$, $w = r(x)\sin\!\left[\theta(x) + C_1 t + C_2\right]$	$ar\theta''_{xx} + 2ar'_x\theta'_x - C_1 r + rg(r^2) = 0$ $ar''_{xx} - ar(\theta'_x)^2 + rf(r^2) = 0$,
$F = uf(u^2 - w^2) + wg(u^2 - w^2)$, $G = ug(u^2 - w^2) + wf(u^2 - w^2)$, $a_1 = a_2 = a$	$u = r(x)\cosh\!\left[\theta(x) + C_1 t + C_2\right]$, $w = r(x)\sinh\!\left[\theta(x) + C_1 t + C_2\right]$	$ar''_{xx} + ar(\theta'_x)^2 + rf(r^2) = 0$, $ar\theta''_{xx} + 2ar'_x\theta'_x + rg(r^2) - C_1 r = 0$
$F = uf(u^2 + w^2) - wg(w/u)$, $G = wf(u^2 + w^2) + ug(w/u)$, $a_1 = a_2 = a$	$u = r(x,t)\cos\varphi(t)$, $w = r(x,t)\sin\varphi(t)$	$\varphi'_t = g(\tan\varphi)$, $\dfrac{\partial r}{\partial t} = a\dfrac{\partial^2 r}{\partial x^2} + rf(r^2)$

where the functions $y = y(\xi)$ and $z = z(\xi)$ are determined by the system of ordinary differential equations

$$k^2[f(ay+bz)y'_\xi]'_\xi + \lambda y'_\xi + g(ay+bz) = 0,$$
$$k^2[f(ay+bz)z'_\xi]'_\xi + \lambda z'_\xi + h(ay+bz) = 0,$$

and the function $\theta = \theta(\xi, t)$ satisfies the heat equation

$$\frac{\partial \theta}{\partial t} = k^2 \frac{\partial}{\partial \xi}\left[f(ay + bz)\frac{\partial \theta}{\partial \xi} \right] + \lambda \frac{\partial \theta}{\partial \xi}.$$

Given $y = y(\xi)$ and $z = z(\xi)$, this equation is linear with respect to θ; its exact solutions can be obtained using the procedure described, for example, in [27].

2°. The system

$$\frac{\partial u}{\partial t} = \frac{\partial}{\partial x}\left[f(t, au + bw)\frac{\partial u}{\partial x} \right] + g(t, au + bw),$$

$$\frac{\partial w}{\partial t} = \frac{\partial}{\partial x}\left[f(t, au + bw)\frac{\partial w}{\partial x} \right] + h(t, au + bw),$$

which generalizes the system of Item 1°, admits exact solutions of the form

$$u = \varphi(t) + b\theta(x, \tau), \qquad w = \psi(t) - a\theta(x, \tau), \qquad \tau = \int f(t, a\varphi + b\psi)\, dt,$$

where $\varphi = \varphi(t)$ and $\psi = \psi(t)$ are determined by the system of ordinary differential equations

$$\begin{aligned} \varphi'_t &= g(t, a\varphi + b\psi), \\ \psi'_t &= h(t, a\varphi + b\psi), \end{aligned}$$

and $\theta = \theta(x, \tau)$ satisfies the linear heat equation

$$\frac{\partial \theta}{\partial \tau} = \frac{\partial^2 \theta}{\partial x^2}.$$

3°. Consider the system

$$\frac{\partial u}{\partial t} = \frac{\partial}{\partial x}\left[f(au + bw)\frac{\partial u}{\partial x} \right] + ug(au + bw) - bwh\left(\frac{w}{u}\right),$$

$$\frac{\partial w}{\partial t} = \frac{\partial}{\partial x}\left[f(au + bw)\frac{\partial w}{\partial x} \right] + wg(au + bw) + awh\left(\frac{w}{u}\right).$$

1. Exact solution:

$$u = br(x, t)\cos^2 \varphi(t), \qquad w = ar(x, t)\sin^2 \varphi(t),$$

where $\varphi = \varphi(t)$ is determined by the separable first-order ordinary differential equation

$$\varphi'_t = \frac{1}{2} a \tan \varphi\, h\left(\frac{a}{b}\tan^2 \varphi\right) \tag{17}$$

and $r = r(x, t)$ satisfies the equation

$$\frac{\partial r}{\partial t} = \frac{\partial}{\partial x}\left[f(abr)\frac{\partial r}{\partial x} \right] + rg(abr). \tag{18}$$

75

The general solution to equation (17) can be written out in implicit form. Equation (18) admits a traveling-wave solution $r = r(kx - \lambda t)$, where k and λ are arbitrary constants.

2. Exact solution:

$$u = br(x,t)\cosh^2\psi(t), \quad w = -ar(x,t)\sinh^2\psi(t),$$

where $\psi = \psi(t)$ is determined by the separable first-order ordinary differential equation

$$\psi_t' = \frac{1}{2}a\tanh\psi\, h\left(-\frac{a}{b}\tanh^2\psi\right)$$

and $r = r(x,t)$ satisfies equation (18).

3. Exact solution:

$$u = -br(x,t)\sinh^2\theta(t), \quad w = ar(x,t)\cosh^2\theta(t),$$

where $\theta = \theta(t)$ is determined by the separable first-order ordinary differential equation

$$\theta_t' = \frac{1}{2}a\coth\theta\, h\left(-\frac{a}{b}\coth^2\theta\right)$$

and $r = r(x,t)$ satisfies equation (18).

4°. Consider the system

$$\frac{\partial u}{\partial t} = \frac{\partial}{\partial x}\left[f(u^2 + w^2)\frac{\partial u}{\partial x}\right] + ug(u^2 + w^2) - wh\left(\frac{w}{u}\right),$$

$$\frac{\partial w}{\partial t} = \frac{\partial}{\partial x}\left[f(u^2 + w^2)\frac{\partial w}{\partial x}\right] + wg(u^2 + w^2) + uh\left(\frac{w}{u}\right).$$

It admits exact solutions in the form

$$u = r(x,t)\cos\varphi(t), \quad w = r(x,t)\sin\varphi(t),$$

where $\varphi = \varphi(t)$ is determined by the separable first-order ordinary differential equation

$$\varphi_t' = h(\tan\varphi), \tag{19}$$

and $r = r(x,t)$ satisfies the equation

$$\frac{\partial r}{\partial t} = \frac{\partial}{\partial x}\left[f(r^2)\frac{\partial r}{\partial x}\right] + rg(r^2). \tag{20}$$

The general solution to equation (19) can be written out in implicit form as

$$\int\frac{d\varphi}{h(\tan\varphi)} = t + C.$$

Equation (20) admits a traveling-wave solution $r = r(kx - \lambda t)$, where k and λ are arbitrary constants.

5°. The system

$$\frac{\partial u}{\partial t} = \frac{\partial}{\partial x}\left[f(u^2 - w^2)\frac{\partial u}{\partial x}\right] + ug(u^2 - w^2) + wh\left(\frac{w}{u}\right),$$

$$\frac{\partial w}{\partial t} = \frac{\partial}{\partial x}\left[f(u^2 - w^2)\frac{\partial w}{\partial x}\right] + wg(u^2 - w^2) + uh\left(\frac{w}{u}\right)$$

admits exact solutions in the form

$$u = r(x,t)\cosh\varphi(t), \quad w = r(x,t)\sinh\varphi(t),$$

where $\varphi = \varphi(t)$ is determined by the separable first-order ordinary differential equation

$$\varphi'_t = h(\tanh\varphi), \tag{21}$$

and $r = r(x,t)$ satisfies the equation

$$\frac{\partial r}{\partial t} = \frac{\partial}{\partial x}\left[f(r^2)\frac{\partial r}{\partial x}\right] + rg(r^2). \tag{22}$$

The general solution to equation (21) can be written out in implicit form as

$$\int \frac{d\varphi}{h(\tanh\varphi)} = t + C.$$

Equation (22) admits a traveling-wave solution $r = r(kx - \lambda t)$, where k and λ are arbitrary constants.

6°. Consider the system

$$\frac{\partial u}{\partial t} = L[u] + b_2 f(a_1 u + b_1 w) + b_1 g(a_2 u + b_2 w),$$

$$\frac{\partial w}{\partial t} = L[w] - a_2 f(a_1 u + b_1 w) - a_1 g(a_2 u + b_2 w),$$

where L is an arbitrary constant-coefficient linear differential operator in x (of any order with respect to the derivatives). It is assumed that $a_1 b_2 - a_2 b_1 \neq 0$.

Multiplying the equations by appropriate constants followed and adding termwise, one obtains two independent equations:

$$\frac{\partial U}{\partial t} = L[U] + (a_1 b_2 - a_2 b_1)f(U), \qquad U = a_1 u + b_1 w;$$

$$\frac{\partial W}{\partial t} = L[W] - (a_1 b_2 - a_2 b_1)g(W), \qquad W = a_2 u + b_2 w.$$

Both equations admit traveling-wave solutions with different wave speeds:

$$U = U(k_1 x - \lambda_1 t), \quad W = W(k_2 x - \lambda_2 t),$$

where k_m and λ_m are arbitrary constants. The corresponding solution to the original system will be a superposition of the two traveling waves.

7°. Consider the system

$$\frac{\partial u}{\partial t} = L[u] + uf(au + bw) - bwg\left(\frac{w}{u}\right),$$

$$\frac{\partial w}{\partial t} = L[w] + wf(au + bw) + awg\left(\frac{w}{u}\right),$$

where L is an arbitrary linear differential operator in x (of any order with respect to the derivatives); the coefficients can depend on x.

1. Exact solution:

$$u = br(x,t)\cos^2\varphi(t), \quad w = ar(x,t)\sin^2\varphi(t),$$

where $\varphi = \varphi(t)$ is determined by the separable first-order ordinary differential equation

$$\varphi_t' = \frac{1}{2}a\tan\varphi\, g\left(\frac{a}{b}\tan^2\varphi\right) \tag{23}$$

and $r = r(x,t)$ satisfies the equation

$$\frac{\partial r}{\partial t} = L[r] + rf(abr). \tag{24}$$

The general solution to equation (23) can be written out in implicit form. Equation (24) admits a t-independent solution, $r = r(x)$. If L is a constant-coefficient linear differential operator, then equation (24) admits a traveling-wave solution $r = r(kx - \lambda t)$, where k and λ are arbitrary constants.

2. Exact solution:

$$u = br(x,t)\cosh^2\psi(t), \quad w = -ar(x,t)\sinh^2\psi(t),$$

where $\psi = \psi(t)$ is determined by the separable first-order ordinary differential equation

$$\psi_t' = \frac{1}{2}a\tanh\psi\, g\left(-\frac{a}{b}\tanh^2\psi\right)$$

and $r = r(x,t)$ satisfies equation (24).

8°. Consider the system

$$\frac{\partial u}{\partial t} = L[u] + uf(u^2 + w^2) - wg\left(\frac{w}{u}\right),$$

$$\frac{\partial w}{\partial t} = L[w] + wf(u^2 + w^2) + ug\left(\frac{w}{u}\right),$$

where L is an arbitrary linear differential operator in x (of any order with respect to the derivatives); the coefficients can depend on x.

Exact solution:

$$u = r(x,t)\cos\varphi(t), \quad w = r(x,t)\sin\varphi(t),$$

where $\varphi = \varphi(t)$ is determined by the separable first-order ordinary differential equation

$$\varphi'_t = g(\tan \varphi) \tag{25}$$

and $r = r(x,t)$ satisfies the equation

$$\frac{\partial r}{\partial t} = L[r] + rf(r^2). \tag{26}$$

The general solution to equation (25) can be written out in implicit form:

$$\int \frac{d\varphi}{g(\tan \varphi)} = t + C.$$

It is noteworthy that equation (26) admits a t-independent solution, $r = r(x)$. If L is a constant-coefficient linear differential operator, then equation (26) admits a traveling-wave solution $r = r(kx - \lambda t)$, where k and λ are arbitrary constants.

9°. Consider the system

$$
\begin{aligned}
\frac{\partial u}{\partial t} &= L[u] + uf(u^2 - w^2) + wg\left(\frac{w}{u}\right), \\
\frac{\partial w}{\partial t} &= L[w] + wf(u^2 - w^2) + ug\left(\frac{w}{u}\right),
\end{aligned}
$$

where L is an arbitrary linear differential operator in x (of any order with respect to the derivatives); the coefficients can depend on x.

Exact solution:

$$u = r(x,t)\cosh \varphi(t), \quad w = r(x,t)\sinh \varphi(t),$$

where $\varphi = \varphi(t)$ is determined by the separable first-order ordinary differential equation

$$\varphi'_t = g(\tanh \varphi) \tag{27}$$

and $r = r(x,t)$ satisfies the equation

$$\frac{\partial r}{\partial t} = L[r] + rf(r^2). \tag{28}$$

The general solution to equation (27) can be written out in implicit form:

$$\int \frac{d\varphi}{g(\tanh \varphi)} = t + C.$$

It is noteworthy that equation (26) admits a t-independent solution, $r = r(x)$. If L is a constant-coefficient linear differential operator, then equation (28) admits a traveling-wave solution $r = r(kx - \lambda t)$, where k and λ are arbitrary constants.

Remark 5. The solutions given in Items 6° to 9° can easily be generalized to the case in which the linear differential operator L depends on n coordinates $x_1, \ldots, x_n$.

$10°$. Consider the system

$$\frac{\partial u}{\partial t} = L[u] + uf\left(\frac{w}{u}\right) - wg\left(\frac{w}{u}\right) + \frac{u}{\sqrt{u^2 + w^2}} h\left(\frac{w}{u}\right),$$

$$\frac{\partial w}{\partial t} = L[w] + wf\left(\frac{w}{u}\right) + ug\left(\frac{w}{u}\right) + \frac{w}{\sqrt{u^2 + w^2}} h\left(\frac{w}{u}\right).$$

Solution:

$$u = r(x,t)\cos\varphi(t), \qquad w = r(x,t)\sin\varphi(t),$$

where the function $\varphi = \varphi(t)$ is determined from the separable first-order ordinary differential equation

$$\varphi'_t = g(\tan\varphi),$$

and the function $r = r(x,t)$ satisfies the linear equation

$$\frac{\partial r}{\partial t} = L[r] + rf(\tan\varphi) + h(\tan\varphi). \tag{29}$$

The change of variable

$$r = F(t)\left[Z(x,t) + \int \frac{h(\tan\varphi)\,dt}{F(t)}\right], \qquad F(t) = \exp\left[\int f(\tan\varphi)\,dt\right]$$

brings (29) to the linear heat equation

$$\frac{\partial Z}{\partial t} = L[Z].$$

$11°$. Consider the system

$$\frac{\partial u}{\partial t} = L[u] + uf\left(\frac{w}{u}\right) + wg\left(\frac{w}{u}\right) + \frac{u}{\sqrt{u^2 - w^2}} h\left(\frac{w}{u}\right),$$

$$\frac{\partial w}{\partial t} = L[w] + wf\left(\frac{w}{u}\right) + ug\left(\frac{w}{u}\right) + \frac{w}{\sqrt{u^2 - w^2}} h\left(\frac{w}{u}\right).$$

Solution:

$$u = r(x,t)\cosh\varphi(t), \qquad w = r(x,t)\sinh\varphi(t),$$

where the function $\varphi = \varphi(t)$ is determined from the separable first-order ordinary differential equation

$$\varphi'_t = g(\tanh\varphi),$$

and the function $r = r(x,t)$ satisfies the linear equation

$$\frac{\partial r}{\partial t} = L[r] + rf(\tanh\varphi) + h(\tanh\varphi). \tag{30}$$

The change of variable

$$r = F(t)\left[Z(x,t) + \int \frac{h(\tanh\varphi)\,dt}{F(t)}\right], \qquad F(t) = \exp\left[\int f(\tanh\varphi)\,dt\right]$$

brings (30) to the linear heat equation

$$\frac{\partial Z}{\partial t} = L[Z].$$

12°. Consider the following nonlinear system consisting of n equations:

$$\frac{\partial u_n}{\partial t} = \frac{\partial}{\partial x}\left[f\left(\sum_{k=1}^{m} a_k u_k\right)\frac{\partial u_n}{\partial x}\right] + u_n g\left(\sum_{k=1}^{m} a_k u_k\right) + u_1 h_n\left(\frac{u_2}{u_1},\ldots,\frac{u_m}{u_1}\right), \quad n = 1,\ldots,m.$$

(31)

Suppose that the condition

$$\sum_{n=1}^{m} a_n h_n(z_2,\ldots,z_m) \equiv 0$$

holds. Then system (31) admits exact solutions of the form

$$u_n = \frac{1}{a_n} r(x,t)\varphi_n(t), \quad n = 1,\ldots,m,$$

where $r(x,t)$ satisfies the equation

$$\frac{\partial r}{\partial t} = \frac{\partial}{\partial x}\left[f(r)\frac{\partial r}{\partial x}\right] + r g(r),$$

and the $\varphi_n(t)$ satisfy the following system of $m-1$ ordinary differential equations and an additional normalization condition:

$$\varphi_n' = \frac{a_n}{a_1} h_n\left(\frac{a_1}{a_2}\frac{\varphi_2}{\varphi_1},\ldots,\frac{a_1}{a_m}\frac{\varphi_m}{\varphi_1}\right), \quad n = 1,\ldots,m-1,$$

$$\sum_{k=1}^{m} \varphi_k = 1.$$

13°. Consider the following nonlinear system consisting of n equations:

$$\frac{\partial u_n}{\partial t} = L[u_n] + u_n f\left(\sum_{k=1}^{m} u_k^2\right) + u_1 g_n\left(\frac{u_2}{u_1},\ldots,\frac{u_m}{u_1}\right), \quad n = 1,\ldots,m.$$

(32)

Suppose that the condition

$$\sum_{n=1}^{m} z_n g_n(z_2,\ldots,z_m) \equiv 0, \quad z_1 \equiv 1, \quad \text{all other } z_m \text{ arbitrary,}$$

holds. Then system (32) admits exact solutions of the form

$$u_n = r(x,t)\varphi_n(t), \quad n = 1,\ldots,m,$$

where $r(x,t)$ satisfies the equation

$$\frac{\partial r}{\partial t} = L[r] + r f(r^2).$$

and the $\varphi_n(t)$ satisfy the following system of $m-1$ ordinary differential equations and an additional normalization condition:

$$\varphi_n' = \varphi_1 g_n\left(\frac{\varphi_2}{\varphi_1}, \ldots, \frac{\varphi_m}{\varphi_1}\right), \quad n = 1, \ldots, m-1,$$

$$\sum_{k=1}^m \varphi_k^2 = 1.$$

NONLINEAR SYSTEMS KLEIN–GORDON TYPE EQUATIONS INVOLVING SECOND DERIVATIVES WITH RESPECT TO t

1°. Consider the system

$$\frac{\partial^2 u}{\partial t^2} = L[u] + b_2 f(a_1 u + b_1 w) + b_1 g(a_2 u + b_2 w),$$

$$\frac{\partial^2 w}{\partial t^2} = L[w] - a_2 f(a_1 u + b_1 w) - a_1 g(a_2 u + b_2 w),$$

where L is an arbitrary constant-coefficient linear differential operator in x (of any order with respect to the derivatives). It is assumed that $a_1 b_2 - a_2 b_1 \neq 0$.

Multiplying the equations by appropriate constants and adding termwise, one obtains two independent equations:

$$\frac{\partial^2 U}{\partial t^2} = L[U] + (a_1 b_2 - a_2 b_1) f(U), \qquad U = a_1 u + b_1 w;$$

$$\frac{\partial^2 W}{\partial t^2} = L[W] - (a_1 b_2 - a_2 b_1) g(W), \qquad W = a_2 u + b_2 w.$$

Both equations admit traveling-wave solutions with different wave speeds,

$$U = U(k_1 x - \lambda_1 t), \quad W = W(k_2 x - \lambda_2 t),$$

where k_m and λ_m are arbitrary constants. The associated solution to the original system will be a superposition of the two waves.

2°. Consider the system

$$\frac{\partial^2 u}{\partial t^2} = L[u] + u f(t, au - bw) + g(t, au - bw),$$

$$\frac{\partial^2 w}{\partial t^2} = L[w] + w f(t, au - bw) + h(t, au - bw),$$

where L is an arbitrary linear differential operator (of any order) with respect to the spatial variables $x_1, \ldots, x_n$, whose coefficients can be dependent on $x_1, \ldots, x_n, t$. It is assumed that $L[\mathrm{const}] = 0$.

1. Solution:

$$u = \varphi(t) + b\theta(x_1,\ldots,x_n,t), \qquad w = \psi(t) + a\theta(x_1,\ldots,x_n,t),$$

where $\varphi = \varphi(t)$ and $\psi = \psi(t)$ are determined by the system of ordinary differential equations

$$\begin{aligned}
\varphi''_{tt} &= \varphi f(t, a\varphi - b\psi) + g(t, a\varphi - b\psi), \\
\psi''_{tt} &= \psi f(t, a\varphi - b\psi) + h(t, a\varphi - b\psi),
\end{aligned}$$

and the function $\theta = \theta(x_1,\ldots,x_n,t)$ satisfies linear equation

$$\frac{\partial^2 \theta}{\partial t^2} = L[\theta] + f(t, a\varphi - b\psi)\theta.$$

2. Let us multiply the first equation by a and the second one by $-b$ and add the results together to obtain

$$\frac{\partial^2 \zeta}{\partial t^2} = L[\zeta] + \zeta f(t, \zeta) + ag(t, \zeta) - bh(t, \zeta), \qquad \zeta = au - bw. \qquad (33)$$

This equation will be considered in conjunction with the first equation of the original system

$$\frac{\partial^2 u}{\partial t^2} = L[u] + uf(t, \zeta) + g(t, \zeta). \qquad (34)$$

Equation (33) can be treated separately. Given a solution $\zeta = \zeta(x,t)$ to equation (33), the function $u = u(x_1,\ldots,x_n,t)$ can be determined by solving the linear equation (34) and the function $w = w(x_1,\ldots,x_n,t)$ is found as $w = (au - \zeta)/b$.

Note three important cases where equation (33) admits exact solutions:

(i) Equation (33) admits a spatially homogeneous solution $\zeta = \zeta(t)$.

(ii) Suppose that the coefficients of L and the functions f, g, h are not implicitly dependent on t. Then equation (33) admits a steady-state, t-independent solution $\zeta = \zeta(x_1,\ldots,x_n)$.

(iii) If the condition $\zeta f(t, \zeta) + ag(t, \zeta) - bh(t, \zeta) = k_1 \zeta + k_0$ holds, equation (33) is linear. If the operator L is constant-coefficient, the method of separation of variables can be used to obtain solutions.

3°. Consider the system

$$\begin{aligned}
\frac{\partial^2 u}{\partial t^2} &= L_1[u] + uf\left(\frac{u}{w}\right), \\
\frac{\partial^2 w}{\partial t^2} &= L_2[w] + wg\left(\frac{u}{w}\right),
\end{aligned}$$

where L_1 and L_2 are arbitrary constant-coefficient linear differential operators (of any order) with respect to x. It is assumed that $L_1[\text{const}] = 0$ and $L_2[\text{const}] = 0$.

1. Solution in the form of the product of two waves traveling at different speeds:

$$u = e^{kx-\lambda t}y(\xi), \quad w = e^{kx-\lambda t}z(\xi), \quad \xi = \beta x - \gamma t,$$

where k, λ, β, and γ are arbitrary constants, and the functions $y = y(\xi)$ and $z = z(\xi)$ are determined by the system of ordinary differential equations

$$\beta^2 y''_{\xi\xi} + 2\lambda\gamma y'_{\xi} + \lambda^2 y = M_1[y] + yf(y/z), \quad M_1[y] = e^{-kx}L_1[e^{kx}y(\xi)],$$
$$\beta^2 z''_{\xi\xi} + 2\lambda\gamma z'_{\xi} + \lambda^2 z = M_2[z] + zg(y/z), \quad M_2[z] = e^{-kx}L_2[e^{kx}z(\xi)].$$

To the special case $k = \lambda = 0$ there corresponds a traveling-wave solution.

2. Periodic multiplicative separable solution:

$$u = [C_1\sin(kt) + C_2\cos(kt)]\varphi(x), \quad w = [C_1\sin(kt) + C_2\cos(kt)]\psi(x),$$

where C_1, C_2, and k are arbitrary constants and the functions $\varphi = \varphi(x)$ and $\psi = \psi(x)$ are determined by the system of ordinary differential equations

$$L_1[\varphi] + k^2\varphi + \varphi f(\varphi/\psi) = 0,$$
$$L_2[\psi] + k^2\psi + \psi g(\varphi/\psi) = 0.$$

3. Multiplicative separable solution:

$$u = [C_1\sinh(kt) + C_2\cosh(kt)]\varphi(x), \quad w = [C_1\sinh(kt) + C_2\cosh(kt)]\psi(x),$$

where C_1, C_2, and k are arbitrary constants and the functions $\varphi = \varphi(x)$ and $\psi = \psi(x)$ are determined by the system of ordinary differential equations

$$L_1[\varphi] - k^2\varphi + \varphi f(\varphi/\psi) = 0,$$
$$L_2[\psi] - k^2\psi + \psi g(\varphi/\psi) = 0.$$

4. Degenerate multiplicative separable solution:

$$u = (C_1t + C_2)\varphi(x), \quad w = (C_1t + C_2)\psi(x),$$

where C_1 and C_2 are arbitrary constants and the functions $\varphi = \varphi(x)$ and $\psi = \psi(x)$ are determined by the system of ordinary differential equations

$$L_1[\varphi] + \varphi f(\varphi/\psi) = 0, \quad L_2[\psi] + \psi g(\varphi/\psi) = 0.$$

Remark 6. The coefficients of L_1 and L_2 as well as the functions f and g in Items 2°–4° can be dependent on x.

Remark 7. If L_1 and L_2 contain only even derivatives, there are solutions of the form

$$u = [C_1\sin(kx) + C_2\cos(kx)]U(t), \qquad w = [C_1\sin(kx) + C_2\cos(kx)]W(t);$$
$$u = [C_1\exp(kx) + C_2\exp(-kx)]U(t), \qquad w = [C_1\exp(kx) + C_2\exp(-kx)]W(t);$$
$$u = (C_1x + C_2)U(t), \qquad w = (C_1x + C_2)W(t),$$

where C_1, C_2, and k are arbitrary constants. Note that the third solution is degenerate.

4°. Consider the system

$$\frac{\partial^2 u}{\partial t^2} = L[u] + au \ln u + uf\left(t, \frac{u}{w}\right),$$

$$\frac{\partial^2 w}{\partial t^2} = L[w] + aw \ln w + wg\left(t, \frac{u}{w}\right),$$

where L is an arbitrary linear differential operator with respect to the coordinates $x_1, \ldots, x_n$ (of any order in derivatives), whose coefficients can be dependent on the coordinates.

Solution:

$$u = \varphi(t)\theta(x_1, \ldots, x_n),$$
$$w = \psi(t)\theta(x_1, \ldots, x_n),$$

where the functions $\varphi = \varphi(t)$ and $\psi = \psi(t)$ are described by the nonlinear system of second-order ordinary differential equations

$$\varphi''_{tt} = a\varphi \ln \varphi + b\varphi + \varphi f(t, \varphi/\psi),$$
$$\psi''_{tt} = a\psi \ln \psi + b\psi + \psi g(t, \varphi/\psi),$$

b is an arbitrary constant, and the function $\theta = \theta(x_1, \ldots, x_n)$ satisfies the steady-state equation

$$L[\theta] + a\theta \ln \theta - b\theta = 0.$$

This equation is linear in the special case $a = 0$.

WEBSITE EQWORLD: THE WORLD OF MATHEMATICAL EQUATIONS. EQARCHIVE: AN EQUATION ARCHIVE

The international scientific-educational website *EqWorld: The World of Mathematical Equations* (web address http://eqworld.ipmnet.ru/) is the world largest specialized online resource devoted to mathematical equations. Listed below are most important sections of the website.

1. *Exact Solutions of Mathematical Equations*
   (http://eqworld.ipmnet.ru/en/solutions.htm). This sections contains reference tables of exact solutions to various classes of ordinary differential, partial differential, integral, functional, and other mathematical equations. It also presents papers of interest on the above topics. Some of the equations considered in Sections 6 and 7 of the present paper, as well as other equations of that type, can be found on the webpage http://eqworld.ipmnet.ru/en/solutions/syspde.htm.

2. *Methods for Solving Mathematical Equations*
   (http://eqworld.ipmnet.ru/en/methods.htm). This section contains papers and lectures on methods for solution of ordinary differential, partial differential, integral, and functional equations. In particular, there are lectures

on the methods of generalized and functional separation of variables (see page http://eqworld.ipmnet.ru/en/methods/meth-pde.htm).

3. *EqArchive: an archive of equations and solutions*
(`http://eqworld.ipmnet.ru/eqarchive/?lang=en`). This section allows any user, once they have registered, to publish a new equation with a solution. All published information is sorted according to a detailed classification and available for viewing. Some of the equations and solutions presented in Sections 3 and 5 of the current paper can be found in EqArchive.

REFERENCES

1. A. D. Polyanin and V. F. Zaitsev, *Handbook of Nonlinear Partial Differential Equations*, Chapman & Hall/CRC Press, Boca Raton, 2004.
2. A. D. Polyanin, V. F. Zaitsev, and A. I. Zhurov, *Methods for the Solution of Nonlinear Equations of Mathematical Physics and Mechanics*, Fizmatlit, Moscow, 2005. (in Russian)
3. A. D. Polyanin and A. V. Manzhirov, *Handbook of Mathematics for Engineers and Scientists*, Chapman & Hall/CRC Press, Boca Raton, 2007.
4. V. A. Galaktionov and S. R. Svirshchevskii, *Exact Solutions and Invariant Subspaces of Nonlinear Partial Differential Equations in Mechanics and Physics*, Chapman & Hall/CRC Press, Boca Raton, 2006.
5. V. A. Galaktionov, *Proc. Roy. Soc. Edinburgh* **125A**(2), 225–448 (1995).
6. S. R. Svirshchevskii, *Phys. Lett. A* **199**, 344–348 (1995).
7. E. A. Saied and R. G. Abd El-Rahman, *Journal of the Physical Society of Japan* **68**(2), 360–368 (1999).
8. V. A. Dorodnitsyn, *Zhurn. vychisl. matem. i matem. fiziki* **22**(6), 1393–1400 (1982). (in Russian)
9. V. A. Galaktionov, *Nonlinear Analys., Theory, Meth. and Applications* **23**, 1595–1621 (1994).
10. A. D. Polyanin and E. A. Vyazmina, *Doklady Mathematics* **72**(2), 798–801 (2005).
11. E. A. Vyazmina and A. D. Polyanin, *Theor. Found. Chem. Eng.* **40**(6), 555–563 (2006).
12. P. G. Estévez, C. Z. Qu, and S. L. Zhang, *J. Math. Anal. Appl.* **275**, 44–59 (2002).
13. A. D. Polyanin and A. I. Zhurov, A. I., *Doklady Physics* **43**(6), 381–385 (1998).
14. N. H. Ibragimov (Ed), *CRC Handbook of Lie Group Analysis of Differential Equations, Vol. 1, Symmetries, Exact Solutions and Conservation Laws*, CRC Press, Boca Raton, 1994.
15. V. F. Zaitsev and A. D. Polyanin, *Doklady Mathematics* **64**(3), 416–420 (2001).
16. S. V. Khabirov, *Mat. Sbornik* **181**(12), 1607–1622 (1990). (in Russian)
17. A. D. Polyanin and E. A. Vyazmina, *Doklady Mathematics* **74**(1), 597–602 (2006).
18. T. Barannyk, "Symmetry and exact solutions for systems of nonlinear reaction-diffusion equations," in *Proc. of Inst. of Mathematics of NAS of Ukraine*, Vol. 43, Part 1, 2002, pp. 80–85.
19. T. A. Barannyk and A. G. Nikitin, "Solitary wave solutions for heat equations," in *Proc. of Inst. of Mathematics of NAS of Ukraine*, Vol. 50, Part 1, 2004, pp. 34–39.
20. R. Cherniha and J. R. King, *J. Phys. A: Math. Gen.* **33**, 267–282, 7839–7841 (2000).
21. R. Cherniha and J.R. King, *J. Phys. A: Math. Gen.* **36**, 405–425 (2003).
22. A. G. Nikitin, *J. Math. Anal. Appl.* **332**(1), 666–690 (2007) (see also `http://arxiv.org/abs/math-ph/0411028`).
23. A. G. Nikitin and R. J. Wiltshire, *J. Math. Phys.* **42**(4), 1667–1688 (2001).
24. A. D. Polyanin, *Foundations of Chemical Engineering* **38**(6), 622–635 (2004).
25. A. D. Polyanin, *Doklady Mathematics* **71**(1), 148–154 (2005).
26. A. D. Polyanin, *Systems of Partial Differential Equations*, From Website *EqWorld: The World of Mathematical Equations*, `http://eqworld.ipmnet.ru/en/solutions/syspde.htm`.
27. A. D. Polyanin, *Handbook of Linear Partial Differential Equations for Engineers and Scientists*, Chapman & Hall/CRC Press, Boca Raton, 2002.

Fractional Differential and Integral Equations of Riemann-Liouville versus Caputo

A.S. Vatsala* and V. Lakshmikantham[†]

*Dept. of Mathematics, University of Louisiana at Lafayette, Lafayette, LA 70504, USA
[†]Dept. of Mathematical Sciences, Florida Institute of Technology, Melbourne, FL 32901, USA

Abstract. Recently fractional differential equations involving Riemann-Loiuville type as well as Caputo type has gained importance due to its application. In this paper we compare and contrast these two types of equations and present the latest development relative to them.

Keywords: Fractional differential inequalities, comparison principle, basic existence results.
PACS: 02.30.Hq, 02.30.Rz

INTRODUCTION

The investigation of the theory of fractional differential equations involving Riemann-Liouville differential operators or of Caputo derivative of order $0 < q < 1$ are important, since such equations are often more useful in the modeling of several physical phenomena [1]-[7], [13]-[17]. Just like integer derivatives the Caputo derivative of a constant k is zero, where as the Riemann-Liouville derivative of a constant k is $\frac{k(t-t_0)^{-q}}{\Gamma(1-q)}$. Thus differential equations involving Caputo derivatives follows almost a similar pattern as that of integer derivatives. Differential equations involving Caputo derivative will have a similar integral representation as that of the integer derivative. Differential equations involving Riemann-Liouville derivatives do not have a similar integral representation as that of the integer derivative. Thus the aim of this paper to compare the study of differential equations involving the Caputo derivative with that of Riemann-Liouville derivative.

PROPERTIES OF FRACTIONAL DERIVATIVES AND INEQUALITIES

Initially, we recall the definition of fractional derivative of Caputo and Riemann-Liouville of order q where $0 < q < 1$. The Caputo derivative of order q where $0 < q < 1$ of a function $m(t)$ is given by

$$^{c}D^{q}m(t) = \frac{1}{\Gamma(p)} \int_{t_0}^{t} (t-s)^{p-1} \frac{d}{ds} m(s) ds.$$

However, the Riemann-Liouville derivative of order q where $0 < q < 1$ of a function $m(t)$ is given by

CP1067, *Applications of Mathematics in Engineering and Economics '34—AMEE '08*, edited by M. D. Todorov
© 2008 American Institute of Physics 978-0-7354-0598-1/08/$23.00

$$D^q m(t) = \frac{1}{\Gamma(p)} \frac{d}{dt} \int_{t_0}^{t} (t-s)^{p-1} m(s) ds.$$

From the definition it is clear that the Caputo derivative exists for functions which are differentiable. The Caputo derivative of a constant is zero whereas the Riemann-Liouville fractional derivative for a constant c is not zero but equals to

$$D^q c = \frac{c(t-t_0)^{-q}}{\Gamma(1-q)}.$$

Also, if $x(t) = (t-t_0)^\omega$, $\omega > 0$, then we get

$$D^q (t-t_0)^\omega = \frac{1}{\Gamma(p)} \left(\frac{d}{dt} \int_{t_0}^{t} (t-s)^{p-1} (s-t_0)^\omega ds \right)$$

$$= (t-t_0)^{\omega-q} \frac{\Gamma(\omega+1)}{\Gamma(1-q+\omega)}$$

after the computation of the integral and differentiation. A simpler result is when $x(t) \equiv 1$,

$$D^q 1 = \frac{(t-t_0)^{-q}\Gamma(1)}{\Gamma(1-q)} = \frac{(t-t_0)^{-q}}{\Gamma(1-q)}.$$

From the definition it follows that

$$^c D^q x(t) = D^q[x(t) - x(t_0)],$$

and

$$^c D^q x(t) = D^q x(t) - \frac{x(t_0)}{\Gamma(1-q)}(t-t_0)^{-q}.$$

In particular, if $x(t_0) = 0$, we have
$$^c D^q x(t) = D^q x(t).$$
Hence, we can see that Caputo derivative is defined for functions for which Riemann-Liouville derivative exists.

Here we recall some of the results developed involving Riemann Liouville derivative. See [10, 11, 12] for details.

Also we introduce some needed notation.

Lemma 1 *Let $m : R_+ \to R$ be locally Hölder continuous such that for any $t_1 \in (0, \infty)$, we have*

$$m(t_1) = 0 \text{ and } m(t) \leq 0 \text{ for } 0 \leq t \leq t_1. \tag{1}$$

Then it follows that

$$D^q m(t_1) \geq 0. \tag{2}$$

We present the proof of this from [12] to see the features of fractional derivative.
Proof: We know that

$$D^q m(t) = \frac{1}{\Gamma(p)} \frac{d}{dt} \int_{0}^{t} (t-s)^{p-1} m(s) ds, \tag{3}$$

where $1 - q = p$. Let $H(t) = \int_0^t (t-s)^{p-1} m(s) ds$. Consider for $h > 0$,

$$H(t_1) - H(t_1 - h) = \int_0^{t_1 - h} [(t_1 - s)^{p-1} - (t_1 - h - s)^{p-1}] m(s) ds$$

$$+ \int_{t_1 - h}^{t_1} (t-s)^{p-1} m(s) ds = I_1 + I_2,$$

say. Since $[(t_1 - s)^{p-1} - (t_1 - h - s)^{p-1}] < 0$ for $0 \le s \le t_1 - h$ and $m(s) \le 0$ by hypothesis, we have $I_1 \ge 0$. Also,

$$H(t_1) - H(t_1 - h) \ge \int_{t_1 - h}^{t_1} (t_1 - s)^{p-1} m(s) ds = I_2.$$

Since $m(t)$ is locally Hölder continuous and $m(t_1) = 0$, there exists a constant $K(t_1) > 0$, such that, for $t_1 - h \le s \le t_1 + h$,

$$-K(t_1)(t_1 - s)^\lambda \le m(s) \le K(t_1)(t_1 - s)^\lambda,$$

where $\lambda > 0$ is such that $\lambda + p - 1 > 0$ and $0 < \lambda < 1$. We then get

$$I_2 \ge -K(t_1) \int_{t_1 - h}^{t_1} (t_1 - s)^{p-1+\lambda} ds = \frac{K(t_1)}{\Gamma(p+\lambda)} h^{p+\lambda}.$$

Hence $H(t_1) - H(t_1 - h) - \frac{K(t_1)}{\Gamma(p+\lambda)} h^{p+\lambda} \ge 0$, for sufficiently small $h > 0$. Letting $h \to 0$, we obtain $H'(t_1) \ge 0$, which implies $D^q m(t_1) = \frac{1}{\Gamma(p)} H'(t_1) \ge 0$ and the proof is complete.

Next we recall some fundamental result relative to strict fractional differential inequalities without proof.

Theorem 1 *Let $v, w : [0, T] \to R$ be locally Hölder continuous, $f \in C([0, T] \times R, R)$ and*

$$(i)\ D^q v(t) \le f(t, v(t)), \quad (ii)\ D^q w(t) \ge f(t, w(t)), \quad 0 \le t \le T,$$

one of the inequalities being strict. Then

$$v(0) < w(0) \tag{4}$$

implies

$$v(t) < w(t), \quad 0 \le t \le T. \tag{5}$$

The next result is for non-strict fractional differential inequalities which requires a one sided Lipschitz type condition.

Theorem 2 *Assume that the conditions of Theorem 3 hold with non-strict inequalities (i) and (ii). Suppose further that*

$$f(t,x) - f(t,y) \leq \frac{L}{1+t^q}(x-y), \quad \text{wherever } x \geq y \text{ and } L > 0. \tag{6}$$

Then $v(0) \leq w(0)$ implies, provided $LT^q \leq \frac{1}{\Gamma(1-q)}$,

$$v(t) \leq w(t), \quad 0 \leq t \leq T. \tag{7}$$

Remark 1 *The conclusion of the above result is true if we replace Riemann-Liouville derivative by Caputo derivative.*

DIFFERENTIAL EQUATIONS WITH CAPUTO VERSUS RIEMANN-LIOUVILLE DERIVATIVE

Consider the initial value problem (IVP) for fractional differential equations which is given by

$$D^q u = f(t,u), \quad u(0) = u_0, \tag{8}$$

where $f \in C([0,T] \times R, R)$, $D^q u$ is the fractional derivative of u and q is such that $0 < q < 1$. Since f is assumed continuous, we considered the IVP (8) is equivalent to the following Volterra fractional integral equation

$$u(t) = u_0 + \frac{1}{\Gamma(q)} \int_0^t (t-s)^{q-1} f(s, u(s)) ds, \quad 0 \leq t \leq T. \tag{9}$$

That is, every solution of (10) is also a solution of (8) and vice versa. Here and elsewhere Γ denotes the Gamma function.

In the initial stages of our study of fractional equations in [10, 12] we assumed this and developed results using the Volterra fractional integral equation (10). This is true only either when $u_0 = 0$, or $D^q u$ (8) replaced by $^c D^q u$. In order to rectify this in [11], we considered the modified version of the differential equation

$$D^q(u - u(0)) = f(t,u), \quad u(0) = u_0, \tag{10}$$

where $f \in C([0,T] \times R, R)$, $D^q u$ is the fractional derivative of u and q is such that $0 < q < 1$ in, as suggested in [5] . Now the results of the previous section can be modified suitably to study the qualitative properties of the solution of 10. However, it is easy to observe that

$$D^q(u - u_0) =^c D^q u.$$

Hence, if the initial condition is specified or well defined, it is useful to consider equation (8) with $D^q u$ replaced by $^c D^q u$.

In order to study the qualitative properties of differential equations using Riemann-Liouville derivative we have to consider

$$D^q x(t) = f(t, x(t)), \quad x(t)(t-t_0)^{1-q}|_{t=t_0} = x^0, \tag{11}$$

and the corresponding integral equation is

$$x(t) = x^0(t) + \frac{1}{\Gamma(q)} \int_{t_0}^{t} (t-s)^{q-1} f(s, x(s)) ds, \tag{12}$$

on $[t_0, T]$ where $x^0(t) = \frac{x^0 (t-t_0)^{q-1}}{\Gamma(q)}$ and $x \in C_p([t_0, T], R]$. Here

$$C_p([t_0, T], R] = [x \in C((t_0, T], R] \text{ and } x(t)(t-t_0)^{1-q} \in C[(t_0, T], R],$$

where $p = 1 - q$. Since $D^q x_0 = \frac{(t-t_0)^{-q} x_0}{\Gamma(1-q)}$.

We can also consider the IVP relative to (11),

$$D^q x(t) = x_0 (t-t_0)^{-q} + f(t, x(t)), \quad x(t)(t-t_0)^{1-q}|_{t=t_0} = x_0, \tag{13}$$

and then the corresponding Riemann Liouville integral will be simply

$$x(t) = x_0 + \frac{1}{\Gamma(q)} \int_{t_0}^{t} (t-s)^{q-1} f(s, x(s)) ds, \tag{14}$$

on the same interval $[t_0, T]$. Thus when one investigates the IVP for fractional differential equations of the R-L type, one has the choice of considering any one of the foregoing formulations. If we consider (11) and (12), (12) is not the usual Volterra integral form which we encounter when we have integer derivative or Caputo derivative in (11). Thus in this situation the study of (12) becomes complicated because of the initial function $x^0(t)$, where as the differential equation in (11) looks similar to what we are accustomed to, except for the initial condition. On the other hand, when we discuss (13) and (14), (13) looks some what different looking, where as (14) is exactly the same type of integral equation we know of. In the monograph [9] (11) and (12) are well documented, where the basic inequalities, existence theories and other qualitative and quantitative properties are investigated.

Here we just recall two integral inequality results from [9] relative to (12) without proof.

Theorem 3 *Let* $v, w \in C_p([t_0, T], \mathbb{R}), f \in C([t_0, T], \mathbb{R})$ *and*

(i) $v(t) \leq \frac{v^0 (t-t_0)^{q-1}}{\Gamma(q)} + \frac{1}{\Gamma(q)} \int_{t_0}^{t} (t-s)^{q-1} f(s, v(s)) ds,$

(ii) $w(t) \geq \frac{w^0(t-t_0)^{q-1}}{\Gamma(q)} + \frac{1}{\Gamma(q)} \int\limits_{t_0}^{t}(t-s)^{q-1}f(s,w(s))ds, \quad t_0 \leq t \leq T,$

one of the inequalities (i) or (ii) being strict. Suppose further that $f(t,x)$ is nondecreasing in x for each t and

$$v^0 < w^0$$

where $v^0 = (t-t_0)^{1-q}v(t)\,|_{t=t_0}$ and $w^0 = (t-t_0)^{1-q}w(t)\,|_{t=t_0}$. Then we have

$$v(t) < w(t), t_0 \leq t \leq T.$$

The next two results are with respect to (13) which we present without proof.

Theorem 4 *Assume that the conditions of Theorem 3 hold with nonstrict inequalities (i) and (ii). Suppose further that*

$$f(t,x) - f(t,y) \leq \frac{L(t-t_0)^{1-q}}{1+(t-t_0)}(x-y),$$

whenever $x \geq y, L > 0$ and $0 < q < 1$. Then $v^0 \leq w^0$ and $L < \Gamma(1+q)$ implies

$$v(t) \leq w(t), t_0 \leq t \leq T.$$

Theorem 5 *Let $v, w \in C_p([t_0,T],R]$, be locally Hölder continuous, $f \in c([t_0,T] \times R, R)$ be continuous and*

$$(i)\ D^q v(t) \leq v_0(t-t_0)^{-q} + f(t,v(t)), \quad v(t)(t-t_0)^{1-q}|_{t=t_0} = v_0$$

$$(ii)\ D^q w(t) \leq w_0(t-t_0)^{-q} + f(t,w(t)), \quad w(t)(t-t_0)^{1-q}|_{t=t_0} = w_0 \quad t_0 < t \leq T,$$

one of the inequalities being strict. Then

$$v_0 < w0 \tag{15}$$

implies

$$v(t) < w(t), \quad t_0 \leq t \leq T. \tag{16}$$

Theorem 6 *Assume that the conditions of Theorem 3 hold with nonstrict inequalities (i) and (ii). Suppose further that*

$$f(t,x) - f(t,y) \leq \frac{L}{1+(t-t_0)^{q-1}}(x-y), \quad \text{wherever } x \geq y \text{ and } L > 0. \tag{17}$$

Then $v_0 \leq w_0$) implies, provided $L(T-t_0)^q \leq \frac{1}{\Gamma(1-q)}$,

$$v(t) \leq w(t), \quad 0 \leq t \leq T. \tag{18}$$

The proof of the above results and their applications will be discussed in our future research work. See [8] for differential and integral inequalities results integer order results.

REFERENCES

1. M. Caputo, *Geophy. J. Roy. Astronom* **13**, 529–539 (1967).
2. W.G. Glöckle and T.F. Nonnenmacher, *Biophy. J.* **68**, 46–53 (1995).
3. K. Diethelm and N.J. Ford, *JMAA* **265**, 229–248 (2002).
4. K. Diethelm and N.J. Ford, *AMC* **154**, 621–640 (2004).
5. K. Diethelm and A.D. Freed. "On the solution of nonlinear fractional differential equations used in the modeling of viscoplasticity," in *Scientific Computing in Chemical Engineering II: Computational Fluid Dynamics, Reaction Engineering, and Molecular Properties*, edited by F. Keil, W. Mackens, H. Vob, and J. Werther, Springer, Heidelberg, 1999, pp. 217–224.
6. Rudolf, Hilfer, *Applications of Fractional Calculus in Physics,* World Scientific, 2008.
7. V. Kiryakova, in *Generalized fractional calculus and applications,* Pitman Res. Notes Math. Ser. Vol 301, Longman-Wiley, New York, 1994.
8. V. Lakshmikantham and S. Leela, *Differential and Integral Inequalities, Vol. I*, Academic Press, New York, 1969.
9. V. Lakshmikantham, S. Leela, and Devi J. Vasundhara, *Theory of Fractional Dynamic Systems,* to be published by Cambridge Scientific Publishers.
10. V. Lakshmikantham and A.S. Vatsala, "Basic Theory of Fractional Differential Equations," in *Nonlinear Analysis: Theory, Methods, and Applications*, to appear in 2008.
11. V. Lakshmikantham and A.S. Vatsala, *Applied Mathematics Letter*, 828–834 2008.
12. V. Lakshmikantham and A.S. Vatsala, *Communication in Applied Analysis* **11**, 395–402 (2007).
13. R. Metzler, W. Schick, H.G. Kilian, and T.F. Nonnenmacher, *J. Chem. Phy.* **103**, 7180–7186 (1995).
14. Keith B. Oldham and Jerome Spanier, *Fractional Calculus*, Dover, 2002.
15. I. Podlubny, *Fractional Differential Equations*, Academic Press, San Diego, 1999.
16. S.G. Samko, A.A. Kilbas, and O.I. Marichev, *Fractional Integrals and Derivatives, Theory and Applications,* Gordon and Breach, Yverdon, 1993.
17. Boris Rubin, *Fractional Integrals and Potentials*, Longman 1996.

Partial Control of a System with Fractal Basin Boundaries

S. Zambrano and M. A. F. Sanjuán

Departamento de Física, Universidad Rey Juan Carlos – Tulipán s/n 28933 Madrid, Spain

Abstract. In this paper we apply the technique of partial control of a chaotic system to a dynamical system with two attractors with fractal basin boundaries, in presence of environmental noise. This technique allows one to keep the trajectories far from any of the attractors by applying a control that is smaller than the amplitude of environmental noise. We will show that the same geometrical horseshoe-like action that gives rise to the existence of fractal basin boundaries will allow us to detect certain sets for the dynamical system considered, the safe sets, that make this type of control possible.

Keywords: Fractal basin boundaries, control, transient chaos.
PACS: 05.45.Gg,05.45.-a

INTRODUCTION

Some dynamical systems are not chaotic but they have an nonattractive invariant set that induces transient chaos [2], usually referred to as a chaotic saddle. There are different mechanisms by which a chaotic attractor is destroyed, giving rise to a chaotic saddle, for example the boundary crisis [5]. Thus, transient chaos can be found in a large variety of contexts [4] so in many different situations it might be desirable to control the system to keep its trajectories close to the chaotic saddle, or simply more convenient than letting them fall in the coexisting attractor. This idea can be framed in the wide field of control of chaotic dynamical systems [16], that has been recently found to have interesting applications in different fields of science and engineering [3].

Considering this, different techniques have been proposed in recent years to control transient chaos. Some of them are inspired in the OGY chaos control scheme [11], based on stabilization of the system around one of the unstable periodic orbits that lie in the chaotic saddle [17, 12, 13] whereas others are focused on accurately perturbing the system in order to avoid "dangerous" regions of the phase space, from which trajectories are expelled to the attractor [14, 6, 4, 1, 18].

There are two main difficulties involving this control task. The first one is the nonattracting nature of the chaotic saddle, from which trajectories typically diverge after a finite amount of time. The second one is that most dynamical systems of interest in practical applications are affected by the presence of noise. We could assume, as an extra difficulty, that our action on the system is bounded to be smaller than the action of noise. In that situation it would seem that it is impossible to sustain transient chaos. However, in a recent paper Aguirre et al [1] showed that this was indeed possible for the simplest dynamical system with a chaotic saddle and escapes to infinity: the slope-three tent map. We recently extended these ideas [19] to the family of dynamical systems for

CP1067, *Applications of Mathematics in Engineering and Economics '34—AMEE '08*, edited by M. D. Todorov
© 2008 American Institute of Physics 978-0-7354-0598-1/08/$23.00

which a horseshoe-like map [15] is the responsible of the appearance of the chaotic sad- dle. We showed that for these systems it was possible to find inside a square Q enclosing the chaotic saddle (where the system considered acts as a horseshoe map) certain *safe sets* that allowed one to achieve this goal. This type of control does not say where the trajectories will exactly go in the vicinity of the chaotic saddle, so we called it *partial control* of a dynamical system.

Chaotic saddles can appear also in systems for which there is more than one attractor. Furthermore, it is quite general to find transient chaos in systems with fractal basin boundaries [7]. For these systems, the dimension of the set that separates the basins of attractions of two or more attractors has a noninteger dimension, which implies a degree of unpredictability which is related with this (fractal) dimension. This phenomenon has been related with the existence of horseshoe like-mapping in a given zones of phase space [7]. Sometimes this relation is found indirectly. For example, by using Melnikov's method [8], it was found that for certain oscillators the appearance of fractal basin boundaries is simultaneous to that of transverse homoclinic points [9] (which imply the existence of a horseshoe on phase space). Considering that our partial control technique applies for dynamical systems that present a horseshoe in phase space, we expect that our technique can be applied in this context to keep the trajectories far from any of the attractors. In fact, in this paper we are going to show that for a paradigmatic map with fractal basin boundaries it is possible to find safe sets analogous to those used in Ref. [19] that can be used to keep the trajectories of the dynamical system far from any of the periodic attractors of the system, even if the control applied is smaller than the amplitude of noise. This is done by making use of the fact that the dynamical system considered behaves approximately like a horseshoe map on a square Q of phase space. Using our technique, we will show that trajectories can be kept inside that square even if the control applied is smaller than the amplitude of noise.

The structure of this paper is the following. First we describe the dynamical system that we are going to use as an example. After this, we are going to state the control problem that we deal with in this paper in a mathematically precise way. After doing this, we will show how the safe sets can be built for this dynamical system, and we will explain why they have the structure that allows one to keep the trajectories far from the attractors using a control smaller than the amplitude of noise. Finally we are going to give a numerical example of application of our technique and we will draw the main conclusions of this work.

DESCRIPTION OF THE SYSTEM AND PROBLEM STATEMENT

The system that we consider here is a two-dimensional map $\mathbf{f}$ that has been thoroughly described in Ref. [7]. The equations of this system are

$$(\theta_{n+1}, x_{n+1}) = \mathbf{f}(\theta_n, x_n) = (\theta_n + a\sin 2\theta_n - b\sin 4\theta_n - x_n \sin \theta_n, -J\cos \theta_n), \quad (1)$$

where the angles θ and $\theta + 2\pi$ are identified. In the following discussion, we will label the points of the trajectories of this dynamical system as $\mathbf{p}_n = (\theta_n, x_n)$, so the relation above can be written as $\mathbf{p}_{n+1} = \mathbf{f}(\mathbf{p}_n)$.

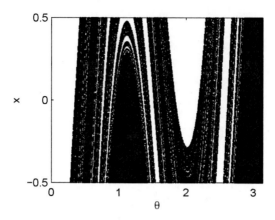

FIGURE 1. Basins of attraction of the fixed points $\mathbf{p}^- = (0, -0.3)$ (white) and $\mathbf{p}^+ = (\pi, 0.3)$ (black). The existence of fractal basin boundaries is clear

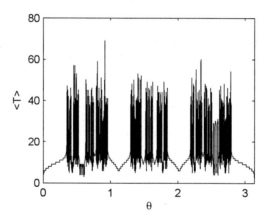

FIGURE 2. Average time $< T >$ needed by a trajectory starting in $x = 0$ and different values of θ to settle into any of the two coexisting periodic attractors

For the values of the parameters that we will use from now on, $a = 1.32$, $b = 0.9$ and $J = 0.3$ the system has two attractive fixed points, $\mathbf{p}^- = (0, -J)$ and $\mathbf{p}^+ = (\pi, J)$. For this values of the parameters, according to Ref. [7] we expect to find fractal boundaries between the basins of attraction of these systems. The basins of attraction of these attractors have been numerically computed and are shown in In Figure 1, where the points that fall after iterations in $\mathbf{p}^-$ are marked in white, whereas those that fall after iterations in $\mathbf{p}^+$ are marked in black. We can notice that the boundaries between the two basins are not smooth, they are fractal. From this figure it is clear that if the initial condition of a trajectory lies in certain regions of the θx plane, its position should be known with a high precision in order to predict whether it will settle into $\mathbf{p}^-$ or $\mathbf{p}^+$.

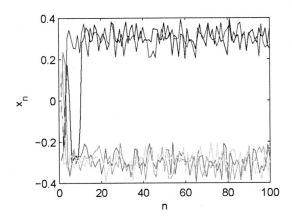

FIGURE 3. Evolution of the x coordinate of a five different trajectories of the map considered affected by noise with $u_0 = 0.1$. Note that after a finite amount of times, the x_n oscillate in the vicinity of the x coordinate of any of the two attractors of the system

The fractal boundary is a zero-measure unstable object, in the sense that trajectories are repelled from it (as long as trajectories that do not start exactly on the boundary will eventually fall in any of the attractors), and it is fractal. These are the typical features of a chaotic saddle. In fact, this system presents a behavior that is typical for systems presenting a chaotic saddle in phase space: the existence of transient chaos. In Figure 2 we represent the average time $< T >$ that a trajectory starting in $\mathbf{p}_0 = (\theta, 0)$, with θ going from 0 to π, needs to fall in the vicinity of $\mathbf{p}^-$ or $\mathbf{p}^+$. In other words, this figure shows the length of the transient needed for a trajectory to settle into a periodic orbit. It is clear from the figure that those average times depend strongly on the initial conditions, and that they can be long. This is a common feature of transient chaos.

Considering all this, what is the typical behavior of a trajectory in this situation? Typically, nearly all the trajectories (except a zero measure set) will fall arbitrarily close to either $\mathbf{p}^-$ or $\mathbf{p}^+$ after a number of iterations. Thus, the goal of our control scheme here will be to keep the trajectories far from these attractors.

As we said in the introduction, the instability of the chaotic saddle is not the only obstacle that we consider when trying to keep trajectories far from the attractor. We must not forget that most dynamical systems found in nature are under the effects of environmental noise. This situation can be modeled by adding a random perturbation to the dynamics of the system considered:

$$\mathbf{p}_{n+1} = \mathbf{f}(\mathbf{p}_n) + \mathbf{u}_n, \qquad (2)$$

where for simplicity we will assume that $\mathbf{u}_n$ is bounded by a positive constant u_0, in such a way that $\|\mathbf{u}_n\| \leq u_0$. The presence of noise in this case can modify the system's dynamics, but not in a substantial way. For moderate noise values, the behavior is quite similar to the behavior described previously: the trajectories will typically wander around for a while before falling in the vicinity of the periodic points. This behavior can be observed for different trajectories in Figure 3. Depending on the initial condition

97

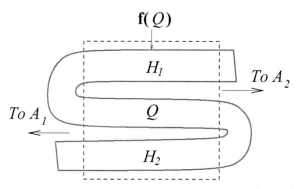

FIGURE 4. Scheme of a map for which fractal basin boundaries arise. We consider that A_1 and A_2 are attractors, in such a way that all points to the left of the square Q fall in A_1 after iterations, and all points to the right of Q fall in A_2 after iterations. Using a typical horseshoe analysis, it can be inferred that the boundaries between the basins of attraction of A_1 and A_2 are fractal. This suggests that the existence of fractal basin boundaries is related with the existence of a horseshoe-like mapping on a square Q of phase space

and on the realization of noise, the system will typically settle into any of the periodic attractors after a transient.

Thus, a natural aim here would be to preserve this transient-like behavior, i. e., to find a way to avoid the settlement into one of the periodic attractors. To do this, we can assume that we can apply each iteration an accurately chosen perturbation $\mathbf{r}_n$ to avoid this phenomenon, so the dynamics of the system considered would be given by the following equations:

$$\begin{cases} \mathbf{q}_{n+1} = \mathbf{f}(\mathbf{p}_n) + \mathbf{u}_n \\ \mathbf{p}_{n+1} = \mathbf{q}_{n+1} + \mathbf{r}_n. \end{cases} \tag{3}$$

We can assume that, as it often occurs when a system needs to be controlled, the accurately chosen perturbations $\mathbf{r_n}$ are bounded by a positive constant r_0, so $||\mathbf{r_n}|| \leq r_0$. In next section we are going to show that for this system it is possible to find the safe sets that will allow us to keep the trajectories far from the attractors even if $r_0 < u_0$.

SAFE SETS AND THE CONTROL STRATEGY

We can say now in a more precise way why we expect to find those safe sets for this dynamical system. In Ref. [19] we showed that for any dynamical system whose geometrical action reminds to that of the horseshoe map on a given square Q (and thus from which nearly all the trajectories escape) it was possible to find certain sets, the safe sets, that allowed to keep the trajectories inside Q with $r_0 < u_0$. This applies for any map that stretches and folds a number of times a given square Q in phase space. From a more technical point of view, these safe sets can be found for any map $\mathbf{f}$ that satisfies the Conley-Moser conditions [10].

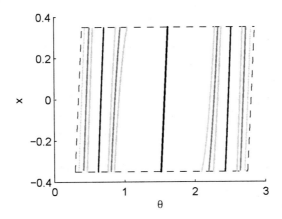

FIGURE 5. Safe sets S^k for the map considered inside the topological square Q (dashed): $k = 0$ (thick black), $k = 1$ (black), $k = 2$ (gray), $k = 3$ (light gray)

As we said in the beginning, the existence of a chaotic saddle in phase space is typically due to the existence of a horseshoe map, so in this particular case in which we have a fractal chaotic set (the fractal basin boundaries) we might find some type of horseshoe of phase space. This intuition is supported by the reasoning made in Ref. [7] by which the typical action of a map giving rise to fractal basin boundaries should be like the one shown in Figure 4.

We can explore this idea in further detail. In that figure we can see the action of a map $\mathbf{f}$ on a square Q of phase space. The map stretches and folds Q in a horseshoe-like way, so all trajectories diverge from Q. On the other hand, we assume that trajectories starting to the left of Q settle into the attractor A_1 after iterations of the map, whereas those starting to the right of the square settle into the attractor A_2. By following a reasoning that reminds to the one used in the construction of the invariant set for a Smale horseshoe map [2], it can be shown [7] that the basins of attraction of A_1 and A_2 inside Q will have an intricate appearance, and that the boundary between these sets will be fractal.

Thus, we expect to be able to find the safe sets that can be found for horseshoe-like maps in a system with fractal basin boundaries due to its underlying geometrical action. In this case, we have that those safe sets $S^k \in \{S^j\}$ would lie in a (topological) square Q between the two attractors of the system considered, $\mathbf{p}^+$ and $\mathbf{p}^-$, i.e., where the map acts like a horseshoe. We expect them to satisfy the following properties [19]:

(i) S^k is part of the preimage of S^{k-1}, and it consists of 2^k vertical curves (from top to bottom of Q).

(ii) Any vertical curve of S^k has two adjacent vertical curves of S^{k+1} *closer* to it than any other curve of S^k.

(iii) The maximum distance between any of the 2^k curves of S^k and its two adjacent curves of S^{k+1}, denoted as δ_k, goes to zero as $k \to \infty$.

(iv) For any point $\mathbf{p} \in S^k$, the distance between $\mathbf{f}(\mathbf{p})$ and the top and bottom sides of Q is $\Delta > 0$.

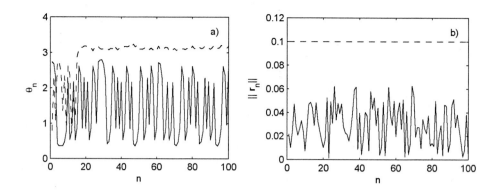

FIGURE 6. (a) Evolution of the θ variable of the system considered when the system is controlled (solid) and under noise (dashed) for $u_0 = 0.1$. The controlled trajectory, contrarily to what happened in absence of control, is kept far from the attractors. (b) Applied control for this time series (solid), which is obviously smaller than the value u_0 (marked with a dashed line)

In fact, safe sets can be found without necessarily building the horseshoe map explicitly, following a procedure that we will detail elsewhere. Those safe sets are shown here up to $k = 3$ in Figure 5. It is easy to see that they satisfy properties $(i) - (iii)$ (property (iv) can be verified mathematically in an easy way, considering that by equation 1 we have that the x variable will always be between $-J$ and J under iterations of the map). In next section we are going to describe briefly how these safe sets can be used as the key ingredient of our partial control strategy and we are going to show a numerical example of application of this technique.

THE CONTROL STRATEGY. EXAMPLE OF APPLICATION

We outline now the strategy that allows to keep the trajectory inside the square Q once the safe sets $\{S^j\}$ have been found, although a more detailed description can be found in Ref. [19]. For simplicity we assume here that u_0 is smaller than the minimum distance Δ described in property (iv) of the safe sets. Given u_0, we have to find the set S^k such that $\delta_k < u_0$, which is always possible by property (iii). Then, put the initial condition $\mathbf{p}$ in any point on S^k. The action of the map will take the trajectory to $\mathbf{f}(\mathbf{p})$, that by definition will lie in one of the 2^{k-1} curves of S^{k-1}. After this, the noise acts. But it is not difficult to see that the fact that any curve of S^{k-1} is surrounded by two adjacent curves of S^k (property (ii)) allows one to use a correction $||\mathbf{r}|| < u_0$ that puts the resulting point $\mathbf{f}(\mathbf{p}) + \mathbf{u} + \mathbf{r}$ back on a point of S^k, and this can be repeated forever. Note that this implies that we can find a value of the control r_0 such that trajectories can be kept inside Q even if $r_0 < u_0$ for any $u_0 > 0$.

As an example, we are going to control here a trajectory of the system affected by noise such that $u_0 = 0.1$. A numerical calculation shows that the adequate safe set where the trajectories can be stabilized is S^3. Thus, we just have to put the initial condition on any of the points of S^3 and apply each iteration the minimum correction $\mathbf{r}_n$ that allows to put the trajectory back on S^3.

A numerical example of application of our control technique is shown in Figure 6. We can see in Figure 6 (a) a controlled trajectory (solid) plotted together to an uncontrolled one (dashed). We can see that the controlled trajectory is kept far from the attractor, whereas the uncontrolled one oscillates around the fixed point. On the other hand, Figure 6 (b) shows clearly the main feature of our partial control scheme: that the amplitude of the control applied each iteration $\|\mathbf{r}_n\|$ is always smaller than u_0, as claimed although the trajectory follows an erratic chaotic-like behavior.

CONCLUSIONS

In this paper we have shown that the technique of partial control of a chaotic system, that typically applies to dynamical systems similar to a horseshoe map, can be applied to a paradigmatic dynamical system presenting fractal basin boundaries. We have shown that the underlying geometrical action of the system considered, that is responsible for the existence of those fractal boundaries, is related to that of a horseshoe map, and thus the safe sets necessary for our partial control strategy can be found. The main consequence of this is that trajectories can be kept far from any of the attractors even in presence of noise. Furthermore, the applied control is smaller than the action of noise. We speculate that an analogous control strategy can be applied to avoid the periodic attractors of other dynamical systems with fractal basin boundaries.

ACKNOWLEDGMENTS

This work was supported by the Spanish Ministry of Education and Science under project number FIS2006-08525 and by Universidad Rey Juan Carlos and Comunidad de Madrid under project number URJC-CM-2007-CET-1601.

REFERENCES

1. J. Aguirre, F. d'Ovidio, and M. A. F. Sanjuán, *Phys. Rev. E* **69**, 016203 (2004).
2. K. T. Alligood, T. D. Sauer, and J. A. Yorke, *Chaos, an introduction to dynamical systems*, Springer-Verlag, New York, 1996.
3. G. Chen, and X. Yu, *Chaos Control: Theory and Applications (Lecture Notes in Control and Information Sciences)*, Springer-Verlag, Berlin, 2003.
4. M. Dhamala, and Y. C. Lai, *Phys. Rev. E* **59**, 1646 (1999).
5. C. Grebogi, E. Ott, and J. A. Yorke, *Phys. Rev. Lett.* **48**, 1507–1510 (1982).
6. T. Kapitaniak, and J. Brindley, *Phys. Lett. A* **241**, 41–45 (1998).
7. S. W. McDonald, C. Grebogi, E. Ott, and J. A. Yorke, *Physica D* **17**, 726 (1985).
8. V. K. Melnikov, *Trans. Moscow Math. Society* **12**, 1 (1963).
9. F. G. Moon, and G. X. Li, *Phys. Rev. Lett.* **55**, 1439 – 1442 (1985).
10. J. Moser, *Stable and Random Motions in Dynamical Systems*, Princeton University Press, Princeton, 1973.
11. E. Ott, C. Grebogi, and J. A. Yorke, *Phys. Rev. Lett.* **64**, 1196 (1990).

12. C. M. Place, and D. K. Arrowsmith, *Phys. Rev. E* **61**, 1357 (2000).
13. C. M. Place, and D. K. Arrowsmith, *Phys. Rev. E* **61**, 1369 (2000).
14. I. P. Schwartz, and I. Triandalf, *Phys. Rev. Lett.* **77**, 4740 (1996).
15. S. Smale, *Bull. Amer. Math. Soc.* **73**, 747–817 (1967).
16. T. Shinbrot, C. Grebogi, E. Ott, and J. A. Yorke, *Nature* **363**, 411 (1993).
17. T. Tél, *J. Phys. A: Math. Gen.* **24**, L1359–L1368 (1991).
18. W. Yang, M. Ding, A. Mandell, and E. Ott, *Phys. Rev. E* **51**, 102–110 (1995).
19. S. Zambrano, M. A. F. Sanjuán, and J. A. Yorke, *Phys. Rev. E* **77**, 055201(R) (2008).

SPECIAL SESSION
ANALYTICAL AND COMPUTATIONAL
TECHNIQUES FOR
NONLINEAR SYSTEMS AND PROCESSES

Numerical Investigation of the Nonlinear Schrödinger Equation with Saturation

M. A. Christou

Dept. of Mathematics and Statistics, University of Cyprus, Nicosia, Cyprus

Abstract. We construct an implicit finite difference scheme to investigate numerically the nonlinear Schrödinger equation with saturation (NLSS). We use Newton's method to linearize the numerical scheme. We examine the propagation, interaction and overtake interaction of soliton solutions of the NLSS. Moreover, we examine the effect of the saturation term on the solution and compare it with the classical case of the cubic nonlinearity without saturation of nonlinearity. We track numerically the conserved properties and the phase shift experienced by the solitons upon collision.

Keywords: Christov difference scheme, Schrödinger, saturation, discrete conservative properties.
PACS: 02.70.Bf, 42.25.Bs, 42.25.Dd, 42.82.Dp

INTRODUCTION

The nonlinear Schrödinger equation (NLSE) models many physical processes. Whilst the main application is in nonlinear optoelectronics (propagation of optical pulses in fibers), this equation is also used as a model in telecommunications, hydrodynamics, nonlinear acoustics, and many other nonlinear systems. The general NLSE in one space dimension is of the form

$$\iota u_t + \alpha_1 u_{xx} + \alpha_2 f(|u|^2)u = 0, \tag{1}$$

where u denotes a smooth solution of the NLSE which vanishes at infinity along with its derivatives, α_1 and α_2 are real scalars and f is a smooth nonlinear function of $|u|^2$.

As is explained in great detail in [4], attention has been paid to higher-order effects resulting from additional terms to the (NLSE), for example, third order dispersion and nonlinear dispersion (self steepening effect). For short pulses and high input peak pulse powers, the field cannot be described by a Kerr-type nonlinearity. Hence at higher field strength, the optically refractive index becomes saturated. Especially in materials with highly nonlinear coefficients, e.g., semiconductor-doped glasses and organic polymers. The nonlinearity which is of the type of $f = \frac{|u|^2}{1+\gamma|u|^2}$, where γ is the saturation parameter, is analogous to systems with resonant interaction. Besides physics, the applications of the model with saturation can be extended to other areas such as economics. An example of such model is acknowledging the market saturation when a product becomes distributed within a market. In color theory, the saturation of a color is determined by a combination of light intensity and how much it is distributed across the spectrum of different wavelengths. In earth sciences the saturation generally refers to the water content in the soil, where the unsaturated zone is above the water table and the saturated

CP1067, *Applications of Mathematics in Engineering and Economics '34—AMEE '08*, edited by M. D. Todorov
© 2008 American Institute of Physics 978-0-7354-0598-1/08/$23.00

zone is below. If we take into account all of the above we arrive to the following model

$$iu_t + \alpha_1 u_{xx} - \alpha_2 u + \alpha_3 \frac{|u|^2}{1+\gamma|u|^2} u = 0. \tag{2}$$

The purpose of this paper is to examine the soliton interactions modelled by equation (2) and to compare them with the results from cubic nonlinear Schrödinger equation ($\gamma = 0$). To do that we constructed a fully implicit nonlinear numerical scheme similar to the one proposed in [1, 2, 3] for a coupled of system of nonlinear Schrödinger equations with cubic nonlinearity.

We investigated the propagation and interaction of soliton solutions using different values of the saturation parameter, γ, and different phase speeds. In the case of interaction we considered the superposition of exact solutions, proposed in [6], situated far from each other in order to avoid intersection in the initial moment. We succeeded full recovery of the initial shapes which concludes the solitonic behavior and proves numerically that the scheme is conservative. Furthermore, we evaluated and compared the phase shift experienced by a soliton during the collision with another one for different values of γ, which is the difference in the actual position of the soliton and the position it would have had reached if no other solitons were present.

NONLINEAR SCHRÖDINGER EQUATION WITH SATURATION

Consider the Nonlinear Schrödinger Equation with Saturation (NLSS) and asymptotic boundary conditions:

$$iu_t + u_{xx} - u + \frac{|u|^2}{1+\gamma|u|^2} u = 0, \quad \text{as} \quad x \to \pm\infty \quad \Re u, \Im u, |u| \to 0. \tag{3}$$

For the last problem an exact solution is available [6, 5]

$$u(x,t) = Ae^{i(\alpha x - \omega t)} \operatorname{sech}\left(\frac{x-ct}{T}\right) \quad \text{where}$$

$$\alpha = \frac{c}{2}, \quad \omega = \frac{c^2}{4} + 1 - \frac{1}{T^2}, \quad A = \sqrt{\frac{2}{T^2 - 2\gamma}}.$$

If we want to define T we can normalize u using $\int\limits_{-\infty}^{\infty} |u|^2 dx = n$. Following [6] we can easily show that (3) can be expressed as variational problem corresponding to Lagrangian of the form

$$L_1 = \frac{i}{2}(u\overline{u_x} - \overline{u}u_x) + \left(\frac{\gamma-1}{\gamma}\right)|u|^2 + |u_x|^2 + \frac{1}{\gamma^2} \int\limits_{0}^{\sqrt{\gamma}|u|^2} \frac{dy}{1+y}. \tag{4}$$

If we use the Lagrangian and the normalization condition we then get the first three conservation laws which corresponds to mass (I_1), momentum (I_2) and energy (I_3)

106

respectively, namely:

$$I_1 = \frac{1}{n} \int_{-\infty}^{\infty} |u|^2 dx,$$

$$I_2 = \iota \int_{-\infty}^{\infty} (u\overline{u}_x - \overline{u}u_x) dx,$$

$$I_3 = \int_{-\infty}^{\infty} \left(2|u_x|^2 + \frac{\gamma-1}{\gamma}|u|^2 + \frac{1}{\gamma^2} \ln(1 + \gamma|u|^2) \right) dx.$$

DIFFERENCE SCHEME

Here we show the construction of the difference scheme for the NLSS and use it for the NLSC too by setting $\gamma = 0$. We first introduce the notation $u = s + ir$ into (3) to get the following system:

$$-r_t + s_{xx} - s + \frac{s^2 + r^2}{1 + \gamma(s^2 + r^2)} s = 0,$$

$$s_t + r_{xx} - r + \frac{s^2 + r^2}{1 + \gamma(s^2 + r^2)} r = 0.$$

(5)

We consider a uniform mesh in the interval $[-L_1, L_2]$

$$x_i = -L_1 + (i-1)h, \quad h = \frac{L_2 + L_1}{N-1},$$

where N is the number of grid points in the interval under consideration and L_1, L_2 are called 'numerical infinities'.

To treat the long-time evolution of the solitons, we construct a conservative scheme following [1, 2, 3], and show the that it reflects the conservative properties of the mass, momentum and energy, namely:

$$-\frac{r^{n+1} - r^n}{\tau} + \frac{1}{2} \left(\frac{s_{i+1}^{n+1} - 2s_i^{n+1} + s_{i-1}^{n+1} + s_{i+1}^n - 2s_i^n + s_{i-1}^n}{h^2} \right) - \frac{1}{2}(s_i^{n+1} + s_i^n)$$

$$+ \frac{\frac{1}{2}[(s_i^{n+1})^2 + (s_i^n)^2 + (r_i^{n+1})^2 + (r_i^n)^2]}{1 + \frac{\gamma}{2}[(s_i^{n+1})^2 + (s_i^n)^2 + (r_i^{n+1})^2 + (r_i^n)^2]} \frac{s_i^{n+1} + s_i^n}{2} = 0,$$

(6)

$$\frac{s^{n+1,k} - s^n}{\tau} + \frac{1}{2} \left(\frac{r_{i+1}^{n+1} - 2r_i^{n+1} + r_{i-1}^{n+1} + r_{i+1}^n - 2r_i^n + r_{i-1}^n}{h^2} \right) - \frac{1}{2}(r_i^{n+1} + r_i^n)$$

$$+ \frac{\frac{1}{2}[(s_i^{n+1})^2 + (s_i^n)^2 + (r_i^{n+1})^2 + (r_i^n)^2]}{1 + \frac{\gamma}{2}[(s_i^{n+1})^2 + (s_i^n)^2 + (r_i^{n+1})^2 + (r_i^n)^2]} \frac{r_i^{n+1} + r_i^n}{2} = 0.$$

Conservative Properties

Consider now the discrete versions of the mass, momentum and energy:

$$\widetilde{I}_1^{\,n+1} = \frac{1}{2}\sum_{i=1}^{N}\left(s_i^{n+1}+r_i^{n+1}+s_i^{n}+r_i^{n}\right),$$

$$\widetilde{I}_2^{\,n+1} = \sum_{i=2}^{N}\left(\frac{s_i^{n+1}+s_i^{n}}{4}-\frac{r_i^{n+1}+r_i^{n}}{4}\right)\left(\frac{s_i^{n+1}-s_{i-1}^{n+1}}{h}+\frac{s_i^{n}-s_{i-1}^{n}}{h}\right),$$

$$+\sum_{i=2}^{N}\left(\frac{r_i^{n+1}+r_i^{n}}{4}-\frac{s_i^{n+1}+s_i^{n}}{4}\right)\left(\frac{r_i^{n+1}-r_{i-1}^{n+1}}{h}+\frac{r_i^{n}-r_{i-1}^{n}}{h}\right).$$

$$\widetilde{I}_3^{\,n+1}=\sum_{i=2}^{N}\left[\left(\frac{s_i^{n+1}-s_{i-1}^{n+1}}{h}\right)^2+\left(\frac{r_i^{n+1}-r_{i-1}^{n+1}}{h}\right)^2+\left(\frac{s_i^{n}-s_{i-1}^{n}}{h}\right)^2+\left(\frac{r_i^{n}-r_{i-1}^{n}}{h}\right)^2\right],$$

$$+\frac{\gamma-1}{2\gamma}\sum_{i=2}^{N}\left[\left(s_i^{n+1}\right)^2+\left(r_i^{n+1}\right)^2+\left(s_i^{n}\right)^2+\left(r_i^{n}\right)^2\right],$$

$$+\frac{1}{2\gamma^2}\ln\left(1+\gamma\left(s_i^{n+1}\right)^2+\gamma\left(r_i^{n+1}\right)^2\right)\left(1+\gamma(s_i^{n})^2+\gamma(r_i^{n})^2\right).$$

They are conserved in the sense that

$$\widetilde{I}_1^{\,n+1}=\widetilde{I}_1^{\,n},\quad \widetilde{I}_2^{\,n+1}=\widetilde{I}_2^{\,n},\quad \widetilde{I}_3^{\,n+1}=\widetilde{I}_3^{\,n}.$$

Linearization of the Scheme

The above conservative scheme, (6), is nonlinear and requires linearization. One should be very careful in selecting the specific linearization, because any 'impurity' of the scheme would spoil the conservative properties. One tested option is to introduce an iterative procedure [3] and to repeat it until convergence. Upon introducing the internal iterations into the scheme we get the following:

$$-2\frac{r^{n+1,k} - r^n}{\tau} + (\frac{s_{i+1}^{n+1,k} - (2+h^2)s_i^{n+1,k} + s_{i-1}^{n+1,k} + s_{i+1}^n - (2+h^2)s_i^n + s_{i-1}^n}{h^2})$$

$$+\frac{2[(s_i^{n+1,k-1})^2 + (s_i^n)^2]s_i^{n+1,k} + 2(s_i^n)^3 + [(r_i^{n+1,k-1})^2 + (r_i^n)^2]s_i^{n+1,k}}{3\vartheta}$$

$$+\frac{[r_i^{n+1,k-1}s_i^{n+1,k-1} + r_i^n s_i^n]r_i^{n+1,k}}{3\vartheta} = 0,$$

$$-2\frac{s^{n+1,k} - s^n}{\tau} + (\frac{r_{i+1}^{n+1,k} - (2+h^2)r_i^{n+1,k} + r_{i-1}^{n+1,k} + r_{i+1}^n - (2+h^2)r_i^n + r_{i-1}^n}{h^2})$$

$$+\frac{2[(r_i^{n+1,k-1})^2 + (r_i^n)^2]r_i^{n+1,k} + 2(r_i^n)^3 + [(s_i^{n+1,k-1})^2 + (s_i^n)^2]r_i^{n+1,k}}{3\vartheta}$$

$$+\frac{[r_i^{n+1,k-1}s_i^{n+1,k-1} + r_i^n s_i^n]r_i^{n+1,k}}{3\vartheta} = 0,$$

where $\vartheta = 1 + \frac{\gamma}{2}[(s_i^{n+1,k-1})^2 + (s_i^n)^2 + (r_i^{n+1,k-1})^2 + (r_i^n)^2]$. The iterations begin from the initial conditions

$$s^{n+1,0} \equiv s^n, \quad r^{n+1,0} \equiv r^n,$$

and are conducted until convergence, in the sense that $\max(N_s, N_r) \leq \varepsilon$ where,

$$N_s = \max_i |\frac{s_i^{n+1,k} - s_i^{n+1,k-1}}{s_i^{n+1,k}}|, \quad N_r = \max_i |\frac{r_i^{n+1,k} - r_i^{n+1,k-1}}{r_i^{n+1,k}}|, \quad \varepsilon = 10^{-12}.$$

After the iterations converge (which is about 12-16 iterations), we complete one time step of the nonlinear scheme with the desired conserved properties.

RESULTS AND DISCUSSION

It is very important to understand the impact of the saturation term on the solution, and the role of the parameter γ and its effect on the shape of the soliton. We can get a clear view of how the soliton looks in the featured example in Figure (1). In the left panel we present the shapes of the stationary propagating solitons obtained for different values of the saturation parameter $\gamma = 0, 2, 10$ and 20. As we increase γ, the soliton becomes shorter and wider. In the right panel of the same figure we show the asymptotic behavior of the solution and observe that the decay is very fast for small values of γ.

We contacted several numerical experiments involving the propagation and interaction of solitons using different γ and different phase speeds c. In all cases, we found that the shapes recover in full and that no amplitude is lost. This demonstrates that the numerical scheme is conservative. Here we present a couple of examples: a propagation of a single soliton without interaction in Figure 2; a head-on collision in Figure 3; and an overtaking interaction in Figure 4.

Another issue we have focused on in our numerical study is the phase shift which is defined as the difference in the actual position of the soliton and the position it would

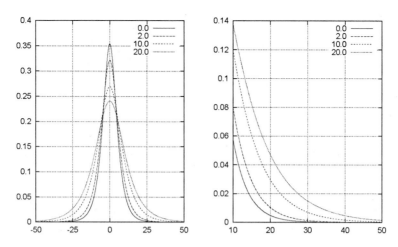

FIGURE 1. Solitons evaluated for different values of γ and $c = 0.2$

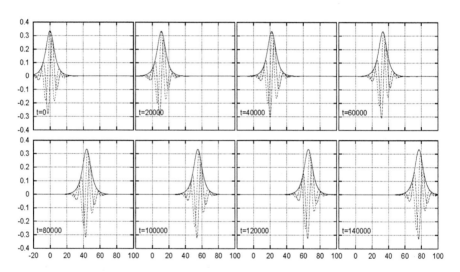

FIGURE 2. Propagation of single soliton for $c = 1.1$ and $\gamma = 1$

have had reached if no other solitons were present. Here we present several different cases obtained using different values of the parameters. When the saturation parameter γ is relatively small, say $0 \leq \gamma \leq 0.5$, then for small values of c the experienced phase shift δ is large but for larger c, δ is becoming smaller and smaller, see Table (1). In the same table we can see that when $\gamma \geq 2$ and $c \geq 4$ then no phase shift is experienced. This can be interpreted that the interaction takes place so quickly that the nonlinear wave does not suffer or experience any change in its course or celerity.

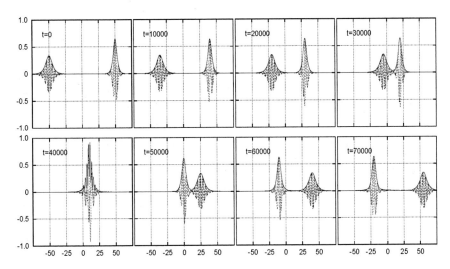

FIGURE 3. Interaction of solitons for $c_1 = 3$, $c_2 = -2$ and $\gamma = 0.5$

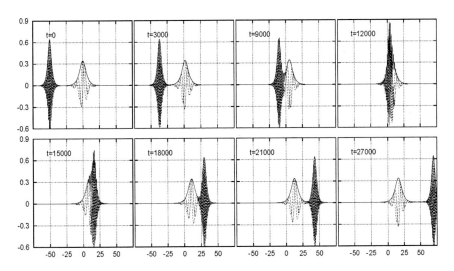

FIGURE 4. Overtake interaction of solitons for $c_1 = 1.2$, $c_2 = 9$ and $\gamma = 0.5$

CONCLUSIONS

In the current work we investigated numerically the nonlinear Schrödinger equation with saturation (NLSS). To do that we constructed an implicit numerical scheme based on the Christov scheme [1, 2, 3] which was linearized using Newton's method. We showed that the scheme conserves integral properties such as mass, momentum and energy.

TABLE 1. Numerically identified phase shifts δ_i using different values of γ

	$\gamma = 0.5$				$\gamma = 0.1$				$\gamma = 0.0$		
c_1	δ_1	c_2	δ_2	c_1	δ_1	c_2	δ_2	c_1	δ_1	c_2	δ_2
0.2	3.8	-0.2	3.8	0.2	7.3	-0.2	7.3	0.2	7.8	-0.2	7.8
0.5	1.6	-0.5	1.6	0.5	2.6	-0.5	2.6	0.5	2.8	-0.5	2.8
0.8	0.9	-0.8	0.9	0.8	1.2	-0.8	1.2	0.8	1.4	-0.8	1.4
2.0	0.2	-2.0	0.2	2.0	0.2	-2.0	0.2	2.0	0.2	-2.0	0.2
4.0	0.1	-4.0	0.1	4.0	0.1	-4.0	0.1	2.5	0.1	-2.5	0.1
8.0	0.0	-8.0	0.0	8.0	0.0	-8.0	0.0	4.5	0.06	-4.5	0.06
0.8	2.3	-0.1	3.5	0.8	3.0	-0.1	3.3	0.8	3.2	-0.1	3.2
0.9	1.1	-0.5	1.8	0.9	1.5	-0.5	1.7	0.9	1.6	-0.5	1.6
1.5	0.4	-1.2	0.5	1.5	0.5	-1.2	0.5	1.5	0.6	-1.2	0.6
2.0	0.1	-4.0	0.1	2.0	0.1	-4.0	0.1	2.0	0.1	-4.0	0.1
2.0	0.0	-8.0	0.0	2.0	0.0	-8.0	0.0	2.0	0.06	-8.0	0.06

	$\gamma = 2.0$				$\gamma = 4.0$				$\gamma = 6.0$		
c_1	δ_1	c_2	δ_2	c_1	δ_1	c_2	δ_2	c_1	δ_1	c_2	δ_2
2.0	0.267	-2.0	0.267	2.0	0.2	-2.0	0.2	2.0	0.16	-2.0	0.16
3.0	0.133	-3.0	0.133	3.0	0.0	-3.0	0.0	3.0	0.0	-3.0	0.0
4.0	0.0	-4.0	0.0	4.0	0.0	-4.0	0.0	4.0	0.0	-4.0	0.0
5.0	0.0	-5.0	0.0	5.0	0.0	-5.0	0.0	5.0	0.0	-5.0	0.0
6.0	0.0	-6.0	0.0	6.0	0.0	-6.0	0.0	6.0	0.0	-6.0	0.0
7.0	0.0	-7.0	0.0	7.0	0.0	-7.0	0.0	7.0	0.0	-7.0	0.0
8.0	0.0	-8.0	0.0	8.0	0.0	-8.0	0.0	8.0	0.0	-8.0	0.0
9.0	0.0	-9.0	0.0	9.0	0.0	-9.0	0.0	9.0	0.0	-9.0	0.0
10.0	0.0	-10.0	0.0	10.0	0.0	-10.0	0.0	10.0	0.0	-10.0	0.0

Using the numerical scheme we studied the propagation and interaction of solitons modeled by the NLSS and calculated the phase shift experienced by the solitons upon collision. We found that when the value of the phase speed is rather large, then the solitons do not experience any significant phase shift. This was actually observed and in the case where $\gamma > 2$.

The current good results suggests the application of the algorithm for the numerical investigation of the coupled nonlinear system namely:

$$\iota u_t + u_{xx} - u - v + \frac{|u|^2 + |v|^2}{1 + \gamma_1 |u|^2 + \gamma_2 |v|^2} u = 0,$$

$$\iota v_t + v_{xx} - u - v + \frac{|u|^2 + |v|^2}{1 + \gamma_1 |u|^2 + \gamma_2 |v|^2} v = 0.$$

ACKNOWLEDGMENTS

The author would like to express his gratitude to the Cyprus Research Promotion Foundation for support through Grant ΦΙΛΟΝΕ/0506/01.

REFERENCES

1. C. I. Christov, S. Dost, and G. A. Maugin, *Physica Scripta* **50**, 449–454 (1994).
2. C. I. Christov and M. G. Velarde, *Int. Journal of Bifurcation and Chaos* **4**, 1095–1112 (1994).
3. C. I. Christov and A. Guran, *Selected Topics in Nonlinear Wave Mechanics*, Birkhauser, 2001.
4. S. Gatz and J. Hermann, *Journal Optical Society of America B*, 2296–2302 (1991).
5. K. H. I. Pushkarov and D. I. Pushkarov, *Rep. on Mathematical Physics* **17**, 37–41 (1980).
6. A. Usman, J. Osman, and D. R. Tilley, *Journal of Physica A* **31**, 8397–8403 (1998).

Analytical Model of the Temperature in UV Cu+ CuBr Laser

S. G. Gocheva-Ilieva*, I. P. Iliev†, K. A. Temelkov**, N. K. Vuchkov**
and N. V. Sabotinov**

*Faculty of Mathematics and Informatics, "Paisii Hilendarski" University of Plovdiv,
24 Tsar Assen Str, 4000 Plovdiv, Bulgaria
†Dept. of Physics, Technical University of Plovdiv, 25 Tzanko Djusstabanov Str,
4000 Plovdiv, Bulgaria
** Metal Vapour Lasers Department, Georgi Nadjakov Institute of Solid State Physics, Bulgarian
Academy of Sciences, 72 Tzarigradsko Shaussee, 1784 Sofia, Bulgaria

Abstract. In this study, a new analytical model for computing the temperature profile in the cross-section of the tube in ultraviolet copper ion excited copper bromide laser is developed. The model is based on the solution of the two-dimensional heat conduction equation under the boundary conditions of the third and fourth kind. Temperature profiles of natural convection are obtained, taking into account the basic laser design parameters, temperature insulation of the tube and the input electric power. Different engineering solutions for effective control of the discharge temperature are proposed and numerically investigated.

Keywords: Temperature profile, ultraviolet, heat conduction equation, boundary conditions of the third and fourth kind.
PACS: 42.55. Lt

INTRODUCTION

The Cu+ copper bromide lasers of ultraviolet (UV) radiation produce a high beam quality, which is used in injection-locking systems, microelectronics, microbiology, photolithography, genetic engineering, etc. In the last decade these lasers were subject of extensive experimental studies and their output characteristics were successfully improved [1, 2].

The gas temperature is an important thermodynamic characteristic of the active laser medium. It takes a basic role in determining the laser lifetime, the laser generation quality, and the distribution of the neutral atoms in the cross-section of the tube. Thermally the lower laser levels are populated, which influences the laser power and the laser beam mode composition. The high temperature can cause thermoionisation instability of the gas discharge. For these reasons, the investigation of the temperature profile remains always in the focus of the experimental and theoretical studies. Some mathematical models are developed especially in [3] and the temperature is found on the basis of an exact solution of the steady-state heat conduction equation under the boundary value of the first kind, where the quartz tube wall temperature is assumed to be constant. In practice, and also in computer modeling and simulations, that method can not be used, because the real values of the

CP1067, *Applications of Mathematics in Engineering and Economics '34—AMEE '08*, edited by M. D. Todorov
© 2008 American Institute of Physics 978-0-7354-0598-1/08/$23.00

temperature of the quartz tube are unknown and will change with variation of the laser geometry, input electric power and other laser parameters and temperature of the surroundings.

In [4], the temperature profile in the case of copper bromide vapor laser with wavelength 510.6 nm and 578.2 nm was carried out by using a new approach. This is based on the boundary value conditions of the third and fourth kind at a given temperature of the surroundings.

In this study, the analogous method is applied to the UV Cu+ CuBr lasers. Analytical model, consisting of the two-dimensional heat conduction equation, subject to nonlinear boundary conditions of the third and fourth kind is derived. At a given air temperature of the surroundings, due to the heat convection and heat radiation, the proposed model allows to take into account the heat exchange processes between the outer surface of the laser tube and its surroundings. The gas temperature profile in the tube and the outer wall temperature are expressed by the exact solution of the obtained problem and are directly dependent on the basic input laser parameters.

The proposed model gives the opportunity to carry out various computer simulations in order to optimize the laser generation, while changing the geometrical design, tube materials, heat insulation, input electric power and laser operating conditions.

DESCRIPTION OF THE MODEL

A subject of study is the determination of the gas heating in the ultraviolet copper ion excited copper bromide vapor lasers. The geometrical design of the cross-section of the laser tube is given in Figure 1. The total consumed power is $P = 1300W$, power after reducing of losses is $P_{in} = 1000W$ and the output laser power P_{out} could rise up to 635mW [1, 2].

The temperature profile of the gas is obtained with the following simplifying assumptions: (i) the temperature profile is determined in a quasi-stationary regime; (ii) the gas temperature does not change substantially in the inter-impulse period; (iii) all the input electric power in the active volume is transformed into heat.

The proposed analytical model is oriented for analyzing and designing of the experiment. It is applicable in computer simulations with a change of the geometric design, the tube material, the heat insulation, the electrical power and the exploitation conditions, for exploration of basic tendencies in changing the temperature profile.

The expected gas temperature T_g satisfies the two-dimensional quasi steady-state heat conduction equation in the cross-section of the laser tube in the form:

$$div\left(\lambda_g grad(T_g)\right) + q_v = 0, \tag{1}$$

In equation (1), due to the radial symmetry, T_g depends only on the variable r along with the radius of the tube, λ_g is the thermal conductivity of the gas (here neon) and q_v is the volume density of the discharge.

Usually, as it was mentioned above, equation (1) is solved in the literature under the boundary condition of the first kind, by giving the constant outer wall temperature of the quartz tube T_3

$$T_g\big|_{w3} = T_3 = const,$$

where $w3$ is the external wall of the quartz tube (see Figure 1). Commonly, λ_g is in the form $\lambda_g = \lambda_0 T^m$. In this case, the equation (1) has the exact solution [3]

$$T_g(r) = \left[T_3^{m+1} + \frac{q_v(m+1)}{4\lambda_0}\left(R^2 - r^2\right) \right]^{1/(m+1)}, \tag{2}$$

where R is the radius of the discharge tube, $0 \le r \le R$. For neon $m = 0.6817$ and $\lambda_0 = 1.0029 \times 10^{-3}$.

Solution (2) of equation (1) for the boundary condition of the first kind is used to model the temperature profile of the copper bromide vapor laser in [5], and also in computer simulations of pure copper vapor lasers in [6].

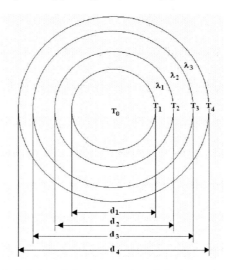

FIGURE 1. The geometrical design of the cross-section of the laser tube, composed of three tubes (outwards from the center): Al_2O_3 tube, quartz tube and zircon dioxide (ZrO_2) insulation tube. The diameters are as follows: d_1 varies from 5.2, 5.8 and 8mm, $d_2 = 18\,mm$, $d_3 = 24.5\,mm$ and $d_4 = 28.5\,mm$

We solve equation (1), subject to mixed boundary conditions of the third and fourth kind, which, for the cylindrical configuration, have the form [7]:

$$T_j = T_{j+1} + \frac{q_l \ln\left(d_{j+1}/d_j\right)}{2\pi\lambda_j}, \quad j = 1, 2, 3, \tag{3}$$

$$Q = \alpha F_4 \left(T_4 - T_0\right) + F_4 \varepsilon c \left[\left(T_4/100\right)^4 - \left(T_{air}/100\right)^4\right]. \tag{4}$$

Boundary conditions (3) express the equation for the continuity of the heat flow on the borders of the three mediums in the composed tube. Here q_l is the power per unit length, $q_l = Q/l_a$, $l_a = 86\,cm$ is the active length [7], λ_j, $j = 1, 2, 3$ are the thermal conductivities of the three tubes, respectively (see Figure 1). The boundary condition (4) shows the heat exchange between the outer surface of the laser tube and the surroundings. The first term on the right hand side of (4) evolves from the Newton-Riemann law for heat exchange by convection. The second term represents the Stefan-Boltzmann law for heat exchange by radiation. The value of Q is equal to the electric power $Q = 1000\,W$, in accordance with assumption (iii), as it was stated above, α is the heat transfer coefficient, F_4 is the outer active surface wall of the tube, ε is the integral emissivity of the material, $c = 5.67\,W/(m^2 K^4)$ is the black body radiation coefficient, and $T_{air} = 300\,K$ is the temperature of the air. In boundary condition (4) there are two unknown values - α and T_4.

Calculation of the Heat Transfer Coefficient α at Natural Convection

For all types of convection the Nusselt criterion holds [7]:

$$Nu = \alpha H/\lambda. \tag{5}$$

For free convection the Grashoff criterion is given by the expression [7]:

$$Gr = g\beta H^3 \left(T_4 - T_{air}\right)/\upsilon^2. \tag{6}$$

For the horizontal tubes at natural convection the two upper criteria can be related to the empirical formula [7]

$$Nu = 0.46 Gr^{0.25} \tag{7}$$

which is valid for $700 < Gr < 7.10^7$. In the previous expressions (5)-(7), α is the heat transfer coefficient of the material, H is a characteristic dimension of the body, (here $H = d_4$), g is the gravitational acceleration, β is the coefficient of cubical heat expansion of the gas, which for the air is $\beta_{air} = 3.14 \times 10^{-3} K^{-1}$, υ is the kinematical viscosity, $\upsilon_{air} = 15.7 \times 10^{-6} m^2 / s$, λ is the thermal conductivity, $\lambda_{air} = 0.0251\, W/(mK)$. The data is correct for air temperature $T_{air} = 300\,K$ [7].

Taking into consideration (5)-(7) we have:

$$\alpha = 0.46\lambda_{air}\left[g\beta_{air}d_4^3\left(T_4 - T_{air}\right)/\upsilon_{air}^2\right]^{0.25}/d_4 \qquad (8)$$

Now, boundary condition (4), represented with respect to the power per unit length q_l, becomes

$$q_l = 0.46\pi\lambda_{air}\left[g\beta_{air}d_4^3\left(T_4 - T_{air}\right)/\upsilon_{air}^2\right]^{0.25}\left(T_4 - T_{air}\right) +$$
$$\pi d_4 \varepsilon c\left[\left(T_4/100\right)^4 - \left(T_{air}/100\right)^4\right] \qquad (9)$$

Here all notations, physical meanings and numerical values of the constants are the same as in equations (3)-(7).

Calculation of the Gas Temperature at Natural Convection

The temperature of the outer tube surface T_4 can be calculated by solving the nonlinear equation (9). It is not difficult to establish that (9) possesses a unique real solution. After that, by means of (3), the temperatures T_j, $j = 3,2,1$ can be calculated.

Now the gas temperature in all internal points in the active discharge medium (inside the internal tube), is given by the generic expression (2) in the form

$$T_g(r) = \left[T_1^{m+1} + \frac{q_v(m+1)}{4\lambda_0}\left(R_1^2 - r^2\right)\right]^{1/(m+1)}. \qquad (10)$$

APPLICATIONS OF THE MODEL

The derived analytical model could be applied for simulation of various laser characteristics.

In all calculations the thermal conductivities of the walls are $\lambda_1 = 2.08\, W/mK$, $\lambda_2 = 1.96\, W/mK$ and $\lambda_3 = 1.78\, W/mK$ [8]. Some numerical results of the gas temperature, calculated for three different internal diameters of the laser tube, respectively $d_1 = 5.2\,mm$, $5.8\,mm$, $8\,mm$ on the walls of the tube are given in Table 1.

In Figure 2 the distribution of the temperature in the cross-section along with the radius of the laser tube for $d_1 = 5.2\,mm$ is presented. It can be observed that the temperature difference in the walls is around $\Delta T = T_1 - T_4 \approx 160^{\circ}C$. This shows that the usually used simple method from [3] (formula (2)), which does not allows for the presence of the walls, is not applicable in this case.

TABLE 1. Temperature in special characteristic points

D_1, mm	T_0, K	T_1, K	T_2, K	T_3, K	T_4, K
5.2	1573.9	838.3	717.4	688.3	672.6
5.8	1566.6	827.7	717.4	688.3	672.6
8.0	1546.7	796.4	717.4	688.3	672.6

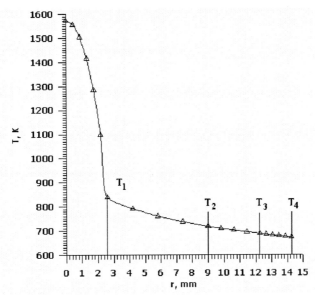

FIGURE 2. Distribution of the temperature in the cross-section of the laser tube for $d_1 = 5.2\,mm$

Figure 3 shows the temperature profile for different values of the internal diameter d_1. The temperature differences are approximately $\Delta T \approx 30^{\circ} - 40^{\circ}C$, so the temperature can not considerably influence the output laser power and the laser beam quality.

Figure 4 illustrates the gas temperature variance with regard to the increase of the input power by 20% (from 1000W to 1200W). It can be seen that the maximal temperature in the center of the laser tube T_0 increases around $70^{\circ}C$ or only 4.5%. This increase is not significant for the laser generation. The computer simulations show that an increase of the input laser power P_{in} within the limits of $\pm 10\%$ do not influence considerably the laser generation.

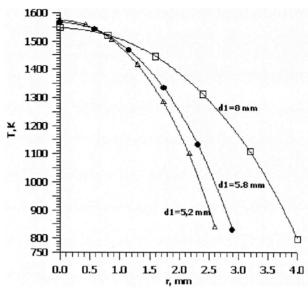

FIGURE 3. Distribution of the temperature in the active volume of the laser source for three different values of the internal diameter d_1, with T_0 (in the center of the tube) and T_1 given in Table 1

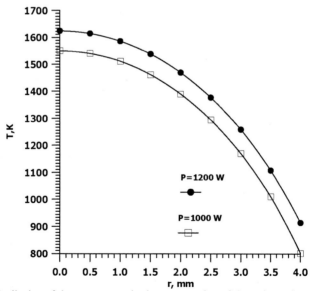

FIGURE 4. Distribution of the temperature in the cross-section of the active volume of the laser for internal diameter $d_1 = 8\,mm$, external diameter $d_4 = 26.4\,mm$ and two useful powers $P_{in} = 1000\,W$ and $P_{in} = 1200\,W$

CONCLUSION

A new approach for solving the heat conduction equation for ultraviolet copper ion excited bromide vapor lasers is proposed. Based on mixed boundary conditions, which describe the heat interaction between the laser tube and the surroundings, the temperature profile in the active medium cross-section is carried out. The method allows preliminary evaluation of the gas temperature for different engineering solutions. In particular the model is used to investigate the influence of the magnitude of the internal diameters on the laser generation and the increase of the laser generation by changing the input power.

Independent of the specific object of study, the proposed method is easily adaptable and can be applied to a wide range of gas lasers, metal vapor lasers and metal compound vapor lasers.

ACKNOWLEDGMENTS

This work is supported by the National Scientific Fund of Bulgarian Ministry of Education and Science under grant VU-MI-205/2006 and Plovdiv University "Paisii Hilendarski" project IS-M4/2008.

REFERENCES

1. N. K. Vuchkov, K. A. Temelkov, P. V. Zahariev, and N. V. Sabotinov, *IEEE J Quant Electronics* **37**(2), 1538-1546 (2001).
2. N. K. Vuchkov, K. A. Temelkov, P. V. Zahariev, and N. V. Sabotinov, *IEEE J Quant Electronics,* **37** (4), 511-517 (2001).
3. M. J. Kushner and B. E. Warner, *J. Appl. Phys.* **54**(6), 2970-2982 (1983).
4. I. P. Iliev, S. G. Gocheva-Ilieva, and N. V. Sabotinov, *Quantum Electronics* **38**(4), 338-442 (2008).
5. D. N. Astadjov, N. K. Vuchkov, and N.V. Sabotinov, *IEEE J. of Quant. Electron.* **24**(9), 1926-1935 (1988).
6. S. I. Yakovlenko, *Quantum Electron.* **30**(6), 501-505 (2000).
7. M. P. Oprev, S. G. Batov, and D. Z. Uzunov, *Heat Technology*, Tekhnika, Sofia, 1978. (in Bulgarian)
8. *Physical Quantities. Handbook*, Energoatomizdat, Moscow, 1991. (in Russian)

On Beam-like Functions with Radial Symmetry

N. C. Papanicolaou[*] and C. I. Christov[†]

[*]*Dept. of Computer Science, University of Nicosia, P.O. Box 24005, 1700 Nicosia, Cyprus*
[†]*Dept. of Mathematics, University of Louisiana at Lafayette, LA 70504-1010, USA*

Abstract. In this work, we introduce a complete orthonormal (CON) set of functions as the eigenfunctions of a fourth-order boundary problem with radial symmetry. We derive the relation for the spectrum of the problem and solve it numerically. For larger indices n of the eigenvalues we derive accurate asymptotic representations valid within $o(n^{-2})$.

Two model fourth order problems with radial symmetry which admit exact analytic solutions are featured: a simple problem involving only the fourth-order radial operator and a constant and the other also involving the second-order radial operator. We show that for both cases, the rate of convergence is $O(N^{-5})$ which is compatible with theoretical predictions. The spectral and analytic solutions are found to be in excellent agreement. With 20 terms the absolute pointwise difference of the spectral and analytical solutions is of order 10^{-7} which means that the fifth order algebraic rate of convergence is fully adequate.

Keywords: Radial beam functions, Galerkin spectral method, fourth-order boundary value problems with radial symmetry, Bessel functions, asymptotic methods.
PACS: 02.70.Hm, 02.60.Lj, 02.30.Mv

INTRODUCTION

In a previous work (see [4] and the literature cited therein) we showed that the set of beam functions was most suitable for solving fourth order boundary value problems with homogeneous boundary conditions.

We now seek to extend this idea to radial-symmetric fourth-order problems in cylindrical coordinates. Chandrasekhar [2] describes sets of functions suitable for fourth order problems with cylindrical and spherical boundaries. His functions satisfy four boundary conditions (the non-slip conditions) and also involve Bessel functions of the second kind Y_n and modified Bessel functions of the second kind K_n. However, Y_n and K_n are highly singular at the origin and thus they cannot be used in the incide of a circular domain. Here, we develop what we call 'Radial Beam Functions'. Our set of functions does not have singularities, and satisfies two homogeneous boundary conditions at the outer boundary $r = \alpha$. Radial Beam Functions (or functions derived in a similar vein) will be very useful for solving biharmonic problems, such as viscous flows in cylindrical pipes (Hagen-Poiseuille flows), and deformation of elastic plates. In this first work, we focus on radial beam functions inside a cylinder of radius $a = 1$.

CP1067, *Applications of Mathematics in Engineering and Economics '34—AMEE '08*, edited by M. D. Todorov
© 2008 American Institute of Physics 978-0-7354-0598-1/08/$23.00

THE STURM-LIOUVILLE PROBLEM

Consider the following eigenvalue problem

$$\mathscr{L}u - \lambda^4 u = L^2 u - \lambda^4 u = 0, \qquad u(1) = \frac{du}{dr}\bigg|_{r=1} = 0, \qquad L \overset{\text{def}}{=} \frac{1}{r}\frac{d}{dr}r\frac{d}{dr}, \qquad (1)$$

which can also be rewritten as follows

$$(L - \lambda^2)(L + \lambda^2)u = (L - \lambda^2)v = 0, \quad (L + \lambda^2)u = v. \qquad (2)$$

The equation for function v is

$$r^2 v''(r) + r v'(r) - \lambda^2 r^2 v(r) = 0, \qquad (3)$$

whose solutions are the the the modified Bessel functions $I_0(\lambda r)$ and $K_0(\lambda r)$.

For the inside of the cylinder we only have to keep the function I_0. Then we have to find the solution of the following nonhomogeneous equation

$$r^2 u''(r) + r u'(r) + \lambda^2 r^2 u(r) = 2Cr^2 I_0(\lambda r). \qquad (4)$$

The particular solution of this equation is $U(r) = C\lambda^{-2} I_0(\lambda r)$, while the homogeneous equation has as solutions the two Bessel functions $J_0(\lambda r)$ and $Y_0(\lambda r)$. Guided by the same considerations, we keep in this note only the function J_0 which is not singular in the origin. Thus the general solution of equation (1) has the form

$$u(r) = AJ_0(\lambda r) + BI_0(\lambda r), \qquad (5)$$

where A and B are two arbitrary constants that has to be identified from the boundary conditions. The latter give us the following system

$$AJ_0(\lambda) + BI_0(\lambda) = 0, \quad AJ_0'(\lambda) + BI_0'(\lambda) = 0.$$

This system can have nontrivial solution only if

$$J_0(\lambda)I_0'(\lambda) - I_0(\lambda)J_0'(\lambda) = 0, \qquad (6)$$

which gives the spectrum. An alternative form can be presented using the properties $J_0'(z) = -J_1(z), \quad I_0'(z) = I_1(z)$ (see, e.g., [5, 1]). Then the spectral equation reads

$$J_0(\lambda)I_1(\lambda) + I_0(\lambda)J_1(\lambda) = 0. \qquad (7)$$

ASYMPTOTIC APPROXIMATION OF THE EIGENVALUES

The solutions λ_n of the spectral equation (7) can be found numerically. We also derive an asymptotic formula which will provide a good approximation for large n.

From [3] we can find the third order asymptotic series expansions for Bessel $J_v(x)$ and modified Bessel functions $I_v(x)$ of order v, for the case of large x. Introducing those into (7) and neglecting the terms of order $O(|\lambda|^{-3})$ we get

$$J_0(\lambda)I_1(\lambda) + I_0(\lambda)J_1(\lambda) \propto \frac{e^\lambda}{\pi\lambda}\left\{(1 - \frac{24}{128\lambda^2} - \frac{3}{8\lambda})\cos(\frac{4\lambda - \pi}{4}) + (\frac{1}{8\lambda} - \frac{3}{64\lambda^2})\right.$$

$$\times \sin(\frac{4\lambda - \pi}{4}) + (1 + \frac{24}{128\lambda^2} + \frac{1}{8\lambda})\cos(\frac{4\lambda - 3\pi}{4}) - (\frac{3}{8\lambda} + \frac{3}{64\lambda^2})\sin(\frac{4\lambda - 3\pi}{4})\right\}$$

$$= \frac{e^\lambda}{\sqrt{2\pi\lambda}}\left\{\cos\lambda(-\frac{1}{4\lambda} - \frac{9}{32\lambda^2}) + \sin\lambda(2 + \frac{1}{4\lambda}) + O(|\lambda|^{-3})\right\}.$$

Then, in the adopted third-order approximation, the equation for the spectrum reduces to

$$\cos\lambda(-\frac{1}{4\lambda} - \frac{9}{32\lambda^2}) + \sin\lambda(2 + \frac{1}{4\lambda}) = 0. \tag{8}$$

To find an approximate solution of equation (8) for large λ, we will stipulate that $\lambda = n\pi + \varepsilon_n + \alpha\varepsilon_n^2$, where ε_n is a small addition to the n-th root of the zeroth-order equation $\sin\lambda = 0$ and α a parameter to be determined. Then,

$$\cos\lambda = \cos(n\pi + [\varepsilon_n + \alpha\varepsilon_n^2]) = (-1)^n(1 - \tfrac{1}{2}\varepsilon_n^2) + O(\varepsilon_n^3) \tag{9}$$

$$\sin\lambda = \sin(n\pi + [\varepsilon_n + \alpha\varepsilon_n^2]) = (-1)^n(\varepsilon_n + \alpha\varepsilon_n^2) + O(\varepsilon_n^3). \tag{10}$$

Introducing into equation (8), keeping only terms up to order $O(\varepsilon_n^2)$, we arrive at

$$(-8\alpha + \tfrac{11}{2} + 132n\pi + 64n^2\pi^2\alpha + 8n\pi\alpha)\varepsilon_n^2 + (-8 + 64n^2\pi^2 + 8n\pi)\varepsilon_n + (-8n\pi - 9) = 0.$$

To achieve a third-order approximation we set the coefficient of ε_n^2 in the last equality equal to zero, and solve for α to obtain

$$\alpha = \frac{-6.5 - 132n\pi}{-8 + 8n\pi + 64n^2\pi^2}, \quad \varepsilon_n = \frac{8n\pi + 9}{-8 + 8n\pi + 64n^2\pi^2} \quad \text{and thus} \tag{11}$$

$$\lambda(n) = n\pi + \frac{8n\pi + 9}{-8 + 8n\pi + 64n^2\pi^2} + \frac{-6.5 - 132n\pi}{-8 + 8n\pi + 64n^2\pi^2}\left(\frac{8n\pi + 9}{-8 + 8n\pi + 64n^2\pi^2}\right)^2 \tag{12}$$

which approximates λ_n up to order $O(\frac{1}{n^3\pi^3})$.

THE SYSTEM OF EIGENFUNCTIONS

The nth eigenfunction of (1) is given by

$$v_n(r) = A_nJ_0(\lambda_nr) + B_nI_0(\lambda_nr) \tag{13}$$

where A_n is arbitrary and $B_n = -\frac{A_nJ_0'(\lambda_n)}{I_0'(\lambda_n)} = \frac{A_nJ_1(\lambda_n)}{I_1(\lambda_n)}$. We will prescribe the value of $A_n = 1$ and thus $B_n = \frac{J_1(\lambda_n)}{I_1(\lambda_n)}$. We will now show the following theorem:

124

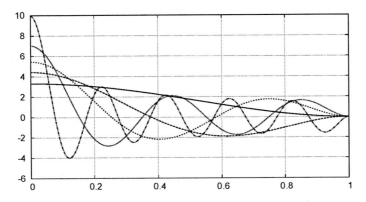

FIGURE 1. The profiles of the radial beam functions $u_1, u_2, u_3, u_5, u_{10}$

Theorem 1. *The set of eigenfunctions $\{v_n(r)\}_{n=1}^{\infty}$ as defined by (13) is orthogonal on $[0,1]$ with respect to the weight $w(r) = r$, i.e.,*

$$< v_n, v_m >_w = \int_0^1 r v_n v_m dr = \begin{cases} 0 & \text{if } n \neq m, \\ c \neq 0 & \text{if } n = m. \end{cases} \tag{14}$$

Proof. Suppose λ_n, λ_m two different eigenvalues of (1). Then

$$\lambda_n^4 < v_n, v_m >_w = \int_0^1 r \lambda_n^4 v_n v_m dr = \int_0^1 [r\mathscr{L}v_n(r)] v_m dr = \int_0^1 \frac{d}{dr} r \frac{d}{dr} \left(\frac{1}{r} \frac{d}{dr} r \frac{d}{dr} \right) v_n v_m dr.$$

Integrating by parts twice and using the boundary conditions we obtain

$$\lambda_n^4 \langle v_n, v_m \rangle_w = \int_0^1 \frac{1}{r} \frac{d}{dr} \left(r v_m'(r) \right) \frac{d}{dr} \left(r v_n'(r) \right) dr. \tag{15}$$

$$\lambda_m^4 \langle u_v, v_m \rangle_w = \int_0^1 \frac{1}{r} \frac{d}{dr} \left(r v_m'(r) \right) \frac{d}{dr} \left(r v_n'(r) \right) dr. \tag{16}$$

Subtracting (15)-(16) we obtain

$$(\lambda_n^4 - \lambda_m^4) < v_n, v_m >_w = 0.$$

But $\lambda_n \neq \lambda_m$ therefore $< v_n, v_m >_w = 0$ when $n \neq m$. $\qquad\square$

Clearly when $n = m$ the number $< v_n, v_n >_w = \int_0^1 r v_n^2(r) \mathrm{d}r = ||v_n||^2$, the square of the L_2 norm of v_n. To normalize $\{v_n(r)\}_{n=1}^\infty$ we simply calculate the norm and divide. Thus,

$$||v_n||^2 \equiv \langle v_n, v_n \rangle_w = \int_0^1 r \left[J_0^2(\lambda_n r) + B_n^2 I_0^2(\lambda_n r) + 2 B_n J_0(\lambda_n r) I_0(\lambda_n r) \right] \mathrm{d}r$$

$$= \frac{1}{2} [J_0^2(\lambda_n) + J_1^2(\lambda_n)] + \frac{1}{2} B_n^2 [I_0^2(\lambda_n) - I_1^2(\lambda_n)].$$

In the above calculations we have used

$$r \frac{d}{dr}\left(\frac{1}{r}\frac{d}{dr} r \frac{d}{dr}(J_0(ar)) \right) = a^3 r J_1(ar), \quad r \frac{d}{dr}\left(\frac{1}{r}\frac{d}{dr} r \frac{d}{dr}(I_0(ar)) \right) = a^3 r I_1(ar),$$

$$\frac{1}{r}\frac{d}{dr} r \frac{d}{dr}(J_0(ar)) = -a^2 J_0(ar), \quad \frac{1}{r}\frac{d}{dr} r \frac{d}{dr}(I_0(ar)) = a^2 I_0(ar),$$

$$J_1(0) = 0, \ I_1(0) = 0, \ J_0(0) = 1, \ I_0(0) = 1, \ J_0(1) \approx 0.765198, \ I_0(1) \approx 1.26607.$$

Thus, our normalized eigenfunctions now read

$$u_n(r) = \frac{\sqrt{2}}{\sqrt{[J_0^2(\lambda_n) + J_1^2(\lambda_n)] + \dfrac{J_1^2(\lambda_n)}{I_1^2(\lambda_n)}[I_0^2(\lambda_n) - I_1^2(\lambda_n)]}} \left(J_0(\lambda_n r) + \frac{J_1(\lambda_n)}{I_1(\lambda_n)} I_0(\lambda_n r) \right).$$

The first couple of the eigenfunctions are presented in Figure 1.

EXPANSIONS INTO SERIES OF RADIAL BEAM FUNCTIONS

In order for us to apply our spectral technique we will need formulas for expanding various expressions into series of radial beam functions.

Expansion of Unity. Unity can be expressed into a series of radial beam functions as follows

$$1 = \sum_{n=1}^\infty o_n u_n(r), \quad o_n = \frac{\sqrt{2}}{\lambda_n} \left([J_0^2(\lambda_n) + J_1^2(\lambda_n)] + \frac{J_1^2(\lambda_n)}{I_1^2(\lambda_n)}[I_0^2(\lambda_n) - I_1^2(\lambda_n)] \right)^{-1/2} J_1(\lambda_n),$$

(17)

In calculating of the above coefficients the following Bessel integrals are used

$$\int x J_0(ax) \mathrm{d}x = \frac{1}{a} x J_1(ax), \quad \int x I_0(ax) \mathrm{d}x = \frac{1}{a} x I_1(ax).$$

(18)

Expansion of the Second-Order Cylindrical Differential Operator. In problems with cylindrical symmetry, the second-order cylindrical operator L from (1) is often involved.

126

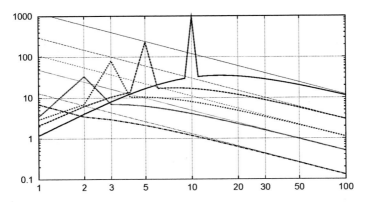

FIGURE 2. The convergence rate of the expansion of second-order operator; From bottom to top: curved lines: the coefficients $g_{1n}, g_{2n}, g_{3n}, g_{5n}, g_{10n}$, accompanying straight lines: the best fits $13n^{-1}$, $49n^{-1}, 110n^{-1}, 300n^{-1}, 1200n^{-1}$

Therefore it is useful to have Galerkin expansions of the form

$$Lu_n(r) = \sum_{m=1}^{\infty} g_{nm} u_m(r). \tag{19}$$

The formula is obtained by evaluating the inner product $< Lu_n, u_m >_w$. Note that all integrals involved can be evaluated symbolically. Thus,

$$g_{nn} = \frac{\lambda_n^2}{2\|v_n\|^2} \left\{ -J_0^2(\lambda_n) - J_1^2(\lambda_n) + B_n^2 \left(I_0^2(\lambda_n) - I_1^2(\lambda_n) \right) \right\}, \tag{20}$$

$$g_{nm} = \frac{\lambda_n^2}{\|v_n\|\|v_m\|} \left\{ -\left(\frac{\lambda_n J_0(\lambda_m) J_1(\lambda_n) - \lambda_m J_0(\lambda_n) J_1(\lambda_m)}{\lambda_n^2 - \lambda_m^2} \right) \right.$$
$$- B_m \left(\frac{\lambda_m I_1(\lambda_m) J_0(\lambda_n) + \lambda_n I_0(\lambda_m) J_1(\lambda_n)}{\lambda_n^2 + \lambda_m^2} \right) + B_n \left(\frac{\lambda_n I_1(\lambda_n) J_0(\lambda_m) + \lambda_m I_0(\lambda_n) J_1(\lambda_m)}{\lambda_m^2 + \lambda_n^2} \right)$$
$$\left. + B_n B_m \left(\frac{\lambda_n I_0(\lambda_m) I_1(\lambda_n) - \lambda_m I_0(\lambda_n) I_1(\lambda_m)}{\lambda_n^2 - \lambda_m^2} \right) \right\} \quad n \neq m. \tag{21}$$

TEST CASES

Simple Fourth-Order BVP. In order for us to test our method, we at first consider the problem

$$\mathscr{L}u = \frac{1}{r}\frac{d}{dr}r\frac{d}{dr}\left(\frac{1}{r}\frac{d}{dr}r\frac{d}{dr} \right)u = 64, \qquad u(1) = \frac{du}{dr}\bigg|_{r=1} = 0, \tag{22}$$

127

which admits the exact analytic solution

$$u_{an}(r) = 1 - 2r^2 + r^4. \tag{23}$$

To apply our technique we expand the sought function into series

$$u_{sp}(r) = \sum_{m=1}^{N} a_m u_m(r). \tag{24}$$

Substituting (17) and (24) into (22), using the basic property of the eigenfunctions (1) and taking successive inner products with $u_n(r)$, $n = 1,2,3,\ldots,N$, yields

$$a_n = 64 \frac{o_n}{\lambda_n^4} = \frac{1}{||v_n||} \frac{128}{\lambda_n^5} J_1(\lambda_n) \ , \quad n = 1,2,3,\ldots,N. \tag{25}$$

From this we may deduce that the convergence rate of the solution coefficients a_n is fifth order algebraic. Indeed, Figure 3 below confirms our assertion.

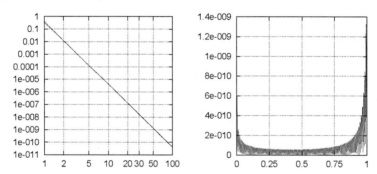

FIGURE 3. Left panel: convergence rate of solution coefficients a_n along with the best fit line $a_n = 0.4 \ n^{-5}$. Right panel: absolute difference between the exact analytic solution and the spectral solution with 100 terms

Furthermore, we observe that the maximum absolute error is of order 10^{-9}.

Fourth-Order BVP involving the Second-Order Radial Differential Operator. We now consider the following problem which also involves the second-order radial differential operator,

$$L^2 u - Lu = -2I_1(1)(\approx -1.13031820798497), \qquad u(1) = u'(1) = 0, \tag{26}$$

and admits the exact analytic solution $u_{an}(r) = -I_0(r) + \frac{1}{2}I_0'(1)r^2 + I_0(1) - \frac{1}{2}I_0'(1)$.

Once again we expand the sought function into a series of radial beam functions $u_{sp}(r) = \sum_{n=1}^{N} a_n u_n(r)$ and employ formulas (1),(17),(20) to obtain the Galerkin system

$$\lambda_n^4 a_n - \sum_{m=1}^{N} g_{mn} a_m = -1.13031820798497 o_n \ , \quad n = 1,2,3,\ldots,N. \tag{27}$$

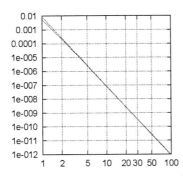
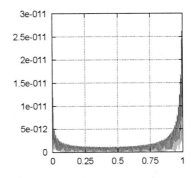

FIGURE 4. Left panel: convergence rate of solution coefficients a_n along with the best fit line $a_n = 0.0086\, n^{-5}$. Right panel: absolute difference between the exact analytic solution and the spectral solution with 100 terms

The matrix multiplying the spectral coefficients $\{a_n\}_{n=1}^{N}$ in (27) is symmetric. We solve (27) using IMSL routine DLSASF for linear systems with a real symmetric coefficient matrix.

We can see in Figure 4 that the convergence rate of the spectral coefficients is once again fifth-order algebraic, and that the overall absolute error when using $N = 100$ terms is $O(10^{-11})$.

CONCLUSIONS

A complete orthonormal set of functions —the so-called Radial Beam functions— is introduced. This is the set of eigenfunctions of a Sturm-Liouville fourth-order boundary problem with radial symmetry. The eigenvalues are found numerically whereas for larger indices n of the eigenvalues, asymptotic representations valid within $o(n^{-2})$ are derived.

The appropriate formulas for expressing unity and the various derivatives of our functions into members of the CON are derived and their convergence rate is verified.

The technique is assessed by applying it to two model fourth order problems with radial symmetry which admit to exact analytic solutions: a simple problem involving only the fourth-order radial operator and a constant and the other also involving the second-order radial operator. In both cases, the convergence rate of the spectral coefficients is $O(N^{-5})$.

The spectral and analytic solutions are found to be in excellent agreement. For the first test problem 200 terms secure a maximum absolute error of order $O(10^{-11})$, whereas for the second test problem this difference is achieved with only 100 terms -a result fully compatible with the theoretical rate of convergence.

The fifth-order algebraic rate of convergence is fully adequate for the problems under consideration. Thus, a new technique suitable for biharmonic fourth-order boundary problems with radial symmetry is introduced. The approach can be easily extended to other fourth-order problems that arise in fluid dynamics and elasticity which involve the bi-Stokesian or even more complex operators.

REFERENCES

1. M. Abramowitz and I. Stegun, *Handbook of Mathematical Functions*, Dover, New York, 1970, 9th edn.
2. S. Chandrasekhar, *Hydrodynamic and Hydromagnetic Instability*, Oxford University Press, Clarendon, London, 1961.
3. A. D. Polyanin and V. F. Zaitsev, *Handbook of Exact Solutions for Ordinary Differential Equations*, Chapman & Hall/CRC, Boca Raton, Florida, USA, 2003, 2nd edn.
4. N. C. Papanicolaou and C. I. Christov, "On the Beam-Functions Spectral Expansions for Fourth-Order Boundary Value Problems: Advantadges and Disadvantages," in *33rd International Conference on Applications of Mathematics in Engineering and Economics, Sozopol, Bulgaria, June 8-14, 2007*, edited by M.D.Todorov, American Institute of Physics CP 946, Melville, New York, 2007, pp. 119–126.
5. M. R. Spiegel and J. Liu, *Mathematical Handbook of Formulas and Tables*, McGraw-Hill, New York, 1999, 2nd edn.

Modelling Hydrodynamic Stability in Electrochemical Cells

J. Pontes*, N. Mangiavacchi†, G. Rabello dos Anjos**, O. E. Barcia‡, O. R. Mattos§ and B. Tribollet¶

* Metallurgy and Materials Engineering Dept. – Federal University of Rio de Janeiro
P.O. Box 68505, Rio de Janeiro, RJ, 21941-972 Brazil
† Group of Environmental Studies of Hydropower Reservoirs (GESAR Group),
State University of Rio de Janeiro,
R. Fonseca Telles 524, Rio de Janeiro, RJ, 20550-013 Brazil
**Group of Environmental Simulations of Hydropower Reservoirs (GESAR Group),
State University of Rio de Janeiro, R. Fonseca Telles 524, Rio de Janeiro, RJ, 20550-013 Brazil
‡ Institute of Chemistry – Federal University of Rio de Janeiro
P.O. Box 68505, Rio de Janeiro, RJ, 21941-972 Brazil
§Metallurgy and Materials Engineering Dept. – Federal University of Rio de Janeiro
P.O. Box 68505, Rio de Janeiro, RJ, 21941-972 Brazil
¶UPR15 – CNRS, Laboratoire Interfaces et Systèmes Electrochimiques
4, Place Jussieu, 75252 Paris Cedex 05, France

Abstract. We review the key points concerning the linear stability of the classical von Kármán's solution of rotating disk flow, modified by the coupling, through the fluid viscosity, with concentration field of a chemical species. The results were recently published by Mangiavacchi et al (*Phys. Fluids*, 19: 114109, 2007) and refer to electrochemical cells employing iron rotating disk electrodes, which dissolve in the 1 M H_2SO_4 solution of the electrolyte. Polarization curves obtained in such cells present a current instability at the beginning of the region where the current is controlled by the the hydrodynamics. The onset of the instability occurs in a range of potentials applied to the cell and disappear above and below this range. Dissolution of the iron electrode gives rise to a thin concentration boundary layer, with thickness of about 4% of the thickness of the hydrodynamic boundary layer. The concentration boundary layer increases the interfacial fluid viscosity, diminishes the diffusion coefficient and couples both fields, with a net result of affecting the hydrodynamic of the problem. Since the current is proportional to the interfacial concentration gradient of the chemical species responsible by the ions transport, the instability of the coupled fields can lead to the current instability observed in the experimental setups. This work presents the results of the linear stability analysis of the coupled fields and the first results concerning the Direct Numerical Simulation, currently undertaken in our group. The results show that small increases of the interfacial viscosity result in a significant reduction of the stability of modes existing in similar configurations, but with constant viscosity fluids. Upon increasing the interfacial viscosity, a new unstable region emerges, in a range of Reynolds numbers much smaller than the lower limit of the unstable region previously known. Though the growth rate of modes in the previously known region is larger than the one of modes in the new region, the amplitude of the concentration unstable modes in this one is very large when compared to the amplitude of the associated hydrodynamic unstable modes. In addition concentration modes are always confined in a rather thin region, leading to the existence of large interfacial concentration gradient. Concentration modes in the new unstable region seem thus, to have a combination of properties sufficient to drive detectable current oscillations. The numerical experiments show that a progressive increase in the interfacial viscosity initially reduces the stability of the flow, but an increase beyond a certain limit restores the stability properties of constant viscosity flows.

Keywords: Rotating disk, hydrodynamic stability, corrosion, chemical oscillations.
PACS: 47.10.ad, 47.15.fe, 82.45.Qr

CP1067, *Applications of Mathematics in Engineering and Economics '34—AMEE '08*, edited by M. D. Todorov
© 2008 American Institute of Physics 978-0-7354-0598-1/08/$23.00

POSING THE PROBLEM

The hydrodynamic field developed close to the axis of a large rotating disk belongs to the restricted class of problems admitting a analytical or semi-analytical similarity solution of the hydrodynamic equations. The angular velocity imposed to the fluid at the surface gives rise to a centrifugal effect and to a radial flow outwards. Continuity requires that the flow be replaced by an incoming one that approaches the disk. Close to the surface, the axial velocity of the incoming flow is reduced and the centrifugal effect appears. This is the region of the hydrodynamic boundary layer. The steady tri-dimensional solution of this flow was found by Von Kármán (1932) and is schematically shown in Figure 1.

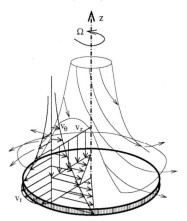

FIGURE 1. Von Kármán's flow close to the axis of a rotating disk

The boundary layer becomes unstable beyond a critical Reynolds number. The first stability studies of Von Kármán's solution are due to Smith (1946)[14], who found the emergence of spirals as shown in Figure 2. The pattern may turn with the disk angular velocity or other. When turning slower than the disk, the spirals arms tend to bend in the upstream direction. The pattern is "pushed" by the mean flow. Patterns bending downstream, or consisting of radial straight lines or either of circumferences are also possible. The coexistence of spirals bending in the up and the downstream directions was observed by Moisy (2004) [11], in a setup of two counter rotating disks.

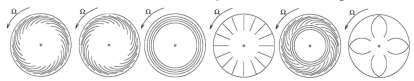

FIGURE 2. Possible perturbation patterns emerging after the first instability of von Kármán's solution

Gregory et al (1955) [4] measured the critical Reynolds number beyond which Von Kármán's solution becomes unstable and proposed the first theoretical approach to the stability problem, not taking into account viscosity effects. Stability studies conducted after the eighties were mainly made in groups interested in the effect of a secondary cross flow, on the stability of a main one, as those found in swept wings. Rotating disk flow presents certain characteristics also found in swept wings. In swept wings the upstream flow is decomposed in a component perpendicular to the wing, responsible by the lift and a cross component that anticipates the transition to turbulence. In rotating disk flow, a secondary radial flow appears in consequence of the imposed azimuthal velocity at the interface. Thanks to Von Kármán's exact solution and to the more controlled conditions found in rotating disks than in wings, rotating disks became a prototype to infer on the instabilization mechanisms occurring in swept wings. Malik (1986) [8] evaluated the first neutral curve of perturbations turning with the angular velocity of the disk (stationary perturbations). The critical Reynolds number was found as 285.36. Lingwood (1995) [7] extended Malik's results by considering non-stationary perturbations, turning with angular velocity different of the disk velocity and found critical Reynolds numbers on the order of 80. In addition, Lingwood addressed the absolute stability problem. The flow becomes absolutely unstable when the a perturbation originated in a point is carried by the angular velocity of the flow and returns to the origin point before being completely damped and excites again the remaining perturbation at that point.

Stability studies conducted after 1990 focused on compressible fluid configurations, having in mind the fact that swept wings operate in the high subsonic regime.

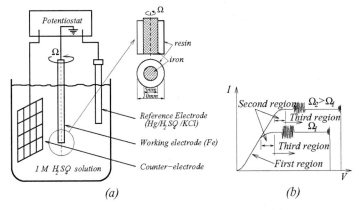

FIGURE 3. An electrochemical cell with a rotating disk electrode (*a*); and the polarization curve showing the three regions (*b*)

The question of rotating disk flow stability appears also in electrochemical problems. Electrochemical cells employing a rotating disk electrode are widely used due to the relatively simple setup and to the fact that the mass flow at the electrode/electrolyte interface is independent of the radial position. The rate of transfer of ions is conveniently controlled by the the angular velocity imposed to the electrode. Figure 3 schematically shows a typical arrangement of these cells, with three electrodes. The counter electrode consists of a platinum screen placed along the cell wall to assure an uniform distribution of the electric potential. Potentials are measured against the reference electrode. The working electrode consists of a 5 mm diameter iron rod covered with a 10 mm diameter resin cast, except in the base, through which the current flows. The electrode is coupled to a variable velocity electric motor. Typical electrode velocities range from 100 to

900 rpm.

Polarization curves experimentally obtained in the dissolution of iron electrodes in H_2SO_4 electrolytes present three regions (Barcia et al, 1992 [1]). The first one occurs at relatively low overvoltages applied to the working electrode. The current is approximately proportional to the overvoltage and depends on the dissolution process. The current is not affected by transport phenomena occurring in the electrolyte, like the angular velocity imposed to the working electrode. Polarization curves show a second region, where the current depends on the applied overvoltage and also on the hydrodynamics, which is defined by the electrode angular velocity. By further increasing the overvoltage, a current *plateau* appears in the polarization curves, where the current depends only on the angular velocity of the electrode, and no longer, on the overvoltage. The overvoltage level of the *plateau* is proportional to $\Omega^{1/2}$.

The hydrodynamic field at the base of the working electrode is approximated by Von Kármán's similarity solution. This hypothesis is justified by the thickness of the hydrodynamic boundary layer, which is proportional to $\Omega^{1/2}$, ranging from 0.5 to 1.3 mm, being thus one order of magnitude smaller than the 10 mm electrode diameter. Dissolution of the iron electrode gives raise to a Fe^{+2} ions concentration boundary layer. The ratio between the hydrodynamic and the concentration boundary layers is given by $\delta_h/\delta_c \approx 2 Sc^{1/3}$ (Levich 1962 [6]), where Sc is the Schmidt number. Typical Schmidt numbers in electrochemical cells is 2000, resulting in concentration boundary layers with thicknesses of order of 4% of the thickness of the hydrodynamic one.

The concentration boundary layer leads to a reduction of the iron diffusion coefficient in the electrolyte. The Stokes-Einstein law postulates that the product of the diffusion coefficient by the electrolyte viscosity is constant, implying in a viscosity increase at the interface. The effect couples both fields.

The cell current may be evaluated through the ions transfer at the electrode/electrolyte interface, which is due to two effects: ions concentration gradient $(i \propto dC/dz|_{z=0})$ and ions migration due to the electric field. Barcia et al (1992) [1] showed that the concentration gradient is the dominant effect, what leads to the conclusion that the current depends on the spatial distribution of a relevant chemical species, that in turn, depends on the hydrodynamic field. Instabilities of the hydrodynamic field and of the interfacial concentration gradient may thus be responsible for the current oscillations observed in the current *plateau*. However, Reynolds numbers attained in electrochemical cells are of order of 20 – 40, being thus clearly smaller than critical experimental values found in the literature and the theoretical values found by Malik and by Lingwood.

This work deals with the coupling of the hydrodynamic field, to the concentration one of a chemical species responsible by the current. Three questions are posed: (a) if the the coupling changes the stability properties of the purely hydrodynamic field, reducing the critical Reynolds number to the range of values attained in electrochemical setups, (b) if possible oscillations of the interfacial concentration gradient are strong enough to drive the current oscillations experimentally observed and (c) what mechanisms lead to the suppression of the current instability beyond a certain overvoltage level. The results herein presented support a positive answer to the first question, suggest that the hydrodynamic instability of the coupled fields is strong enough to drive the current oscillation, but still do not explain the instability suppression.

GOVERNING EQUATIONS

The problem is governed by the continuity and the Navier-Stokes equations, coupled through the viscosity, to the transport equation of the relevant chemical species. These equations, written in the frame attached to the surface of the rotating disk read:

$$\left.\begin{array}{ll} \operatorname{div}\mathbf{v} = 0 & \dfrac{D\mathbf{v}}{Dt} = -2\mathbf{\Omega}\times\mathbf{v} - \mathbf{\Omega}\times(\mathbf{\Omega}\times\mathbf{r}) - \dfrac{1}{\rho}\operatorname{\mathbf{grad}}p + \dfrac{1}{\rho}\operatorname{div}\tau \\ \dfrac{D\mathscr{C}}{Dt} = \operatorname{div}(\mathscr{D}\operatorname{\mathbf{grad}}\mathscr{C}) & \end{array}\right\} \tag{1}$$

where Ω is the angular velocity of the rotating disk electrode, $\mathscr{C}$ and $\mathscr{D}$ are, respectively, the concentration and the diffusion coefficient of the representative chemical species and τ is the viscous stress tensor for a newtonian fluid with the viscosity depending on the concentration of the chemical species.

The evolution equations are rewritten in dimensionless form. Variables having units of length or its reciprocal (radial and axial coordinates, perturbation wave number along the radial direction) are made dimensionless with the length used to measure the thickness of the boundary layer, $(\nu_\infty/\Omega)^{1/2}$, where ν_∞ is the bulk viscosity of the fluid. Velocity components are divided by the local imposed azimuthal velocity $r_e\Omega$, pressure is divided by $\rho(r_e\Omega)^2$, viscosity is divided by the bulk viscosity, ν_∞, time and the eigenvalue of the linearized problem are made dimensionless with the time required by a particle, turning with the azimuthal velocity $r_e\Omega$, to move a distance equal to the reference length, $(\nu_\infty/\Omega)^{1/2}$. Here, r_e is the dimensional coordinate along the radial direction at which the stability analysis will be carried out. The dimensionless concentration of the chemical species is defined by:

$$C = \frac{\mathscr{C} - \mathscr{C}_\infty}{\mathscr{C}_S - \mathscr{C}_\infty} \tag{2}$$

where $\mathscr{C}_S$ and $\mathscr{C}_\infty$ are, respectively, the concentration of the chemical species at the electrode surface and in the bulk. We define also the Reynolds and the Schmidt numbers by the relations:

$$R = r_e\left(\frac{\Omega}{\nu_\infty}\right)^{1/2} \quad \text{and} \quad Sc = \frac{\mathscr{D}_\infty}{\nu_\infty}. \tag{3}$$

The Reynolds number may be seen as a dimensionless distance to the disk axis. At this point, we assume that the viscosity depends on the dimensionless concentration of the chemical species according to:

$$\nu = \nu_\infty \exp(mC) \tag{4}$$

where m is a dimensionless parameter depending on the electrochemical characteristics of the system (electrode material, type of electrolyte, applied potential), but not on the concentration of the chemical species. In particular, this parameter defines the interface viscosity, given by $\nu = \nu_\infty \exp(m)$. This equation is based on a thermodynamic model proposed by Esteves et al (2001) [2]. We also assume the Stokes-Einstein law:

$$\mathscr{D}\nu = \mathscr{D}_\infty \nu_\infty \tag{5}$$

where $\mathscr{D}_\infty$ is the bulk diffusion coefficient. Using the bulk coefficients to rewrite equations (5) and (4) in dimensionless form, we obtain:

$$\mathscr{D}v = 1 \qquad \text{and} \qquad v = \exp(mC). \qquad (6)$$

Equations (1) are rewritten as follows, in dimensionless form:

$$\left.\begin{array}{l} \operatorname{div}\mathbf{v} = 0 \qquad \dfrac{D\mathbf{v}}{Dt} = -2\mathbf{e}_z \times \mathbf{v} - \mathbf{e}_z \times (\mathbf{e}_z \times r\mathbf{e}_r) - \mathbf{grad}\,p + \dfrac{1}{R}\operatorname{div}\tau \\[2mm] \dfrac{DC}{Dt} = \dfrac{1}{R\,Sc}\operatorname{div}(\mathscr{D}\,\mathbf{grad}\,C) \end{array}\right\}. \qquad (7)$$

A destabilizing potential of the coupling between the hydrodynamic and the chemical species fields can be seen in equation (7): the Reynolds number is amplified by the Schmidt number, which takes here the value $Sc = 2000$, typical for electrochemical cells.

THE BASE STATE

Base State Equations

The base state is the von Kármán similarity solution for a fluid with the viscosity depending on the concentration field, which is assumed to vary along the axial coordinate only.

$$\left.\begin{array}{lll} \bar{v}_r = r\Omega F(z) & \bar{v}_\theta = r\Omega G(z) & \bar{v}_z = (v_\infty\Omega)^{1/2}H(z) \\[2mm] p = \rho v_\infty\Omega P(z) & \mathscr{C} = \mathscr{C}_\infty + (\mathscr{C}_S - \mathscr{C}_\infty)C(z) \end{array}\right\}. \qquad (8)$$

All variables in equation (8) are dimensional, except the axial dependent profiles, F, G, H, C and the axial coordinate z. Boundary conditions for F, G, H and P are $F = G = H = P = 0$ at the disk surface ($z = 0$), $F = H' = 0$ and $G' = -1$ as $z \longrightarrow \infty$. The dimensionless concentration profile, C, varies from 1, at $z = 0$, to 0, as $z \longrightarrow \infty$.

Introducing equation (8) and the constitutive equations of the stress tensor for a newtonian fluid with variable viscosity in equations (1), together with equations (4) and (5), we obtain the ordinary nonlinear system for the axial profiles F, G, H, P and C:

$$\left.\begin{array}{ll} 2F + H' = 0 & F^2 - (G+1)^2 + HF' - vF'' - v'F' = 0 \\[2mm] 2F(G+1) + HG' - vG'' - v'G' = 0 & \\[2mm] P' + HH' - vH'' - 2v'H' = 0 & Sc\,HC' - \dfrac{C''}{v} + \dfrac{v'}{v^2}C' = 0 \end{array}\right\} \qquad (9)$$

where prime denotes derivatives with respect to the dimensionless axial coordinate z. The viscosity v and its derivatives are written in dimensionless form.

Evaluation of the Viscosity at the Electrode/Electrolyte Interface

Solving equations (9) requires specification of two parameters: the bulk Schmidt number and the parameter m appearing in equation (4), which ultimately defines the

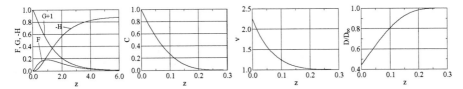

FIGURE 4. Stationary dimensionless velocity, concentration, viscosity and diffusion, profiles, F, G, H and C for $v_0/v_\infty = 2.255$ and $Sc = 2000$

electrolyte viscosity at the interface with the electrode. At this point, we assume that the limit current density at the interface is proportional to the concentration gradient of the relevant chemical species generated by the dissolution of the electrode. Ions migration due to the potential gradient is neglected. The current density is given by the relation (Barcia, 1992 [1]):

$$\frac{i}{n\mathscr{F}\dfrac{1}{Sc}\dfrac{1}{v_0/v_\infty}(\mathscr{C}_\infty - \mathscr{C}_s)\sqrt{v_\infty\Omega}} = \left.\frac{dC}{dz}\right|_{z=0} \tag{10}$$

where i is expressed in [A/cm^2], n is the valence number of the chemical species ($n = 2$), $\mathscr{F} = 96500$ C/mol, is the Faraday constant, $\mathscr{C}_s = 2.0 \times 10^{-3}$ mol/cm^3 is the dimensional concentration of the species at the electrode interface (saturated condition),

Quantity $\mathscr{C}_\infty = 0$ mol/cm^3 and v_0 is the fluid viscosity at the interface. The limit current density obtained experimentally is $i = 0.8810$ A/cm^2 at 900 rpm (Geraldo, 1998 [3]). An initial value is specified for m, equations (9) are solved. The ratio v_0/v_∞ is evaluated and the dimensionless normal derivative of the concentration at the interface, $dC/dz|_{z=0}$, is obtained from the profiles. These figures are introduced in equation (10), leading to a value for the current density. The value initially specified for m is corrected and the procedure is repeated until convergence to the experimental value of i is reached. At convergence we obtain $v_0/v_\infty = 2.255$.

Base State Profiles

The velocity, concentration, viscosity and diffusion coefficient profiles, F, G, H, C, v and $\mathscr{D}$, obtained for $v_0/v_\infty = 2.255$ and $Sc = 2000$ are shown in Figure 4. The thin concentration boundary layer, results in velocity profiles close to the ones obtained for the constant viscosity case. However, the derivatives of the velocity profiles are strongly affected inside the concentration boundary layer, as shown in Figure 5.

STABILITY OF THE BASE STATE

Stability Equations

We turn now to the question of the stability of the base state with respect to small disturbances. The hydrodynamic and chemical fields are written as a sum of the base state plus a perturbation:

$$v_r = \bar{v}_r + \tilde{v}_r \qquad v_\theta = \bar{v}_\theta + \tilde{v}_\theta \qquad v_z = \bar{v}_r + \tilde{v}_z$$
$$p = \bar{p} + \tilde{p} \qquad \mathscr{C}_T = \bar{\mathscr{C}} + \tilde{\mathscr{C}} \tag{11}$$

where the perturbation, in dimensional form, is given by:

$$\begin{pmatrix} \tilde{v}_r \\ \tilde{v}_\theta \\ \tilde{v}_z \\ \tilde{p} \\ \tilde{\mathscr{C}} \end{pmatrix} = \begin{pmatrix} r_e \Omega f \\ r_e \Omega g \\ r_e \Omega h \\ \rho v_\infty \Omega \pi \\ (\mathscr{C}_s - \mathscr{C}_\infty) c \end{pmatrix} \exp\left[i(\alpha r + \beta R\theta - \omega t)\right] + cc. \tag{12}$$

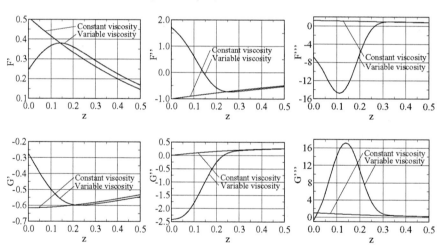

FIGURE 5. The first three derivatives of the dimensionless velocity profiles, F and G, for the constant and variable viscosity cases, with $v_0/v_\infty = 2.255$ and $Sc = 2000$

Here ω is a complex number, with $\Re(\omega)$ and $\Im(\omega)$ being, respectively, the frequency and the rate of growth of the perturbation. The functions f, g, h, π and c depend on the axial coordinate z and the parameters α and β are the components of the perturbation wave-vector along the radial and azimuthal directions. For a given time, the phase of the perturbation is constant along branches of a logarithmic spiral, with the branches curved in the clockwise direction if β/α is positive and counter-clockwise, if negative. The structure turns counter-clockwise if $\Re(\omega)/\beta$ is positive and clockwise, if negative.

The base state and the perturbation variables are rewritten in dimensionless form, introduced in equations (7) and nonlinear terms are dropped. Perturbation variables are not, strictly speaking, separable since the resulting equations for the profiles f, g, h, π and c still contain the radial coordinate r. In order to overcome the problem it is necessary to make the *parallel flow* assumption, where it is assumed that variations of the above profiles are small as far as $\Delta r/r \ll 1$. This approximation holds whenever the stability analysis is carried sufficiently far from $r = 0$. If variations of the profiles with r are small, this coordinate can be assumed as constant. The dimensionless constant r is the Reynolds number at which the stability analysis is carried. This is the parallel flow hypothesis. Adoption of this hypothesis in rotating disk flow [9, 15, 7], is made by replacing r by R.

To conclude, terms of order R^{-2} are dropped, leading to a generalized complex non-symmetric eigenvalue-eigenfunction problem in the form:

$$\begin{pmatrix} A_{11} & A_{12} & A_{13} \\ A_{21} & A_{22} & A_{23} \\ A_{31} & & A_{33} \end{pmatrix} \begin{pmatrix} h \\ \eta \\ c \end{pmatrix} = \omega R \begin{pmatrix} B_{11} & & \\ & B_{22} & \\ & & B_{33} \end{pmatrix} \begin{pmatrix} h \\ \eta \\ c \end{pmatrix} \tag{13}$$

where $\eta = \alpha g - \beta f$, missing elements in the matrices are zero and the operators A_{ij} and B_{ij} are given by:

$$
\begin{aligned}
A_{11} &= a_{114}D^4 + a_{113}D^3 + a_{112}D^2 + a_{111}D + a_{110} & A_{12} &= a_{121}D + a_{120} \\
A_{13} &= a_{132}D^2 + a_{131}D + a_{130} \\
A_{21} &= a_{211}D + a_{210} & A_{22} &= a_{222}D^2 + a_{221}D + a_{220} & A_{23} &= a_{231}D + a_{230} \\
A_{31} &= a_{310} & A_{33} &= a_{332}D^2 + a_{331}D + a_{330}
\end{aligned}
$$

$$B_{11} = D^2 - \bar{\lambda}^2 \qquad B_{22} = 1 \quad B_{33} = iSc$$

where $D^n = d^n/dz^n$. By defining $\lambda^2 = \alpha^2 + \beta^2$, $\bar{\alpha} = \alpha - i/R$ and $\bar{\lambda}^2 = \alpha\bar{\alpha} + \beta^2$ we obtain for the coefficients a_{ijk}:

$$a_{114} = iv \qquad a_{113} = i(2v' - H)$$

$$a_{112} = iv'' - iv\left(\lambda^2 + \bar{\lambda}^2\right) + R\left(\alpha F + \beta G\right) - i\left(H' + F\right)$$

$$a_{111} = -iv'\left(\lambda^2 + \bar{\lambda}^2\right) + iH\bar{\lambda}^2$$

$$a_{110} = i\bar{\lambda}^2\left(v'' + v\lambda^2\right) - R\left(\alpha F + \beta G\right)\bar{\lambda}^2 - R\left(\bar{\alpha}F'' + \beta G''\right) + iH'\bar{\lambda}^2$$

$$a_{121} = 2\left(G + 1\right) \qquad\qquad a_{120} = 2G'$$

$$a_{132} = R\left(\bar{\alpha}F' + \beta G'\right)\gamma \qquad a_{131} = \left[2R\left(\bar{\alpha}F'' + \beta G''\right) + 6i\bar{\lambda}^2 F\right]\gamma + 2R\left(\bar{\alpha}F' + \beta G'\right)\gamma'$$

$$a_{130} = \left[R\bar{\lambda}^2\left(\alpha F' + \beta G'\right) + R\left(\bar{\alpha}F''' + \beta G'''\right) + 4i\bar{\lambda}^2 F'\right]\gamma$$
$$\qquad + \left[2R\left(\bar{\alpha}F'' + \beta G''\right) + 6i\bar{\lambda}^2 F\right]\gamma' + R\left(\bar{\alpha}F' + \beta G'\right)\gamma''$$

$$a_{211} = 2\left(G + 1\right) \qquad\qquad a_{210} = -iR\left(\alpha G' - \beta F'\right)$$

$$a_{222} = iv \qquad a_{221} = i\left(v' - H\right) \qquad a_{220} = -iv\lambda^2 + R\left(\alpha F + \beta G\right) - iF$$

$$a_{231} = iR\left(\alpha G' - \beta F'\right)\gamma \qquad\qquad a_{230} = iR\left[\left(\alpha G'' - \beta F''\right)\gamma + \left(\alpha G' - \beta F'\right)\gamma'\right]$$

$$a_{310} = RSc C' \qquad\qquad a_{332} = -\frac{1}{\bar{v}} \qquad\qquad a_{331} = \frac{1}{\bar{v}}\left(\frac{\bar{v}'}{\bar{v}} + \frac{1}{\bar{v}}C'\gamma + Sc\,\bar{v}H\right)$$

$$a_{330} = iRSc\left(\alpha F + \beta G\right) - \frac{1}{\bar{v}}\left\{-\bar{\lambda}^2 + \frac{1}{\bar{v}}\left[\left(2\frac{\bar{v}'}{\bar{v}}\gamma - \gamma'\right)C' - \gamma C''\right]\right\}$$

where $\gamma = dv/dC$. Boundary conditions of the problem require non-slip flow, vanishing axial component of the velocity and saturation concentration of the chemical species at the electrode surface. These conditions are already fulfilled by the base-state and cannot be modified by the perturbation. In consequence we require $h = \eta = c = 0$ in $z = 0$. Moreover, continuity requires that $h' = 0$ at the interface. In $z \longrightarrow \infty$ we require that the perturbation vanishes $(h = h' = \eta = c = 0)$.

Stationary Neutral and Unstable Level Curves

The neutral stability curves for stationary disturbances $[\Re(\omega) = 0]$, obtained by solving the eigenvalue-eigenfunction problem for $Sc = 2000$ and $v_0/v_\infty = 2.255$, are presented in Figure 6. This figure shows that the coupling enlarges the unstable region of constant viscosity fluids to a wider range of wave-numbers and to a critical Reynolds number of order of 50% of the critical wave number of constant viscosity fluids. Figure 6 shows also the existence of a new family of much more unstable modes. Critical Reynolds numbers of this region are in the range of the ones attained in experimental setups. We refer to the new modes as *chemical* and to modes of the former family as *hydrodynamic*.

The enlarged unstable hydrodynamic region suggests that modes inside this region might possibly have larger growth rates (ω_i) than unstable modes of constant viscosity fluids. Similarly, we could expect that the narrow region of unstable chemical modes

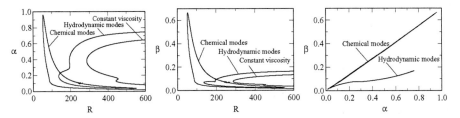

FIGURE 6. Neutral stability curves of stationary perturbations, in the $\alpha \times R$, $\beta \times R$ and $\alpha \times \beta$ planes

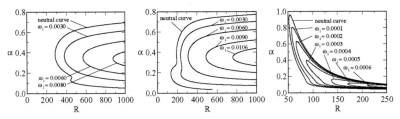

FIGURE 7. Neutral and level curves with positive growth rates (ω_i) of fluids with constant viscosity (on left) and in the unstable region of hydrodynamic (at the center) and chemical modes (on right) of the variable viscosity case

could not allow for the existence of modes with large growth rates. This is the case, indeed. Figure 7 shows neutral and level curves evaluated for positive rates of growth for fluids with constant viscosity on left, for the unstable region of hydrodynamic and chemical modes at the center and right, respectively.

An analysis of Figure 7 shows that growth rates of hydrodynamic modes of variable viscosity fluids attaint values up to 30% larger than in the case of constant viscosity case. However, growth rates of chemical modes are more than one order of magnitude smaller than those of hydrodynamic modes.

Marginally Stable Eigenmodes – Hydrodynamic and Chemical Unstable Regions

Despite of the low growth rate, unstable modes in the chemical region show a combination of properties that seem capable to drive the current oscillations observed in experimental setups. Figure 8 shows the hydrodynamic and chemical concentration components of a mode on the neutral curve of the chemical region (first row) and on the neutral curve of hydrodynamic region (second row). The amplitude of the chemical component is of same order of the hydrodynamic components in the second case but significantly larger in the case of modes in the chemical region. In addition, the chemical component is always confined to a narrow region close to the interface, leading to high values of the interfacial concentration gradient. This gradient seems sufficiently strong to drive detectable current oscillations.

141

Effects of Decoupling the Transport of the Chemical Species and of Variation of the Interfacial Viscosity

Numerical experiments conducted by the authors Mangiavacchi at al, 2007 [10] show that the coupling, and not the viscosity stratification is responsible for the existence of the new chemical region of unstable modes, for the enlargement and reduction of the critical Reynolds number of the hydrodynamic region. The experiment was done by setting operators $A_{13} = A_{23} = 0$ in equation (13).

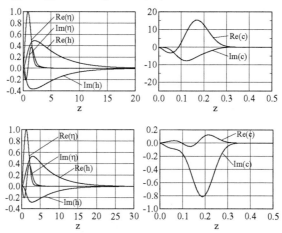

FIGURE 8. Real and imaginary parts of one modes on the neutral curve of the chemical region, at the point $R = 156.99$, $\alpha = 0.20124$, $\beta = 0.13767$, in the first row and of a mode on the neutral curve of the hydrodynamic region, at the point $R = 135.48$, $\alpha = 0.18726$, $\beta = 0.067182$ (second row)

The authors also showed that a progressive increase of the interfacial viscosity results firstly, in a decrease of the overall stability, but, as the viscosity increases the destabilizing effect is inverse and the stability properties of constant viscosity fields are restored. The $v_0/v_\infty = 2.255$ is not the most unstable one. An increase by 5% of the interfacial viscosity enlarges the area of the unstable hydrodynamic region to practically its maximum. A narrow new unstable region already exists with a 2% increase in the interfacial viscosity, but with a critical Reynolds number of order of 1100. As the interfacial viscosity increases the area of the unstable chemical region increases and moves to inner radius. The area attains a maximum at $v_0/v_\infty \approx 1.5$. Further increasing the interfacial viscosity reduces the area of the unstable chemical region. The region disappears close to $v_0/v_\infty = 2.8$. New increases in the interfacial viscosity reduce the hydrodynamic unstable region to the one of constant viscosity fluids.

DIRECT NUMERICAL SIMULATION

The linear stability analysis identifies the distance from the electrode axis, below which perturbations introduced in Von Kármán's solution are damped. Perturbations with geometry within a range that is enlarged as the distance of the axis increases are amplified.

The linear analysis identifies the stability limit but does answer questions on the unstable modes first selected. Selection is made through the nonlinear coupling, when the amplitude of unstable modes is sufficiently large to trigger interaction mechanisms. The nonlinear interaction leads to the selection of one or the few modes of the first emerging unstable pattern. The study of pattern selection is done by following the evolution and the interaction of a small number of prescribed modes, close to a bifurcation point. An alternative method consists in doing a three-dimensional direct numerical simulation. This is the method we currently follow. A FEM code has been developed [5, 12, 16], with the following characteristics: pressure and diffusive terms are treated by a Galerkin method, convective terms are treated in a semi-lagrangean form in order to assure the necessary stability to the code. Time is discretized according to a first order forward differences scheme, leading to a system in the form:

$$\left. \begin{array}{cc} M\left(\dfrac{\mathbf{v}_i^{n+1} - \mathbf{v}_d^n}{\Delta t}\right) + \dfrac{1}{R}K\mathbf{v}^{n+1} - Gp^{n+1} = 0 & \qquad D\mathbf{v}^{n+1} = 0 \\[3mm] M_c\left(\dfrac{c_i^{n+1} - c_d^n}{\Delta t}\right) + \dfrac{1}{R\,Sc}K_c c^{n+1} = 0 & \end{array} \right\} \qquad (14)$$

where M and M_c are mass operators, K and K_c are the momentum and concentration diffusion operators, G and D are the gradient and divergent operators. The hydrodynamics is solved first and the results are used to evaluate the concentration of the chemical species. Upon rewriting the hydrodynamic equations in matrix form, we obtain:

$$\begin{bmatrix} B & -\Delta t\,G \\ D & 0 \end{bmatrix} \begin{bmatrix} \mathbf{u}^{n+1} \\ \mathbf{p}^{n+1} \end{bmatrix} = \begin{bmatrix} \mathbf{r}^n \\ 0 \end{bmatrix} + \begin{bmatrix} \mathbf{bc}_1 \\ \mathbf{bc}_2 \end{bmatrix} \qquad (15)$$

where $B = M + (\Delta t/R)\,K$. The LHS matrix in equation (15) is factorized as follows:

$$\begin{aligned} \begin{bmatrix} B & -\Delta t\,G \\ D & 0 \end{bmatrix} &= \begin{bmatrix} B & 0 \\ D & \Delta t\,DB^{-1}G \end{bmatrix} \begin{bmatrix} I & -\Delta t\,B^{-1}G \\ 0 & I \end{bmatrix} \\[2mm] &\approx \begin{bmatrix} B & 0 \\ D & \Delta t\,DB_L^{-1}G \end{bmatrix} \begin{bmatrix} I & -\Delta t\,B_L^{-1}G \\ 0 & I \end{bmatrix} . \end{aligned} \qquad (16)$$

The resulting linear systems are solved by a projection method. The application of the projection method decouples the velocity and pressure fields computations, resulting in two symmetric and positive definite matrices that are solved by the preconditioned conjugate gradient method. A reverse Cuthill-McKee reordering and an incomplete Cholesky preconditioning scheme are applied to solve the linear systems efficiently. Cubic tetrahedral elements are employed, pressure is evaluated at the vertices and the velocity, at the vertices and the centroid of the element (mini-element). Variables from the previous time step are used to evaluate the chemical species field, decoupling the hydrodynamics from the scalar variable transport.

Figure 9 shows results of the FEM simulations: the velocity profiles for constant viscosity, first figure, and for variable viscosity, second figure. The results agree with the analytical von Kármán solution [13]. The boundary layer for the variable viscosity

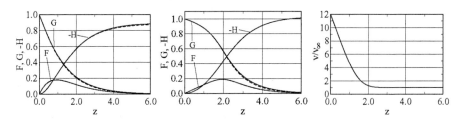

FIGURE 9. Steady hydrodynamic profiles for constant viscosity fluids (on left) and for fluids with the viscosity depending on the axial coordinate z (at the center). Full lines refer to the FEM solution and dashed lines to the solution of Von Kármán's equations. The right panel shows the steady viscosity profile used to evaluate the velocity profiles in the central panel

is thicker than the constant viscosity case, and presents an additional inflection point, increasing thus the instability of the profile.

The FEM code is now validated and will be employed for the simulation of the nonlinear evolution of the most unstable modes found in the linear analysis.

REFERENCES

1. O. E. Barcia, O. R. Mattos, and B. Tribollet, *J. Eletrochem. Soc.* **139**, 446–453 (1992).
2. M. J. C. Esteves, M. J. E. M. Cardoso, and E. Barcia, *Ind. Eng. Chem. Res.* **40**, 5021–5028 (2001).
3. A. B. Geraldo, O. E. Barcia, O. R. Mattos, F. Huet, and B. Tribollet, *Electrochim. Acta* **44**, 455–465 (1998).
4. N. Gregory, J. T. Stuart, and W. S. Walker, *Phil. Trans. Roy. Soc. London* **A-248**, 155–199 (1955).
5. T. Hughes and A. Brooks, *A Theoretical Framework for Petrov-Glaerkin Methods with Discontinuous Weighting Functions: Application to the Streamline Upwind Procedure*, Wiley, 1982.
6. V. G. Levich, *Physicochemical Hydrodynamics*, Prentice Hall, Englewood Cliffs, NJ, 1962.
7. R. J. Lingwood, *J. Fluid Mech.* **299**, 17–33 (1995).
8. M. R. Malik, *J. Fluid Mech.* **164**, 275–287 (1986).
9. M. R. Malik, Wilkinson, and S. A. Orzag, *AIAA J.* **19**(9), 1131–1138 (1981).
10. N. Mangiavacchi, J. Pontes, O. E. Barcia, O. E. Mattos, and B. Tribollet, *Phys. Fluids* **19**, 114109 (2007).
11. F. Moisy, O. Doaré, T. Passuto, O. Daube, and M. Rabaud, *J. Fluid Mech.* **507**, 175–202 (2004).
12. O. Pironneau, *Finite Element Methods for Flows*, Wiley, 1989.
13. H. Schlichting and K. Gersten, *Boundary Layer Theory*, Springer, Berlin, 1999.
14. N. Smith, "Exploratory investigation of laminar boundary layer oscillations on a rotating disk," Technical Report TN-1227, NACA, Dec. 1946.
15. S. Wilkinson and M. R. Malik, *AIAA J.* **23**, 588 (1985).
16. O. C. Zienkiewicz and R. L. Taylo, *The Finite Element Method for Fluids Dynamics*, Wiley, 2000, 5th edn.

Use of Particular Analytical Solutions for Calculations of Nonlinear PDEs

A. Porubov*, D. Bouche[†] and G. Bonnaud**

*Institute of Problems of Mechanical Engineering, Russia
[†]CMLA, ENS de Cachan, Cachan, France
**CEA, INSTN, Centre de Saclay, France

Abstract. It is shown, how analytical solutions of non-linear partial differential equations (PDE) may be applied for a design of their numerical solutions are considered. First, explicit particular exact and asymptotic solutions predict important features of the waves behavior even outside their formal applicability realized in numerics. Thus a prediction may be done whether an arbitrary initial condition splits into the train of localized waves, and each of these waves in numerical solution is described by the analytical travelling wave solution. This happens both for the bell-shaped and kink-shaped localized waves. Second application concerns the study of numerical dispersion and dissipation using the method of differential approximation. This method allows us to study defects of a discrete scheme by an analysis of a non-linear PDE obtained from the scheme. In particular, exact solutions of this equation help us to improve the scheme diminishing unwanted effects.

Keywords: Nonlinear equation, exact solution, method of differential approximation.
PACS: 04.30.Nk, 04.20.Jb, 02.70.Bf, 05.45.Yv

INTRODUCTION

Analytical solutions of most of non-linear partial differential equations (PDE) is difficult to obtain since the latter are usually non-integrable. As a rule only particular solutions (e.g., travelling wave solutions) may be found mainly in the 1D case. At the same time, numerical solutions of nonlinear PDE may be obtained in the general 3D case. However, a solution of a non-linear equation is very sensitive to the values of the equation coefficients and to the initial conditions; this may be missed using only numerical modeling. Moreover, some numerical results look rather unusual, and its difficult to decide whether they are correct or caused by a defect of a scheme. Therefore numerical results require a confirmation, that may be done through comparing with analytical solutions. But may even particular solutions predict and explain real behavior of a numerical solution? Is it possible to improve a scheme for a nonlinear equation using particular analytical solutions?

To answer these questions, one has to note that full comparison with analytical solutions is impossible, and only some features of the numerical solution may be checked. In particular, this is the case of localization when an initial disturbance evolves into a sequence of localized bell-shaped travelling waves or transformation to a travelling shock wave with permanent slope and velocity. This happens due to the balances between various factors kept in the governing PDE. There exist analytical solutions that arise due to the balances, and their finding may help to find the conditions of the wave localization. These are the solutions that keep their shape on propagation: bell-shaped

CP1067, *Applications of Mathematics in Engineering and Economics '34—AMEE '08*, edited by M. D. Todorov
© 2008 American Institute of Physics 978-0-7354-0598-1/08/$23.00

145

solitary wave (balance between nonlinearity and dispersion) and shock wave or a kink (balance between nonlinearity and dissipation). The conditions of the existence of the exact solution allows us to predict the type of the initial condition that may produce or not the localized waves in numerical solution, in particular, it may be the suitable sign of the amplitude. Also the resulting waves of permanent form in the numerical solution may be compared with analytical travelling wave solutions to confirm the validness of the numerical solution.

Some recently obtained results for the bell-shaped solutions [3, 4] illustrate above mentioned ideas. Thus, the exact solutions were employed in [5, 6, 7] to predict an appearance of localized strain waves governed by the equation

$$v_{tt} - a\,v_{xx} - c_1\,(v^2)_{xx} - c_2\,(v^3)_{xx} + \alpha_3\,v_{xxtt} - \alpha_4\,v_{xxxx} = 0. \tag{1}$$

The exact travelling wave solutions to equation (1) give rise to the conditions of the wave localization in dependence of the equation coefficients. In particular, the conditions were found when the waves of opposite signs of the amplitude, or compression and tensile strain waves, may coexist. Numerical simulations on an arbitrary pulse evolution, performed in [5, 6, 7], confirmed theoretical results both qualitatively and quantitatively. Namely, both the positive and negative amplitude inputs give rise to the localized waves provided that theoretical restrictions on the equation coefficients are satisfied. Each generated solitary wave is found to propagate according to the exact solution. No no localization of the waves was observed when the analytical conditions of the localized wave existence are not met. The validness of the single wave solution prediction was found in [7] even in the presence of the solitary waves interaction.

Asymptotic solutions [3, 4, 8, 9] allow us to describe even more complicated wave behavior. In particular, the selection of localized waves was studied in [3, 8] for the equation

$$u_t + u\,u_x + d\,u_{xxx} = \varepsilon f(u), \tag{2}$$

where f is defined as

$$f(u) = -\left(a_1 u - a_2 u^2 + a_3 u^3\right), \tag{3}$$

a_1, a_2, a_3 are positive constants and ε is a small parameter. The asymptotic solution of equation (2) was obtained in [8] in the form of a travelling solitary wave whose parameters slowly vary with $T = \varepsilon\,t$. An analysis of the solution allowed to reveal the selection of the wave when an increase and/or decrease in the amplitude happens by some finite value fully defined by the equation coefficients. Based on this particular solution, numerical simulations were performed in [8] where it was found first that theoretical domains of the values of the initial wave amplitudes define localization ro a decay of the solitary waves. Those waves that are not decayed, evolve into the waves described by the asymptotic solution. An arbitrary input splits into a train of solitary waves each evolving according to the asymptotic solution. Also the solitary waves interaction does not affect a selection of solitary waves. Finally, it happened that the selection of the solution to the theoretical value of the amplitude is realized even when the parameter ε is not small, i.e., outside the formal applicability of the asymptotic solution. Similar findings were obtained in [9] for another equation governing the strain waves in a medium with microstructure. These analytical results may be also obtained using the conservation laws like in Ref. [1].

The nonlinearity, dispersion and dissipation in the problems mentioned above are caused by physical factors. At the same time, the scheme itself possesses dispersive and/or dissipative properties that affect the numerical solution of the physical problem. It turns out that these features of the discrete equations may be analyzed using the solution of the PDEs, and this will be considered further in the paper. However, the first section will be devoted to the use of the kink-shaped solution of both the single and coupled PDEs for the prediction of the wave behavior in numerical solutions. Later these results will be extended to the description of the dispersive features of the schemes using the method of differential approximation. An approach will be developed to add artificial nonlinear terms in the scheme to suppress its dispersion. The choice of nonlinear terms is based on the analytical results obtained in the beginning of the paper for the single and coupled PDEs.

EXACT SOLUTIONS VS NUMERICAL SOLUTIONS

Kink-shaped Solutions to the Single Equations

Let us begin with the celebrated Burgers equation,

$$u_t + u\, u_x + b\, u_{xx} = 0, \tag{4}$$

Its known shock-wave solution (or a kink) at the boundary conditions $u_\infty = 0, u_{-\infty} > 0$ reads

$$u_0 = \frac{u_{-\infty}}{2}\left\{1 - \tanh[p(X - V\,t)]\right\}. \tag{5}$$

where p and V are defined by the boundary conditions as

$$V = \frac{u_{-\infty}}{2},\ p = -\frac{u_{-\infty}}{4b}. \tag{6}$$

This solution accounts for a smooth transition from the undisturbed state to the state $u_{-\infty}$. The solution requires specific initial condition in the form of (5) at $t = 0$. However, an initial shock with a slope different from that of (5) nevertheless evolves into the latter as shown in Figure 1. It happens because the solution (5) arises as a result of the simultaneous action of nonlinearity that tends shock wave to break and dissipation that smoothes the profile of the shock. These numerical results are possible to find analytically since the Burgers equation is integrable. Only numerical simulations allow one to study the same processes for nonintegrable equations. Now its to be noted that the particular travelling wave solution (5) predicts both qualitatively and quantitatively the shape and the velocity of the wave arising from rather arbitrary input.

The Korteweg-de Vries-Burgers (KdVB) equation generalizes the Burgers equation by an addition of the dispersion term

$$u_t + u\, u_x + b\, u_{xx} + s\, u_{xxx} = 0, \tag{7}$$

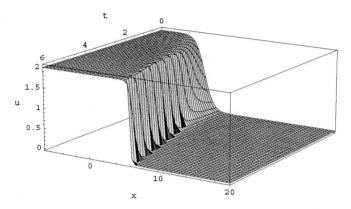

FIGURE 1. Transformation of the initial shock to the Burgers shock wave (5)

that destroys the integrability. Nevertheless, equation (7) possesses known exact travelling kink-shaped solution,

$$u = 12\, s\, k^2\, \text{sech}^2[k(x - Vt)] - \frac{12\, b\, k}{5}\{1 - \tanh[k(x - Vt)]\}, \tag{8}$$

whose parameters are *uniquely defined* by the coefficients of the KdVB equation,

$$k = -\frac{b}{10s},\, V = \frac{6b^2}{25\, s}.$$

and the boundary conditions cannot be satisfied like (6) for any $u_{-\infty}$. As a result, another type of profile arises in the numerical solution shown in Figure 2. Instead of a smooth shock a shock with a bump on the upper side of the wave front appears. This wave propagates keeping its shape and velocity but it is not described by the exact solution, the last accounts for a smooth shock like the solution (5).

One can suggest that deviations in the shape presented in Figure 2 are caused by the absence of free parameters in the solution (8). As noted before, the shock wave of permanent shape and velocity exists due to a balance between nonlinearity and dissipation. The balance in equation (7) may be recovered if an additional nonlinear term is to suppress the dispersion term $s\, u_{xxx}$ like the nonlinear term in the Burgers equation compensates the influence of dissipation term $b\, u_{xx}$ on the profile of the shock. Then an improved equation reads

$$u_t + u\, u_x + b\, u_{xx} + s\, u_{xxx} + \gamma\, u_{xx}^2 = 0, \tag{9}$$

Direct substitution of equation (5) into equation (9) confirms the existence of this exact smooth kink-shaped solution provided that

$$\gamma = -\frac{s}{2(b - 2p\, s)},\, p = \frac{b \pm \sqrt{b^2 + 2s\, u_{-\infty}}}{4s},\, V = \frac{u_{-\infty}}{2} \tag{10}$$

148

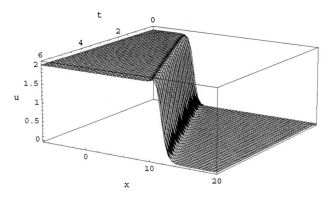

FIGURE 2. Evolution of the initial shock of the KdVB equation

Numerical simulations of equation (9) confirms the suggestion about the role of the free parameters in the exact solution. Now the evolution to the exact solution takes place similar to that shown in Figure 1.

Coupled Equations

The isothermal coupled compressible Navier-Stokes equations (NSE) read

$$\rho_t + (\rho\, u)_x = 0, \tag{11}$$

$$(\rho\, u)_t + (\rho\, u^2 + a^2\, \rho)_x - \nu(\rho\, u_x)_x = 0. \tag{12}$$

Its exact travelling wave solution should satisfy the following boundary conditions,

$$\rho \rightarrow \rho_{\pm\infty} \quad \text{at } x \rightarrow \pm\, \infty, \tag{13}$$

$$u \rightarrow 0 \text{ at } x \rightarrow \infty,\, u \rightarrow u_{-\infty} \text{ at } x \rightarrow -\, \infty. \tag{14}$$

Transformation to the phase variable $\theta = x - V\, t$ gives rise to the relationship between ρ and u following from equation (11)

$$\rho = \frac{\rho_\infty V}{V - u}. \tag{15}$$

Substitution of the last expression into equation (12) yields the equation for u whose solution is obtained by direct integration in the form

$$u = \frac{u_{-\infty}}{2}\, \{1 - \tanh[k\, (X - V\, t)]\}, \tag{16}$$

where V and k are defined by

$$k = \frac{u_{-\infty}}{2\, \nu},\, V = \frac{u_{-\infty} + \sqrt{4a^2 + u_{-\infty}^2}}{2}. \tag{17}$$

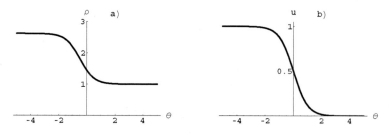

FIGURE 3. Typical profiles of the exact solution to the isothermal Navier-Stokes equations

Typical profiles for u and ρ are shown in Figure 3. This solution also possesses free parameters defined by the boundary conditions, then it should arise in an unsteady process like the solution (5) of the Burgers equation. The solution obtained does not exist for the NSE modified by dispersion,

$$\rho_t + (\rho\, u)_x = \gamma_1 \rho_{xxx} + \gamma_2 (\rho\, u)_{xxx}, \tag{18}$$

$$(\rho\, u)_t + (\rho\, u^2 + a^2\, \rho)_x - v(\rho\, u_x)_x = \gamma_3 \rho_{xxx} + \gamma_4 (\rho\, u)_{xxx}. \tag{19}$$

unless $\gamma_1 = -\gamma_2 V$, $\gamma_3 = -\gamma_4 V$. This is similar to non-existence of the solution (5) for the KdVB equation since the velocity is fixed by the equation coefficients. Then the idea of adding terms to compensate the dispersion may be employed again. The modification is not so obvious as for the KdVB equation, one can try it as

$$\rho_t + (\rho\, u)_x = \gamma_1 \rho_{xxx} + \gamma_2 (\rho\, u)_{xxx} + \alpha_1 \rho + \alpha_2 \rho^2 + \alpha_3 \rho^3, \tag{20}$$

$$(\rho\, u)_t + (\rho\, u^2 + a^2\, \rho)_x - v(\rho\, u_x)_x = \gamma_3 \rho_{xxx} + \gamma_4 (\rho\, u)_{xxx} + \beta_1 \rho + \beta_2 \rho^2 + \beta_3 \rho^3. \tag{21}$$

In this case the exact solution (15), (16) exists provided that

$$\alpha_1 = \frac{4k^2}{u_{-\infty}^2} \left(6V[u_{-\infty} - V] - u_{-\infty}^2\right)(\gamma_1 + \gamma_2\, V),$$

$$\alpha_2 = \frac{12k^2}{\rho_{-\infty}\, u_{-\infty}^2} \left(V[2V - 3u_{-\infty}] + u_{-\infty}^2\right)(\gamma_1 + \gamma_2\, V),$$

$$\alpha_3 = -\frac{8k^2\, (V - u_{-\infty})^2}{\rho_{-\infty}^2\, u_{-\infty}^2}(\gamma_1 + \gamma_2\, V),$$

with V and k defined by (17). Expressions for $\beta_i, i = 1 \div 3$, are obtained by changing γ_1 by γ_3 and γ_2 by γ_4 in corresponding expressions for α_i.

IMPROVEMENT OF THE DIFFERENCE SCHEMES

As noted before, a discrete model itself may possess internal dispersive and/or dissipative properties caused by a method of discretization. It gives rise to non-physical deviations in the numerical solution. The method of differential approximation [11, 2] allows us to study dispersive and dissipative features of a discrete equation by analyzing the *differential* equation called the differential approximation (DA) of the scheme.

Method of Differential Approximation

Consider first the schemes for the Burgers equation. According to the method of differential approximation the Taylor expansions of the discrete function is substituted into the *difference* scheme. However, instead of the strict continuum limit, some terms proportional to the powers of the spacial, $\triangle x$, and the temporal, $\triangle t$, steps are left. Truncating these series to some order, another differential equation appears, and it is called the DA. In particular, according to the Lax-Wendroff (LW) scheme, the differential approximation is written as [11]

$$u_t + u\,u_x + bu_{xx} = \frac{\triangle t^2}{24}(u^4)_{xxx} - \frac{\triangle x^2}{12}(u^2)_{xxx}, \qquad (22)$$

when only terms proportional to the square of the steps are taken into account. Certainly, the Burgers equation arises in the strict continuum limit when both $\triangle x$ and $\triangle t$ tend to zero.

For the NSE the DA is written for the LW scheme in the form [11]

$$w_t + f(w)_x = \frac{\triangle t^2}{6}\left(f_{ww}\,f_x^2 + f_w(f_w\,f_x)_x\right)_x - \frac{\triangle x^2}{6}f_{xxx}, \qquad (23)$$

with

$$w = \begin{pmatrix} \rho \\ \rho u \end{pmatrix}, \quad f = \begin{pmatrix} \rho u \\ \rho u^2 + a^2\rho \end{pmatrix},$$

f_w, f_{ww} denote differentiation with respect to w. Again the NSE arise in the strict continuum limit.

Improvement of the Scheme Dispersion for the Burgers Equation

One can see that DA for a dicretization of a *non-linear* equation is also non-linear and nonintegrable equation whose analysis is difficult. A simplification may be done linearizing the r.h.s. parts proportional to the steps. Thus, for the DA of the Burgers equation (22) it may be done around the value of $u = u_{-\infty}$. Then equation (22) turns out the KdVB equation (7) with

$$s = \frac{\left(\triangle x^2 - \triangle t^2 u_{-\infty}^2\right)u_{-\infty}}{6}. \qquad (24)$$

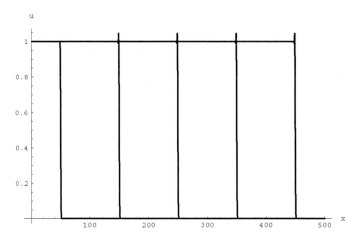

FIGURE 4. Temporal evolution of the Burgers shock wave (5) according to the LW scheme

Now it is clearly seen that the last term in equation (22) accounts for dispersion, and also $s > 0$ if the standard criterium of stability holds. Numerical simulation of the Burgers equation by the LW scheme yields the evolution shown in Figure 4. One can see a bump arising at the upper part of the wave front like it happens for the evolution of a shock for the KdVB equation shown in Figure 2. It was found in previous section that addition of a nonlinear term to equation (7) suppresses the bump which leads to the formation of the smooth shock shown in Figure 1. It may be used now to avoid the defect of the LW scheme. First, the solution of the DA with linearized r.h.s. may be improved by adding the nonlinear term $\gamma\, u_{xx}^2$ like equation (7). Then the coefficient γ is calculated using equations (10), (24). After that the LW scheme is modified by the addition of the discrete analog of this nonlinear term. It turns out that even central discretization is enough to achieve the smooth profile shown in Figure 5. Note that the value of γ should be defined using the analytical solutions, any deviations in it yield non-suppression of the bump at the shock wave profile.

Isothermal Navier-Stokes equations

Application of the procedure developed in the previous subsection first yields the linearization of the r.h.s. of equations (23) yields equations (18), (19) with

$$\gamma_1 = (a^2 - u_{-\infty}^2)u_{-\infty}\Delta t^2/3, \quad \gamma_2 = (a^2 + 3u_{-\infty}^2)\Delta t^2/6 - \Delta x^2/6, \qquad (25)$$

$$\gamma_3 = [(a^4 - 3\,u_{-\infty}^4 + 2a^2\,u_{-\infty}^2)\Delta t^2 - a^2\Delta x^2]/6,$$

$$\gamma_4 = [4(a^2 + u_{-\infty}^2)\Delta t^2 - \Delta x^2]u_{-\infty}/6 \quad (26)$$

As found in a previous section, an addition of nonlinear terms is needed to achieve the suitable shock wave solution for the DA. The coefficients of the terms introduced like in

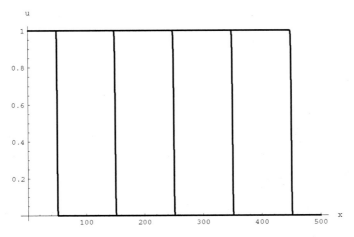

FIGURE 5. Temporal evolution of a shock wave profile according to the LW scheme improved by an addition of artificial non-linear term described by equation (9)

equations (20), (21), are calculated using equations (25), (26), definitions of γ_i and the boundary conditions. Then, the corresponding discretizations of the artificial nonlinear terms should be added into the LW scheme to obtain the smooth shock wave propagation for the Navier-Stokes equations.

CONCLUSIONS

Particular exact solutions may predict localization of a wave (or corresponding balances governed by an equation) that happens in an unsteady numerical solution. The conditions of localization may be obtained in an explicit form that allows us to predict the domain of the values of the coefficients of the governing equations when the balance is realized and localization happens. It helps to describe physical factors responsible for the localization of the wave. In elasticity, these are the elastic properties of a material of a wave guide [3, 5].

Factors affecting the balances may not be of physical nature but be caused by the defects of a scheme. In this case the particular exact solution may be used to improve the scheme. In the last case the conditions of the existence of exact solution help us to add suitable artificial additional terms to the scheme to avoid these defects, in particular, the scheme dispersion.

Asymptotic solutions also may be used to analyze the dispersive (and dissipative) features of a scheme, in particular, the bump shown in Figure 4. It is possible to establish a link between the height of the bump and the scheme steps using the asymptotic solution. Various schemes were considered in our previous work [10] where besides the Lax-Wendroff scheme, the Warming-Beam, Mac-Cormack schemes with dispersion were studied for the Burgers equation. Also the schemes with dissipate defects were considered.

153

Our future work is focused on the study of the schemes used for the full Navier-Stokes equations.

ACKNOWLEDGMENTS

This work of one of the authors (AVP) was supported by the CEA/DAM grant 0611S/DIR and by the grant of the Russian Science Support Foundation.

REFERENCES

1. C.I. Christov and M.G. Velarde, *Physica D* **86**, 323–347 (1995).
2. S.I. Mukhin, S.B. Popov, and Yu. P. Popov, *Numerical Math. and Math. Phys.* **6**, 45–53 (1983).
3. A.V. Porubov, *Amplification of Nonlinear Strain Waves in Solids,* World Scientific, Singapore, 2003.
4. A.V. Porubov, *Rendiconti del Seminario Matematico dell'Universita' e Politecnico di Torino* **65**, 217–230 (2007).
5. A.V. Porubov and G.A. Maugin, *Intern. J. Non-Linear Mech.* **40**, 1041–1048 (2005).
6. A.V. Porubov and G.A. Maugin,*Phys. Rev. E.* **74** 046617 (2006).
7. A.V. Porubov and G.A. Maugin,*Journal of Sound and Vibration* **310/3**, 694–701 (2008).
8. A.V. Porubov, V.V. Gursky, and G.A. Maugin, *Proc. Estonian Acad. Sci., Phys. Math.* **52**, 85–93 (2003).
9. A.V. Porubov and F. Pastrone, *Intern. J. Non-Linear Mech.* **39**, 1289–1299 (2004).
10. A.V. Porubov, D. Bouche, and G. Bonnaud, *International Journal of Finite Volumes* **5**, 1–16 (2008) (see http://www.latp.univ-mrs.fr/IJFV/.)
11. Yu. Shokin, *The Method of Differential Approximation*, Springer, Berlin, 1983.

On Numerical Simulation of Propagation of Solitons in Microstructured Media

A. Salupere, L. Ilison and K. Tamm

Center for Nonlinear Studies, Institute of Cybernetics at Tallinn University of Technology,
Dept. of Mechanics, Tallinn University of Technology,
Akadeemia tee 21, 12618, Tallinn, Estonia

Abstract. Wave propagation in microstructured media is simulated numerically making use of two different models. In the first case a Korteweg–de Vries type equation is used for modeling 1D wave motion in granular materials. In the second case a Boussinesq type equation is applied for modeling 1D wave motion in Mindlin type microstructured solids. Both equations are integrated numerically under localized initial conditions by employing the discrete Fourier transform (DFT) based pseudospectral method. Emergence of trains of solitons is demonstrated in both cases.

Keywords: Solitons, Korteweg-de Vries type equation, Boussinesq type equation, pseudospectral method, granular materials, microstructured solids.
PACS: 05.45.Yv, 02.60.Cb

INTRODUCTION

Microstructured materials (alloys, ceramics, functionally graded materials, etc) are characterised by various scales of microstructure. This circumstance should be taken into account when wave propagation in such materials is modeled [1, 2, 3]. The scale-dependence involves dispersive as well as nonlinear effects and it is widely known, that the balance between these two effects may result in emerging of solitary waves and solitons.

Two examples of microstructured materials are considered in the present paper. In the first case one-dimensional wave propagation in dilatant granular materials is studied. Corresponding model equation

$$u_t + uu_x + \alpha_1 u_{xxx} + \beta \left(u_t + uu_x + \alpha_2 u_{xxx} \right)_{xx} = 0 \tag{1}$$

is derived by Giovine and Oliveri [4]. Here variable u is bulk density, x — space coordinate, t — time; α_1 and α_2 are macro- and microlevel dispersion parameters, respectively, and β is a parameter involving the ratio of the grain size and the wavelength. Equation (1) consists of two Korteweg–de Vries (KdV) operators: the first describes the motion in the macrostructure and the second (in the brackets) — the motion in the microstructure. Equation (1) is clearly hierarchical in the Whitham's sense — the parameter β controls the influence of the microstructure [5]. Due to that kind of hierarchy the equation (1) could by called hierarchical Korteweg–de Vries (HKdV) equation.

In the second case a Boussinesq-type [6] equation

$$v_{tt} - bv_{xx} - \frac{\mu}{2} \left(v^2 \right)_{xx} - \delta \left(\beta v_{tt} + \gamma v_{xx} \right)_{xx} + \delta^{3/2} \frac{\lambda}{2} \left(v_x^2 \right)_{xxx} = 0 \tag{2}$$

CP1067, *Applications of Mathematics in Engineering and Economics '34—AMEE '08*, edited by M. D. Todorov
© 2008 American Institute of Physics 978-0-7354-0598-1/08/$23.00

is used in order to model one-dimensional wave propagation in microstructured solids. Here v is deformation, x — space coordinate, t — time, b, μ, δ, β, and γ are material parameters (see [7, 8] for details). In order to derive equation (2) the Mindlin model [9] of continua with microstructure and hierarchical approach by Engelbrecht and Pastrone [1] is applied. This equation is referred below as the MEP equation.

The main goal of the present paper is to simulate numerically the emergence of trains of solitons by means of two different material models. For this reason model equations (1) and (2) are integrated numerically under localised initial conditions. Results are analysed in terms of solitonics, i.e., the goal is to understand whether or no solitary waves that emerge from initial sech^2-type pulse propagate at constant speed and amplitude and interact elastically. Interaction of solitary waves is called elastic, if they restore their speed and amplitude after interactions. During elastic interaction solitons may experience phase shift. In turn, if interactions between solitary waves are elastic they are called solitons.

NUMERICAL TECHNIQUE

In the present paper the discrete Fourier transform (DFT) based pseudospectral method (PsM) [10, 11] is used for numerical integration of model equations, i.e., for numerical simulation of wave propagation. Let us introduce the DFT and the inverse DFT (IDFT) as follows:

$$U(k,t) = Fu = \sum_{j=0}^{N-1} u(j\Delta x,t)\exp\left(-\frac{2\pi i jk}{N}\right),\tag{3}$$

$$u(j\Delta x,t) = F^{-1}U = \frac{1}{N}\sum_{k} U(k,t)\exp\left(-\frac{2\pi i jk}{N}\right).\tag{4}$$

Here $u(x,t)$ is a periodic function (with space period 2π, i.e., $0 \leq x \leq 2\pi$), N – the number of space grid points, $\Delta x = 2\pi/N$ – the space step, i – the imaginary unit, $k = 0, \pm1, \pm2, \ldots, \pm(N/2 - 1), -N/2$ are wavenumbers, F denotes the DFT and F^{-1} the IDFT. Space derivatives of function $u(x,t)$ can now be calculated making use of properties of the DFT [12, 13]:

$$\frac{\partial^n u(x,t)}{\partial x^n} = F^{-1}\left[(ik)^n Fu\right].\tag{5}$$

If the length of the space period for $u(x,t)$ is not 2π, but $2m\pi$, then one must use quantity k/m instead of k in last formulae:

$$\frac{\partial^n u(x,t)}{\partial x^n} = F^{-1}\left[\left(\frac{ik}{m}\right)^n Fu\right].\tag{6}$$

In a nutshell the idea of the PsM is very simple. If a PDE is given in a general form

$$u_t = \Phi(u, u_x, u_{xx}, \ldots)\tag{7}$$

156

then one can formally reduce the latter to ODE

$$u_t = \Psi(u) \tag{8}$$

making use of formulae (6), which can be considered as numerical differential operators. The ODE (8) can now be integrated with respect to time variable t by standard ODE solvers. The method is called pseudospectral because integration with respect to time is carried out in physical space and the Fourier transform related quantities are used only for calculation of space derivatives.

The HKdV equation (1) includes mixed partial derivative term βu_{xxt}. Therefore the usual PsM algorithm cannot be applied directly and one has to introduce suitable change of variables. At first the HKdV equation is rewritten in the form

$$(u + \beta u_{xx})_t + (u + 3\beta u_{xx})u_x + (\alpha_1 + \beta u)u_{xxx} + \beta\alpha_2 u_{xxxxx} = 0 \tag{9}$$

and a variable

$$\phi = u + \beta u_{xx} \tag{10}$$

is introduced. Making use of the DFT and its properties the last expression can be rewritten in form

$$\phi = F^{-1}(Fu) + \beta F^{-1}(-k^2 Fu) = F^{-1}\left[(1 - \beta k^2) Fu\right]. \tag{11}$$

In turn, the variable u and its space derivatives can be expressed from (11) in terms of variable ϕ:

$$u = F^{-1}\left[\frac{F\phi}{1 - \beta k^2}\right], \qquad \frac{\partial^n u}{\partial x^n} = F^{-1}\left[\frac{(ik)^n F\phi}{1 - \beta k^2}\right]. \tag{12}$$

It is clear, that in order to avoid division by zero, one can consider only such values for parameter β which result in $1 - \beta k^2 \neq 0$. Finally equation (9) is rewritten in the form

$$\phi_t = -(u + 3\beta u_{xx})u_x - (\alpha_1 + \beta u)u_{xxx} - \alpha_2\beta u_{xxxxx}. \tag{13}$$

where the variable u and its space derivatives could be expressed in terms of ϕ according to formulae (12). Therefore equation (13) can be considered as an ODE with respect to variable ϕ and could be integrated numerically making use of standard ODE solvers.

The MEP equation (2) includes a mixed partial derivative term $\delta\beta v_{xxt}$ and therefore it is rewritten in the form

$$(v - \delta\beta v_{xx})_{tt} - bv_{xx} - \frac{\mu}{2}(v^2)_{xx} + \delta\gamma v_{xxxx} + \delta^{3/2}\frac{\lambda}{2}(v_x^2)_{xxx} = 0 \tag{14}$$

and new variable

$$\phi = v - \delta\beta v_{xx} = F^{-1}(Fv) - \delta\beta F^{-1}(-k^2 Fv) = F^{-1}[(1 + \delta\beta k^2)Fv] \tag{15}$$

is introduced. From the latter variable v and its space derivatives can be expressed in terms of variable ϕ:

$$v = F^{-1}\left[\frac{F\phi}{1 + \delta\beta k^2}\right], \qquad \frac{\partial^n v}{\partial x^n} = F^{-1}\left[\frac{(ik)^n F\phi}{1 + \delta\beta k^2}\right]. \tag{16}$$

Finally we can rewrite equation (14) in the form

$$\phi_{tt} = \left[bv + \frac{\mu}{2}v^2 - \delta\gamma v_{xx} - \delta^{3/2}\frac{\lambda}{2}\left(v_x^2\right)_x \right]_{xx} \qquad (17)$$

where variable v and its space derivatives are expressed through new variable ϕ making use of expressions (16).

RESULTS AND DISCUSSION

Emergence of trains of solitons from localised initial excitation will be discussed in the present section in case of the HKdV as well as in case of the MEP equations. For this reason both equations are integrated numerically by PsM under sech^2-type initial conditions and periodic boundary conditions.

The HKdV Equation

The HKdV (1) is integrated under initial condition

$$u(x,0) = A\operatorname{sech}^2\frac{x-x_0}{\delta}, \qquad \delta = \sqrt{\frac{12\alpha_1}{A}}, \qquad 0 \le x \le 16\pi, \qquad x_0 = 8\pi, \qquad (18)$$

where A is the amplitude and δ the width of the initial pulse. It is clear that the latter is the analytical solution of KdV equation that corresponds to the first KdV operator in equation (1).

Numerical experiments are carried out for $0 < \alpha_1 < 1$, $0 < \alpha_2 < 1$ and $\beta = 111.11, 11.111, 1.111, 0.111, 0.0111$. These particular values of β are selected in order to to avoid division by zero in expressions (12). The number of space grid points is $n = 1024$ and the length of the time interval is $t_f = 100$. Ilison and Salupere [14, 15, 16] have demonstrated that depending on values of material parameters α_1, α_2 and β solutions of different types can emerge from initial sech^2-shape wave. Here we demonstrate a case where the solution type is a train of KdV solitons. This solution type appears for $\alpha_2 < \alpha_1$ in cases of $\beta = 111.11$ and $\beta = 11.111$, i.e., in cases when the influence of the microstructure is relatively strong but microlevel dispersion parameter is smaller than that of the macrolevel.

In the present paper the following values of material parameters are considered: $\alpha_1 = 0.4$, $\alpha_2 = 0.01$ and $\beta = 111.11$. Amplitude of the initial pulse is $A = 4$. Time-slice plot (Figure 1) and pseudocolor plot (Figure 2) demonstrate the formation of train of solitons in the beginning of integration interval and subsequent interactions between emerged solitons. In Figure 3 the initial wave profile at $t = 0$ and waveprofile at $t = 18.7$ (just before emerged solitons start to interact) is plotted against space coordinate x. In this figure eight solitons can be clearly detected. However, the formation of soliton train is not finished at $t \approx 18.7$ — Figure 2 demonstrates that trajectory of the ninth soliton appears only for $t > 35$. Analysis of numerical results demonstrate that the amplitude of

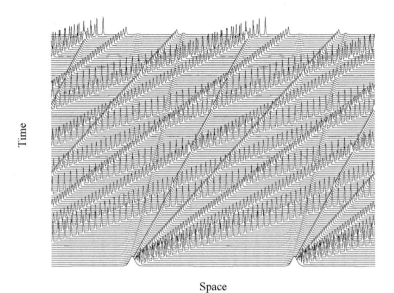

FIGURE 1. Emergence of train of KdV solitons in case of the HKdV model — time-slice plot over two space periods

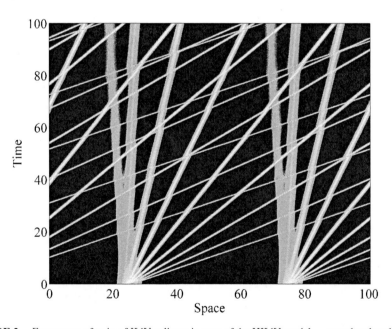

FIGURE 2. Emergence of train of KdV solitons in case of the HKdV model — pseudocolor plot over two space periods

159

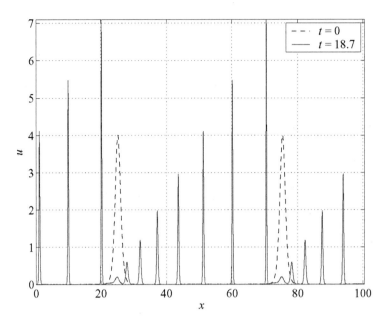

FIGURE 3. Initial wave and train of KdV solitons at $t = 18.7$ over two space periods in case of the HKdV model

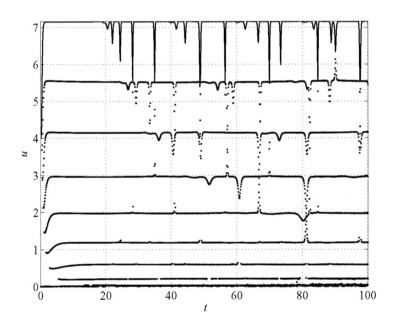

FIGURE 4. Wave profile maxima against time in case of the HKdV model

the ninth soliton is of order of 0.025 and therefore it is hard to distinguish it in time-slice plot in Figure 1 or in amplitude curves in Figure 4. Furthermore, a very small amplitude tail can be detected besides the soliton train. The behaviour of the emerged solitons is practically identical to these emerged in case of KdV equation (see Figures 1–4): (i) the higher solitons in the train (three higher solitons in the present case) are higher than the initial pulse, (ii) the higher the soliton the higher its speed, (iii) if two solitons interact then the amplitude (height) of the higher soliton decreases during interactions, (iv) after interactions solitons restore their initial amplitude and speed. For these reasons the considered type of solution is called train of KdV solitons.

For other sets of material parameters (from the same domain, i.e., $\alpha_2 < \alpha_1$ and $\beta = 111.11$ or $\beta = 11.111$) the tail that forms besides the train of solitons can be more distinctive, but it does not influence the behavior of emerged solitons essentially — it only causes small amplitude oscillations of amplitudes of solitons. Nevertheless, the character of the interactions remains practically elastic because of the amplitudes oscillate with respect to the initial level and speeds of solitons are not altered.

The MEP Equation

In [17] we studied head on collisions of solitary waves over long time intervals. Notwithstanding that interactions were not completely elastic we concluded that the behavior of the solution was very close to solitonic. Here the MEP equation is integrated under initial conditions

$$v(x,0) = A \operatorname{sech}^2 B(x - x_0), \qquad 0 \leq x \leq 200\pi, \qquad x_0 = 100\pi. \qquad (19)$$

In the numerical experiment, discussed below, values $A = 1$ and $B = 0.05$ are the pulse amplitude and width respectively, $c = 0$ is the initial phase velocity and $b = 0.719$, $\mu = 2.083$, $\delta = 0.25$ $\beta = 45.040$, $\gamma = 9.375$, $\lambda = 2.083$ are the equation parameters. The time-space behavior of the solution can be observed in Figures 5–9. The MEP equation is of Boussinesq type and therefore contrary to the HKdV equation, waves going to the left as well as to the right can emerge. Figures 5 and 6 demonstrate that two solitons that propagate to the right and two solitons that propagate to the left form in the present case. Because of the fact that the phase velocity of the initial wave is zero, solitons going to the right and solitons going to the left have equal amplitudes and equal speeds. In Figures 7 and 8 some typical wave profiles are plotted against space coordinate x. Time moment $t = 418$ corresponds to the beginning of the integration interval before interactions, at $t = 1128$ and $t = 4525$ interactions between the two highest solitons take place, at $t = 1845$ interaction between the higher and the lower solitons and at $t = 2704$ between the two lower solitons take place. In Figure 9 amplitudes of solitons are plotted against the time. It is clear, that interactions between emerged solitons are not completely elastic — Figures 5–8 demonstrate that each interaction produces small amplitude radiation. The more interactions have taken place the more distinctive the radiation is (see Figures 7 and 8). During interactions the amplitudes of waves increase and after interactions they are almost restored on the initial level. However, amplitudes of waves are not constant between interactions but oscillate with respect to the initial level.

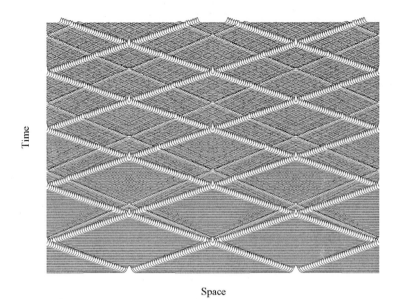

FIGURE 5. Emergence of trains of solitons in case of the MEP model — time-slice plot over two space periods

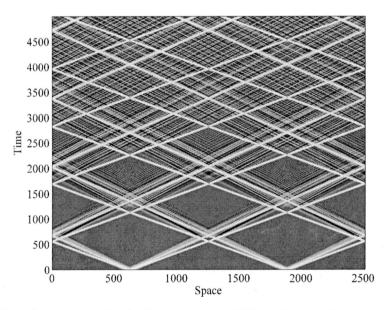

FIGURE 6. Emergence of trains of solitons in case of the MEP model — pseudocolor plot over two space periods

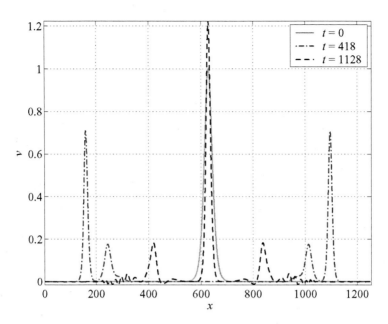

FIGURE 7. Single wave profiles in case of the MEP model

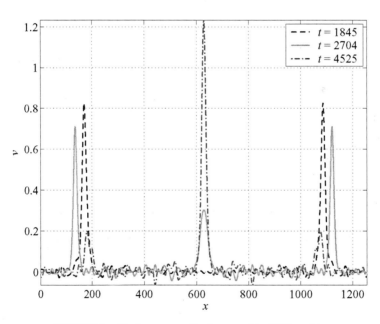

FIGURE 8. Single wave profiles in case of the MEP model

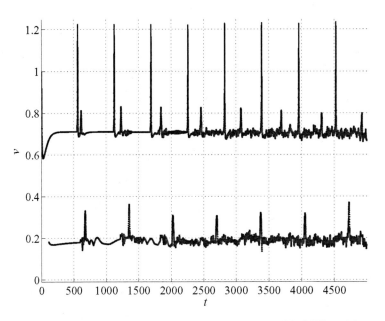

FIGURE 9. Amplitudes of solitons against time in case of the MEP model

These oscillations are caused by the radiation produced during interactions. Therefore the more interactions have taken place the more distinctive are the oscillations of soliton amplitudes (see Figure 9). The speed of waves is not altered during interactions. To sum up, one can say that interactions between emerged localized waves are almost elastic and thereof we call them solitons.

Of course, for different values of parameters of the MEP equation and of the initial pulse the number of emerged solitons can be higher than two. If other parameters (except the width B_0) are fixed, then the smaller the parameter B_0 (the wider the initial pulse) the larger the number of solitons in the train.

CONCLUSIONS

In the present paper propagation and interaction of solitons is studied in case of the HKdV as well as in case of the MEP model. Corresponding equations are integrated numerically under sech^2-type initial conditions making use of the PsM. Analysis of numerical results demonstrates that emerged solitary waves practically restore their amplitudes and speeds throughout interactions. Therefore the interactions are practically elastic and the emerged solitary waves can be called solitons in both considered cases.

ACKNOWLEDGMENTS

Authors of this paper thank Professor Jüri Engelbrecht for helpful discussions and Dr. Pearu Peterson for Python related scientific software developments and assistance. The research was supported, in part, by Estonian Science Foundation Grant No 7035 (A.S., L.I. and K.T.) and EU Marie Curie Transfer of Knowledge project MTKD-CT-2004-013909 under FP 6 (A.S.).

REFERENCES

1. J. Engelbrecht and F. Pastrone, *Proc. Estonian Acad. Sci. Phys. Math.* **52**, 12–20 (2003).
2. V. I. Erofeev, *Wave Processes in Solids with Microstructure*, World Scientific, Singapore, 2003.
3. A. V. Porubov, *Amplification of Nonlinear Strain Waves in Solids*, World Scientific, Singapore, 2003.
4. P. Giovine and F. Oliveri, *Meccanica* **30**, 341–357 (1995).
5. G. Whitham, *Linear and Nonlinear Waves*, John Wiley & Sons, New York, 1974.
6. C. Christov, G. Maugin, and A. Porubov, *Comptes Rendus Mécanique* **335**, 521–535 (2007).
7. J. Janno and J. Engelbrecht, *J. Phys. A: Math. Gen.* **38**, 5159–5172. (2005).
8. J. Janno and J. Engelbrecht, *Inverse Probl.* **21**, 2019–2034 (2005).
9. R. D. Mindlin, *Arch. Rat. Mech. Anal.* **16**, 51–78 (1964).
10. H.-O. Kreiss and J. Oligeri, *Tellus* **24**, 199–215 (1972).
11. B. Fornberg, *A Practical Guide to Pseudospectral Methods*, Cambridge University Press, Cambridge, 1998.
12. E. Oran Brigham, *The Fast Fourier Transform and Its Applications*, Prentice Hall Int., London, 1988.
13. R. N. Bracewell, *The Fast Fourier Transform and Its Applications*, McGraw-Hill, New York, 1978.
14. L. Ilison and A. Salupere, Propagation of localised perturbations in granular materials, Research Report Mech 287/07, Institute of Cybernetics at Tallinn University of Technology (2007).
15. L. Ilison, A. Salupere, and P. Peterson, *Proc. Estonian Acad. Sci. Phys. Math.* **56**, 84–92 (2007).
16. L. Ilison and A. Salupere, *Math. Comput. Simulat.* (2008), (submitted).
17. A. Salupere, K. Tamm, and J. Engelbrecht, *Int. J. Non-linear Mech.* **43**, 201–208 (2008).

On the Stability of a Free Viscous Film

S. Tabakova*,† and S. Radev†

*Dept. of Mechanics, Technical University of Sofia, branch Plovdiv, Bulgaria
†Laboratory of Physico-Chemical Hydrodynamics, Institute of Mechanics, BAS, Sofia, Bulgaria

Abstract. The free films have applications in some technological processes including liquid-liquid or liquid-gas disperse systems, as well as in the production of semiconductor micro-materials from melt. The problem of film stability during these processes is of major interest. In the present work a free thin film, attached on a rectangular frame surrounded by an ambient gas, is considered. The film is assumed viscous and under the action of the capillary forces. The linear stability problem is studied, when a small perturbation is imposed on the steady-state solution. An evolutionary non-linear system for the longitudinal velocity and film thickness is obtained from the long-wavelength approach. This system is solved numerically for the perturbed flow by the method of differential Gauss elimination. The results show, that for a large range of Re and wetting angles α, the film steady shapes are stable with respect to antisymmetrical disturbances.

Keywords: Thin film, linear stability, differential Gauss elimination, eigenvalue search.
PACS: 47.15.gm, 47.20.Dr, 47.20.Gv

INTRODUCTION

Macroscopic thin liquid films are important for different branches of science and engineering. When the films are subjected to various mechanical or thermal actions, they exhibit interesting dynamic phenomena such as wave propagation, rupture phenomena creating holes, etc. [1]. In the case of free thin films, usually, the disparity of the length scales favors the application of the long-wavelength model reducing the full set of governing equations and boundary conditions to a simplified, highly nonlinear, evolution set of equations. One nonlinear evolution PDE of 4th order for film thickness is obtained, if the film surfaces are immobile (with zero tangential velocities) [2]. However, at mobile film surfaces (with zero shear stress), a set of 2 or 3 nonlinear evolution PDE of 2nd order for film thickness and longitudinal velocity (or velocities in 2D case) is found: in [3], [4], [5] for laterally periodic and in [6], [7] for laterally confined film by a solid boundary. The linear and nonlinear stability at the presence of the van-der-Waals forces is discussed in [2] - [4]. In [6] the non-linear evolution is studied numerically, only when the inertia, viscosity and capillarity are taken into account. The results obtained there give an insight of the stable and unstable film solutions. However, it is difficult to study the whole spectrum of process parameters and only several cases are considered. For all of them, the found solutions are steady.

In the present work, the linear stability analysis of a free film laterally bounded is studied in 1D case. The perturbed flow equations are solved by the numerical method of differential Gauss elimination. It is shown that for all physically possible wetting angles and combinations of Reynolds number and Weber number the steady film shape is stable with respect to antisymmetrical disturbances.

CP1067, *Applications of Mathematics in Engineering and Economics '34—AMEE '08*, edited by M. D. Todorov
© 2008 American Institute of Physics 978-0-7354-0598-1/08/$23.00

FORMULATION OF THE PROBLEM

A thin free film symmetrically attached on a rectangular frame with respect to the central plane $z = 0$ (in a Cartesian coordinate system (x,y,z) connected with the frame $x = \pm a$, $y = \pm b$ as shown in Figure 1) is considered. The film fluid is assumed Newtonian viscous with constant density ρ and dynamic viscosity μ. Since the film is thin enough, the capillary forces are important, but the gravity force is negligible. As the mean film thickness εa is much smaller than a, then $\varepsilon \ll 1$. The free symmetrical surfaces are defined as $z = \pm h/2$, where $h(x,y,t) = O(\varepsilon)$ represents the film shape.

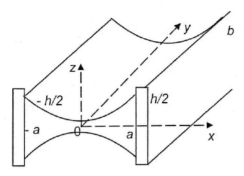

FIGURE 1. Free thin film sketch

Similarly to long-wavelength model [2] - [7] the velocity and pressure are developed in asymptotic expansions on the coordinate $z = O(\varepsilon)$.

For the zero order term of longitudinal velocity and film thickness a dynamic system is obtained from the Navier-Stokes equations, the kinematic and dynamic boundary conditions on the interfaces [5], [6]:

$$\frac{\partial h}{\partial t} + \nabla \cdot (h\mathbf{u}) = 0, \tag{1}$$

$$\rho \left\{ \frac{\partial \mathbf{u}}{\partial t} + \mathbf{u} \cdot \nabla \mathbf{u} \right\} = \frac{1}{h} \nabla \cdot \hat{\mathbf{T}}, \tag{2}$$

where $\mathbf{u} = (u,v)$ is the longitudinal velocity (mean velocity with respect to film cross-section) vector, $\nabla = \left(\frac{\partial}{\partial x}, \frac{\partial}{\partial y} \right)$ is the plane gradient, $\hat{\mathbf{T}} = -\mathbf{P} + \mathbf{T}$ is the film stress tensor, $\mathbf{P} = -0.5\sigma \left[h\nabla^2 h\mathbf{I} + 0.5(\nabla h)^2\mathbf{I} - \nabla h \otimes \nabla h \right]$ is the pressure tensor with $\mathbf{I}$ standing for the identity tensor and the viscous stress tensor given as $\mathbf{T} = 2\mu h \left\{ (\nabla \cdot \mathbf{u})\mathbf{I} + 0.5 \left[\nabla \mathbf{u} + (\nabla \mathbf{u})^T \right] \right\}$.

The upper system (1), (2) for the film thickness h and longitudinal velocity $\mathbf{u}$ is nonlinear and coupled. Due to the projection performed on the mean plane, this system is 2D in general. However, if the frame is much longer in y direction ($b \gg a$) and if the end effects in y direction are neglected, then the system is 1D, i.e., depends only on (x,t). Here after, the 1D system in dimensionless form with the corresponding boundary and initial conditions is considered. The following characteristic scales are used: a for length, U for velocity (capillary), a/U for time and εa for film thickness.

The 1D dimensionless variant of the system (1), (2) reads:

$$\frac{\partial h}{\partial t} + \frac{\partial}{\partial x}(uh) = 0, \tag{3}$$

$$\frac{\partial u}{\partial t} + u\frac{\partial u}{\partial x} = \frac{\varepsilon}{\text{We}}\frac{\partial^3 h}{\partial x^3} + \frac{4}{\text{Re}\, h}\frac{\partial}{\partial x}(h\frac{\partial u}{\partial x}), \tag{4}$$

where $\text{Re} = \rho a U/\mu$ is the Reynolds number, $\text{We} = \text{Re} \cdot \text{Ca} = 2\rho a U^2/\sigma$ is the Weber number, Ca is the capillary number.

The boundary conditions for h and u on the frame are the following:

$$u(\pm 1, t) = 0, \tag{5}$$

$$\frac{\partial h}{\partial x}(\pm 1, t) = \pm\tan\alpha, \tag{6}$$

where $\pi/2 - \alpha$ is the wetting angle with the frame.

The mass conservation condition during the film dynamics is expressed as:

$$\int_{-1}^{1} h\, dx = W, \tag{7}$$

where $W = \text{const}$ is the initial film volume.

LINEAR STABILITY

The system (3)-(7) has an analytical steady solution, given by:

$$h_s(x) = 0.5(x^2 - 1/3)\tan\alpha + W, \qquad u_s(x) \equiv 0. \tag{8}$$

Since the film thickness is positive, $h_s(x) > 0$, an upper bound for the wetting angle is found from (8): $\alpha < \arctan(3W)$. Further in our analysis, $W = 2$ is taken, without influencing the results generality and thus $\alpha < 1.40565$.

If on the steady-state solution (8) small disturbances $\tilde{h}(x,t)$, $\tilde{u}(x,t)$ are imposed, then the film thickness h and velocity u are expressed as:

$$h = h_s + \tilde{h}, \qquad u = \tilde{u} \tag{9}$$

Inserting (9) in (3)-(6) and neglecting 2nd order disturbances, the following linearized system is obtained:

$$\frac{\partial \tilde{h}}{\partial t} + \frac{\partial(h_s\tilde{u})}{\partial x} = 0, \tag{10}$$

$$\frac{\partial \tilde{u}}{\partial t} = \frac{\varepsilon}{\text{We}}\frac{\partial^3 \tilde{h}}{\partial x^3} + \frac{4}{\text{Re}\, h_s}\frac{\partial}{\partial x}(h_s\frac{\partial \tilde{u}}{\partial x}), \tag{11}$$

168

$$\tilde{u}(\pm 1, t) = 0, \tag{12}$$

$$\frac{\partial \tilde{h}}{\partial x}(\pm 1, t) = 0 \tag{13}$$

The mass conservation condition (7) requires $\int_{-1}^{1} \tilde{h}\,dx = 0$, which follows directly from (10) and (12).

The disturbances $\tilde{h}$ and $\tilde{u}$ are sought in the form:

$$\tilde{h} = H(x)\exp(\omega t), \qquad \tilde{u} = V(x)\exp(\omega t), \tag{14}$$

where $\omega = \omega_r + i\omega_i$ is the disturbance growth rate. Further, our goal is to find ω_r. If it is positive, the steady-state solution (8) is unstable, while negative, the solution is stable.

Using (14) the system (10)-(13) is transformed into a system for H and V:

$$H = -\frac{(h_s V)'}{\omega}, \tag{15}$$

$$h_s(h_s V)^{IV} - 4\Omega(h_s V')' + P\Omega^2 h_s V = 0, \tag{16}$$

$$V(\pm 1) = 0, \tag{17}$$

$$H'(\pm 1) = 0, \tag{18}$$

the latter being equivalent to

$$(h_s V)''(\pm 1) = 0, \tag{19}$$

where $\Omega = \frac{We}{\varepsilon Re}\omega$, $P = \frac{\varepsilon Re^2}{We}$. Then (16) with (17) and (19) represents an eigenvalue problem for Ω, which has no analytical solution for $\alpha > 0$.

In the next section a numerical method for solving the non-linear eigenvalue problem (16), (17) and (19) is proposed.

NUMERICAL MODEL

Since the eigenvalue problem is non-linear, special methods are appropriate for its solution [8]: the eigenvalue search by secant method and the differential Gauss elimination method applied at each step of secant search. In the following, both methods are described briefly.

Differential Gauss Elimination Method

New variables are introduced by:

$$\mathbf{v} = (v_1, v_2) = (h_s V', (h_s V)''' - \Omega V'), \quad \mathbf{w} = (w_1, w_2) = (h_s V, (h_s V)'' - \Omega V). \tag{20}$$

Then the 4th order ODE (16) is transformed into a linear ODE system of 1st order:

$$\mathbf{v}' = \mathbf{K}\mathbf{v} + \mathbf{G}\mathbf{w}, \tag{21}$$

169

$$\mathbf{w}' = \mathbf{D}\mathbf{v} + \mathbf{E}\mathbf{w}, \tag{22}$$

where

$$\mathbf{K} = \begin{pmatrix} -\frac{h_s'}{h_s} & 0 \\ -\frac{2\Omega h_s}{h_s^2} & 0 \end{pmatrix}, \mathbf{G} = \begin{pmatrix} -\frac{h_s''+\Omega}{h_s} & 1 \\ -\frac{\Omega}{h_s^2}(3h_s''-3\Omega+P\Omega h_s) & \frac{3\Omega}{h_s} \end{pmatrix}, \mathbf{E} = \begin{pmatrix} \frac{h_s'}{h_s} & 0 \\ 0 & 0 \end{pmatrix}$$

and $\mathbf{D} = \mathbf{I}$, as $\mathbf{I}$ is the identity matrix.

Equations (17) and (19) are the homogeneous boundary conditions for $\mathbf{v}$ and $\mathbf{w}$, which are assumed symmetrical with respect to $x = 0$ (hence H is antisymmetrical). Further, only the half interval $x \in [0,1]$ is considered. Then the boundary conditions (17), (19) become:

$$\mathbf{v}(0) = \mathbf{0}, \tag{23}$$

$$\mathbf{w}(1) = \mathbf{0}. \tag{24}$$

The unknown vector $\mathbf{v}$ is sought in the form [8]:

$$\mathbf{v}(x) = \mathbf{A}(x)\mathbf{w}(x), \tag{25}$$

where

$$\mathbf{A}(x) = \begin{pmatrix} A_{11}(x) & A_{12}(x) \\ A_{21}(x) & A_{22}(x) \end{pmatrix} \tag{26}$$

is an unknown matrix, whose elements will be found after solving the following differential equation:

$$\mathbf{A}' = \mathbf{G} - \mathbf{A}\mathbf{E} + (\mathbf{K}\text{-}\mathbf{A}\mathbf{D})\mathbf{A}. \tag{27}$$

This equation is obtained when (25) is substituted in (21), (22). The boundary condition for $\mathbf{A}$ is homogeneous obtained from (23):

$$\mathbf{A}(0) = \mathbf{0} \tag{28}$$

The solution of (27), (28) represents the first part of the differential Gauss elimination method or the direct differential Gauss elimination.

The homogeneous boundary condition (24) makes the matrix elements $\mathbf{A}(1)$ infinite, when Ω becomes equal to an eigenvalue. In order to prevent from this singularity, an inverse differential Gauss elimination starting at $x = 1$ and directing towards $x = 0$ is applied.

Similarly to the direct elimination, the variables

$$\overline{\mathbf{v}} = (\overline{v}_1, \overline{v}_2) = (w_1, w_2), \quad \overline{\mathbf{w}} = (\overline{w}_1, \overline{w}_2) = (v_1, v_2), \tag{29}$$

are solutions of the differential equations:

$$\overline{\mathbf{v}}' = \overline{\mathbf{K}}\mathbf{v} + \overline{\mathbf{G}}\mathbf{w}, \tag{30}$$

$$\overline{\mathbf{w}}' = \overline{\mathbf{D}}\mathbf{v} + \overline{\mathbf{E}}\mathbf{w}, \tag{31}$$

$$\overline{\mathbf{K}} = \mathbf{E}, \quad \overline{\mathbf{G}} = \mathbf{D}, \quad \overline{\mathbf{D}} = \mathbf{G}, \quad \overline{\mathbf{E}} = \mathbf{K},$$

with homogeneous boundary conditions

$$\overline{\mathbf{v}}(1) = 0, \tag{32}$$

$$\overline{\mathbf{w}}(0) = 0. \tag{33}$$

The vector $\overline{\mathbf{v}}$ is represented as:

$$\overline{\mathbf{v}}(x) = \mathbf{B}(x)\overline{\mathbf{w}}(x), \tag{34}$$

where the unknown matrix $\mathbf{B}$ given by

$$\mathbf{B}(x) = \begin{pmatrix} B_{11}(x) & B_{12}(x) \\ B_{21}(x) & B_{22}(x) \end{pmatrix} \tag{35}$$

is solution of the differential equation

$$\mathbf{B}' = \overline{\mathbf{G}} - \mathbf{B}\overline{\mathbf{E}} + (\overline{\mathbf{K}} - \mathbf{B}\overline{\mathbf{D}})\mathbf{B} \tag{36}$$

with boundary condition

$$\mathbf{B}(1) = \mathbf{0}. \tag{37}$$

The direct and inverse eliminations are matched at some point $x_c \in (0,1)$, which leads to the following condition:

$$\mathbf{Q}_c(x_c) \cdot [v_1(x_c), v_2(x_c), w_1(x_c), w_2(x_c)]^T = \mathbf{0}, \tag{38}$$

where

$$\mathbf{Q}_c = \begin{pmatrix} \mathbf{I} & -\mathbf{A}(x_c) \\ \mathbf{B}(x_c) & -\mathbf{I} \end{pmatrix}$$

The existence of non-trivial solutions of $\mathbf{v}$ and $\mathbf{w}$ requires the determinant of $\mathbf{Q}_c$ to vanish, which gives the characteristic equation for Ω:

$$F_c(\Omega) = \det(\mathbf{Q}_c) = 0. \tag{39}$$

Both systems (27), (28) and (36), (37) are numerically integrated by the Runge-Kutta method to find the matrices $\mathbf{A}$ and $\mathbf{B}$.

Eigenvalues Search Method

At $\alpha = 0$ the eigenvalue problem (16), (17) and (19) has the analytical solution:

$$\Omega_{1,2} = \frac{-\pi^2(2k+1)^2(2 \mp \sqrt{4-P})}{4P}, \quad \Omega_{3,4} = \frac{-\pi^2 k^2(2 \mp \sqrt{4-P})}{P}, \tag{40}$$

where $k = \pm 1, \pm 2, \ldots$

At given P, the eigenvalue with the greatest real part corresponds to $\Omega_0 = \Omega_1$ at $k = -1$, obtained by (40). If instead, the eigenvalue problem written in the form (21)-(24) is

171

considered, its analytical solution at $\alpha = 0$ coincides with $\Omega_{1,2}$ given by (40). Therefore, at $\alpha > 0$ it is expected that the eigenvalue search in the form (21)-(24) assures the greatest (with the greatest real part) eigenvalue.

At $\alpha > 0$ the characteristic equation (39) is nonlinear. Then, it is difficult to find its roots in the parametric space (α, P). However, the knowledge of the eigenvalue Ω_0 can be used as an initial approximation in an iterative method for Ω search at any $\alpha > 0$.

If $F_c(\Omega)$ is developed in series at Ω_0

$$F_c(\Omega) = F_c(\Omega_0) + (\Omega - \Omega_0)F_c'(\Omega_0) + ..,$$ (41)

and if $|\Omega - \Omega_0|$ is assumed small enough to neglect the nonlinear terms in (41), then from $F_c(\Omega) = 0$ it follows:

$$\Omega = \Omega_0 - \frac{F_c(\Omega_0)}{F_c'(\Omega_0)}.$$

Since $F_c(\Omega)$ is found numerically, then $F_c'(\Omega)$ is approximated by finite differences. Finally, the iterative formula of the secant method is obtained:

$$\Omega_{n+1} = \Omega_n - F_c(\Omega_n)\frac{(\Omega_n - \Omega_{n-1})}{[F_c(\Omega_n) - F_c(\Omega_{n-1})]},$$ (42)

where $n = 1, 2, ...$

The real and imaginary parts Ω_r and Ω_i are sought simultaneously as in [8].

NUMERICAL RESULTS AND DISCUSSION

In our numerical calculations the matching of the direct and inverse Gauss elimination is done at $x_c = 0.5$, although the same results could be obtained with any $x_c \in (0, 1)$.

At given $P > 0$ and $0 \leq \alpha < 1.40564$, only 4-5 iterations are sufficient to find the correspondent eigenvalue Ω that satisfies (39) with accuracy $O(10^{-7})$. When α is closer to its upper bound more iterations are necessary to reach this accuracy.

Since the secant method is faster converging, when the initial approximation is closer to the root, an outer iteration for α is performed for a fixed value of P. The correspondent analytical solution Ω_0 is used as an initial approximation. Afterwards α is increased with a variable step, becoming smaller as α approaches its upper bound. The initial step in the iterative procedure (42) for Ω at a given P, when increasing α, is found by a gradient method, starting with $10^{-3}(1+i)$ at $\alpha = 0$.

The real and imaginary parts of the eigenvalue solution at $0 \leq \alpha < 1.40564$ and $P \in [0.1, 10^3]$ are plotted in Figure 2. The curves at $P < 0.1$ coincide with that at $P = 0.1$.

CONCLUSIONS

The linear stability of a free thin film with respect to exponentially growing perturbation is studied at different wetting angles and different Re and We. Special numerical model to find the eigenvalues of time perturbation is developed, based on the differential Gauss

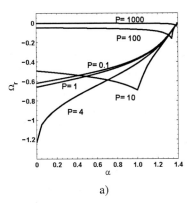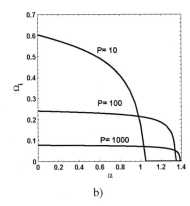

FIGURE 2. Eigenvalues as functions of α and P: a) real Ω_r b) imaginary Ω_i

elimination. A search algorithm of the eigenvalues for a large range of Re, We and α is applied.

The linear stability analysis shows that the steady shape at different wetting angles is stable with respect to antisymmetrical disturbances (symmetrical velocity disturbances). The case of symmetrical disturbances as well as the general disturbances case will be studied in future. The non-linear evolution analysis in [6], also shows that the steady-state solution is reached for the considered cases of $\alpha = 1.37$, Ca $= \varepsilon$ (correspondent to $P \equiv$ Re according to the present notation) and $1 \leq \text{Re} \leq 100$. As a continuation of the present work, the van-der-Waals forces will be included in the linear stability model. The non-linear analysis in [7] shows that at the van-der-Waals forces action the film reaches a stable shape at some process parameters, while at others it ruptures.

REFERENCES

1. A. Oron, S. H. Davis and S. G. Bankoff, *Rev. Mod. Phys.* **69**, 931–980 (1997).
2. M. Prevost, D. Gallez, *J. Chem. Phys.* **84**, 4043–4048 (1986); **85**, 4757–4758(E) (1986).
3. T. Erneux, S. H. Davis, *Phys. Fluids* **A 5**(5), 1117–1122 (1993).
4. A. De Wit, D. Gallez, and C. I. Christov, *Phys. Fluids* **6**(10), 3256–3266 (1994).
5. D. Vaynblat, J. R. Lister, and T. P. Witelski, *Phys. Fluids* **13**(5), 1130–1140 (2001).
6. L. Popova, G. Gromyko, and S. Tabakova, *Mathematical Modelling and Analysis* **8**(1), 291–311 (2003).
7. L. Popova, G. Gromyko, and S. Tabakova, *Differential Equations* **39**(7), 1037–1043 (2003).
8. M. A. Goldshtik and V. N. Shtern, *Hydrodynamic Stability and Turbulence*, Nauka, Novosibirsk, 1977. (in Russian)

Prediction of Properties of Nonlinear Processes

P. Volkov

Yugora State University, Tchechov Str., bl. 16, 628012 Khanty-Mansiïsk, Russia

Abstract. Determination of the reference length and velocity in the similarity theory in terms of a parameter varying within wide limits enables the obtaining of a prediction of the properties of nonlinear processes depending on the value of this parameter. As an example, an analysis of convective processes is carried out in connection with the body force parameter variation. Prediction data are confirmed by the bubble emersion process. The variation of convective processes versus the pressure is predicted. A variant of the model medium (the water vapor), in which the processes are similar to those occurring in a channel with water under a high pressure is presented under the normal conditions.

Keywords: Boundary-value problem, degeneracy, regularization, FEM, similarity theory, benchmark, maps of regimes.
PACS: 47.55-t; 47.56+r

INTRODUCTION

It is practically impossible to obtain "a priori" the answer concerning the behavior of nonlinear systems under new conditions. The nonlinearity factor plays the main positive role in the description of processes by mathematical methods; it is simultaneously an insurmountable obstacle for obtaining the prediction. Right for this reason, the reduction of mathematical models with a subsequent computational experiment is the main way for obtaining the predictions. In connection with the determination of reduction ways, the development of methods for a preliminary prediction of the behavior of nonlinear systems is along with the well-posed ness problem of obtained problems of an immense practical importance. In the present work, the techniques of obtaining trustworthy knowledge about the behavior of the processes under consideration in new conditions are studied by the examples of nonlinear convective processes in the mechanics of deformable media.

The systems of equations governing the motions of liquids and gases are as a rule nonlinear integrodifferential and involve partial derivatives [1, 2]. Various hypotheses leading to a simplification of the original system are employed on the way of designing numerical algorithms. The medium incompressibility hypothesis having a sufficiently clear physical substantiation has gained the most widespread acceptance. As a result, the models of incompressible media were derived: the systems of Euler and Navier-Stokes, which are basic for the modeling of flows of liquids and gases [3]. In the present work, we restrict ourselves to the discussion of the Navier-Stokes model and the Oberbeck-Boussinesq model derived from it for non-isothermal processes.

The investigation of properties of particular (approximate) solutions of the models of incompressible media leads to ambiguous conclusions. Several solutions agree well

CP1067, *Applications of Mathematics in Engineering and Economics '34—AMEE '08*, edited by M. D. Todorov
© 2008 American Institute of Physics 978-0-7354-0598-1/08/$23.00

with the data of laboratory experiments. There are, however, a number of paradoxes (due to Stokes, Whitehead et al [3]), which point to the absence of a strong solution (the functions whose substitution into the equations and boundary conditions reverts them into identities). Theoretical investigations of the existence of solutions for the Navier-Stokes system reveal fundamental difficulties. For the boundary-value problems with a given velocity vector at the boundaries, the existence theorems for generalized (approximate!) solutions were proved [4, 5], the existence of strong solution in the general case was not proved so far. The strong solution existence was proved in the class of problems of flows through channels (with a given pressure on separate parts of the boundary of a region occupied with fluid) [6, 7].

The advent of computers stimulated the development of computational technologies for solving various boundary-value problems. Benchmark tests were developed for the purpose of verifying the numerical techniques. The structure of isothermal flows was studied by the example of water flow in cavity (a rectangular prism with a lid moving at constant velocity) [8]. Non-isothermal gas flows were studied at the air motion in a cube with different temperatures of the walls [9]. Despite an apparent simplicity of experiments these two flow kinds give a wide spectrum of the phenomena observed in nature: from stratified laminar ones to vortex turbulent ones. Experimental data are enough to verify the mathematical models, including high parameter values of problems.

An analysis of the data of numerical calculation of the Navier-Stokes system showed that the numerical algorithms in velocity-pressure variables have a poor convergence [10]. As a result, the techniques using the stream function and vorticity have enjoyed a development. The application of various numerical schemes to the solution of the driven cavity problem leads to considerable discrepancies in velocity profiles [11, 12]. The development of a high-accuracy method for solving the Navier-Stokes equations in primitive variables [13] have not brought the desired result in terms of the comparison with experimental data [8]. A peculiarity of the models of incompressible media is to be noted. The pressure field is determined with the accuracy up to an additive function of time. This leads to a conclusion about the independence of kinematics parameters of pressure magnitude. Since the motion is the result of a nonlinear interaction a special research is necessary on the possibility of applying the models of incompressible media for prediction. Summarizing the research results one can conclude that a trustworthy prediction of fluid flows in the regions bounded by solid walls is impossible when using the models of incompressible media.

Speaking about the prediction of processes we will distinguish between two aspects: a) the indication of directions in which the qualitative changes will occur; b) quantitative characteristic of deviations from some state. Qualitative differences in convective processes can be determined depending on the value of some parameter from the similarity theory. The quantitative changes may be obtained from the maps of flow regimes for some process (benchmark) having a fairly complete experimental investigation and an adequate mathematical description. In the present work, convective processes will be analyzed under the body force parameter variation. The process of bubble emersion in the fluid, for which the maps of flow regimes were constructed [14, 15], will be used as a benchmark. A prediction for the variation of convective processes in deposit strata with a liquid under a high pressure will also be given. This task is topical in connection with the prediction of the recovery of natural media from underground strata containing water, oil or gas.

NAVIER-STOKES SYSTEM, BOUNDARY-VALUE PROBLEMS

The motion equations for a viscous incompressible fluid with density ρ, viscosity v for the velocity $\mathbf{u}(\mathbf{x}, t) = (u, v, w)$ and the pressure p have the form

$$\frac{\partial \mathbf{u}}{\partial t} + (\mathbf{u} \cdot \nabla)\mathbf{u} = -\frac{1}{\rho}\nabla p + v\Delta\mathbf{u} - \mathbf{F} \tag{1}$$

$$\nabla \cdot \mathbf{u} = 0. \tag{2}$$

Here $\mathbf{x}$ are the Eulerian coordinates, t is the time, $\mathbf{F} = g\mathbf{k}$ is the body force.

In terms of the type of boundary conditions specification at the boundary Γ of region Ω occupied by the fluid, one distinguishes between two basic statements: a) with a given velocity vector on Γ; b) the flows through channels; for example, the pressure and the tangent velocity components are specified at the parts of the Γ_1 boundary (the inlet) and Γ_2 (the outlet), and the velocity vector is specified at the remaining boundary parts.

A direct discretization of (1)–(2) for problem a) leads to a slowly convergent iterative process in nonlinearity. As a reason for this, one usually points to the necessity of an exact treatment of the nonlinear term. Since it is impossible to eliminate the scheme viscosity in finite-difference techniques, the convergence and accuracy depend on the technique of discretization of nonlinear terms and of the iterative process itself.

The attempts to solve consistently (1)-(2) by the finite element method (FEM) having a strict mathematical substantiation have led to an ill-conditioned matrix of the system of algebraic equations. As a result, one had to refuse the classical FEM variant and to go over to the use of approximating functions of different orders for the velocities and the pressure. This technique enables the solution of problems and it "naturally" introduces in the solution an extra error from a cruder approximation of the function. Thus, the FEM indeed acquires a "scheme viscosity" in this form.

We note in summary a paradoxical fact. Numerical models from which the solutions were obtained contain the "scheme viscosity" as a consequence of the approximation of equations, boundary conditions, and the technique for splitting up the nonlinearity in finite-difference and finite-element techniques. The history of FEM application shows that only after the incorporation of the procedure for deterioration of the accuracy of the entire solution one succeeded in obtaining the convergent algorithms for the method, in which the integral conservation laws were satisfied originally on the elements. To elucidate the reasons for such a behavior of numerical solutions we investigate the spectral properties of boundary-value problems for the Navier—Stokes system.

The algorithms using the variational Bubnov—Galerkin methods are the most efficient for this purpose. The well-known Galerkin method combined with the finite-element model belongs to such algorithms. The given approach assumes a search for a piecewise polynomial approximate solution on the elements provided that the difference between the approximate and exact solutions must be orthogonal to the functions employed for approximation. The availability of the residual minimization procedure over the entire element volume and, as a consequence, the absence of scheme viscosity makes this technique a unique tool for investigating the incompressible fluid models under consideration, as this takes place also in many other areas of computational mathematics.

A classical application of FEM for (1)–(2) with polynomials of equal degree for all functions and the use of an implicit finite-difference approximation in time (Δt is the time step) leads to a system of linear algebraic equations for the vectors of unknowns $\{\Phi\}$ in grid nodes

$$\left(\frac{1}{\Delta t}[C] + [K]\right)\{\Phi^{n+1}\} = \left(\frac{1}{\Delta t}[C]\right)\{\Phi^n\} - \{F\} \tag{3}$$

where $[C], [K]$ are the damping and stiffness matrices, $\{F\}$ is the vector of forces for the FEM system. At the derivation of (3), the coefficient in the nonlinear term in (1) has been taken from the nth time layer. System (3) is the original one for analyzing the well-posedness of boundary-value problems for (1)–(2), which bases on the investigation of the spectrum structure of the stiffness matrix [K]. A full realization of the described approach in the system of symbolic computations "Mathematica" starting from the grid generation procedure to solving system (3) gives a possibility of performing the analytic computations completely in rational fractions as well as the approximate floating-point computations at a user-specified mantissa length.

The well-posedness was investigated at the example of four problems. These were the water flows in the square and cubic regions due to the moving lid, the flow between two rotating circular cylinders (the Couette flow), and the air motion in a square region due to heated walls.

The application of the analytic FEM at the derivation of the system of linear algebraic equations for the Navier-Stokes and Oberbeck-Boussinesq systems with a given velocity vector on the boundaries leads to the system of equations (3) with matrix [K] having the multiple zero eigenvalue. The approximate FEM variant shows the presence in the spectrum of the same number of the numbers in which the number of zeroes after the decimal point is determined by the given mantissa length. With increasing mantissa length up to (20÷50) this group of numbers tends to zero. These facts characterize the degeneration of system (3), which also points to the boundary-value problem degeneracy.

Such a behavior of spectra (the presence of a group of small numbers) is observed for triangular simplex-elements of the first and second orders, rectangular Hermitean fourth-order elements preserving the first derivatives (fluxes) at the element boundaries, curvilinear isoparametric second-order elements as well as the simplex elements of the first and second orders (tetrahedrons) for the three-dimensional tasks. So for two-dimensional problems with linear, quadratic, and Hermitean elements the number of zeroes was eight (for each of three problems), for the isoparametric elements there were 17 zeroes, and for three-dimensional ones there were over 100 zeroes. (On coarse grids, the number of zeroes is less; the above presented numbers correspond to sufficiently fine grids.)

Thus, the opinion of the necessity of preserving the fluxes across the element boundaries and a high degree of approximation to ensure a good convergence does not obtain its confirmation. The problem is related to the fundamental property of degenerate boundary-value problems with a given velocity vector at the boundaries for the incompressible fluid equations. It makes no sense to search for the explanation in the influence of nonlinearity because the linear problem also possesses this property, which explains apropos the origin of the Stokes paradox for the cylinder. This investigation indeed points to a source of problems and the direction for their resolution. In the Stokes problem involving the parabolic-type equations for velocity components and the condition for the solenoidal velocity vector, only one reason may lead to the degeneracy of boundary-value problems – this is the condition for solenoidal velocity field. Thus, we arrive at a necessity of the regularization of the incompressibility condition.

The following formula for regularization of (2) proved to be successful for practical application

$$\operatorname{div} \mathbf{u} = \tau \, \nabla(\nabla p + \mathbf{F}) \tag{4}$$

which had appeared for the first time in the scheme of splitting in terms of physical processes for the Navier-Stokes equations. Here τ is a parameter responsible for the compressibility of particles along their trajectories [16].

Together with the natural boundary condition for the pressure $(\nabla p + \mathbf{F})\mathbf{n} = 0$, where $\mathbf{n}$ is the normal to the boundary surface, formula (4) has produced good results in terms of the convergence of the numerical techniques [17, 18]. Only one zero has remained in the regularized problem spectrum. This zero corresponds to the circumstance that the pressure is determined with the accuracy up to a constant [19, 20].

For the boundary-value problems of flows through channels (problem b)), the pressure is determined unambiguously. There is a **complete** absence of zeroes in the stiffness matrix spectrum for linear, quadratic, and Hermitean elements. These tasks do not need regularization, which agrees completely with the investigations of the existence and uniqueness of the strong solution [6, 7]. The solution of problems of flow through channels in the Stokes approximation (the Poiseuille flow in a channel and in a channel with a backward-facing step) leads to the unique solution when using various types of finite elements. The solution of the same problems with regard for the nonlinear term in the governing system was obtained only under the preservation of fluxes at the boundaries between elements (at the Hermitean elements). These investigations confirm that the iterative process convergence for well-posed nonlinear problems is reached by increasing the accuracy. However, a drastic improvement of the convergence is obtained at the regularization of the incompressibility condition (4). The solution is now obtained on the same grids both on linear and quadratic finite elements.

PREDICTION FROM THE MAPS OF FLOW REGIMES

A successful choice of the process as a benchmark facilitates the solution of the prediction problem. For example, a bubble in the fluid is an object which is unique in its simplicity and which manifests the interaction of practically all significant forces in the classical fluid mechanics: inertia, viscous friction, capillary forces, and gravitation. Right these circumstances enable the use of the bubble emersion process as a benchmark and for the prediction of changes in nonlinear convective processes under different conditions.

Investigation of the bubble emersion process drew considerable attention of both the theoreticians and experimentalists by virtue of its originality. Series of experiments on the emersion of bubbles of various sizes in various liquids were conducted [21]. The graphs of bubble emersion velocity were plotted versus the volume, and the emersion features were indicated. Figure 1a) shows a small part of investigation results. The data in Figure 1a) give the idea of some peculiarities of bubble emersion, but they give no chance to obtain the predictions for the emersion processes under new conditions. Let us show how one can reach this without applying nearly no efforts.

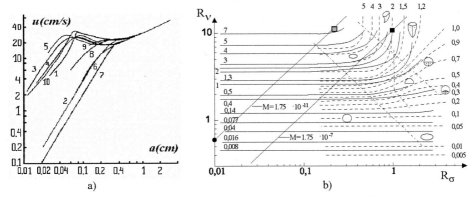

FIGURE 1. a) Bubble emersion speed u as a function of the equivalent radius a in liquids [21].

1 - Water,	$M= 1.8 \cdot 10^{-10}$;	6 - 62% solution of syrup in water, $M=1.55 \cdot 10^{-4}$;
2 - Mineral oil,	$M= 1.45 \cdot 10^{-2}$;	7 - 68% solution of syrup in water, $M=2.12 \cdot 10^{-3}$;
3 - Varsol,	$M= 4.30 \cdot 10^{-10}$;	8 - 56% solution of glycerin in water, $M= 1.75 \cdot 10^{-7}$;
4 - Turpentine,	$M= 2.40 \cdot 10^{-10}$;	9 - 42% solution of glycerin in water, $M=4.18 \cdot 10^{-8}$;
5 - Methyl alcohol,	$M=0.89 \cdot 10^{-10}$;	10 - 13% solution of glycerin in water, $M=1.17 \cdot 10^{-8}$.

b) Map of flow regimes. $R_\sigma = a/\delta_\sigma$. $R_v = a/\delta_v$. $M = g\rho^3 v^4/\sigma^3$. Solid lines are the contours of constant values of the Froude number from the calculation of the full Navier-Stokes equations; the dashed lines are the experimental data for different liquids.

■ – data for a liquid in terrestrial conditions ($u \approx 22$ cm/s); ● – data for the bubble of the same volume ($a = 0.25$ cm, $u \approx 0.02$ cm/s) in the same liquid in microgravitation conditions, ☐ – data for the bubble in microgravitation conditions, which has the same Archimedes' force as on the Earth ($a \approx 5.3$ cm, $u \approx 2.2$ cm/s).

One can draw two important conclusions from Figure 1a). First, the bubble emersion process is determined by two independent parameters (the liquid, a is the bubble size). Second, there is at least one parameter to be determined: u, the emersion velocity. This means in summary that one can construct in a plane different maps of flow regimes by taking two dimensionless criteria as the coordinates and plotting the contours of the third one by recalculating them from experimental data. From the viewpoint of the similarity theory, this will enable the identification of the effect of various mechanisms determining the emersion [22].

Let us demonstrate this as follows. Let us choose two dimensionless complexes computed from g: $R_\sigma = a/\delta_\sigma$, $R_v = a/\delta_v$ as the coordinates. Here a is the radius of a sphere equal to the bubble in its volume, $\delta_\sigma = (\sigma/g\rho)^{1/2}$ is the capillary constant, $\delta_v = (v^2/g)^{1/3}$ [23], g is the gravity acceleration, σ is the surface tension coefficient. We will consider the Froude number Fr $= u^2/ga$ the parameter to be determined. According to the definition of R_σ and R_v, we have $R_\sigma / R_v = (g\rho^3 v^4/\sigma^3)^{1/6} = M^{1/6}$ (M is the Morton number), and, consequently, the data for a specific liquid will be presented by the points of a straight line in these coordinates. Figure 1b) shows the map of flow regimes with the Froude number contours computed from experimental data of [14, 15]. A comparative analysis of the information content of the data representation in coordinates R_σ and R_v was already discussed in [14]. We will show here the advantages of Figure 1b) for predicting the convective processes under new conditions.

Consider the emersion of bubbles in the 56% solution of glycerin in water (M = $1.75 \cdot 10^{-7}$) – Figure 1a). The black square in Figure 1b) corresponds to the emersion of a bubble whose radius a equals the capillary constant. Let us assume that exactly the same experiment was conducted on board of a space vehicle, where the quasi-static component of the residual acceleration is by four orders of magnitude lower than the terrestrial value of g. In this case, the data of experiments are depicted by points of the straight line with $M = 1.75 \cdot 10^{-11}$. The black circle in Figure 1b) indicates the location corresponding to the emersion of the bubble of the same volume as in the case of the bubble marked by a black square. A much lower value of the emersion velocity (the Froude number) is an expected effect of the Archimedean force magnitude reduction. Consider the emersion of a bubble of a larger size (the grey square), whose Archimedean force is the same as in the case of the bubble marked by a black square. Its emersion velocity is much lower than on the Earth! A comparison of the Reynolds numbers (Re=ua/v) shows that for space conditions it is nearly twice higher than the terrestrial one.

One can establish exactly in the same way the character of the variation of capillary forces. The value of the Weber number (We= $\rho u^2 a/\sigma$) in space is much lower than its terrestrial value by a factor of nearly 5. This means that under space conditions, the influence of capillary forces increases (and/or the pressure head contribution drops). The grey square lies in the region of ellipsoidal bubbles with attached flow around them, and the black one is for a bubble in the form of a "cup" with a separation zone in the trailing part. All of this and the straight line itself with M= $1.75 \cdot 10^{-11}$ point to that the inertial properties of the liquid in space shift towards the ideal fluid model. The presence of such changes can explain a number of peculiarities in the data of experiments in the area of spatial materials science [15, 24].

PREDICTION ON THE SIMILARITY THEORY

From the viewpoint of the similarity theory there is no need at the first stage to consider the motion equations. It is sufficient to write down the physical constants determining the state of a system under consideration and make the complete list of dimensionless parameters [22].

Let the liquid with parameters ρ, v, and σ which is located in the field of a force characterized by the quantity g contain a bubble. The pressure jump on the free surface and in the liquid is Δp (it determines with σ the bubble volume, and hence, the linear size a in the equilibrium state). Thus, we have five physical constants determining the bubble emersion process: ρ, v, σ, g, Δp. Depending on the technique of passing to dimensionless quantities, one can obtain a different number of dimensionless parameters. If we take L as the reference length (for example, a) and U as the reference velocity (for example, the bubble emersion velocity u), then we obtain four dimensionless parameters, which are well known in fluid mechanics and which have a physical background:

$$\mathrm{Re} = UL/v, \mathrm{We} = \rho\, U^2 L/\sigma, \mathrm{Fr} = U^2/gL, \mathrm{Eu} = \Delta p/\rho\, U^2. \tag{5}$$

The Reynolds number Re is responsible for the balance of inertia and viscosity (skin friction) forces, the Weber number We is responsible for the balance of forces of the dynamic pressure and elasticity of the bubble surface, the Froude number Fr is responsible for the dynamic properties of the liquid in the force field, the Euler number Eu is responsible for the balance of pressure and dynamic pressure forces. According to the similarity theory, there are two independent dimensionless parameters. This means that there are dependencies between parameters in (5), and, therefore, the analysis of the dynamic properties of the liquid is not obvious.

Accounting for a need in obtaining the prediction of processes in connection with the variations of the quantity g (from g_0 – terrestrial value to $10^{-4}g_0$ – the microgravitation), we accept as L and U the quantities depending on g:

$$L = (\sigma/g\rho)^{1/2} = \delta_\sigma, \quad U = (\sigma g/\rho)^{1/4}.$$

In this case, $\mathrm{We} \equiv 1$, $\mathrm{Fr} \equiv 1$, the Reynolds number will revert to $\mathrm{Re}_g = M^{-1/4}$, the Euler number will revert to $\mathrm{Eu}_g = \Delta p/(\sigma g\rho)^{1/2}$. There are now two dimensionless complexes, and, hence, this is the minimum set of independent parameters. The diminution of parameter g from g_0 to $10^{-4}g_0$ leads to an increase in the quantities Re_g and Eu_g under equal remaining conditions. This means that the viscous friction contribution to the development of flows and to the dynamic pressure role reduces. This finally leads to the enhancement of the role of capillary forces. These conclusions coincide completely with the results of Section 2 obtained experimentally and confirmed by direct computations of the Navier-Stokes equations.

PREDICTION OF THE FILTRATION UNDER THE PRESSURE IN A CHANNEL, DETERMINATION OF SIMILAR PROCESSES

Consider a model problem the solution of which can elucidate the processes occurring in the porous underground strata through which the water, oil or gas are filtered. Let the liquid (with parameters ρ, v, and σ in the force field g) flows out of a channel with the orifice h, at the inlet and at the outlet of which the pressures $p_s + \Delta p$ and p_s are given. Thus, the filtration process is determined by seven physical constants:

$$\rho, v, \sigma, g, p_s, \Delta p, h. \tag{6}$$

We suppose that the pressure magnitude p_s can be very high (in the underground strata with the oil, it is of the order of 200 atm.). Under high pressures, the information about the rheological laws is scarce, and the matter with the prediction problem is even more complex [25]. From the viewpoint of the Navier-Stokes model, the additive constant in the pressure must not affect the motion. Therefore, any information is important, which indicates the qualitative peculiarities in convective processes in connection with the pressure magnitude variation.

Let us make use of the similarity theory. Let us choose the length and velocity scales depending on the p_s value:

$$L = p_s/\rho g, \ U = (p_s/\rho)^{1/2}.$$

As a result, Fr $\equiv 1$, Eu $\equiv 1$, and the Reynolds and Weber numbers have taken the following form:

$$\text{Re} = (p_s/\rho)^{3/2}/\nu g, \ \text{We} = p_s^2/\rho g \, \sigma. \tag{7}$$

It is necessary to add the remaining two parameters to (7)

$$\Delta = \Delta p/p_s, \quad H = \rho g h/p_s. \tag{8}$$

There is now a minimum set of four independent dimensionless parameters in (7) and (8). The pressure variation from the atmospheric one to the value p_s leads to the enhancement of the inertial properties of the liquid (the reduction of the internal friction influence) and to a substantial weakening of the influence of capillary forces. The latter conclusion is noted in [25] from the data of the results of geological investigations, but belongs to the category of the ones to be verified and substantiated.

The obtained conclusions point to the changes in convection, they are, however, insufficient yet to make a decision about the mathematical model choice. It is possible to elucidate this if one chooses the parameters of the medium in ground-based conditions for which the convective processes are similar to those in strata with a liquid under a high pressure.

Two physical processes are similar if the numerical values of all dimensionless parameters are the same [22]. In our case, the parameters must be the same in (7) and (8), which we will denote by the subscripts "s" and "g":

$$\text{Re}_s = \text{Re}_g, \quad \text{We}_s = \text{We}_g, \ \Delta_s = \Delta_g, \ H_s = H_g.$$

As a result, we have the following dependencies between the physical constants of media in the stratum and on the surface and the geometric scales:

$$(p_s/\rho_s)^{3/2}/\nu_s g_s = (p_g/\rho_g)^{3/2}/\nu_g g_g, \quad p_s^2/\rho_s g_s \, \sigma_s = p_g^2/\rho_g g_g \, \sigma_g \tag{9}$$

$$\Delta p_s/p_s = \Delta p_g/p_g, \quad \rho_s g_s h_s/p_s = \rho_g g_g h_g/p_g. \tag{10}$$

The system of four algebraic equations (9) and (10) for seven parameters with subscript "g" is nonlinear and practically little applicable. There are, however, as a rule additional considerations concerning several parameters. For an exact treatment of the stratum porous space it is necessary to have equal geometric sizes of the samples. This will be at $h_s = h_g$. Since the stratum lies at the depth of 2-3 km it follows that the value of the gravity acceleration changes by fractions of the percent. That is we can take $g_s = g_g$. Thus, we obtain from the second equation (10) the relation

$$\rho_g/p_g = \rho_s/p_s. \tag{11}$$

At the given pressure and density values for the liquid in stratum one can determine the model medium density on the surface $\rho_g = p_g \rho_s/p_s$, which will be (in the given case) by two orders of magnitude lower than the liquid density in the stratum.

With regard for (11) we obtain from equations (9):

$$\nu_g = \nu_s, \quad \sigma_g = \sigma_s p_g/p_s. \tag{12}$$

182

The last relation in (12) points to that the surface tension of the model medium is less than unity, which corresponds to the gaseous medium. It follows from (11) and the first relation (12) that

$$v_g \rho_g = v_s \rho_s p_g / p_s,$$

which is the dynamic viscosity of the medium on the surface is by two order lower than for the liquid in stratum. Right such a relation in orders of quantities will be for the air and water [3].

The temperature is known to be high in underground strata (~$100°C$ [25]). The presence of temperature jumps ΔT at the boundaries of the channel and liquid gives rise to heat fluxes. Their consideration introduces at least two more physical parameters: the coefficients of volumetric expansion β and the thermal diffusivity k_T. As a result, similar processes must have equal Prandtl numbers $Pr = v/k_T$ and Boussinesq numbers $Bu = \beta \Delta T$. Since the Prandtl number decreases with increasing temperature, the conditions in the stratum lead to a diminution of the difference in Prandtl numbers for the water in stratum and gas on the surface. So the water has $Pr = 7$ at $T=20°$ C, and at $T=100°$ C it already has $Pr \approx 1.8$, the air at $T=20°$ C has $Pr = 0.7$. The addition of water vapors increases the Prandtl number. One can obtain the equality $\beta_g \Delta T_g = \beta_s \Delta T_s$ for the given value of β_g by choosing ΔT_g. These conclusions already make it possible to select the mathematical model for predicting the convection in the underground stratum. The model must account for compressibility effects [26]. As the first step one can take the mathematical models of incompressible fluids, which are regularized with regard for weak compressibility along the motion trajectories [16, 18], and watch the compressibility influence on the dynamics of flows in channels [27-28].

CONCLUSIONS

The approach proposed for the computation of reference length and velocity scales from the value of the physical parameter of the system, which is varied within wide limits, enables the obtaining of the predictions for variation of convective processes by using the conclusions of the similarity theory.

The maps of benchmark processes in dimensionless criteria and the determination of model medium parameters on the Earth on which the processes are similar to those which occur under the conditions under study enable the specification of the mathematical model. A preliminary analysis of the well-posedness of boundary-value problems enables one to avoid many problems on the way of numerical solution and the interpretation of obtained data.

The determination of model media and conditions for which the state is similar to the one in the process under investigation enables the creation of the laboratory benches for prediction.

One can estimate the error between the parameter values of the model medium, which are required by the similarity theory, and the parameters of real media at the computation of benchmarks.

The method of finite elements is a universal exact method for solving the well-posed boundary-value problems, it allows the parallelization both at the region segmentation and at the solution of the system of linear algebraic equations.

ACKNOWLEDGMENTS

Researches of modeling the processes in underground layers have been supported by Governor of Hanty-Mansijsk autonomous region: Grant No. НУ 06.4/06-ЮГУ-234.

REFERENCES

1. L.D. Landau and E.M. Lifschitz, *Mechanics of Contnua*, Gostekhizdat, Moscow, 1954. (in Russian)
2. C. A. Truesdell, *First Course in Rational Continuum Mechanics*, Academic Press, New York, 1977.
3. G.K. Batchelor, *An Introduction to Fluid Dynamics*, Cambridge University Press, London, 1967.
4. O.A. Ladyzhenskaya, *Mathematical Questions of the Viscous Incompressible Fluid Dynamics*, Nauka, Moscow, 1970. (in Russian)
5. R. Temam, *Navier-Stokes Equations: Theory and Numerical Analysis*, North-Holland, Amsterdam, 1984, 3rd edn.
6. V.V. Ragulin, *Continuum Dynamics* **27**, Novosibirsk, 78-92 (1976).
7. B.G. Kuznetsov, N.P. Moshkin, and Sh. Smagulov, *Numerical Methods of the Viscous Fluid Dynamics*, Inst. Theor. Appl. Mech. of the USSR Acad. Sci. Siberian Branch, Novosibirsk, 1983, pp. 203-207.
8. K.R. Koseff and R.L. Street, *Trans. ASME/ J. Fluids Engng.* **106**, 390-398 (1984).
9. W.H. Leong, K.G.T. Hollands, and A.P. Brunger, *Int. J. Heat Mass Transfer* **42**, 1979-1989 (1999).
10. P. Roache, *Computational Fluid Dynamics*, Hermosa, Albuquerque, New Mexico, 1976.
11. S. Thakur and W. Shyy, *Numer. Heat Transfer* **24**, N. B. 31-55 (1993).
12. Kh. Miloshevich, A.D. Rychkov, and Yu.I. Shokin, *Modeling of Jet Flows in Steelmaking Vessels*, Published by the Siberian Branch of the Russian Academy of Sciences, Novosibirsk, 2000.
13. C.I. Christov, R.S. Marinova, *Computational Technologies* **6**(4), 81-108 (2001).
14. P.K. Volkov, *J. Engng. Phys.* **66**(1), 93-123 (1994).
15. P.K. Volkov, in *Russian Science at the Dawn of New Century*, edited by academician V.P. Skulachev, Nauchnyi mir, Moscow, 2001, pp. 8-17.
16. P.K. Volkov, *Vychisl. Tekhnol.* **9**(2), 20-30 (2004).
17. P.K. Volkov and A.V. Pereverzev, *Mat. Model.* **15**(3), 15-28 (2003).
18. P.A. Anan`ev and P.K. Volkov, *Computational Mathematics and Mathematical Physics* **45**(7), 1245-1258 (2005).
19. P.A. Anan`ev, P.K. Volkov, and A.V. Pereverzev, *Mat. Model. (RAS)* **16**(7), 68-76 (2004).
20. P. Volkov, A. Pereverzev, and P. Ananiev, in *Proc. 8th World Multiconference on Systemics, Cybernetics and Informatics*, July 18-21, 2004 Orlando, Florida, Vol. 9, pp.331-336.
21. W.L. Haberman and R.K. Morton, *Proc. Amer. Soc. Civil. Engrs.* **49** (387), 1-25 (1954).
22. L.I. Sedov, *Mechanics of Continuum*. Vol. 1. Nauka, Moscow, 1973 (English translation: Wolters-Noordhoff, Groningen, 1971).
23. S.S. Kutateladze and V.E. Nakoryakov, *Heat Exchange and Waves in Gas-Liquid Systems*. Nauka, Novosibirsk, 1984. (in Russian)
24. V.S. Avduevsky, I.V. Barmin, S.D. Grishin et al, *Problems of Space Production*. Moscow, 1980.
25. M. Muskat, *Physical Principles of Oil Production*, Mc Graw-Hill Book Co., New York, 1949.
26. P.K. Volkov, *Doklady Physics*, **52**(11), 625-629 (2007).
27. P.A. Anan`ev and P.K. Volkov, *High Temperature* **44**(3), 425-434 (2006).
28. P.A. Anan`ev and P.K. Volkov, *Fluid Dynamics* **41**(6), 871 – 880 (2006).

SPECIAL SESSION

TEACHING ENVIRONMENTS FOR

MATHEMATICS COURSES

Graphing Parametric Curves Using Spreadsheets

A. Didenko, D. Allison and G. Miller

The Petroleum Institute,P.O. Box 2533, Abu Dhabi, The United Arab Emirates

Abstract. Traditionally, using only pencil and paper the graphing of parametric functions was a time-consuming task. Microsoft Excel is a convenient tool that can be used in the classroom to facilitate and enhance this procedure. A simple, but effective, spreadsheet is described.

Keywords: Mathematics education, technology, parametric functions, Microsoft Excel spreadsheet.
PACS: 01.50.H-, 01.50.ht, 01.50.hv

INTRODUCTION

Sophisticated educational technology is available, but at a price. In many parts of the world there are schools and tertiary level institutions that are not provided with sufficient funds to purchase either instructional math software such as Maple or Mathematica, or to equip a dedicated mathematics teaching lab with PCs. Often an instructor only has a single classroom PC loaded with a standard home computing/business software item such as MS Office available. However, the Excel component of this package can be a useful instructional aid for the teaching of college level mathematics [1], [2], [3]. In a previous paper [4], the authors described how Excel can be effectively deployed in the classroom as a graphing tool for pre-calculus and introductory calculus courses. In this paper we continue the theme, highlighting the ability of this software to rapidly and effectively generate graphs of parametrically defined functions.

THE EXCEL SPREADSHEET

In this section of the paper the authors describe an in-house developed Excel spreadsheet that can serve as a graph generator for parametric equations, suitable for pre-calculus or introductory calculus laboratories. The spreadsheet is a cost-effective and easy to use teaching tool.

The authors wish to emphasize that the development of the spreadsheet did not require any software programming skills (in Visual Basic for example). They set out to develop a tool that is easy to construct and utilize. Only the standard features of the Excel package were needed, such as cell referencing and table calculations. Key cells for data entry were designed to incorporate pop-up comments/instructions to aid the user.

The spreadsheet features a list of pre-set parametric functions, typical of cases found in most standard calculus sequence textbooks. These are accessible via a pull-down menu and are accessed in the examples below.

Figure 1 shows the graphical output corresponding to when

CP1067, *Applications of Mathematics in Engineering and Economics '34—AMEE '08,* edited by M. D. Todorov

$$x(t) = \cos t \quad \text{and} \quad y = \sin 2t.$$

By moving an Excel 'slider' control the value of the parameter t can be altered, and the point corresponding to the t-value moves along the curve.

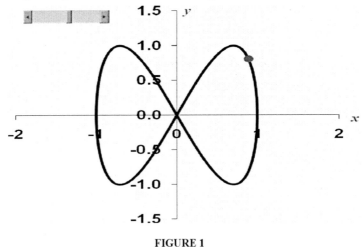

FIGURE 1

The software is able to handle more complicated parametric curves, as illustrated in the following figure. Figure 2 shows the graphical output corresponding to when

$$x(t) = \cos t - \cos 80t \sin t \quad \text{and} \quad y = 2 \sin t - \sin 80t$$

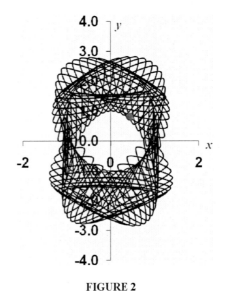

FIGURE 2

Figure 3 represents the case of a *cycloid*.

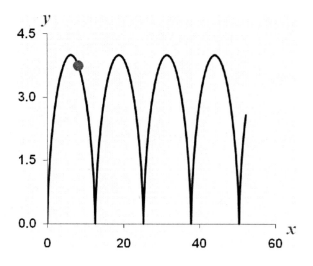

FIGURE 3

The parametric equations

$$x(t) = \frac{1}{2}\cos t - \frac{1}{4}\cos(2t) \quad \text{and} \quad y(t) = \frac{1}{2}\sin t - \frac{1}{4}\sin(2t)$$

produce a heart-shaped sub-region of the famous Mandelbrot subset (see Figure 4).

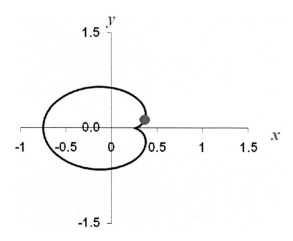

FIGURE 4

If one considers a point on a circle whose radius is *n*, and lets the circle roll (without slipping) on the outside of a fixed circle whose radius is *m*, the curve that the point traces is called an *epicycloid*, and its parametric equations are given by:

$$x(t) = (m+n)\cos t - n\cos\left(\frac{(m+n)}{n}t\right), \quad y(t) = (m+n)\sin t - n\sin\left(\frac{(m+n)}{n}t\right).$$

Notice now that the curve is dependent upon two parameters *m* and *n* in addition to *t*.

By means of two additional sliders integer values for *m* and *n* can be inputted. In Figure 5, we can see the graphical output corresponding to the case of *m* = 4 and *n* = 1.

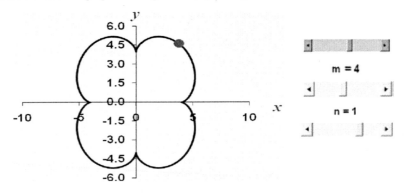

FIGURE 5

Figure 6 shows the parametric curve corresponding to *m* = 7 and *n* = 1. Note that the cusps are clearly evident in the output.

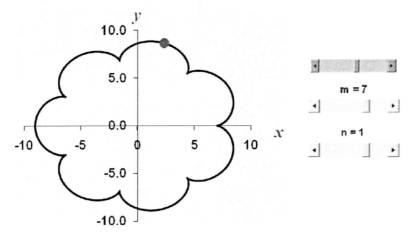

FIGURE 6

190

AN APPLIED EXAMPLE - BREAKING THE SOUND BARRIER

Parametric equations occur in the modeling of how a shock front develops when an aircraft passes through the sound barrier [5]. In this section we describe how the modeling can be done using the Excel spreadsheet.

In what follows, one unit equals the distance traveled by sound in one second.

The parametric equations of the position (x, y) at time t (seconds) of a sound wave emitted from a point (a, b) at a time $t = c$ are given by:

$$x(t) = a + (t - c)\cos\theta, \qquad y(t) = b + (t - c)\sin\theta,$$

where we restrict the angular parameter $0 \leq \theta < 2\pi$.

Consider the case of an airplane flying due East at a constant speed of v units per second. Let its position at time t be $(vt, 0)$.

Three different values of the airplane's speed are considered:

$$v \in \{0.8 \ units/sec, \ 1.0 \ units/sec, \ 1.4 \ units/sec\}$$

Mach 0.8 (Below the Speed of Sound)

When $v = 0.8$ (Mach 0.8) the position of the airplane at time t is $(0.8t, 0)$. The position of the airplane and the position of a number of sound waves at various different times are then found.

$t = 0$

At time $t = 0$ the airplane is at the origin $(0, 0)$.

After 5 seconds a sound wave emitted from the airplane has position (x, y) given by
$$x(t) = 0 + (5 - 0)\cos\theta, \quad y(t) = 0 + (5 - 0)\sin\theta$$
i.e. the parametric curve representing the position of this sound wave is a circle, center the origin, with a radius equal to 5 units.

$t = 1$

At time $t = 1$ the airplane has position $(0.8, 0)$.

After 4 seconds the position of a sound wave emitted by the airplane from this point is

$$x(t) = 0.8 + 4\cos\theta, \quad y(t) = 4\sin\theta.$$

$t = 2$

The airplane is at $(1.6, 0)$.

After 3 seconds the position of an emitted sound wave is

$$x(t) = 1.6 + 3\cos\theta, \ y(t) = 3\sin\theta.$$

$t = 3$

The airplane is at $(2.4, 0)$.

After 2 seconds the position of an emitted sound wave is
$$x(t) = 2.4 + 2\cos\theta, \ y(t) = 2\sin\theta.$$

$t = 4$

Airplane is at $(3.2, 0)$.

The position of a sound wave after 1 second is $x(t) = 3.2 + \cos\theta, \quad y(t) = \sin\theta.$

191

$t = 5$

The airplane is at (4, 0).

Using the Excel spreadsheet the positions of the airplane (points) with the corresponding sound waves can be graphed (Figure 7). The sound wave positions lie on circles, which can be produced by generating incremental values of θ between 0 to 2π.

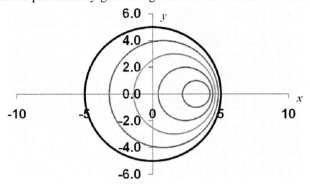

FIGURE 7

Mach 1 (At the Speed of Sound)

Now $v = 1.0$.

Repeating the above procedure we have:

$t = 0$

The position of the airplane is (0, 0).

After 4 seconds the position of an emitted sound wave is $x(t) = 4\cos\theta, \quad y(t) = 4\sin\theta$.

$t = 1$

The position of the airplane is (1, 0).

After 3 seconds the position of an emitted sound wave is

$$x(t) = 1 + 3\cos\theta, \quad y(t) = 3\sin\theta.$$

$t = 2$

The airplane's position is (2, 0).

After 2 seconds the wave is at $x(t) = 2 + 2\cos\theta, \quad y(t) = 2\sin\theta$.

$t = 3$

The airplane's position is (3, 0).

After 1 second the wave is at $x(t) = 3 + \cos\theta, \quad y(t) = \sin\theta$.

$t = 4$

The airplane is at position (4, 0).

Plotting the position points of the airplane with the corresponding sound wave curves we get Figure 8.

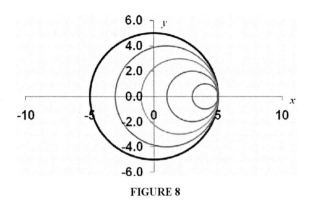

FIGURE 8

Note that all the sound waves intersect at the airplane's $t = 4$ position. When this happens we say that the airplane has reached the "sound barrier".

Mach 1.4 (Breaking the Sound Barrier)

Now the airplane's speed is $v = 1.4$, which is faster than the speed of sound.

$t = 0$

The position of the airplane is (0, 0).

After 5 seconds the position of an emitted sound wave is $x(t) = 5\cos\theta, \quad y(t) = 5\sin\theta$.

$t = 1$

The position of the airplane is (1.4, 0).

After 4 seconds the position of an emitted sound wave is

$$x(t) = 1.4 + 4\cos\theta, \quad y(t) = 4\sin\theta.$$

$t = 2$

The airplane's position is (2.8, 0).

After 3 seconds the wave is at position $x(t) = 2.8 + 3\cos\theta, \quad y(t) = 3\sin\theta$.

$t = 3$

The airplane's position is (4.2, 0).

After 2 seconds the wave is at $x(t) = 4.2 + 2\cos\theta, \quad y(t) = 2\sin\theta$.

$t = 4$

The airplane is at (5.6, 0).

The position of a sound wave after 1 second is $x(t) = 5.6 + \cos\theta, \quad y(t) = \sin\theta$.

$t = 5$

The airplane is at position (7, 0).

Plotting the position points of the airplane with the corresponding sound wave curves we get FIGURE 9.

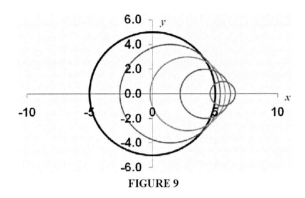

FIGURE 9

The sound waves intersect each other to form a shock front. The sound barrier has been broken by the airplane, and this is signaled by a loud "sonic boom".

The shock front can be modeled by the two straight lines, whose parametric equations are respectively [5]:

$$x(t) = 7 - \sqrt{0.96t}, \quad y(t) = t$$

and

$$x(t) = 7 - \sqrt{0.96t}, \quad y(t) = -t.$$

These two lines are superimposed on the sound wave diagram to illustrate the shock front (Figure 10).

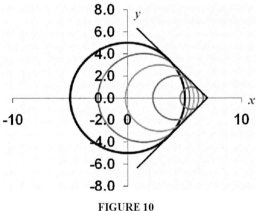

FIGURE 10

REFERENCES

1. B.S. Gottfried, *Spreadsheet Tools for Engineers Using Excel,* McGraw-Hill Science Engineering, New York , 2005.
2. S.C. Bloch, *Excel for Engineers and Scientists,* John Wiley & Sons, New York, 2003, 2nd edn.
3. E. Neuwirth and D. Arganbright, *The Active Modeler: Mathematical Modeling with Microsoft Excel,* Duxbury Press, 2003.

4. A. Didenko, D. Allison, and G. Miller, "Using Spreadsheets as Graphing Tools" in *The 6th UAE Mathday*, edited by A.Didenko, Conference CD Proceedings, ISBN 978-9948-03-668-5, The Petroleum Institute, Abu Dhabi, UAE, 2008.
5. R.T. Smith and R.B. Minton, *Calculus*, McGraw-Hill Science Engineering, New York, 2002.

Promoting Creative Engagement with SpicyNodes

M. Douma[a], G. Ligierko[a] and I. Angelov[b]

[a]Institute for Dynamic Educational Advancement (IDEA),
616 Great Falls Road, Rockville, MD USA
[b]University of Plovdiv, FMI, 236 Bulgaria Blvd., 4000 Plovdiv, Bulgaria

Abstract. The Information Age has posed new challenges to the presentation and communication of knowledge. In their attempts to handle the abundance of information in an effective way, people have come to rely on maps. Maps are a useful tool for visualization and the arrangement of concepts as they mark the current location in the context of their surroundings. This paper presents SpicyNodes, an improved method of organizing and visually presenting web-based information using radial tree maps. The project encourages creative thought on two levels. For those who create, it provides a platform for organizing and displaying information in a creative manner, and sharing those methods with others. For users, it allows the exploration of complex systems while gaining the perspective of a broader context. By presenting information in a natural and intuitive way, such radial maps can help people think in novel and creative ways. It may also aid and find application in the learning process of kids and students with learning disabilities.

Keywords: SpicyNodes, radial tree, radial map, layout, animated, interactive, creativity, creative engagement, learning disabilities, tree, network, hierarchy.
PACS: 01.40.-d; 01.50.-i

INTRODUCTION

Creative insight is often triggered by novel interpretations of information and the recognition of relationships between seemingly disparate pieces of information. While the abundance of information available today facilitates an environment of creativity, the increasingly complex interrelationships between pieces of information are often lost because of the sheer volume of the world's knowledge repository.

More specifically, the burgeoning of the Internet has presented new challenges in developing ways to both display information and to make it accessible to billions of people around the globe. Creative designers, artists, and writers seek new ways to humanize the information-saturated world, making information and ideas more accessible and engaging through the interplay of how information is logically structured and how information is visually presented.

This paper presents SpicyNodes, an advanced method of organizing and visually presenting web-based information using radial tree maps that encourages creative thought on two levels. For those who create, such radial maps provide a platform for organizing and displaying information in a creative manner, and sharing those

CP1067, *Applications of Mathematics in Engineering and Economics '34—AMEE '08,* edited by M. D. Todorov
© 2008 American Institute of Physics 978-0-7354-0598-1/08/$23.00

methods with others. For their viewers, radial maps allow the exploration of complex systems while gaining the perspective of a broader context.

OBTAINING INFORMATION

Throughout the history of civilization, many human needs have remained constant, but the need for information has increased exponentially. As the sum of human knowledge grew, so did the challenge of classifying, exploring, organizing and searching systems of information. Yet until the late 20th century, the most advanced system widely available for storing and retrieving information was the classified written index, which although adequate, is fundamentally a static, hierarchical system.

Today, the enormous quantity of information on the Internet is organized in many ways, from simple sets of web pages; to complex, database-driven e-commerce sites; to ever-changing news sites. Users have an expectation that the Internet will provide them with information at their fingertips, but it often does not provide the specific, useful information they are seeking. In order to make information truly accessible and drive creative thought, the gap in organizing online data needs to be filled.

In traditional methods, information seekers are like real world explorers, hacking their way through the undergrowth with only a vague sense of where they are in relation to the jungle as a whole. In contrast, natural systems offer a helicopter ride above the information jungle, allowing views of the whole landscape along with opportunities to land in clearings for detailed exploration. The effectiveness of naturalistic tools and models comes from their emulation of the way the human mind perceives and stores information.

SPICYNODES: A FAMILIAR ENVIRONMENT

In order to aid and accelerate the browsing process, the project uses geometrical layouts, animation, and interactivity. By employing simple and yet expressive geometrical forms, and designing the elements so that they behave like their counterparts in the real world, the radial map creates a familiar and intuitive environment for the user.

With an appropriate interactive design, a system guides, encourages, and permits a user to take a certain course of action or creates a state of mind that fosters creativity [1]. In contrast, an inappropriate interactive design may distract, discourage, and prohibit a user from taking a certain course of action, thus obstructing creativity.

SpicyNodes' interactivity materializes in the wide variety of options with which it responds to user input. Contextual controls can be utilized by the user to perform various actions on the active node. Typically, the available options are to (a) display more information about the node, e.g., panning through text; (b) go to a related web page; (c) go to another nodemap, allowing multiple nodemaps to be linked together. Optionally, these contextual controls can be configured to evoke more suggested actions or to include interaction sounds.

The project employs an environment suited to depict both hierarchies and networks. When exploring hierarchies, a "tree" metaphor is used, and the user is always shown "where they are" and where "home" is. In the project's network mode, networks are displayed using graphical interfaces that are similar to those used in hierarchies, but there is no "home," and the user can browse endlessly. The creator can also use a mixture of "hierarchy" and "network" structures.

"Explorational" Interface, Creative Panorama

Such radial maps allow users to search efficiently for information – even if they are not sure it exists. Because SpicyNodes shows both the big picture and the detail simultaneously, users can explore a site's information architecture while keeping track of where they are within the whole structure – in other words, without getting lost. In fact, such a radial map might be better described as an "explorational" GUI rather than a "navigational" one.

This approach allows a kind of visualization that lets users see the structures and substructures of information in each section of a web site or database. Also, in each area users are able to perceive the information's relative density, which provides a panorama presenting meta-models in a new way.

This results in numerous opportunities for creative insights. Creative insights fall into three categories: selective encoding, selective combination and selective comparison [2]. Selective encoding involves distinguishing irrelevant from relevant information; selective combination refers to taking "selectively encoded" information and combining it in a novel way, whereas selective comparison indicates a novel way of relating new information to old information.

Radial maps facilitate the creative process by accomplishing several of the same creative goals as keyword searching, but in a visual manner. Users can discover unnoticed connections, explore in an iterative manner to focus on a topic, and serendipitously find new insights.

Intuitive Arrangement

In order to interpret a situation, people rely on information-rich representations. Examples include graphical representations, metaphors, and spatial positioning of objects. Spatial relationships among freely positioned objects may be viewed as linearly ordered, spatially arranged, or grouped together. Spatial positioning of objects provides an effective means to express emerging relationships among the positioned objects [1].

This project achieves its usability and user-friendly model by seeming to base nodes and connections in a space that is familiar and intuitive to humans. The nodes exert gravitation-like attraction to each other, node relationships and movement take place in a two-dimensional pseudo-Newtonian space. When focused on the periphery, related nodes (children) radiate from their immediate parent in a cascading manner, just as the sun swirls in our galaxy, planets orbit the sun, moons orbit planets, satellites orbit moons, and so on.

Information is easiest to understand if ancillary elements are hidden from view. Designers should be aware of this "precognitive" phenomenon so they can reduce distraction and allow people to maximize their creative opportunities [3].

Similar to our experience of walking around, where we only see our immediate surroundings; this implementation of radial maps provides an option to limit the vista so that only nearby nodes have large, legible labels. To reduce local complexity in different areas of a map, the system conceals the detail of sections of the whole graph that are not immediately relevant to the section the user is browsing.

Intuitive Motion

Human-computer interaction studies have concluded that perception affects the creative potential. Systems that require extra effort to find or notice things tend to be distracting and the time spent on concentration takes away from actual cognition. People cannot function productively below their proprioceptive control limit. A model of eye–hand coordination indicates that movement might give important proprioceptive feedback [3].

Thus, this project's movement subsystem is designed to mimic movement in the real world as closely as possible. When users click a node, it gathers speed and then decelerates as it settles into its new orbit. This naturalistic mode of movement is aesthetically satisfying, and allows the user a much greater sense of actually working with the interface rather than simply using it. The way in which nodes jump from orbit to orbit is similarly elegant: instead of tracing a line based on rectangular coordinates within the graph plane, they follow a route based on polar coordinates. To correct a common problem of earlier radial tree implementations, constraints are placed on the motion of nodes so that the whole layout does not appear to flip in a way that is disorienting to the user. Instead, nodes arc from orbit to orbit, as a spacecraft would when maneuvering under the influence of a planet's gravity.

SpicyNodes employs systems for panning and rotation based on physical models of motion, thereby creating a living, organic space and providing intuitive ways for the user to rotate and move a scene. To create a lifelike interaction, we use artificial motion rules, inspired by, but not directly based on, traditional Newtonian motion. Since a mouse can only move horizontally and vertically, the project utilizes a novel method of interaction using centers of mass, momentum, and friction – similar to pushing plates around a dining room table – so that the user can both rotate and pan using only two dimensions of mouse motion.

Additional Navigational Aides

Complex working environments are problematic for users who need to keep track of information, how to remember it, and how to use it efficiently. The human memory works best when certain environmental cues serve as reminders. The absence of such environmental cues can cause memory challenges for the user [4].

With this project, a breadcrumb display serves as an environmental cue. Users always know where they are because at any moment they can learn their location by looking at the route map (e.g., Home > Fruit > Sweet > Apples).

In most styles, colors are widely used to group the nodes, either by major group, or by the distance from the center. This helps users keep track of nodes as they refocus from one node to another.

This project also incorporates traditional search methods to "jump" to a node. For example, in small data sets, misspellings of proper names are a common problem. SpicyNodes incorporates a text search engine based on a fuzzy search algorithm that is designed to detect phonetic and typographical misspellings, thereby incorporating the most flexible aspect of traditional navigation technology – searchability – into its new model of an explorational interface.

FOSTERING CREATIVITY

Theoretical Origins

The project is designed to facilitate creative thought in users. Several models of creativity informed the approach.

Contextualizing Information

Providing information on a need-to-know basis helps avoid information overload and increases opportunities for learning on demand. Projects should avoid producing more out-of-context information since human attention is a scarce resource and help people attend to the information that is the most relevant for their task at hand [5]. Humans want to act—they do not want to study large information spaces (e.g., help information, design rationale) in the abstract [6].

This project developed an efficient presentation of a large tree graph in a small space that allows users to appreciate context as well as grasping detail. Users do not have to scan back and forth between pages on a website as they would when using a traditional sitemap.

Creativity factors

Creative individuals typically share certain characteristics, described by Sternberg and Lubart [7]. But not everyone is inherently creative. To foster creativity, a tool must be convenient for highly creative people, while also useful for unimaginative users. Radial maps meet those criteria.

TABLE 1. Fostering a willingness to be creative

Quality	How radial maps help
Willing to redefine difficult problems, analyze solutions to problems, and surmount obstacles	Radial maps simplify complex concepts, starting broad, and converging on a narrowing focus.
Willing to work hard in finding something they love to do	The burden of finding information is reduced because information is structured.
Willing to tolerate ambiguity	Navigating a network structure, users always feel they have discrete paths, even though there may be multiple equivalent paths.
Willing to take sensible risks	By keeping all visualizations in a single web page, latency is minimal, and there is no "cost" to getting information from a wrong node.

Willing to see things from multiple points of view	Radial visualizations present a web of ideas and concepts simultaneously.
Willing to keep a perspective on oneself	A radial layout inherently shows the user a perspective of "where they are" in a broader network of nodes.
Believing in one's ability to be creative	Radial layouts are easy to use and place minimal burden on the user, building confidence that they will find what they seek.
Appreciating the aesthetic qualities of an experience [8]	Well designed visuals are beautiful and exciting to view.

Csikszentmihalyi's Flow Model

SpicyNodes is an apt tool for navigation and visualization and its usage will likely evoke in users the sensation of "flow."

Flow is a concept introduced by the psychologist Mihaly Csikszentmihalyi, a professor at the University of Chicago, to describe a particular state of "intense emotional involvement" [9] which people experience when heavily focused on challenging activities. When in flow, people feel it easy to achieve peak performance, following a naturally acquired sensation of conceptual progression.

The work of Csikszentmihalyi has been found to have considerable impact on human-computer interactions. Professors Donna Hoffman and Thomas Novak from the Vanderbilt University defined the flow experience as "the state occurring during network navigation, which is (1) characterized by a seamless sequence of responses facilitated by machine interactivity, (2) intrinsically enjoyable, (3) accompanied by a loss of self-consciousness, and (4) self-reinforcing" [10]. Hoffman and Novak also believe that flow is "a central construct when considering consumer navigation on commercial Web sites" [11].

SpicyNodes has a feature set that completely matches the Hoffman-Novak definitions. When changing focus between the nodes, the user will smoothly extend his/her route as prompted by the environment. The experience of frictionless goal achievement should bring about pleasure and satisfaction, and it may eventually result in total absorption in the activity. By intuitively steering the user to further findings, SpicyNodes will increasingly start feeling as an inventive tool, natural and delightful to operate. Any design featuring SpicyNodes will not only spur exploratory behavior but will also provide utter fulfillment.

Van Oech Model

In 1986, Roger von Oech argued that there are four stages in the creative cycle, each of which requires distinct, although imaginary, roles. The paradigm offers flexibility for fact- and value-based thinkers alike. It is as outlined in Table 2 [12].

In the development phase of the project, all four Oech roles were used. The collaborating designers drew on new and uncommon places in the outside world, and experimented with new metaphors (explorer). After reviewing all the pertinent information gathered as explorers, the designers' most difficult task was working with the basic concept of a radial layout and inventing various implementations (artist). The designers evaluated ideas and simplified the designs to arrive at the core elements. The designers had to be self-critical, and balance new, risky ideas against familiarity (judge). They faced many challenges, and took the offensive to balance visual simplicity against clear information design (warrior).

TABLE 2. van Oech model

Role	Description
Explorer	Finds problems. Employs curiosity and willingness to look for ideas in out-of-the-way places, and applies principles learned from other seemingly unrelated topics.
Artist	Finds ideas. Take this material and experiments with it in a variety of ways.
Judge	Finds solutions. Helps determine if an idea is any good, and if it's what is wanted and needed.
Warrior	Takes action. Takes the offensive and devises strategies to overcome barriers to make the idea a reality.

More importantly, the Oech model also applies to end users. Most people have skills in at least one of the roles necessary for creativity to flourish. Using a computer tool is normally a lone activity, with no collaboration. This project is designed to facilitate each Oech role, and to help bolster roles where users are weak. With radial maps, users can easily jump from idea to idea (explorer). Users can manipulate and interact with a radial map, adjusting the physical space, and choosing a wide range of paths though the web of information (artist). Users can quickly determine which nodes seem relevant from labels that contextually appear, and by viewing preview information about a node, without going to a new web page (judge). Accessible creative tools eliminate barriers, and encourage the user to take action and devise a strategy for finding the information they seek (warrior).

Freedom to Organize and Explore Ideas

Computers can help stimulate creative thought by providing support for techniques like brainstorming, divergent thinking and lateral thinking, as well as by offering a number of ways for users to visualize the problem, such as sketching and concept mapping [13]. This project is based on the premise that most people, regardless of how spatially oriented they are, are inherently good at navigating their environment. The approach is a series of maps depicting how concepts are arranged. And by putting the viewer in the map, just like being at a crossroads or intersection, the user always has a choice of where to go. In the process, users have the opportunity to see new relationships between pieces of information, which in turn sparks creative thought.

Design Elements

In the course of research on how user interfaces influence creativity, some design principles have emerged that enable a user to be not only be more productive but also more innovative when using software [14]. An interface compliant with these guiding rules facilitates effective searching of intellectual resources, improved collaboration, and accelerated discovery processes. These principles are especially applicable to composition tools – environments that provide the means for physical representations.

TABLE 3. Summary of key design elements

Element	Description
Familiar environment	Layouts and interactions that are reminiscent of the real world.
Intuitive arrangement	Radial layout, similar to orbits, with the periphery shrunk or hidden.
Intuitive motion	User can drag and rotate the nodes, as if he or she were pushing a dinner plate around a table with a finger.
Animation & Interactivity	The slow-in, slow-out animation technique enlivens the interface; the node interactivity enhances the user experience.
Breadcrumbs	Explicitly delineate a user's path back home.
Color coding	Optional grouping by color coding.
Keyword searches	Full text searching can find content within a node, while fuzzy searching matches commonly misspelled words or phonetic misspellings.

RESULTS AND DISCUSSION

Creative Engagement of Design Team

More than twenty designers from all around the world participated in creating SpicyNodes, and were tasked with finding modern and effective ways of using radial maps in the 21st century.

The designers were specifically instructed to develop two designs each, one more "conventional" and familiar to users, and the other more "experimental." As a result, the designers could indulge their creativity, and often the experimental designs drew on more interesting metaphors and were thus more intuitive than the conventional designs.

The development of this project into a complete and universal radial layout is an ongoing creative and collaborative process. During the core design phase, over a dozen designers and artists within and outside of our organization were invited to collaborate. This was a self-selected community of designers who share an interest in new interfaces and information design. In keeping with the suggested means of stimulating creativity [15], they had enough latitude to shake themselves free of the established practices. As a result, they contributed diverse original ideas for sample visualizations, use cases, interactions, and so on. Some of these are displayed by Figure 1.

Creative Engagement by Users

SpicyNodes promotes creative engagement by both creators and end users. Individual creators are encouraged to customize their implementation using many dimensions of variability. End users who explore nodemaps embark on a creative journey during their brief use of the interactive nodemap, and chart their own course through a sea of information.

Individual Creativity

Each information structure, such as a web site, has its own identity and culture. Its appearance and functionality results from either purposeful design or an organic process. In marketing terms, a web site's appearance is an integral part of its "brand," an aspect that visitors come to recognize and appreciate. In artistic terms, this project provides an empty canvas, as well as a rich palette of colors and fundamental shapes that creators can assemble into unique representations of information.

Interfaces typically provide standardized styles from which the user can view, use, and choose. This method, however, inhibits creativity rather than enhances it. Creativity support tools should offer control of and flexibility in the creative processes [13].

This project offers creators a graphical interface that provides an intuitive method for personalizing the appearance of their nodemaps. With a few mouse clicks, creators can select the layout of their nodemaps, as well as the shapes, motions, and navigational methods they would like to incorporate into their unique version of SpicyNodes. Further, such radial maps allow creators to select or design colors by manipulating several gradient scales, such as warm-cool, harmonious-opposing, and muted-vivid. These "skins" (Figure 1) allow creators to match or complement the appearance of their web sites. The wide variety of choices in color schemes, backgrounds, and complementary fonts enhance the user experience and foster creative thought while maintaining the project's consistent functionality.

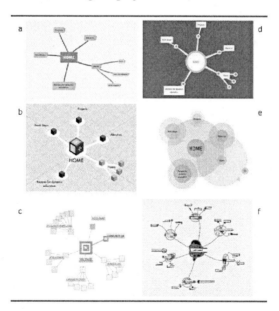

FIGURE 1. Examples of the widely varied skins that users can implement to style their nodes. (a) "folders," (b) "boxes," (c) "squares," (d) "subway," (e) "circles," and (f) "scribble"

In an educational setting, the process of authoring nodemaps (similar to constructing concept maps) is itself educational. Students can enjoy creating and linking nodes, and decorating them with colors and graphics.

Online Interchange

This project is designed to promote community-based collaborations and promote shared achievements. Creative activity grows out of the relationship between an individual and the world of his or her work, as well as from the ties between an individual and other human beings [16].

Community collaborations foster social creativity, which thrives in environments that support so-called meta-design [17]. Meta-design is defined as the necessity for systems to be open in a way that allows users to modify them. The perspective of meta-design is that problems cannot be anticipated at system design time and result in gaps between users' needs and systems' functionality. These gaps, referred to as mismatches, cause breakdowns that serve as potential sources for new insights, new knowledge, and new understanding. [18]. Meta-design calls for a shift in focus from finished products to solutions that enable users to handle mismatches they encounter. In this way, better systems evolve while users can engage their creativity and act as designers.

This project features an open architecture and is meta-design compliant. The creators can upload their work in online galleries for display, and the showcased pieces are then shared with the community. This embraces different perspectives and encourages novel approaches that enable new, original solutions.

By offering such opportunities, SpicyNodes also serves as a system that supports the creation and evolution of so-called active boundary objects [18] – artifacts that facilitate shared understanding and activate information relevant to a specific task. Active boundary objects create awareness of others' work, afford individual reflection and exploration, enable simultaneous, parallel and serial co-creation, allow participants to build on the work of others, and provide mechanisms to help draw out individual and collective knowledge and perspectives.

FIGURE 2. An example of SpicyNodes "skinned" implementation. It's been prepared and uploaded by a designer outside the developing team and star-rated for achievement by the supporting community. Such collaboration is indicative of open community interchange

End Users' Unique Interactive Journey

In compliance with the recommended design principles that facilitate creativity, and in contrast to guided navigation, this project sports an interface characterized as "explorational."

SpicyNodes' interface gives users an incentive to explore and find their route themselves. It appeals to their curiosity and need for discovery by offering a variety of intertwined paths. Although not all of these paths immediately lead a user to where he or she needs to go, the journey enables users to discover new relationships among the objects represented as nodes. Some of these discoveries may appear serendipitously, while others may be the result of a logical sequence of actions; either way, the underlying process is highly inventive.

APPLICATIONS

SpicyNodes is currently utilized for education purposes. Using it as a tool for representing a complex informational structure, an interactive family tree of the ancient Greek Gods was built and published on the Web[19]. This web site demonstrates the power and versatility of the technology.

This project's Greek Gods implementation is well suited for use by museums. Its historical context can be expanded to include genealogies of medieval dynasties (Merovingian, Tudor, Bourbon, and so forth), intellectual and emotional relationships between significant figures (like the influence of John Locke over David Hume in philosophy or the liaison of Percy Bysshe and Mary Shelley in literature), the cultural background of important events (such as the ascent of the British Empire during the Elizabethan Age and its reflection in the works of Shakespeare and other playwrights), and many others. For natural sciences, SpicyNodes can visualize patterns, such as interrelation of minerals based on their ingredients, the transformation of rocks in view of their geological formation, maps of the constellations and other cosmic objects such as nebulae, quasars, pulsars, and so on.

Site maps comprise another current successful application of the project. With its intuitive navigation, SpicyNodes reveals to visitors the underlying model of data organization which, coupled with its "focus + context" functionality, makes moving back and forth as natural as it feels in the physical world. In this way, site maps become meaningful to even less experienced users and allows them to grasp the notion of site navigation quickly and easily.

The project also has potential for use in libraries. Whether electronic or book-based, every library needs a comprehensive and user-friendly cataloguing system. Because of the complexity and enormity of the content stored in a library, cataloguing is inherently problematic. Radial maps can allow users to navigate knowledge-dense areas, many of which often overlap. For reference purposes, the radial tree's searching features will be very useful.

Some research evidence indicates that graphic organizers like SpicyNodes enhance the performance of students with learning disabilities and could further apply learning tools like SpicyNodes for exceptional children and special education [20]

DELIVERY MEDIA

The project was developed using Adobe's (formerly Macromedia's) Flash technology, and is written in the ActionScript language. Its current form is determined by the abilities of Flash, with all of its pros and cons. Recent researches show Flash as the most popular visualization technique currently on the Web and it has begun penetrating the mobile devices and consumer electronics markets as well. SpicyNodes is currently deliverable as a Web-based Flash application. The animated radial map is rendered in a browser capable of displaying Flash, version 7 and above, images.

The ubiquity and the versatility of Flash open the doors wide for radial map applications. If used to display mobile site maps, for example, radial maps will be an apt implementation for a small screen owing to its intelligent visualizing abilities. They can also serve as visual environment for PDAs, dealing easily with tasks such as scheduling, reminding, mind-mapping, and so forth. In this way, its information organizing functionality will find a natural use.

CONCLUSION

Today's repository of knowledge is greater than ever before, and the Internet has the potential to make enormous quantities of information accessible to any person. Yet the methods of information retrieval most often used on the Web have inherent limitations that can be overcome by organizing information using radial maps, such as those employed by SpicyNodes. When information is accessible in a natural, intuitive way, people can spend more time thinking about that information in novel and creative ways. This can lead to breakthroughs in critical thinking and advances in virtually every field, including, but not limited to, the realm of personal satisfaction.

REFERENCES

1. Y. Yamamoto and K. Nakakoji, *Interaction design of tools for fostering creativity in the early stages of information design*, Research Center for Advanced Science and Technology, University of Tokyo, 2005
2. R. J. Sternberg, *Creativity or creativities?*, Department of Psychology, Center for the Psychology of Abilities, Competencies, and Expertise, Yale University, USA, 2005
3. T. Selker, *Fostering motivation and creativity for computer users*, MIT Media Lab, USA, 2005
4. T. T. Hewett, *Informing the design of computer-based environments to support creativity*, Department of Psychology, Drexel University, Philadelphia, USA, 2005
5. H. A. Simon, *The Sciences of the Artificial*, The MIT Press, Cambridge, MA, 1996, 3rd edn.
6. T. P. Moran and J. M. Carroll (Eds), *Design Rationale: Concepts, Techniques, and Use*, Lawrence Erlbaum Associates, Inc., Hillsdale, NJ, 1996.
7. R.J. Sternberg and T. Lubart, *American Psychologist* **51**(7), 677-688 (1996).
8. P. Ahmed, *European Journal of Innovation Management* **1**, 3-43.
9. M. Csikszentmihalyi, *Creativity: Flow and the Psychology of Discovery and Invention*, HarperCollins Publishers, New York, NY, 1999.
10. D. Hoffman, T. Novak, and Y. Yung, *Modeling the structure of the flow experience among web users*, 1997, Available at http://club.telepolis.com/ohcop/sikszen.html
11. D. Hoffman and T. Novak, *Journal of Marketing* **80**(4), (1996).
12. R. van Oech, *A Kick in the Seat of the Pants. Using your Explorer, Artist, Judge and Warrior to be More Creative*, Harper and Row, New York., 1986
13. H. Johnson and L. Carruthers, *Supporting Creative and Reflective Processes*, HCI Laboratory, Department of Computer Science, University of Bath, UK, 2006.
14. M. Resnick, B. Myers, K. Nakakoji, B. Shneiderman, R. Pausch, T. Selker, and M. Eisenberg, *Design Principles for Tools to Support Creative Thinking. Report of Workshop on Creativity Support Tools*, Oct. 2005.

15. G. Fischer, *Distances and Diversity: Sources for Social Creativity*, Center for LifeLong Learning and Design (L3D); Department of Computer Science, University of Colorado, Boulder, USA, 2005.
16. H. Eden, G. Fischer, A. Gorman, and E. Scharff, *Transcending the Individual Human Mind – Creating Shared Understanding through Collaborative Design*, Center for LifeLong Learning & Design and Institute of Cognitive Science, Department of Computer Science, University of Colorado, Boulder, USA, 2000.
17. G. Fischer, E. Giaccardi, Y. Ye, and A.G., Mehandjiev, *Meta-Design: A Manifesto for End-User Development*, Dept. of Computer Science, University of Colorado at Boulder; Sutcliffe N., 2004.
18. G. Fischer, E. Giaccardi, H. Eden, M. Sugimoto, and Y. Ye, *Beyond Binary Choices: Integrating Individual and Social Creativity*, Center for LifeLong Learning and Design (L3D), 2005.
19. M. Douma, *Genealogy of the Greek Gods*, WebExhibits, 2006,
http://webexhibits.org/greekgods/index.html
20. S. V. Horton, *Journal of Learning Disabilities* **23**(1), 12-22 (1990).

SpicyNodes Radial Map Engine

M. Douma[a], G. Ligierko[a] and I. Angelov[b]

[a]*Institute for Dynamic Educational Advancement (IDEA),*
616 Great Falls Road, Rockville, MD USA
[b]*University of Plovdiv, FMI, 236 Bulgaria Blvd., 4000 Plovdiv, Bulgaria*

Abstract. The need for information has increased exponentially over the past decades. The current systems for constructing, exploring, classifying, organizing, and searching information face the growing challenge of enabling their users to operate efficiently and intuitively in knowledge-heavy environments. This paper presents SpicyNodes, an advanced user interface for difficult interaction contexts. It is based on an underlying structure known as a radial map, which allows users to manipulate and interact in a natural manner with entities called nodes. This technology overcomes certain limitations of existing solutions and solves the problem of browsing complex sets of linked information. SpicyNodes is also an organic system that projects users into a living space, stimulating exploratory behavior and fostering creative thought. Our interactive radial layout is used for educational purposes and has the potential for numerous other applications.

Keywords: Radial, map, layout, nodes, searching, browsing, navigating, information, concepts, relations, hierarchy, network, organic, interactive, animated.
PACS: 01.40.-d; 01.50.-i

INTRODUCTION

While the burgeoning of the Internet has put an abundance of information at our fingertips, the information is often organized in such a way that the user becomes the passive recipient of data. Because of the sheer volume of the world's knowledge repository, we face challenges in displaying complex information and making it accessible to billions of people around the globe.

This paper presents SpicyNodes, an improved user interface for difficult interaction contexts. Users manipulate and interact with entities called "nodes" – pieces of information – which they can manipulate in an intuitive and natural manner. This technology solves the problem of browsing complex sets of linked information.

SpicyNodes is based on an underlying structure called a "radial map," which has been explored in various formats for over a century. As a platform for creating node maps, SpicyNodes makes this approach come alive by (a) making node maps easy to create, allowing authors to focus on concepts rather than on software development; (b) allowing the user to navigate to and focus his or her attention on a given node while automatically adjusting the layout to accommodate the user's point of view; (c) animating the display so that it is smooth and intuitive when the user changes his or her focus; (d) giving node authors a myriad of ways to personalize the appearance of the nodes; and (e) placing the user in an organic setting where he or she can create and

CP1067, *Applications of Mathematics in Engineering and Economics '34—AMEE '08*, edited by M. D. Todorov
© 2008 American Institute of Physics 978-0-7354-0598-1/08/$23.00

interact with nodes, as well as see the relationships between nodes that are visualized as links.

THE PROBLEM: FINDING INFORMATION

As the sum of human knowledge has increased, so has the challenge of constructing, exploring, classifying, organizing, and searching systems of information.

Today, people have an expectation that the Internet will provide them with information at their fingertips, but it often does not provide the specific, useful information they are seeking. In order to make information truly accessible, the gap in organizing online data needs to be filled.

The process of searching for digital information can be discussed in the context of three broad categories:

Navigating Maps

People rely on maps to find things in the real world, and consult, for example, subway maps or road maps. Maps are also a useful visualization of concepts.

In large, dynamic networks such as the Internet, users also benefit from a sense of context – a class of information that tells them their location in a larger sea of information, and, therefore, the relevance and importance of the main class of information for which they have been searching.

Navigating the Internet using current models can be like looking at a map through a cardboard tube. Although the browser can pick out tiny details of the current area of focus, the larger map itself is obscured.

There are three main approaches to show an individual where they are on a map. First, the "You are here" approach utilizes a symbol or box to show where the user is located. This is only useful for large maps that can either be physically manipulated by the user or toward which the user can move closer, such as a subway map.

Second, "overview + detail" (also known as "detail-in-context") is a more complex system, and uses two scales simultaneously, one broad and one magnified. Typically, the whole information space is depicted, with the region of interest in a separate area. This approach is analogous to the key on a map and can be cumbersome when attempting to cross-reference two views.

Third, "focus + context" utilizes the entire visual space and employs effects to emphasize the region of interest in the same window. Like looking through a magnifying glass, the user views data of interest at the focal point in full size and detail, while the area around the focal point (the context) illustrates how the focal information relates to the entire data structure. In this most complex approach, the overview and detail are combined, serving to both conserve space and to provide an intuitive sense of context.

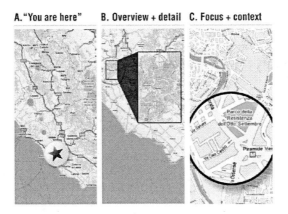

FIGURE 1. (a): In "You are here," the region of interest is marked. (b): In [Overview + Detail], the region of interest is in a separate area. (c): In [Focus + Context], effects are employed to emphasize the region of interest in the same window

These approaches are part of a continuum. For example, if "focus + context" is implemented with a sharp-edged lens, the appearance is quite similar to "overview + detail." The distinction between the techniques can be relative, especially when using add-on effects like folding, transparency, or inset or offset of detail [1].

Overall, "focus + context" approaches more closely resemble real life, where, as people navigate the world, they see the objects near them in detail, while the periphery is small or beyond their sight [2]. In order to stress the focus and enhance so-called Degree-of-Interest (DOI) [3], the display can employ text formatting, enlarged paths, color-coding, and so forth.

Browsing Topics

Browsing is useful when users do not know exactly what they are looking for, or where to find it. Users can browse with the purpose of "zooming in" on a narrow concept, honing in through a cascade of increasing specificity through a hierarchy. Also, users can browse in a less-directed manner, jumping between related concepts, like hyperlinks on web pages, through a network. Hierarchies and networks can be and are often combined in web sites.

Hierarchies

The simplest linear search involves a linear approach and is manageable with only a small number of items. An increasing number of topics makes linear search impractical. Over time, this has led to the development of tree-like structures, such as a book classification system in a library or a table of contents that organizes the information fragments in a book. By dividing topics into a cascading set of categories and subcategories, it becomes possible for people to quickly find the topics they want.

When people do not know exactly what they seek, an exploratory or "wandering" approach can be more effective, and can be categorized as browsing networks.

A thesaurus is one example of network browsing, as it is based on the idea of jumping among related items. On a broader level, the Internet is based on using hyperlinks to link together billions of web pages. The drawback is that general browsing often results in the "lost-in-space" syndrome, losing the context or looping through the same topics, especially when the author uses a slightly different or unexpected term for naming the concept.

Searching Keywords

When people are not certain where information can be found, they search by keyword(s) in indices, reference tables, or search engines. Searching keywords is a powerful and flexible way of obtaining information, yet the method has serious limitations.

Well-formed searches should produce results that a creative information seeker will consider very useful. In order to be useful for the creative process, keyword searches should be constructed to have at least one of the following goals [4]:

TABLE 1. Fundamental goals for keyword searching

Type	Description
Generative goals	The goal of the search can be to learn about a topic area or generate an idea, rather than find a specific document or fact.
Cross-context	Searches may extend across domains or collections to enable the searcher to gain alternative perspectives or discover unnoticed connections.
Exploratory and iterative	Searchers issue multiple, successive queries as their information need evolves. In the process, they collect useful bits of information, follow promising threads and identify new sources.
Serendipity and non-linearity	Serendipitous findings can provide valuable insight for the creative searcher.

Common search engines typically don't meet the fundamental goals for keyword searching. Because they work well only when users know what they are looking for, search engines do not meet generative goals. Indeed, people often guess at keywords or search for an approximation, then read web pages to discover what search terms to use to find the information they need. Likewise, because keywords searches work best when users are searching over the billions of web sites that comprise the Internet[5], they do not meet exploratory and iterative goals within small repositories of content.

Currently, research into alternative search engines that address these issues delves into artificial intelligence and natural language processing; clustering of results; giving recommendations and participating in social groups; and various improvements on metasearch algorithms and the visual browsing of results.

THE SOLUTION: INTERACTIVE RADIAL TREE

Radial trees comprise entities known as nodes that stand for concrete or abstract notions. The Swiss mathematician Leonhard Euler was the first to use nodes for this purpose. Node layouts were initially arranged in a radial manner in response to the problem of drawing free trees, or trees without a specified root [6]. Further research elaborated on their features[7].

Radial graph layouts typically have a focus node located at the center of the viewport and nodes connected to it that radiate outward. The siblings occupy uniformly separated rings standing for the levels in the tree. Each node is given a sector of the ring proportionate to the node's angular size and the number and size of its children.

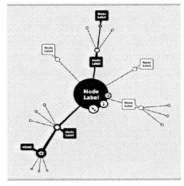

FIGURE 2: Schematic of radial tree implementation

Radial trees are useful designs for organizing, displaying and accessing information in difficult, knowledge-heavy environments. Animation further boosts their efficiency because it gives the user the sense of moving through information in a way that mimics natural human movement.

These improvements result in the increased usability of the layout and in enhanced user experience. The combination of natural animation and a radial map provides a useful and versatile tool for finding information.

Explorational Features

Radial maps show both the big picture and the contiguous details simultaneously, allowing users to explore a site's navigational architecture while keeping track of where they are within the whole structure. In this way, the system not only prevents a user from getting lost, but also allows him or her to search efficiently for information – even if the user is not sure if such information exists. Because of this feature set, a radial map might be better described as an "explorational" graphical user interface rather than a "navigational" one.

Previous and Similar Solutions

When first implemented, hierarchical structures mirrored an object familiar to humans from their natural surroundings: the tree. As simple hierarchical structures, trees normally comprise a set of nodes with directional relations. The direction indicates parent and children nodes. However, linear tree-like views become problematic when the nodes are too numerous to be visualized in a single view. Web sites tackle this problem by employing sitemaps.

Circular Depictions

Circular tree models are useful in displaying a hierarchy of any size within the finite area of a circle. Focus is acquired by enlarging the portion of the space currently in view. The number of outer leaf nodes tends to increase exponentially in large hierarchical structures. By employing an algorithm of distributing nodes in a full 360° circumference by siblings' count and importance, the angle between edges depends on the weight of the nodes, or the count of their sub-nodes, thus making optimal use of the available space. This is illustrated in Figure 3.

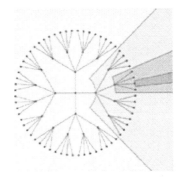

FIGURE 3. The hierarchical structure in 2D Euclidean space

Hyperbolic Projections

In hyperbolic space[8], the circumference of a circle is always exponentially more than 2pi.r and the angle polar coordinate in a hyperspace can have a range greater than 360°. The main advantages of using hyperbolic space is that many items can be fit into the periphery, and hyperbolic space can be warped in real time, moving distortion from the center of view. Hyperbolic circles have a circumference that increases exponentially with respect to radius. Therefore, a hyperbolic circle can devote more room to nodes than a Euclidian circle. In a hyperbolic visualization structure, every node is allocated a hemisphere that lies directly on the parental hemisphere's tangent plane.

A disadvantage of hyperbolic projections is that the warping of the space on the periphery tends to increase with the size of the radius. The hyperbolic tree model is practical only in a finite number of radii around the center. The exact number varies with different purposes, but is usually fewer than 10 units.

Cone Tree

The Cone tree [9] (Figure 4) displays hierarchies in three dimensions. Branches from any node are spread out in a cone. This allows a layout that is denser than traditional 2D diagrams, though nodes may be obscured.

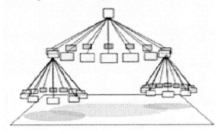

FIGURE 4. The Xerox PARC Cone tree displays hierarchical data in three dimensions

TreeMap

A TreeMap [10] (Figure 5) represents hierarchies as a series of embedded boxes, where each box contains its children, usually by alternating vertical and horizontal layers. The relative size of each box can be made proportional to a property, where the parent's property is the sum of the children.

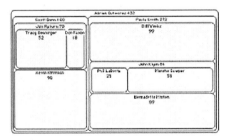

FIGURE 5. A TreeMap displays the hierarchy as a series of boxes

IMPLEMENTATION

This system seeks to fill the gap by helping people browse web sites in ways that match how human minds naturally seek information, while representing and communicating the relationship between pieces of information. Using static radial maps as the starting point, SpicyNodes introduces an environment where path-interconnected nodes represent hierarchical or network information. The development phase involved software engineering in Adobe Flash (ActionScript v2), with algorithms that drew from geometry, computer graphics, parallel processing, and physics. We developed sample nodemaps on a range of topics, such as browsing the art collection in a museum, finding information in a city web site, and browsing a music encyclopedia. The final product is an online service, SpicyNodes, which is customizable in terms of (a) the nodes and their logical arrangement; (b) the color scheme; (c) the animations, motion and interactivity; and (d) the shapes, lines, and labels. Creators can participate in the project's online community by posting their work in the online gallery.

Intuitive Motion

SpicyNodes' movement subsystem is designed to mimic motion in the real world as closely as possible, and uses movement to give important proprioceptive feedback[11]. When users click a node, it gathers speed and then decelerates as it settles into its new orbit. To correct a common problem of earlier radial tree implementations, constraints are placed on the motion of nodes so that the whole layout does not appear to flip in a way that is disorienting to the user.

A technique for animating layout transitions, which takes place when the focus node changes, has been the subject of recent research[12]. The layout adjusts by moving the new focus node to the fore in orbital circles, instead of pushing it on a straight-line path. The animation calculates the nodes' new positions by applying linear interpolation to their polar coordinates

Motion is controlled by applying the sidely used slow-in, slow-out timing is a technique widely used for visualization in order to "humanize" machine-generated output. To keep the user from getting lost, geometric transformations should retain a familiar view and retain visual memory[13]. To create a lifelike interaction, we use artificial motion rules, inspired by, but not directly based on, traditional Newtonian motion. Since a mouse can only move horizontally and vertically, the system utilizes a method of interaction using centers of mass, momentum, and friction – similar to pushing plates around a dining room table – so that the user can both rotate and pan (3D manipulation: x-transform, y-transform, and theta-transform) using only two dimensions of mouse motion.

Radial Layout Algorithms

SpicyNodes' navigation is based on an animated radial tree representation that uses a layout algorithm with rearrangement constraints to prevent user disorientation. The layout engine is based on several fundamental algorithms that generate the layouts, and allow layouts to be rearranged to change the current "focus" from node to node. The core layout algorithm is common to many radial tree representations, and is based on sibling counts.

As the user navigates through the nodemap, the engine maintains an intuitive layout by providing a visual fusion of a typical tree exploration with a history based navigation system. The layout algorithm uses constraints to transform the typical radial layout into a layout with radial clusters along a path of nodes between the currently selected focus and the root node. The path of radial clusters is almost straight so the user has a constant orientation to the root node. The comparison of both layouts is presented in Figure 6. The algorithm to achieve "clusters along a path" reinitiates the balancing recursion for each of the nodes placed along the path. Path neighbors are excluded from the recursion and are positioned using a fixed angle.

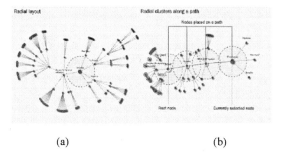

(a)   (b)

FIGURE 6. Two radial layouts. (a): A common radial layout with one center of recursive branches balancing. (b): A radial layout with a number of centers of branches balancing, all of them placed in centers of nodes forming a path between the currently selected node and the root node

Rotation and Translation

To facilitate intuitiveness, SpicyNodes employs systems for panning and rotation that allow a user to rotate and move a scene[14].

Panning and rotation are not two separated handles, but form a compact display manipulation tool that can be used by simply dragging the mouse cursor on the viewport background. Which effect - panning or rotation - affects the display more intensively depends upon the direction of the cursor motion (Figure 7).

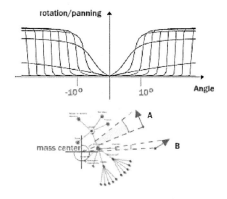

FIGURE 7. The relation between the rotation (moment) and the panning (linear velocity) depends on a modified arctangent function. In (a), where the angle between the center of mass and the cursor motion vector (in a short period of time) is relatively large, the rotation has stronger effect. In (b), where this angle is relatively small, panning has stronger effect and the stage moves much faster than it rotates, or it does not rotate at all

Changing Focus

The currently "selected" node is the "focus." It is typically centered and is visually dominant. When a new node is selected, the focus changes, and the entire radial layout shifts to adjust the viewport accordingly. The adjacent elements consequently re-group to fit the altered layout.

The first step of the rearrangement procedure consists of building a new layout based on the template tree but rotated to a temporary new root node, which is exactly the new focus. The process of swapping layouts using the tree template is presented in Figure 8. The second step is adjusting the whole tree alignment in the viewport. The last step is preparation of keyframes for the animation that follows the change of focus.

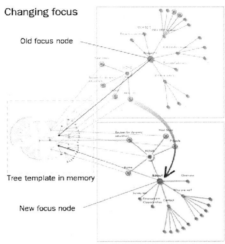

FIGURE 8. The first step of changing focus consists of making a copy of a subset of nodes of the template tree. For most layouts, some branches of the tree template that are distant to the current focus node are invisible (the light gray nodes in the tree template)

Displaying Labels

When labeling a radial tree, it is important to avoid overlapping node labels while ensuring that each label correctly references its node. Because rotation and zoom are an integral part of SpicyNodes, the labeling system has to fit to the changing coordinates' space in the displayed tree. Labeling is not haphazard, but instead is governed by geometrical rules that dictate the positions of labels relative to the nodes they describe. The novel layout algorithm of SpicyNodes described above marries the model of movement and users' perceptions of the node space, and allows dense label layouts without collisions.

The layouts and transitions are expansions on work by Yee, et al [12] In some visualizations, text is displayed. Often several sentences or paragraphs (with scrolling) are displayed for the focus node, and 1-5 words are displayed for peripheral nodes. The text causes the nodes to be rectangular because words tend to be wider than high.

This is particularly problematic on the top and bottom portions of an orbit because the words overlap. Overlap is prevented several ways: (a) Repulsion - a force-directed algorithm adjusts the geometrically calculated positions; (b) Oval warp - all positions are warped from a square grid to a wider grid; and cursor interactions have a reverse warp. When labels are displayed outside of a node, the are positioned radially away from the center of view.

Information About the Nodes

SpicyNodes' interactivity materializes in the wide variety of options with which it responds to user input. Typically, the available options are to (a) display more information about the node, e.g., panning through text; (b) go to a related web page; (c) go to another nodemap, allowing multiple nodemaps to be linked together[15]. Optionally, these contextual controls can be configured to evoke more suggested actions.

Nodes can be expanded with one click to display detailed information (Figure 9). Among the contextual controls available in any setting, there always exists a "Show Info" option that, when activated, brings forth a text field. This field contains associated information that is either auxiliary or explanatory to the node in focus. Analogous to similar resources, the field usually displays related and referenced items at the bottom. In most cases, the text field is animated consistently with the overall implementation of the radial map.

FIGURE 9. Simple text-only additional information is displayed in a popup dialog for any node

Other Navigation Aids

Full text searching can find content within a node, while fuzzy searching matches commonly misspelled words or phonetic misspellings

Breadcrumbs explicitly delineate a user's path back home. Users always know where they are because at any moment they can learn their location by looking at the route map (e.g., Home > Fruit > Sweet > Apples). Also, colors are widely used to group the nodes, either by major group, or by their distance from the center.

CONCLUSION

Today's repository of knowledge is greater than ever before, and people increasingly face the challenge of finding the specific, useful information that they look for. SpicyNodes is an emerging technology that seeks to fill the gap in organizing online data by illustrating the relationships between items and concepts in an intuitive graphical manner. Overcoming certain shortcomings in similar prior technologies, our radial map optimizes the user experience. SpicyNodes advancement comes from employing a pure "focus + context" technique to mimic the human brain's perspective, giving users an intuitive navigating experience, and creating a naturally feeling living environment. SpicyNodes' at-a-glance visualization of hierarchical or network relationships combined with the interactive fluidity of natural slow-in and slow-out animation reorganizes complex sitemaps into straightforward, relationship-oriented navigation that allows users to explore related and relevant information without having to access a multitude of web pages or multiple sites. Because SpicyNodes makes better use of the available screen space by enlarging or rearranging nodes, it can contain more information than a linear or grid sitemap while utilizing a fraction of the space.

REFERENCES

1. M.S.T. Carpendale and C. Montagnese, *A Framework for Unifying Presentation Space*, University of Calgary, 2001.
2. J. Stasko and E. Zhang, "Focus+context display and navigation techniques for enhancing radial, space-filling hierarchy visualization," in *Proc. of Information Visualization*, 2000, pp. 57-65.
3. G. W. Furnas, "Generalized Fisheye Views," in *Proc. of the Conference on Human Factors in Computing Systems, CHI'86*, ACM Press, 1986, pp. 16-23.
4. B. Kules, "Supporting Creativity with Search Tool," Report of Workshop on Creativity Support Tools, Univ. of Maryland, USA, 2005.
5. A. Langville and C. Meyer, *Google's PageRank and Beyond: The Science of Search Engine Rankings*, Princeton University Press, 2006.
6. P. Eades, "Drawing Free Trees," Bulletin of the Institute for Combinatorics and its Applications, Vol. 5, 1992, pp. 10-36.
7. G. Di Battista, P. Eades, R. Tamassia, and I. G. Tollis, *Drawing: Algorithms for the Visualization of Graphs*, Prentice Hall, Upper Saddle River, NJ, 1999.
8. T. Munzner, *Radial and Hyperbolic Trees, Interactive Visualization of Large Graphs and Networks*, Stanford University, 2000.
9. G. G. Robertson, J. D. Mackinlay, and S. K. Card, "Conetrees: Animated 3d visualizations of hierarchical information," in *Proc. of the ACM SIGCHI Conferencion Human Factors in Computing Systems*, 1991, pp 189–194.
10. B. B. Bederson, B. Shneiderman, and M. Wattenberg, *ACM Transactions on Graphics (TOG)* **21**, 833-854 (2002).
11. T. Selker, *International Journal of Human-Computer Studies* **63**, 410-421 (2005).
12. K. Yee, D. Fisher, R. Dhamija, and M. Hearst, "Animated exploration of dynamic graphs with radial layout," in *Proc. of the IEEE Symposium on Information Visualization 2001*, IEEE Computer Society 43, 2001.
13. H. Lam, R. Rensink, and T. Munzner, "Effects of 2D geometric transformations on visual memory," in *Proc. of the 3rd Symposium on Applied Perception in Graphics and Visualization (Boston, MA, July 2006)*, ACM Press, 2006, pp. 119-126.
14. M. Hancock, F. Vernier, and D. Wigdor, "Rotation and Translation Mechanisms for Tabletop Interaction," IEEE International Workshop on Horizontal Interactive Human-Computer Systems, ISBN: 0-7695-2494-X, 2006.
15. M. Douma and H. Dediu "Interactivity techniques: Practical suggestions for interactive science web sites," in *E-learning in Virtual Science Centers*, edited by R. Subramaniam, Hershey, IDEA Group., PA, 2005.

Construction of Mathematical Problems by Students Themselves

M. Rodionov[*] and S. Velmisova[†]

[*]*Penza State Pedagogical University, Penza, Russia*
[†]*Ulyanovsk State University, Ulyanovsk, Russia*

Abstract. The article illustrates how a mathematical task can be developed by means of transformations of its conditions to construct a chain of tasks progressing in difficulty. Within the consideration of several intermediate tasks several new theorems are proved.

Keywords: Geometric configuration, triangle, polygon, tetrahedron, square, ratio, interval length, bisector, median, generalization of results, combination of interconnections.
PACS: 01.40.gb

INTRODUCTION

There are a lot of methodical approaches that motivate and stimulate the students to show their creative initiative in constructing mathematical problems. One of these approaches consists in transformation of a task which leads to a new task. The initiative is most effective when the result is expressed by the student in terms of conjectures and hypotheses. Within the process of consistent task transformation the students supported by the teacher can construct cycles/series of tasks (dynamical tasks) that are based on a common idea and cover a broad part of a mathematical course.

To illustrate this approach we have selected the topic AREAS of the elementary geometry college course. Consistent constructions of appropriate series of tasks and their solutions are represented below.

Let us consider an initial task.

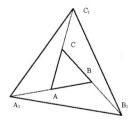

FIGURE 1.

CP1067, *Applications of Mathematics in Engineering and Economics '34—AMEE '08*, edited by M. D. Todorov
© 2008 American Institute of Physics 978-0-7354-0598-1/08/$23.00

Initial Task (Figure 1). Extend the edges of the triangle ABC as it is shown in Figure 1 locating the points A_1, B_1, C_1 so that $AA_1=AB$, $BB_1=CB$, $CC_1=AC$. Find the ratio of both triangles $A_1B_1C_1$ and ABC areas.

Solution. Notice that the triangles A_1AC and ABC have equal areas because they have one common vertex and equal altitudes. Similarly, the triangles A_1CC_1, CC_1B, BC_1B_1, AB_1B, AA_1B_1 have the same area as ABC. Hence, the ratio of these triangles areas and of ABC area equals 7.

To transform the task one should understand which elements may by varied or generalized. Among such elements we may consider vertices, directions that identify the points A_1, B_1, C_1, lengths of segments, space dimension and also the question of interest which depends on the way of the task transformation.

After identifying these elements and ratios together with the students, the teacher offers them to point out several concepts:

Firstly, transformations of the triangle could result in the following: rectilinear triangle, rectilinear polygon, quadrangle, arbitrary polygon, rectilinear tetrahedron, arbitrary pyramid, simplex, spherical triangle.

Secondly, the segments can be selected at bisectors of triangle angles, medians, triangle altitudes, lines with arbitrary angles between them and triangle sides.

Thirdly, the lengths of triangle sides can be in different ratios with respect to the lengths of constructed segments or be equal to the lengths of other sides.

Fourthly, the task may be formulated in the 3-dimensional space or n-dimensional space.

So we can vary any element or combination of elements listed above and obtain series of tasks different in complexity, didactic and developmental opportunities. Note that each separate result can serve for further dynamical development of the initial task. Our experience shows that such development demonstrates the sense and value of the initial problem and outlines its characteristics which are not revealed in direct and/or trivial analysis.

Here we present some directions of the initial task proposed by the students during practical classes.

First Transformation Direction

One may construct a sequence of comprehensive triangles and obtain the following task:

Task 1.1 (Figure 2) There is a triangle ABC. The points A_1, B_1, C_1 are such that $AA_1=AB$, $BB_1=CB$, $CC_1=AC$. The points A_2, B_2, C_2 are such that $A_1A_2=A_1B_1$, $B_1B_2=C_1B_1$, $C_1C_2=A_1C_1$. Find the ratio of the triangles $A_2B_2C_2$ and ABC areas.

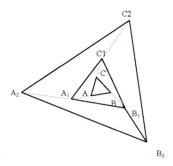

FIGURE 2.

Second Transformation Direction

Let us change the lengths of the segments constructed on the rays from the points A, B, C outside the triangle. In this case the task can be formulated in the following way:

Task 2.1 (Figure 3). Given a triangle ABC. The points A_1, B_1, C_1 are such that
   a) $AA_1 = n\ AB$, $BB_1 = n\ CB$, $CC_1 = n\ AC$
   b) $AA_1 = n\ AB$, $BB_1 = m\ CB$, $CC_1 = k\ AC$
Find the ratio of the triangles $A_1 B_1 C_1$ and ABC areas.

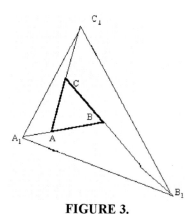

FIGURE 3.

Having solved the problem we obtained that the areas ratio is equal to
   a) $3n(n+1)+1$;
   b) $m(n+1)+n(k+1)+k(m+1)+1$.

Further questions concerning the task development are:
- Could the points A_1, B_1, C_1 be taken on the edges of the triangle?

223

- Could the task be generalized by changing the number of the figure vertices in question?
- Is it possible to change the dimension of the problem?

Task 2.2. ABC is an arbitrary triangle. The points A_1, B_1, C_1 are such that C_1 belongs to the extension of BA in the direction from B to A, B_1 belongs to the extension of AC in the direction from A to C, A_1 belongs to the extension of CB in the direction from C to B and

a) $BC_1 = m\, AB$, $AB_1 = m\, AC$, $CA_1 = m\, CB$;

b) $BC_1 = k\, AB$, $AB_1 = m\, AC$, $CA_1 = n\, CB$.

Find the ratio of the triangles $A_1B_1C_1$ and ABC areas.

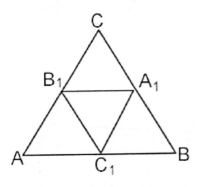

FIGURE 4.

Solution. Task 2.1 a) is a particular case of Task 2.2. a) with $m = n + 1$. If $m < 1$ then the points A_1, B_1, C_1 belong to the edges of ABC. Let us denote the areas of triangles $A_1B_1C_1$ and ABC as S_1 and S respectively. As the ratio of triangles with common angle equals the ratio of products of the segments lengths which form the angle we receive that S_1: $S = 1 - 3m(1-m)$, where m is any positive number. As this ratio never vanishes it follows that the points A_1, B_1, and C_1 are not located on one line.

Case b) requires deeper analysis. The ratio of the two triangles areas can be transformed into

$S_1 : S = 1 - m(1 - k) - n(1 - m) - k(1 - m)$ or

$S_1 : S = (1 - m)(1 - n)(1 - k) + m \cdot n \cdot k$.

Let us assume that neither of the points A_1, B_1, C_1 (see Figure 4) coincides with a vertex of the triangle ABC. Then the last expression can be written as

$S_1 : S = 1 + n'm'k'$, where $n' = n/(1-n)$, $m' = m/(1-m)$, $k' = k/(1-k)$.

The ratio of interest is zero (all the points - A_1, B_1, and C_1 lie on one line) in case when $n' \cdot m' \cdot k' = -1$. The ratios of the length may take either positive or negative values. So, the points A, B, and C lie on a straight line if and only if $n' \cdot m' \cdot k' = -1$.

Let us consider particular cases when A_1, B_1, C_1 are the basis points of bisectors or of medians.

Task 2.3 (Figure 4). Let the points A_1, B_1, C_1 be the midpoints of triangle ABC sides. Find the ratio of the triangles $A_1B_1C_1$ and ABC areas.
One can use the result obtained in Task 2.1 a) and take into account that $m=1/2$ to get S_1: $S = 1/4$.

Task 2.4. In an arbitrary triangle ABC A_1, B_1, C_1 are the basis points of bisectors. Find the ratio of the triangles $A_1B_1C_1$ and ABC areas.

Solution. Let us denote the sides lengths of ABC as $BC = a$, $AC = b$, $AB = c$. Apply the bisector property in a triangle:

$AB_1 : B_1C = c : a$, $CA_1 : A_1B = b : c$, $BC_1 : C_1A = a : b$.

Then

$m = CA_1 : CB$, $n = BC_1 : BA$, $k = AB_1 : AC$, i.e. $m = \dfrac{c}{b+c}$, $n = \dfrac{a}{c+a}$, $k = \dfrac{b}{a+b}$.

Using the result in Task 2.2 b) we can calculate the ratio

$$S_1 : S = \left(1 - \frac{c}{b+c}\right)\left(1 - \frac{a}{c+a}\right)\left(1 - \frac{b}{a+b}\right).$$

Hence, $S_1 : S = \dfrac{2abc}{(a+b)(c+a)(b+c)}.$

As $(a+b) \cdot (b+c) \cdot (c+a) \geq 8$, we conclude that $S_1 : S \leq \dfrac{1}{4}$.

Having compared the areas of triangles AB_1C_1, C_1BA_1, and A_1B_1C we can see that the area of $A_1B_1C_1$ exceeds the smallest area of the triangles in question. Let's denote $AB_1 : B_1C = n$,

$CA_1 : A_1B = m$, $BC_1 : C_1A = k$ and consider all possible cases.

- $n \geq m \geq k \geq 1$.

  In this case

  $$S_{CA_1B_1} : S_{A_1B_1C_1} = \frac{m(1+k)}{1+mnk}$$

  and $1 + m \cdot n \cdot k - m(1+k) \geq 1 + m^3 - m(1+m) = (1+m)(m-1)^2 \geq 0$

  or $S_{CA_1B_1} : S_1 \leq 1$, i.e., $S_1 \geq S_{CA_1B_1}$;

- $n \geq 1, k \leq 1$.

  It is not difficult to check that

  $1 + n \cdot m \cdot k - m \cdot (1+k) = m \cdot k \cdot (n-1) + (1-m) \geq 0$ and $S_1 \geq S_{CA_1B_1}$.

Task 2.5. The three points A_1, B_1, C_1 are on the sides of the triangle ABC correspondingly. Prove that the area of $A_1B_1C_1$ is not less than the area of AB_1C_1, C_1A_1B, A_1B_1C.

If we consider the situation described in the Task1.1 we note that the lines AA_1, BB_1, and CC_1 intersect the sides B_1C_1, A_1C_1, A_1B_1 in points A_2, B_2, C_2 so that

$C_1A_2 : A_2B_1 = B_1C_2 : C_2A_1 = A_1B_2 : B_2C_1 = 1:2$.

After this analysis several more tasks of this type may be formulated.

The third group of students proposed an arbitrary polygon to be considered.

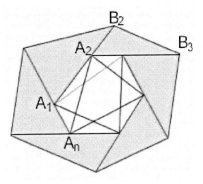

FIGURE 5.

Task 3.1 (Figure 5). Each edge $A_K A_{K+1}$ of a convex polygon $A_1...A_n$ ($n>4$) is extended so that $A_{K+1}B_K = A_K A_{K+1}$. Prove that the area of the constructed polygon is no greater than 5 times than the area of the initial one.

Solution: It is necessary to prove that the sum of the n additional triangles areas shaded in the figure is not greater than $4S$ (S is the area of the initial polygon).

Each of these triangles is divided by its median into two triangles with equal areas.

Hence, its area is two times greater than the area of the triangle bounded by two adjacent sides and a diagonal of the initial polygon (corresponding triangles marked with black lines). But the sum of 'black' triangles areas obviously does not exceed $2S$ (see Figure 5).

We may obtain the next specification of the task if we consider rectilinear n-gon as a case of a convex *n*-gon.

Task 3.2. Each edge $A_K A_{K+1}$ of a rectilinear *n*-gon $A_1...A_n$ is extended as follows: $A_{K+1}B_K = A_K A_{K+1}$. Find the ratio of the polygons $B_1...B_n$ and $A_1...A_n$ areas.

Further development of the considered task allows to formulate it in case of rectilinear n-gon.

Task 3.3. All sides of a rectilinear triangle *ABC* are extended with a segment of equal length: *AB* in the direction from *A* to *B*, *BC* in direction from *B* to *C*, *CA* in direction from *C* to *A*. The new points are vertices of a new triangle $B_1C_1A_1$. Find the ratio of $B_1C_1A_1$ and *ABC* areas.

Task 3.4. Points C_1, A_1, B_1 belong respectively to the extensions of *AB* in the direction from *A* to *B*, *BC* in direction from *B* to *C*, *CA* in direction from *C* to *A* such that $BB_1 = n\cdot AB$, $CC_1 = n\cdot BC$, $AA_1 = n\cdot CA$. Find the ratio of $B_1C_1A_1$ and *ABC* areas.

Task 3.5. The points C_1, A_1, B_1 are located on the sides of the triangle *ABC* so that $AC = n\cdot AB$, $BA_1 = m\cdot BC$, $CB_1 = k\cdot CA$ (*n, k, m* are less than 1). The lines AA_1, BB_1 and CC_1 cross each other in points *M, N,* and *P,* respectively. Find the ratio of *MNP* and *ABC* areas.

The next group of students proposed to consider various types of quadrangles instead of triangles. As a result a wide range of problems with different levels of difficulty have been constructed.

Task 4.1. The edges *AB, BC, CD,* and *DA* of a convex quadrangle *ABCD* are extended: *AC* in direction of *B* with $BB_1 = BC$, *DC* in direction of *C* with $CC_1 = DC$, *AD* in direction of *D* with $DD_1 = DA$, *BA* in direction of *A* with $AA_1 = AB$. Find the ratio of the quadrangles $A_1B_1C_1D_1$ and *ABCD* areas.

(Answer. $S_1 : S = 5$ where S_1 is the area of $A_1B_1C_1D_1$ and S is the area of *ABCD*)

Further generalization of the problem was generated by substituting numerical values or considering particular cases of quadrangles. Here are some examples.

Task 4.2. On the edges AB, BC, CD, and DA (or their extensions) of a convex quadrangle $ABCD$ the points A_1, B_1, C_1, D_1 are selected so that: $AA_1 = n \cdot AB$, $BB_1 = n \cdot BC$, $CC_1 = n \cdot CD$, $DD_1 = n \cdot DA$. Find the ratio of the quadrangles $A_1B_1C_1D_1$ and $ABCD$ areas. *($S_1 : S = 1 - 2n(1 - n)$)*

Task 4.3. (Figure 6) On the edges AB, BC, CD, and DA of a parallelogram $ABCD$ the points A_1, B_1, C_1, D are selected so that
$$AA_1 = n \cdot AB, \ BB_1 = m \cdot BC, \ CC_1 = n \cdot CD, \ DD_1 = m \cdot DA. \ \text{(Figure 6)}.$$
Prove that the points of the lines AC_1, BD_1, CA_1, DB_1 intersection form a parallelogram and the ratio of the constructed and initial parallelogram areas is equal to $\dfrac{n \cdot m}{(1+n)(1+m)+1}$.

Task 4.4. The middle points of a parallelogram edges are connected with the vertices of the corresponding opposite edge. Find the ratio of both constructed octagon and initial parallelogram areas.

Task 4.5. The diagonals AC and BD of quadrangle $ABCD$ are extended in directions of C and D in such a way that $CC_1 = n \cdot AC$, $DD_1 = m \cdot BD$. Find the ratio of quadrangles ABC_1D_1 and $ABCD$ areas. *(Answer: $S_1 : S = (n+1)(m+1)$.)*

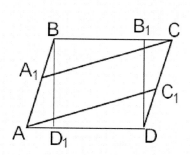

FIGURE 6.

Task 4.6. The diagonals AC and BD of quadrangle $ABCD$ are extended in such a way that $CC_1 = n \cdot AC$, $AA_1 = m \cdot AC$, $BB_1 = k \cdot BD$, $DD_1 = l \cdot BD$. Find the ratio of quadrangles $A_1B_1C_1D_1$ and $ABCD$ areas.

NOTE. Further ways for developing the task may be concerned with spherical geometry. This kind of work was fulfilled by students in the form of individual tasks. The initial task may be the following:

Task 5.1. (Figure 7) Edges of tetrahedron $ABCD$ are extended: BA in direction from B to A, CB in direction from C to B, DC in direction from B to C, AD in direction from A to D. The new points A_1, B_1, C_1, and D_1 belong to these extensions with properties $AA_1 = AB$, $BB_1 = BC$, $CC_1 = CD$, $DD_1 = DA$. Find the ratio of the pyramids $A_1B_1C_1D_1$ and $ABCD$ volumes.

Task 5.2. Edges of tetrahedron $ABCD$ are extended: BA in direction from B to A, CB in direction from C to B, DC in direction from B to C, AD in direction from A to D. The new points A_1, B_1, C_1, D_1 belong to these extensions with properties
$$AA_1 = n \cdot AB, \ BB_1 = n \cdot BC, \ CC_1 = n \cdot CD, \ DD_1 = n \cdot DA.$$
Find the ratio of the pyramids $A_1B_1C_1D_1$ and $ABCD$ volumes.

Tasks 2.3 may have spatial analogues:

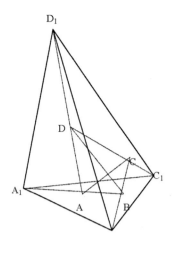

Task 5.3. Bisector plane of dihedral angle splits the opposite edge in the ratio which equals the ratio of the sides squares.

Task 5.4. Bisector plane of dihedral angle divides the planes into triangles which areas are proportional to the areas of the planes which bound this dihedral angle.

Next step may include fourth parameter in the consideration of spherical triangles and n-dimensional simplexes. The students get opportunity to continue the work on the task and combine still unused material.

FIGURE 7.

CONCLUSION

With the help of this example we tried to demonstrate how a "creative laboratory" can be "developed" within the framework of traditional mode of teaching and learning mathematics. We are convinced that such investigations allow not only to understand the material but also to develop students' creativity.

REFERENCES

1. M. A. Rodionov, The Motivation of Mathematical Studies and the Ways to form it, Saransk, 2001. (in Russian)
2. M. A. Rodionov and N. V. Sadovnikov, *Interaction of Theoretical and Practical Aspects of Tasks Usage in Mathematical Teaching. Tutorial for Teachers and Students, The Training Appliance*, Penza, 1997. (in Russian)
3. E.E. Moise, *Elementary Geometry from Advanced Standpoint*, Mass, Reading, 1964.

Computer-supported Classes in Mathematics for Engineering Students

Y. D. Stoynov

Faculty of Applied Mathematics and Informatics, Technical University of Sofia, 1000 Sofia, Bulgaria

Abstract. A combined method for teaching and learning selected topics of Mathematics for engineering students is presented. Some points in approaching these topics using CAS DERIVE are discussed and CAS-support of students' learning is shown.

Keywords: Computer algebra, mathematics teaching and learning, engineering students.
PACS: 01. 40. Di, 01.40.gb, 01.50.ht

INTRODUCTION

In this paper we present seminars in Mathematics for engineering students (at the Faculty of Automatics and Faculty of Mechanical Engineering). During the teaching and learning process we combine traditional methods with computer algebra applications. The aims of this combined procedure are:

1. To make the students familiar with CAS *DERIVE*.
2. To help the students to assimilate selected topics introduced in lectures and seminars.
3. To provide a possibility for students with gaps in their secondary education and in Mathematics-1 (their number is large) to fill them to be able to acquire the material and above all to successfully solve problems in Mathematics-2.

In view of removed from the curriculum laboratory classes in Mathematics(now we have only 3 school hours for a lecture and 2 school hours for a seminar weekly) we decided to proceed to computer-supported lectures and seminars. We chose some topics which teaching was significantly supported by using CAS:
- extremum of a function of 2 variables;
- solving Ordinary Differential Equations;
- calculation of double integrals.

TEACHING ACTIVITIES

In the beginning of the classes we give the students general knowledge in the work with *DERIVE*-command line, dropping menus, executing of a function- an interface

CP1067, *Applications of Mathematics in Engineering and Economics '34—AMEE '08*, edited by M. D. Todorov
© 2008 American Institute of Physics 978-0-7354-0598-1/08/$23.00

similar to many of widely used programs. After that we go into beforehand selected problems from the chosen topics.

Activity 1

One of the basic notions in Mathematics-2 is a partial derivative. Differentiation, whose learning begins from high school on and continuous in Mathematics-1 , creates problems for a large part of the students. *DERIVE* can eliminate this problems. The new thing are mixed partial derivatives, whose finding is difficult for some students even with *DERIVE*, because they do not know what z_{xy} means. We write the definition of the z_{xy}: $z_{xy}=(z_x)_y, z_{yx}=(z_y)_x$ and begin to explain how to find it. We illustrate the theorem of the equality of mixed derivatives.

Some Examples

PROBLEM 1. For the function $z=x^2+\sin(x^3-4xy)+\operatorname{atan}(1/(xy))$ find:

a) z_x, z_y

b) $z_{xx}, z_{xy}, z_{yx}, z_{yy}$

c) make sure that $z_{xy}=z_{yx}$

PROBLEM 2. For the function $z=\exp(x^2+2xy)$ prove that $z_{xxy}=z_{yxx}=z_{xyx}$

The student has to execute the following succession of commands for **PROBLEM 1**:

1.) Enter the expression for the function in the command line and push the button ***Enter.***

2.) Push the button ∂ (for differentiation) and fill the respective fields in the dialogue window:

    2.1. the variable with respect to which he will differentiate.

    2.2. the order of differentiation.

    2.3. push the button ***OK.***

The symbol for derivative, applied to the respective function appears on the screen. The student checks for errors and execute the command with the button = *(Simplify)*. In this way one of the first partial derivatives is found.

To find the partial derivative with respect to the other variable, the student has to click with the mouse on the expression of the function and go to 2). A list with the problems and the succession of the commands, which are used to solve problems, is presented to the students. After the sense of the symbols $z_{xx}, z_{xy}, z_{yx}, z_{yy}$ is explained the student go to Problems 1b and 1c. These solutions are called solutions-components (that is why the respective problems are called preparatory) for the solutions of the problems in which extremum is sought. One approach and his realization to problems in which we seek extrema with *DERIVE* are considered in [1]. For us it is important the student to follow right the algorithm of the solution.

A Typical Problem

Find the local extrema of the function $z = x^3+y^3-3xy$ and make sure that your conclusion is correct as you make the graph.

First Step:

Finding the stationary points.

In this step we reiterate the definition of stationary points and the students have to find z_x and z_y with *DERIVE*. Then we explain how to solve systems.

#1: $x^3 + y^3 - 3 \cdot x \cdot y$

#2: $\dfrac{d}{dx} (x^3 + y^3 - 3 \cdot x \cdot y)$

#3: $3 \cdot x^2 - 3 \cdot y$

#4: $\dfrac{d}{dy} (x^3 + y^3 - 3 \cdot x \cdot y)$

#5: $3 \cdot y^2 - 3 \cdot x$

#6: $\text{SOLVE}\left(\left[3 \cdot x^2 - 3 \cdot y,\ 3 \cdot y^2 - 3 \cdot x\right],\ [x,\ y],\ \text{Real}\right)$

#7: $[x = 0 \wedge y = 0,\ x = 1 \wedge y = 1]$

There are also complex solutions, but we do not need them.

Second Step:

Finding the Hessian.

In this step the students have to find the expression $\Delta = z_{xx} \cdot z_{yy} - (z_{xy})^2$. The mixed derivatives have already been explained. We reiterate this if necessary and make the students to find the derivatives which we need. Then we explain how to find Δ.

#9: $x^3 + y^3 - 3 \cdot x \cdot y$

#10: $\left(\dfrac{d}{dx}\right)^2 (x^3 + y^3 - 3 \cdot x \cdot y)$

#11: $6 \cdot x$

#12: $\left(\dfrac{d}{dy}\right)^2 (x^3 + y^3 - 3 \cdot x \cdot y)$

#13: $6 \cdot y$

#14: $\dfrac{d}{dy} \dfrac{d}{dx} (x^3 + y^3 - 3 \cdot x \cdot y)$

#15: -3

#16: $\Delta = \left[\left(\dfrac{d}{dx}\right)^2 (x^3 + y^3 - 3 \cdot x \cdot y)\right] \cdot \left[\left(\dfrac{d}{dy}\right)^2 (x^3 + y^3 - 3 \cdot x \cdot y)\right] - \left[\dfrac{d}{dy} \dfrac{d}{dx} (x^3 + y^3 - 3 \cdot x \cdot y)\right]^2$

#17: $\Delta = 36 \cdot x \cdot y - 9$

Third step:

Finding the local extrema.

We substitute the values of x and y found in the first step in Δ and see that there is no extremum in the point $M_1(0,0)$ and there is in the point $M_2(1,1)$. After that we find $z_{xx}(1,1)$ and see that it is positive. We conclude that there is a local minimum in $M_2(1,1)$. The graph shows that our conclusion is right.

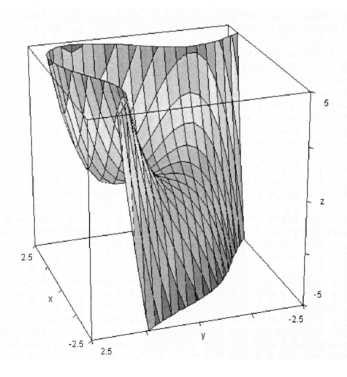

Several problems are given to the students to assimilate the algorithm.

Activity 2

The next topic is Differential equations. Our purpose is to make the students to rationalize the following mathematical concepts: general, singular and particular solution and the fact that the first two satisfy only the equation and the third satisfies an additional initial condition.

Finding General Solutions

PROBLEM 1. Determine the type of the equation: $xdy+(y-\ln(x)y^2)dx=0$ and find the general solution.

The students solve the first part on a list of paper by hand. They have to reduce the problem in the form $y'+(1/x)y=(\ln(x)/x)y^2$ and guess that this is Bernoulli's equation. They are told that the general solution of the equation of the type $y'+p(x)y=q(x)y^m$ can be solved with the function *Bernoulli_ode_gen(p,q,m,x,y,c)* and begin to solve the problem by determining $p(x)=1/x$, $q(x)=\ln(x)/x$, $m=2$. After that they enter *Bernoulli_ode_gen(1/x,ln(x)/x,2,x,y,c)* in the command line and push the button $=$(Simplify).We underline the fact that the constant C (an arbitrary constant), which appears in the formula of the solution and can take innumerable values, shows the difference between the general and the particular solution and for every particular value of this constant we have different particular solution.

The student is given a list on which are written the type of the equation and the respective *DERIVE* function, which solves it. For example, the general solution of the linear equation $y'+p(x)y=q(x)$ is solved with *Linear1_gen(p,q,x,y,c)*.

In the following problems we want the student to determine the type of the equation by the respective reducing (by hand on a list of paper) and then to find the general solution using the necessary function. The first part is similar to what is needed in the exam. The second is done by CAS *DERIVE* and many difficulties for students, for whom solving integrals is not easy are eliminated. Here we give some additional problems.

PROBLEM 2. $xydy-(x^2+y^2)dx=0$

PROBLEM 3. $(1/x)dy+(y-2x^2)dx=0$

Finding solution of the Bernoulli's equation can be done directly or first reducing it (by hand) to linear and then using the respective function. In this way we overcome the difficulty in solving integrals, emphasize on knowledge necessary for the exam and can go to find particular solution. The two ways have to give (of course) the same result.

Finding Particular Solutions

It can be done by clicking on the general solution and then with the button **_SUB_** we fill the initial values for x and y, then solve with respect to C with the command **_Solve_** and substitute this value in the general solution. Using *DERIVE* we have the possibility to plot the integral portrait. We underline the fact that the general solution is presented by the family of innumerable number of integral curves and a given particular solution is presented by a curve from this family. Consider the following problem:

PROBLEM 4. For the equation $y' = \sqrt{(1-y^2)}$ find the general solution, the particular solution which satisfy the initial condition $y(0)=0$ and check for singular solutions.

The student determine the type and find the general solution with *Separable_gen(1,* $\sqrt{(1-y^2)},x,y,c$*)*. Then we substitute the initial values and find C with **_Solve_**. We plot the integral portrait for some values of C (including the C from the particular solution). Plotting the particular solution for several times (with different color) we see that it is presented by a curve from the integral portrait. The eventual singular solutions could be $y=1$ and $y=-1$, because we divide by $\sqrt{(1-y^2)}$ when we solve the equation by hand and we make sure that they are solutions by substituting them in the equation.

We plot on the same graph these two solutions and see that they touch the curves of the general solution and therefore are singular solutions. This can be seen from the following graph.

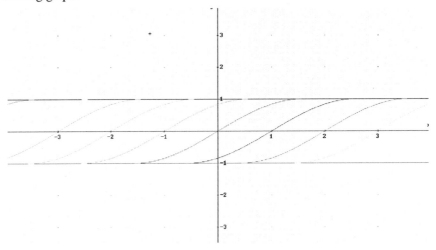

Solving differential equations with *DERIVE* is taken up in [2].

Activity3

The final topic that we consider are double integrals.

The main point here is how to reduce a double integral to two simple integrals. This depends on the type of the domain of integration and as we have a source to plot curves it can help to solve such problems.

PROBLEM 1. The domain $D=\{y \geq x^3,\ x \geq y, x \geq 0\}$ is given. Plot the curves, shade the domain and solve $\iint\limits_{D}(x+y-1)dxdy$.

First we plot the curves. For many of our students it is difficult to plot them by hand and this is one of the main obstacles to solve the integral.

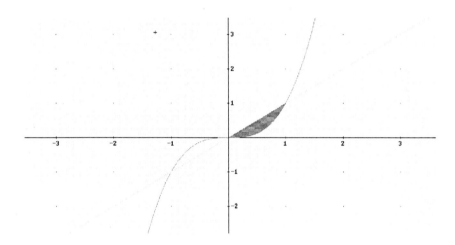

#19: $x + y - 1$

#20: $\displaystyle\int_{x^3}^{x} (x + y - 1)\ dy$

#21: $\displaystyle\int_{0}^{1}\int_{x^3}^{x} (x + y - 1)\ dy\ dx$

#22: $\qquad\qquad\qquad -\dfrac{3}{140}$

#23: $\displaystyle\int_{y}^{y^{(1/3)}} (x + y - 1)\ dx$

#24: $\displaystyle\int_{0}^{1}\int_{y}^{y^{(1/3)}} (x + y - 1)\ dx\ dy$

#25: $\qquad\qquad\qquad -\dfrac{3}{140}$

235

After we have solved the geometric part of the problem we go to determine the limits of variables x and y. This can be done in two ways because the integral can be solved by integrating first with respect to x and then with respect to y or first with respect to *y* and then with respect to *x*. The student is asked to do it in both ways and then calculate the integral. The result must be the same. Here we stress the theorem about the change of the order of integration in double integral.

Some of the problems are solved by the students by hand after they plotted and shaded the domain .

CONCLUSION

We want to stress that CAS *DERIVE* can help significantly in explaining important ideas from the material in Mathematics 2 and make it more attractive to students.

REFERENCES

1. M.D. Todorov and E.A.Varbanova, "Application of DERIVE in the Investigation of Explicit Functions of Two Variables", in *Proc. of 25th Summer School Application of Mathematics in Engineering and Economics, Sozopol, June 1998*, edited by B.I.Cheshankov and M.D. Todorov, Heron Press, Sofia, 1999, pp.212-218.
2. E.A.Varbanova and Y.D. Stoynov, "DERIVE-Approach to Teaching and Learning First Order Ordinary Differential Equations", in *Proc.of 29th International Conference Application of Mathematics in Engineering and Economics, Sozopol, June 2002*, edited by M.S. Marinov and G.P. Venkov, Bulvest 2000, Sofia, 2003, pp. 308-314.
3. Y.D. Stoynov, "Computer algebra in mathematics education", in *Proc. of the Third Conference "Education in Higher Mathematics in Technical and Economics Universities,"* Bansco, September, 2007.
4. E.A. Varbanova, "A CAS supported environment for learning and teaching calculus," in *Proc. of the 1st KAIST Intern. Symposium on Teaching*, CBMS Issues in Mathematics Education, Vol. 14, 2007.
5. B.Kutzler and V. Kokol-Voljc, *Introduction to Derive 5 , The Mathematical Assistant for Your PC*, Texas Instruments 2000.
6 J.Berry and B.Lowe, *Learning Differential Equations Through DERIVE*, Studentliteratur, Lund, Sweden, 1998.

NUMERICAL METHODS
IN MATHEMATICAL MODELING

Modeling and Applications of the Cylindrical Couette Flow of a Rarefied Gas

D. Dankov and V. Roussinov

Institute of Mechanics, Bulgarian Academy of Sciences,
Acad. G. Bonchev Str., bl. 4, 1113 Sofia, Bulgaria

Abstract. The cylindrical Couette flow of a rarefied gas is studied in the case when the inner cylinder is rotating while the outer cylinder is at rest. Velocity, density and temperature profiles are investigated by a Direct Monte Carlo Simulation method and a numerical solution of the Navier-Stokes equations is found. The results prove good agreement between flow macro-characteristic values obtained by the two methods.

Keywords: Fluid dynamics, kinetic theory, rarefied gas, DSMC.
PACS: 51.20.+d

INTRODUCTION

Fluid transport in micro and macro channels yields the necessity to study flow in a cylindrical coordinate system. Note that Couette cylindrical flow is a fundamental problem in the rarefied gas kinetic theory [1]. As such, its modeling and numerical solving is a basic test of program packages, which we plan to use in studying more complex rarefied gas flows in a channel [2,3].

The design of adequate mathematical models of real processes and phenomena is one of the most important tasks of the studies. Since analytical methods lack, various numerical schemes are developed to solve the problems posed. One of the methods of checking solution plausibility is to compare results found by employing different methods of mathematical modeling.

In the present paper, we compare results found by using the Direct Simulation Monte Carlo Method with those calculated by employing the finite difference method. Both methods are used to solve the Navier-Stokes equations for Knudsen number 0.02 and for different rotation velocities of the inner cylinder.

The numerical results show good agreement for velocity and density profiles. Small differencec in the temperature profile are due to the difference between the two models: Navier-Stokes equations follow from the Boltzmann equation while the DSMC method accounts for effects which are a priory eliminated from the Boltzmann equation.

CP1067, *Applications of Mathematics in Engineering and Economics '34—AMEE '08*, edited by M. D. Todorov

FORMULATION OF THE PROBLEM AND METHODS OF METHODS

We study a rarefied gas between two coaxial cylinders with equal temperatures T_0. The inner cylinder has radius R_1 and the outer – R_2. The inner cylinder rotates with a constant velocity V_1 and the outer one is static – Figure 1.

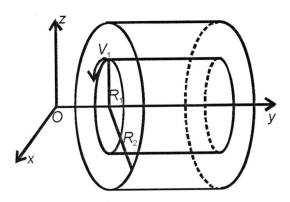

FIGURE 1. Flow geometry

Direct Simulation Monte Carlo (DSMC) Method

The gas considered is simulated as a stochastic system of N particles [4,5]. All quantities used are non-dimensional, so that the mean free path at equilibrium is equal to 1. The basic steps of simulation are as follows:

A. The time interval $[0;T]$ over which the solution is found, is subdivided into subintervals with step Δt.

B. The space domain is subdivided into cells with sides $\Delta y, \Delta z$.

C. Gas molecules are simulated in gap G using a stochastic system of N points (particles) having position $y_i(t), z_i(t)$ and velocities $\overrightarrow{\xi_i(t)}$.

D. N_m particles are located in the mth cell at any given time. This number varies during the computer simulation by the following two stages:

Stage 1. Binary collisions in each cell are calculated, whereas particles do not move. Collision modeling is realized using Bird's scheme "no time counter".

Stage 2. Particles move with new initial velocities acquired after collisions, and no external forces act on particles. No collisions are accounted for at this stage.

E. Stage 1 and Stage 2 are repeated until $t = T$.

F. Flow macro-characteristics (density, velocity, temperature) are calculated as time-averaged when steady regime is attained.

G. Boundary conditions are diffusive over the cylinders and periodical along axis Oy.

The Finite Difference Method for the Navier-Stokes Equations

The Navier-Stokes equations, which describe rarefied gas flow in a cylindrical coordinate system, have following form:

$$\frac{\partial \rho}{\partial t} + \frac{1}{r}\frac{\partial}{\partial r}(r\rho u) = 0 \tag{1}$$

$$\rho\left(\frac{\partial u}{\partial t} + u\frac{\partial u}{\partial r} - \frac{v^2}{r}\right) = -\frac{\partial P}{\partial r} + \frac{4}{3}\mu\frac{\partial}{\partial r}\left[\frac{1}{r}\frac{\partial}{\partial r}(ru)\right] + \frac{2}{3}\left(2\frac{\partial u}{\partial r} - \frac{u}{r}\right)\frac{\partial \mu}{\partial r} \tag{2}$$

$$\rho\left(\frac{\partial v}{\partial t} + u\frac{\partial v}{\partial r} + \frac{uv}{r}\right) = \mu\frac{\partial}{\partial r}\left[\frac{1}{r}\frac{\partial}{\partial r}(rv)\right] + r\frac{\partial}{\partial r}\left(\frac{v}{r}\right)\frac{\partial \mu}{\partial r} \tag{3}$$

$$\rho c_V\left(\frac{\partial T}{\partial t} + u\frac{\partial T}{\partial r}\right) = \frac{1}{r}\frac{\partial}{\partial r}\left(kr\frac{\partial T}{\partial r}\right) - \frac{P}{r}\frac{\partial}{\partial r}(ru) +$$

$$\frac{4}{3}\mu\left[\left(\frac{\partial u}{\partial r}\right)^2 - \frac{u}{r}\frac{\partial u}{\partial r} + \left(\frac{u}{r}\right)^2\right] + \mu\left[r\frac{\partial}{\partial r}\left(\frac{v}{r}\right)\right]^2 \tag{4}$$

$$P = \rho RT \tag{5}$$

where μ is the viscosity which is the function of the temperature [6]. Non-dimensional complexes, the implicit numerical scheme and slip boundary conditions for velocity and temperature are as in [6, 7].

NUMERICAL RESULTS

We study four cases at fixed Knudsen number 0.02, where the radius of the outer cylinder is twice larger than that of the inner one, and different rotation velocities of the inner cylinder are considered: $V_1 = 0.3; 0.5; 0.7; 1$. The outer cylinder is stationary. The modeling particles number for DSMC method is 3200000.

The results show very good agreement regarding density and z velocity component – Figure 1 and 2. Slight differences of order 0.002 are observed in the temperature profile. They are due to the approaches of flow modeling, which are a priori different. However, the good agreement and the unconditional convergence of the DSMC method is a numerical proof of the stability of the finite difference method.

An object of further studies is flow stability when:

- varying the Knudsen number;
- employing different boundary condition for velocity and temperature;
- considering rotation of the outer cylinder;
- considering simultaneous rotation of both cylinders, in one and the same direction and in opposite directions.

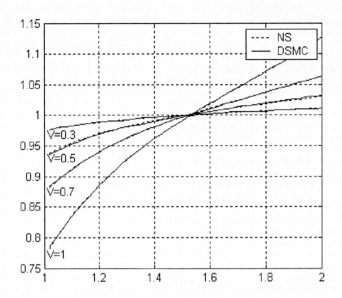

FIGURE 2. Density profile for inner cylinder rotating velocity $V_1 = 0.3; 0.5; 0.7; 1$

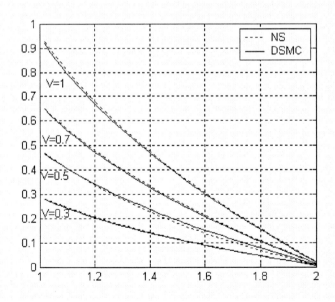

FIGURE 3. The z velocity profile for inner cylinder rotating velocity $V_1 = 0.3; 0.5; 0.7; 1$

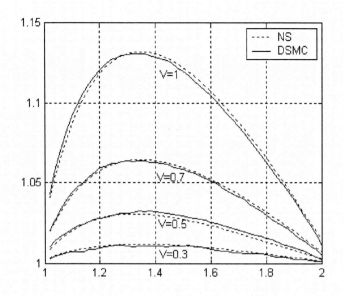

FIGURE 4. Temperature profile for inner cylinder rotating velocity $V_1 = 0.3; 0.5; 0.7; 1$

ACKNOWLEDGMENTS

This work is supported by the Bulgarian National Science Foundation, under Grant MM 1404/2004.

REFERENCES

1. C. Cercignani and F. Sernagiotto, *Physics of Fluids* **10**, 1200 (1967).
2. K. W. Tibbs, F. Baras, and A. L. Garcia, *Phys, Rev. E* **56**, 2282-2283 (1997).
3. K. Aoki, H. Yoshida, T. Nakanishi, and A. Garcia, *Phys. Rev. E* **68**, 016302 (2003).
4. G. A. Bird, *Molecular Gas Dynamics and the Direct Simulation of Gas Flows*, Oxford University Press, Oxford, 1994.
5. C. Cercignani, *Rarefied Gas Dynamics, From Basic Concepts to Actual Calculations*, Cambridge University Press, Cambridge, 2000.
6. S. K. Stefanov, V. M. Roussinov, and C. Cercignani, *Physics of Fluids* **14**, 2255-2269 (2002).
7. S. K. Stefanov, V. M. Roussinov, and C. Cercignani, *Physics of Fluids* **14**, 2270-2288 (2002).

Model and H_∞ Controller for a Distributed Parameter System

B. Gilev

Technical University of Sofia, Faculty of Applied Mathematics and Informatics
1000 Sofia, Bulgaria

Abstract. This paper presents a mathematical model of inductor electromagnetic and heating processes. There is introduced a controller as well, designed to provide a uniform heating supply. The mathematical model is used for a coupled electromagnetic-thermal analysis, since the electromagnetic analysis calculates the density of the heating source and the thermal analysis, that follows, utilizes the obtained heating source density to calculate a bounded heat transfer problem. The finite element method (FEM or FE method) is applied to reduce the problem to a linear system of ODEs and then, by the use of the control theory and the Glover-Doyle algorithm, it is demonstrated how to design a controller to establish a min temperature gradient change in the radial direction (a uniform heating mode). It is well-known that processes, such as the semi-solid forming and hot casting, require the max achievable uniform heating for a min period of time.

Keywords: Induction heating, Maxwell equation, heat transfer, robust control.
PACS: 84.30.-r, 44.05.+e, 41.20.-q

INTRODUCTION

The multitude of mathematical models that exist, both numeric and analytic, for analysis of the induction heating processes, is due to the proposed various geometric and environmental (including non-linear) conditions [2, 4, 8]. Usually, only one of the two primary optimization tasks - uniform heating and heating for hardening, is solved [2, 7, 10]. The solutions are often obtained for a specific shape of the induction coil or pre-set process parameters, determined after a corresponding numeric optimization [2].

The input voltage Ud of the assigning inverter drive [6] is the selected controlled variable. The transfer function

$$I = k\,U_d, \tag{1}$$

is linear, where the nominal value of k is known, and I is the amplitude of the loading coil current of the inverter.

The electric current amplitude is the main quantity that controls the variations in the magnetic field intensity H_z of a cylindrical ferromagnetic body, placed inside the inductor device. An electromagnetic analysis of the inductor calculates the closed form relation between H_z and I [9, 5].

Once is known, the estimation of the heating source density F, in the heat-transfer equation (HTE), is facilitated. The FE method reduces the HTE to a system of linear ODEs. The state variables y_1, y_2, y_3 and y_4, are the temperatures in four equidistant points, located on the radius, in direction from the center to the surface of the heated component.

CP1067, *Applications of Mathematics in Engineering and Economics '34—AMEE '08*, edited by M. D. Todorov
© 2008 American Institute of Physics 978-0-7354-0598-1/08/$23.00

The so formed model includes a range of coefficients, usually set to pre-defined nominal values. However, in practice, these coefficients are temperature dependent and alter their values during the induction heating process. Therefore, this model is suitable for robust control, i.e., it is feasible to design a controller to give required output quantities when model parameters change within certain limits.

ELECTROMAGNETIC ANALYSIS

The Maxwell equations are used to describe the electromagnetic processes of an inductor, placed in a cylindrical ferromagnetic body of radius R and length l. On certain assumptions and a cylindrical coordinate system (r, θ, z), Maxwell's equations lead to Bessel's equation [9, 5]

$$\frac{d^2 H_z}{dr^2} + \frac{1}{r}\frac{dH_z}{dr} - k^2 H_z = 0, \quad \text{for} \quad k^2 = i\sigma\omega\mu\mu_0 - \omega^2\mu\mu_0\varepsilon\varepsilon_0, \tag{2}$$

where $i^2 = -1$, σ is the electrical conductivity of the material, ε, μ – the electrical and magnetic permeability of the material, ε_0, μ_0 – the electrical and magnetic constants, ω – the frequency of the inductor current, H_z – the complex amplitude of the intensity of the electrical field.

It is assumed, in this case, that H_z has a sinusoidal time behavior

$$H_z(t) = H_z\, e^{i\omega t}.$$

The equation (2) is solved with the following boundary condition [9, 5, 8]

$$H_{z0}\, l = I\, n, \tag{3}$$

where l is the length of cylindrical body, n – number of inductor coils, I – the complex amplitude of the current in the coil of the inductor, H_{z0} – magnetic field intensity on the surface of the cylindrical body $(H_{z0} = H_z(R))$.

Finally the solution of the problem (2)-(3) is given by the equation [9, 5]

$$H_z(r) = \frac{I\, n}{l\, I_0(kR)}\, I_0(kr), \tag{4}$$

where $I_0(\cdot)$ is a Bessel function.

After H_z is estimated, Maxwell's equations are used once more, for calculating the density of the heating source

$$F(r) = \frac{1}{2}\sigma |E_\varphi|^2 = \frac{1}{2}\sigma \left| \frac{1}{\sigma + i\varepsilon\varepsilon_0}\frac{dH_z}{dr} \right|^2 = \frac{1}{2}\sigma \left| \frac{1}{\sigma + i\varepsilon\varepsilon_0}\frac{k\, I\, n}{l\, I_0(kR)} I_1(kr) \right|^2, \tag{5}$$

where E_φ is the complex amplitude of the electrical intensity; and $I_1(\cdot)$ – Bessel's function $(\dot{I}_0(x) = I_1(x))$.

THERMAL ANALYSIS

The heat transfer equation

$$C\rho \frac{\partial U}{\partial t} = \frac{\lambda}{r} \frac{\partial}{\partial r}\left(r\frac{\partial U}{\partial r}\right) + F(r), \text{ for } r \in [0,R] \text{ and } t \geq 0 \tag{6}$$

is solved in polar coordinates for the boundary conditions:

$$\frac{\partial U}{\partial r}(0,t) = 0; \quad \frac{\partial U}{\partial r}(R,t) = -\frac{\alpha}{\lambda}U \tag{7}$$

and the initial condition

$$U(r,0) = 0, \tag{8}$$

where U is the temperature of the heated body, ρ – density of the heated body, C – thermal capacity, λ – coefficient of thermal conductivity, α – release coefficient, $F(r)$ – the density of the heating source is calculated by formula (5).

The first boundary condition assumes symmetry, while the second one reflects the heat release at the boundary.

The FE method reduces equations (6)-(8) to a system of ODEs. The interval $[0,R]$ is divided into equal sub-intervals $[r_{k-1}, r_k]$, for $k = 2,...,n$. Within each element $[r_{k-1}, r_k]$, the unknown function $U(r,t)$ is approximated with a piecewise function $y(r,t)$, i.e.,

$$y(r,t) = \sum_{j=1}^{2} y_j^k(t)N_j^k(r), \text{ for } r \in [r_{k-1}, r_k]. \tag{9}$$

where $N_1^k = \frac{r_k - r}{r_k - r_{i-1}}$ and $N_2^k = \frac{r - r_{k-1}}{r_k - r_{k-1}}$.

We choose 3 elements, because if we increase the number of the elements, the result does not change significantly. Then the approximate solution, in the range $[0,R]$, is

$$y(r,t) = \sum_{k=1}^{3} \sum_{j=1}^{2} y_j^k(t)N_j^k(r), \text{ for } r \in [0,R]. \tag{10}$$

The Galerkin method is applied, i.e.,

$$\int_{r_{k-1}}^{r_k} \left(C\rho\, r \frac{\partial U}{\partial t} - \lambda \frac{\partial}{\partial r}\left(r\frac{\partial U}{\partial r}\right) - r\, F(r)\right)N_i^k(r)dr = 0, \text{ for } i = 1,2 \text{ and } k = 2,3,4 \tag{11}$$

Integrating by parts and substituting the exact solution $U(r,t)$, with the approximate one $y(r,t)$ from (11), results in

$$\sum_{j=1}^{2} \frac{\partial y_j^k(t)}{\partial t}K_{ij}^k + \sum_{j=1}^{2} y_j^k(t)M_{ij}^k = L_i^k, \text{ for } i = 1,2 \text{ and } k = 2,3,4, \tag{12}$$

246

where

$$K_{ij}^k = \int_{r_{k-1}}^{r_k} C\rho\, r N_i^k N_j^k dr, \text{ for } i = 1,2 \text{ and } j = 1,2; \tag{13}$$

$$M_{ij}^k = \int_{r_{k-1}}^{r_k} \lambda\, r \frac{\partial N_i^k}{\partial r} \frac{\partial N_j^k}{\partial r} dr, \text{ for } i = 1,2 \text{ and } j = 1,2; \tag{14}$$

$$L_i^k = \lambda_k\, r_k N_i^k(r_k)\frac{\partial U}{\partial r}(r_k) - \lambda_{k-1}\, r_{k-1} N_i^k(r_{k-1})\frac{\partial U}{\partial r}(r_{k-1}) + B_i^k, \tag{15}$$

$$\text{where } B_i^k = \int_{r_{k-1}}^{r_k} r F(r)\, N_i^k(r)dr, \text{ for } i = 1,2.$$

Global variables are assigned to the state space variables as follows:

$$y_1(t) = y_1^2(t),\ y_2(t) = y_2^2(t) = y_1^3(t),\ y_3(t) = y_2^3(t) = y_1^4(t),\ y_4(t) = y_2^4(t). \tag{16}$$

The second and the third as well as the fourth and the fifth equations are summed to receive the following

$$\begin{pmatrix} K_{11}^2 & K_{12}^2 & & \\ K_{21}^2 & K_{22}^2 + K_{11}^3 & K_{12}^3 & \\ & K_{21}^3 & K_{22}^3 + K_{11}^4 & K_{12}^4 \\ & & K_{21}^4 & K_{22}^4 \end{pmatrix} \begin{pmatrix} \dot{y}_1(t) \\ \dot{y}_2(t) \\ \dot{y}_3(t) \\ \dot{y}_4(t) \end{pmatrix} \tag{17}$$

$$+ \begin{pmatrix} M_{11}^2 & M_{12}^2 & & \\ M_{21}^2 & M_{22}^2 + M_{11}^3 & M_{12}^3 & \\ & M_{21}^3 & M_{22}^3 + M_{11}^4 & M_{12}^4 \\ & & M_{21}^4 & M_{22}^4 \end{pmatrix} \begin{pmatrix} y_1(t) \\ y_2(t) \\ y_3(t) \\ y_4(t) \end{pmatrix} = \begin{pmatrix} L_2^2 \\ L_2^2 + L_1^3 \\ L_2^3 + L_1^4 \\ L_2^4 \end{pmatrix}.$$

Solving for the boundary conditions (7) and transforming the free terms of (17), yields:

$$\begin{pmatrix} K_{11}^2 & K_{12}^2 & & \\ K_{21}^2 & K_{22}^2 + K_{11}^3 & K_{12}^3 & \\ & K_{21}^3 & K_{22}^3 + K_{11}^4 & K_{12}^4 \\ & & K_{21}^4 & K_{22}^4 \end{pmatrix} \begin{pmatrix} \dot{y}_1(t) \\ \dot{y}_2(t) \\ \dot{y}_3(t) \\ \dot{y}_4(t) \end{pmatrix} \tag{18}$$

$$+ \begin{pmatrix} M_{11}^2 & M_{12}^2 & & \\ M_{21}^2 & M_{22}^2 + M_{11}^3 & M_{12}^3 & \\ & M_{21}^3 & M_{22}^3 + M_{11}^4 & M_{12}^4 \\ & & M_{21}^4 & M_{22}^4 + R\alpha \end{pmatrix} \begin{pmatrix} y_1(t) \\ y_2(t) \\ y_3(t) \\ y_4(t) \end{pmatrix} = \begin{pmatrix} B_1^2 \\ B_2^2 + B_1^3 \\ B_2^3 + B_1^4 \\ B_2^4 \end{pmatrix},$$

where the coefficients are calculated by formulae (13), (14) and (15).

A MODEL WITH UNCERTAIN PARAMETERS

The nominal value of the constant k from (1) is

$$k = 10.135 \ 1.777.$$

The indeterminateness in this coefficient is in the interval $\pm 35\%$. It follows, from (1), that

$$l^2 = 10.135^2 1.777^2 (1+d_1) U_d^2, \text{ where } d_1 \in [-0.35, 0.35].$$

This is a transfer function of the inverter with uncertain parameters.

Parameters C, λ and α, included in the inductor model, use to vary during heating and take part in the calculation of the K_{ij}, M_{ij} and B_i coefficients. Parameters C and λ are included only once, as multipliers in the numerators of and respectively.

The indeterminateness of parameter C is given by the formula

$$C = \frac{C_{nom}}{(1+a_1)} = \frac{650}{(1+a_1)}, \text{ where } a_1 \in [-0.3, 0.3].$$

As the nominal value of the parameter C is included in the calculation of the elements K_{ij}, then correction of the matrix $K = \{K_{ij}\}$ is made in accordance with the formula

$$\frac{1}{(1+a_1)} K, \text{ where } a_1 \in [-0.3, 0.3].$$

On the other hand λ is corrected with $\pm 45\%$, i.e.,

$$\lambda = \lambda_{nom}(1+b_1) = 45(1+b_1), \text{ where } b_1 \in [-0.45, 0.45].$$

The nominal value of the parameter λ is included in the calculation of the elements M_{ij}, and its influence is given by

$$(1+b_1)M, \text{ where } b_1 \in [-0.45, 0.45].$$

The parameter α is corrected as follows

$$\alpha = \alpha_{nom}(1+c_1) = 25(1+c_1), \text{ where } c_1 \in [-0.4, 0.4].$$

Finally the uncertain model of the system is

$$\frac{K}{(1+a_1)} \begin{pmatrix} \dot{y}_1 \\ \dot{y}_2 \\ \dot{y}_3 \\ \dot{y}_4 \end{pmatrix} + \left\{ (1+b_1)M + \begin{pmatrix} 0 & 0 & 0 & 0 \\ 0 & 0 & 0 & 0 \\ 0 & 0 & 0 & 0 \\ 0 & 0 & 0 & R\,\alpha_{nom}(1+c_1) \end{pmatrix} \right\} \begin{pmatrix} y_1 \\ y_2 \\ y_3 \\ y_4 \end{pmatrix} \quad (19)$$

$$= \bar{B}\, 10.135^2 1.777^2 (1+d_1) U_d^2$$

where $a_1 \in [-0.3, 0.3]$, $b_1 \in [-0.45, 0.45]$, $c_1 \in [-0.4, 0.4]$, $d_1 \in [-0.35, 0.35]$, $K = \{K_{ij}\}$, $M = \{M_{ij}\}$ and vector $\bar{B}$ results from $B = \{B_i\}$, after calculating $B = l^2 \bar{B}$.

The values of the parameters, which remain unchanged during the time of heating are:

$$\omega = 2\pi\,4000\ s^{-1}, \quad R = 0.008\ m, \quad \rho = 7.83\ 10^3\ kg/m^3.$$

The nominal values of the uncertain parameters are the following values:

$$\alpha_{nom} = 25\ W\,m^{-2}C^{-1}, \quad \lambda_{nom} = 45\ W\,m^{-1}C^{-1}, \quad C_{nom} = 650\ J\,kg^{-1}C^{-1}.$$

The nominal matrixes K and M are:

$$K = \begin{bmatrix} 3.044 & 3.044 & 0 & 0 \\ 3.044 & 24.0356 & 9.0133 & 0 \\ 0 & 9.0133 & 48.0711 & 15.0222 \\ 0 & 0 & 15.0222 & 33.0489 \end{bmatrix} \quad \text{and}$$

$$M = \begin{bmatrix} 22.5 & -22.5 & 0 & 0 \\ -22.5 & 90 & -67.5 & 0 \\ 0 & -67.5 & 180 & -112.5 \\ 0 & 0 & -112.5 & 112.5 \end{bmatrix}$$

The vector $\bar{B}$ is

$$\bar{B} = 10^{-3} \begin{bmatrix} 0.00005 \\ 0.0021 \\ 0.0446 \\ 0.1028 \end{bmatrix}, \text{ for } \mu = 15 \text{ and } \bar{B} = 10^{-5} \begin{bmatrix} 0.0034 \\ 0.1033 \\ 0.6266 \\ 0.6842 \end{bmatrix}, \text{ for } \mu = 1.$$

The so formed matrices are included in the substitutions (19).

The uncertain model is obtained on the basis of the formula (19) implementing the function SYSIC from the Robust Control Toolbox (2006) and author's code. As a results one find an uncertain state space object. These system is described by the equation

$$(y_1\ y_2\ y_3\ y_4)^T = G\,u$$

where $u = U_d^2$ and G is the plant transfer function.

H_∞ SYNTHESIS

This synthesis is done in the frequency domain.

The block-diagram of the close-loop system that includes the uncertain model G, the feedback and controller K, as well as the elements reflecting the performance requirements W_p and W_u, is shown in Figure 1. The signals ref, y_i and u are respectively, reference output, obtained outputs and control signal.

In H_∞ synthesis is used the object $Gnom$, where the uncertain parameters are fixed at their nominal values.

The controller K is synthesized by an open-loop system P, obtained from the closed-loop system (Figure 1), after the controller K is removed. The inputs of this system are ref and u, and the outputs are e_u, e_{y_1}, e_{y_2}, e_{y_3}, e_{y_4} and $ref - y_4$.

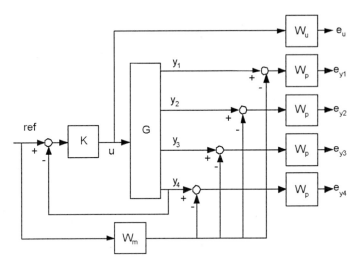

FIGURE 1. Block-diagram of the closed-loop system with filters

If the open system P and the controller K are closed by lower fraction linear transform $F_l(\cdot,\cdot)$ we obtain the closed system $F_l(P,K)$ from Figure 1.

The H_∞ synthesis is searching for a suboptimal controller K_{sub} that fulfills

$$||F_l(P,K_{\text{sub}})(j\omega)||_\infty < \gamma, \text{ where } \gamma > \gamma_{\text{min}}, \tag{20}$$

where the γ_{min} designates a number that fulfills

$$\min_K ||F_l(P,K_{\text{sub}})(j\omega)||_\infty = ||F_l(P,K_{\text{sub}})(j\omega)||_\infty = \gamma_{\text{min}}.$$

The synthesis of the suboptimal controlling device is performed through the use of the Glover-Doyle algorithm [3], as follows:

1. Select an initial approximation of γ and obtain the solutions $X_\infty > 0$ and $Y_\infty > 0$ (positively determined matrices) of the Riccati equations

$$A^T X_\infty + X_\infty A + C_1^T C_1 + X_\infty(\gamma^{-2} B_1 B_1^T - B_2 B_2^T)X_\infty = 0$$

$$A Y_\infty + Y_\infty A^T + B_1^T B_1 + Y_\infty(\gamma^{-2} C_1^T C_1 - C_2^T C_2)Y_\infty = 0,$$

where the controlled system is described by the following system of linear DEs:

$$\left.\begin{array}{l} \left(\begin{array}{c} \dot{x}_1 \\ \dot{x}_2 \end{array}\right) = A\left(\begin{array}{c} x_1 \\ x_2 \end{array}\right) + (B_1\, B_2)\left(\begin{array}{c} u_1 \\ u_2 \end{array}\right) \\[12pt] \left(\begin{array}{c} y_1 \\ y_2 \end{array}\right) = (C_1\, C_2)\left(\begin{array}{c} x_1 \\ x_2 \end{array}\right) + \left(\begin{array}{cc} D_{11} & D_{12} \\ D_{21} & D_{22} \end{array}\right)\left(\begin{array}{c} u_1 \\ u_2 \end{array}\right) \end{array}\right\}. \tag{21}$$

The variables u_2 and y_2, which are used for closing the feedback are designated with index 2, while the variables u_1 and y_1 stay free and are designated with index 1. The bounded with $u_2(y_2)$ and $u_1(y_1)$ state space variables are designated respectively with x_2 and x_1. The constraints, which are put over controlled system (21) are given in [3].

2. If matrices X_∞ and Y_∞ are obtained, synthesize a suboptimal controlling device, to fulfill (21), according to the formula

$$K_c(s) = -Z_\infty L_\infty (sI - A_\infty)^{-1} F_\infty, \text{where}$$

$$F_\infty = -B_2^T X_\infty, \quad L_\infty = -Y_\infty C_2^T, \quad Z_\infty = (I - \gamma^{-2} Y_\infty X_\infty)^{-1} \text{ and}$$

$$A_\infty = A + \gamma^{-2} B_1 B_1^T X_\infty + B_2 F_\infty + Z_\infty L_\infty C_2.$$

The H_∞ synthesis of the controller is an iterative procedure, where an initial approximation of γ is selected and a further bisection by γ is done.

The H_∞ controller is synthesized in MATLAB environment, using the standard command HINFSYN and private source code. A controller of 24th order is obtained. The order of this controller is very high, which eventually makes difficult its real time implementation. That is why the order of the controller is reduced to 10.

The H_∞ synthesis is done for several performance weighting functions that ensure a good balance between system performance and robustness. On the basis of the experimental results, we choose the performance and the control weighting functions, Figure 1

$$W_p = \frac{s+50}{s+2.5} \text{ and } W_u = 10^{-2}.$$

The reference model is

$$W_m = \frac{1}{10s+1}.$$

The G-model with uncertain parameters is used for the simulation. Thus, four families of curves, representing the various values of the uncertain parameters, are obtained, as shown in Figure 2.

The plots on Figure 2 prove that the synthesized controller establishes effectively a uniformly heated state of the system for a relatively short time.

CONCLUSION

The new point in this work is the contemporary approach of the robust control for achieving uniform heating of a piece in inductor. For this purpose the nonlinear problem for the induction heating is linearized by the finite elements method and by fixing the uncertain parameters to their nominal values. Based on the obtained linear model by using the synthesis we develop the robust controller, which provides the desired regime. The operation of the controller is tested over the model with uncertain parameters and the results show that the family lines, which correspond to various values uncertain parameters are getting closer quickly, which shows the robustness of the control.

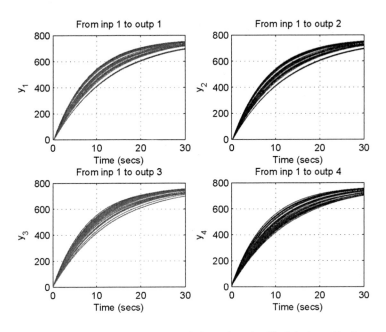

FIGURE 2.   The temperature variations along 4 radii of the heated body

REFERENCES

1. Balas, R. Chiang, A. Packard, and M. Safonov, *MATLAB, Robust Control Toolbox, User's Guide,* The MathWorks, Inc., 2006.
2. O. Bodart, A.-V. Boureau, and R. Touzani, *Applied Mathematical Modeling* **25**(8), 697–712 (2001).
3. J.C. Doyle, B. Francis, and A. Tannenbaum, *Feedback Control Theory,* Macmillan Publishing Company, N.Y., 1992.
4. M. Feliachi and G. Develey, *IEEE Trans. on Magnetics* **27**(6), 5235–5237 (1991).
5. B. Gilev, G. Kraev, and G.I. Venkov, "A Double Layer Model of the Electromagnetic and Thermal Processes in Induction Heating of Ferromagnetic Material," in *Applications of Mathematics in Engineering and Economics'33,* edited by M. D. Todorov, AIP CP946, Melville, New York, 2007, pp. 153–162.
6. N. Hinov, G. Kraev, B. Gilev, and D. Vakovsky, *Facta Universitatis (NIS), SER.: Elec. Energ.* **19**(1), 99–107 (2006), [Online: http://factaee.elfak.ni.ac.yu/fu2k61/hinov.html].
7. E. Rapoport, Y. Pleshivtseva, *Optimal Control of Induction Heating Processes,* Taylor&Francis, Boca Ration, London, New York, 2006.
8. A.E. Slukhotski, *Induction Heating Devices,* Energoizdat, Leningrad, 1991. (in Russian)
9. L.A. Veinstein, *The Electromagnetic Waves,* Radio Svyaz, Moscow, 1988. (in Russian)
10. F. Young, *IEEE Transactions on Magnetics* **13**(6), 1776–1785 (1977).

Increasing of the Accuracy in the Computation of the Concentrations in Diffusion Models with Localized Chemical Reactions

B. Jovanovic[*], J. Kandilarov[†] and L. Vulkov[†]

[*]University of Belgrade, Faculty of Mathematics, 11000 Belgrade, Serbia
[†]Dept. of Numerical Analysis, FNSE, University of Rousse, 8 Studentska str.,
7017 Rousse, Bulgaria

Abstract. In the present work Richardson extrapolation to accelerate the convergence of finite difference schemes for reaction-diffusion problems with localized nonlinear reaction terms is used. A rigorous rate of convergence analysis is given. Numerical experiments confirm the theoretical results. Conditions for the coefficients of the differential operator that guarantee well posedness of the boundary value problems are also discussed.

Keywords: Richardson extrapolation, local jump conditions, difference schemes, convergence.
PACS: 01.30.Cc, 31.15.xf, 44.05.+e, 68.35.Fx

INTRODUCTION

The basic physical system considered is that of linear diffusion and adsorption-desorption but nonlinear localized chemical reactions:

$$-\frac{d}{dx}\left(D(x)\frac{dU}{dx}\right) + K(x)U - \sum_{s=1}^{S} Q_s(U)\delta(x - \xi_s) = F(x), \quad x \in \Omega \equiv (0,1). \quad (1)$$

Here $U(x,t) \in R^{+N}$ is a vector with positive valued components representing the concentrations of N different species, D is a matrix of diffusion coefficients, K is a matrix representing adsorption-desorption of linear bulk reaction, Q is the rate term due to reactions taking place at the active site ξ_s, $F(x)$ is the incident flux and δ is the Dirac-delta function. In this paper we assume that the bulk processes are decoupled so that D and K are diagonal.

To determine the solution U of (1) we prescribe appropriate boundary conditions that may be expressed in the general form

$$(\alpha_i \frac{d}{dx} + d_i)U(x_i) = g_i, \quad (2)$$

$i = 0$ with $x_0 = 0$ and $i = 1$ with $x_1 = 1$. Here, as with the bulk, we assume that the boundary conditions are decoupled so that α_i and d_i are diagonal and $g_i, i = 0, 1$ are specified constant vectors. Hence the only coupling in the equations governing the various species occurs through the reaction term, which is generally nonlinear.

CP1067, Applications of Mathematics in Engineering and Economics '34—AMEE '08, edited by M. D. Todorov
© 2008 American Institute of Physics 978-0-7354-0598-1/08/$23.00

It is possible to rewrite (1) in an alternative form if we assume that D, K, F are piecewise continuous functions and U is continuous:

$$-\frac{d}{dx}\left(D(x)\frac{dU}{dx}\right) + K(x)U = F(x), \quad x \in \Omega = \bigcup_{s=1}^{S+1} \Omega_s, \tag{3}$$

$$[U]_{x=\xi_s} = 0, \quad \left[D\frac{dU}{dx}\right]_{x=\xi_s} = Q(U(\xi_s)), \quad s = 1, \ldots, S. \tag{4}$$

Here $\Omega_s = (\xi_{s-1}, \xi_s)$, $s = 1, \ldots, S+1$, $\xi_0 = 0$, $\xi_{S+1} = 1$ and

$$[v]_{x=\xi_s} \equiv \lim_{\varepsilon \to 0+} (v(\xi_s + \varepsilon) - v(\xi_s - \varepsilon)).$$

Physically, these equations can be used to represent localized reactions that occur at chemically active parallel line-defects on a two-dimensional surface, or they can be used in the modeling of influence of membrane-bound enzymes in the living organisms [2, 12]. Also, some processes in electrodynamics, fluid mechanics [5], generate own, intrinsic for the concrete phenomena localized sources and can be described by similar equations. In the present paper we consider only 1D model equations.

The immersed interface method (IIM) of second order was applied to 1D nonlinear parabolic problems and 2D problems in the case of line interface and interface relations of the form (4) by Kandilarov and Vulkov [9], for 2D parabolic problems with curve interface [8], for system of two localized reactions [10] and for elliptic problems by B. Jovanovic, J. Kandilarov and L. Vulkov [6].

Richardson extrapolation is often used to improve the numerical methods [4, 7, 11]. We first explain how to implement the Richardson method for (1)-(2) on the scalar case. Then we extend the algorithm to nonlinear problems. System of linear equations are also discussed. Numerical experiments are reported in the last section.

SCALAR CASE

The main ideas and results of the paper will be discussed on the case of one equation.

Linear Reactions

We explain the idea of Richardson extrapolation on the simplest problem of a single reaction

$$\begin{aligned}
-u'' + k(x)u &= f(x), & x &\in (0, \xi) \cup (\xi, 1), \\
[u]_{x=\xi} &= 0, & [u']_{x=\xi} &= qu(\xi), \\
u(0) &= 0, & u(1) &= 0.
\end{aligned} \tag{5}$$

Let us introduce the following function spaces:

$$Q_\xi^m = C^m([0, \xi]) \cap C^m([\xi, 1]) \quad \text{and} \quad CQ_\xi^m = Q_\xi^m \cap C^m([0, 1]), \quad m = 1, 2, \ldots.$$

Following [11] one can prove that if $k(x) \geq 0$ and $f, k \in Q_\xi^m$, then there exists unique solution $u \in Q_\xi^{m+2}$ of problem (5).

We also denote $\{u(x)\} := \frac{u(x-0)+u(x+0)}{2}$, $u_x := \frac{u(x+h)-u(x)}{h}$ and $u_{\bar{x}} := \frac{u(x)-u(x-h)}{h}$.

Suppose (for simplicity of the exhibition) that ξ is rational number. Then on $(0,1)$ one can construct uniform mesh ω_h with step h for which ξ is a node. We approximate the problem (5) by the difference scheme

$$-u_{\bar{x}x}^h + k(x)u^h + q\delta_h(x-\xi)u^h = \tilde{f}(x), \quad x \in \omega_h$$
$$u^h(0) = 0, \quad u^h(1) = 0, \tag{6}$$

where

$$\delta_h(x-\xi) = \begin{cases} 1/h, & x = \xi, \\ 0, & x \in \omega_h \setminus \{\xi\}, \end{cases} \quad \tilde{f}(x) = \{f(x)\} = \begin{cases} \frac{f(x-0)+f(x+0)}{2}, & x = \xi, \\ f(x), & x \in \omega_h \setminus \{\xi\}. \end{cases}$$

The scheme (6) could be written in the form:

$$-u_{\bar{x}x}^h + k(x)u^h = f(x), \quad x \in \omega_h, \ x \neq \xi,$$
$$u_x^h - u_{\bar{x}}^h = qu^h + h(ku^h - \{f\}), \quad x = \xi, \tag{7}$$
$$u^h(0) = 0, \quad u^h(1) = 0.$$

We will construct the expansion of the difference solution u^h along the power of h

$$u^h = u_0 + h^2 u_1 + h^4 u_2 + \dots, \tag{8}$$

where $u_i = u_i(x)$, $i = 0, 1, 2, \dots$ are unknown functions with independent variable $x \in (0,1)$.

Let $U(x)$ be continuous on $[0,1]$, arbitrary smooth on each subintervals $[0, \xi]$ and $[\xi, 1]$ function. Using the Taylor formula we get:

$$U_{\bar{x}x} = U'' + \frac{h^2}{12}U^{(4)} + \frac{h^4}{360}U^{(6)} + \dots, \quad x \in \omega_h, \ x \neq \xi,$$

and

$$U_x - U_{\bar{x}} = [U'] + h\{U''\} + \frac{h^2}{6}[U'''] + \frac{h^3}{12}\{U^{(4)}\} + \frac{h^4}{120}[U^{(5)}] + \dots, \quad x = \xi.$$

We find from (5)-(7)

$$-\left(u_0'' + \frac{h^2}{12}u_0^{(4)} + \frac{h^4}{360}u_0^{(6)} + \dots\right) - h^2\left(u_1'' + \frac{h^2}{12}u_1^{(4)} + \dots\right) - h^4\left(u_2'' + \dots\right) - \dots$$
$$+ k(x)\left(u_0 + h^2 u_1 + h^4 u_2 + \dots\right) = f(x), \quad x \neq \xi. \tag{9}$$

It follows from (5), (6) and (8)

$$\left([u'_0] + h\{u''_0\} + \frac{h^2}{6}[u'''_0] + \frac{h^3}{12}\{u^{(4)}_0\} + \frac{h^4}{120}[u^{(5)}_0] + \dots\right)$$

$$+h^2\left([u'_1] + h\{u''_1\} + \frac{h^2}{6}[u'''_1] + \dots\right) + h^4\left([u'_2] + \dots\right) + \dots \tag{10}$$

$$= q(u_0 + h^2 u_1 + h^4 u_2 + \dots) - h\{f\}$$

$$+hk(x)(u_0 + h^2 u_1 + h^4 u_2 + \dots), \qquad x = \xi.$$

Now, comparing in (9) and (10) the coefficient at the equal (even) powers of h, we get:

$$\begin{aligned}
-u''_i + k(x)u_i &= f_i(x), & x \in (0,\xi) \cup (\xi,1) \\
[u_i]_{x=\xi} &= 0, & [u'_i]_{x=\xi} = qu_i(\xi) - \kappa_i, \\
u_i(0) &= 0, & u_i(1) = 0,
\end{aligned} \tag{11}$$

where $i = 0, 1, 2, \dots$, and

$$f_0(x) = f(x), \qquad\qquad\qquad \kappa_0 = 0,$$

$$f_1(x) = \frac{1}{12}u^{(4)}_0(x), \qquad\qquad \kappa_1 = \tfrac{1}{6}[u'''_0]_{x=\xi},$$

$$f_2(x) = \frac{1}{360}u^{(6)}_0(x) + \frac{1}{12}u^{(4)}_1(x), \qquad \kappa_2 = \tfrac{1}{120}[u^{(5)}_0]_{x=\xi} + \tfrac{1}{6}[u'''_1]_{x=\xi},$$

and so on. Therefore, $u_0(x) = u(x)$ and the next members $u_1(x)$, $u_2(x)$, ... and so on, are solution of the problem (11) – a problem of type (1), (2).

From (10), comparing the coefficients at the equal (odd) powers of h, we also obtain:

$$-\{u''_i(\xi)\} + k(\xi)u_i(\xi) = \{f_i(\xi)\}, \qquad i = 0, 1, 2, \dots$$

It follows from (11) that these requirements are also fulfilled.

Nonlinear Reactions

The typical is the scalar case with a single nonlinear reaction

$$\begin{aligned}
-u'' + k(x)u &= f(x), & x \in (0,\xi) \cup (\xi,1), \\
[u]_{x=\xi} &= 0, & [u_x]_{x=\xi} = q(u(\xi)), \\
u(0) &= 0 & u(1) = 0.
\end{aligned} \tag{12}$$

For the nonlinearity we assume:

$$0 < q'(\xi) < \infty, \qquad \xi \in R^1. \tag{13}$$

We study the difference scheme

$$-u^h_{\bar{x}x} + k(x)u^h = f(x), \qquad x \neq \xi, \tag{14}$$

$$u^h_x - u^h_{\bar{x}} = q(u^h) + hau^h - h\frac{f^+ + f^-}{2}, \; x = \xi.$$

In the last section we have assumed that the expansion (8) a priori holds. Starting with this assumption we found the members of the expansion. In the next theorem we prove that a finite expansion really exists

Theorem 1 *Suppose that the solution of (12)* $u \in Q^7_\xi$. *Then there exists* $h_0 > 0$, *such that for* $h < h_0$ *the expansion holds*

$$u^h(x) = u(x) + \sum_{j=1}^{2} h^{2j} v_j(x) + h^5 \eta^h(x), \quad x \in \omega_h, \tag{15}$$

where the functions $v_j \in Q^{7-2j}_\xi$ *do not depend on h and* η^h *are uniformly bounded with respect to* $\overline{\omega}_h$.

Outline of the proof: We consider the auxiliary problems (10) but with new interface conditions:

$$[u'_0]_\xi = q(u_0(\xi)), \qquad [u'_1]_\xi = q'(u_0(\xi))u_1(\xi) - \frac{1}{6}[u'''_0]_\xi,$$

$$[u'_2]_\xi = q'(u_0(\xi))u_2(\xi) - \frac{1}{6}[u'''_1]_\xi - \frac{1}{120}[u^v_0]_\xi.$$

Using formulas (6)-(7) we insert (15) in (14) to obtain for η^h the difference problem

$$-\eta^h_{\bar{x}x} + a\eta = \sum_{j=0}^{2} \chi^h_j, \qquad x \neq \xi,$$

$$\eta^h_x - \eta^h_{\bar{x}} - q'(\xi_1)\eta^h - ha\eta^h = -\left(\frac{1}{7!}[u^{(7)}_0] + \frac{1}{5!}[u^{(5)}_1] + \frac{1}{3!}[u^{(3)}_2]\right) \tag{16}$$

$$+u_1 u_2 q''(u_0) + \frac{1}{3!}u^3_1 q'''(\xi_2) + O(h^2), \quad x = \xi,$$

where $\xi_1, \xi_2 \in (u^h, \sum_{i=0}^{2} h^{2i}u_i)$ and $\chi^h_j \leq \max |u^{(7-2j)}_j(x)|$. It follows from the assumption (14) that for the problem (16) the maximum principle holds. Therefore $|\eta^h| \leq \|\frac{\tilde{F}}{\tilde{D}}\|$, where $\tilde{F}$ is the right-hand side of (16) and $\tilde{D} = a$, $x \in \omega_h \setminus \{\xi\}$, $\tilde{D} = q'(\xi_1) + ha$, $x = \xi$.

In a similarly way could be shown that $|[u^{(7-2i)}_i]|$ are bounded as solutions of corresponding problems. □

The expansion (15) permit us to perform the Richardson extrapolation. Let u^h be the solution of (14) and $u^{h/2}$ be the corresponding one with two times smaller mesh step. Then the extrapolated solution is [11]

$$U^h = \frac{4}{3}u^{h/2} - \frac{1}{3}u^h, \quad x \in \overline{\omega}_h. \tag{17}$$

Theorem 2 *Assume that the conditions in Theorem 1 are fulfilled. Then for the extrapolated solution (17) the estimate holds*

$$\max_{\overline{\omega}_h} |U^h(x) - u(x)| \leq Ch^4,$$

where the constant C is independent of h.

SYSTEMS OF EQUATIONS

There are only a few papers on Richardson extrapolation for nonlinear elliptic equations [1, 11]. For systems of nonlinear equations such papers are lacking. The difficulty is in theoretical analysis of the convergence rate.

In our case the first question is the well posedness of problem (1), (2) for $N \geq 2$. If the analysis relies on inverse-monotonicity method the first step is to formulate sufficient conditions (generalizations of (13)) in order the maximum principle holds for (1), (2), i.e., the solution takes its maximal (minimal) value at the boundary $x = 0$ or $x = 1$. Following results for system of ODEs with continuous coefficients, see [13] (in the system (1), some of the coefficients are Dirac-delta functions) we will discuss the linear case for a single reaction:

$$Q(U) = QU, \quad Q = \{q_{ij}\}_{i,j=1}^{N}, \ q_{ij} - \text{constants}.$$

First we assume that $D(x) = \{d_i(x)\}_{i=1}^{N}$ and $K = \{k_i(x)\}_{i=1}^{N}$ are diagonal matrices with elements that satisfy

$$0 < d_{i0} \leq d_i(x) \leq D_{i0}, \ 0 \leq k_{i0} \leq k_i(x) \leq K_{i0}, \ i = 1, \ldots, N, \tag{18}$$

where $d_{i0}, D_{i0}, k_{i0}, K_{i0}$ are constants.

Proposition 1 *Suppose that additional to (18) the coefficient matrix Q is diagonally dominant with strictly positive diagonal entries, and non-positive off diagonal entries. Then for problem (1), (2) the MP holds.*

But the system (1) may be analyzed with weakened hypothesis of Q. A such one is the following

Proposition 2 *Suppose in additional to (18) Q has strictly positive diagonal entries and the matrix*

$$\Gamma := \begin{pmatrix} 1 & -\|q_{12}/q_{11}\| & \cdots & -\|q_{1N}/q_{11}\| \\ -\|q_{21}/q_{22}\| & 1 & \cdots & -\|q_{2N}/q_{22}\| \\ \vdots & \vdots & \cdots & \vdots \\ -\|q_{N1}/q_{NN}\| & -\|q_{N2}/q_{NN}\| & \cdots & 1 \end{pmatrix}$$

is inverse monotone of Q, i.e., all entries of Γ are nonnegative. Then for problem (1), (2) the MP holds.

On the base of these linear criteria one can obtain similar ones for the nonlinear case.

TABLE 1. Grid refinement analysis for **Example 1**: $K = 3$, $\xi = 0.5$ and $\beta_1 = \beta_2 = 1$

N	$\|E_N\|$	$\|E_{2N}\|$	Ratio	$\|E_N\|_{rich}$	Ratio
8	9.3733e-05	2.3792e-05	4.00	1.8307e-08	15.96
16	2.3792e-05	5.9489e-06	4.00	1.1617e-09	15.76
32	5.9489e-06	1.4873e-06	4.00	7.2552e-11	16.01
64	1.4873e-06	3.7183e-07	4.00	4.5541e-12	15.93

NUMERICAL EXPERIMENTS

We test various numerical examples.

Example 1 *Single equation and linear local own source:*

$$(\beta u_x)_x = f(x) + q\delta(x - \xi)u, \ x \in (0,1).$$

We choose $\beta_1 = \beta_2 = 1$, $\xi = 0.5$, $q = 3$ and the exact solution of the form:

$$u(x,t) = \begin{cases} \exp(-\frac{x}{\xi}) - x(1 - \xi) + 3, & \text{if} \quad 0 \leq x \leq \xi; \\ \exp(-\frac{x}{\xi}) - \xi(1 - x) + 3, & \text{if} \quad \xi \leq x \leq 1. \end{cases}$$

The boundary conditions are found from the exact solution. In Table 1 we present the mesh refinement analysis. With $\|E_N\|_\infty$ we denote the infinity norm of the error when the mesh parameter $h = 1/N$. The ratio $\|E_N\|_\infty / \|E_{2N}\|_\infty$ near 4 indicates second order of convergence for the scheme without extrapolation. With $\|E_N\|_{rich}$ we denote the error, obtained after Richardson extrapolation. Ratio near 16 confirms the fourth order of convergence for the proposed method.

Example 2 *Single equation and nonlinear local own source:*

$$(\beta u_x)_x = f(x) + \delta(x - \xi)q(u), \ x \in (0,1).$$

Let the local source term be in the form $q(u) = -\exp(-u)$ and the exact solution be

$$u(x,t) = A \begin{cases} \sinh(x/\alpha_1)/\sinh(\xi/\alpha_1), & \text{if} \quad 0 \leq x \leq \xi, \\ \sinh((1 - x)/\alpha_2)/\sinh((1 - \xi)/\alpha_2), & \text{if} \quad \xi \leq x \leq 1, \end{cases}$$

where $\alpha_i^2 = \beta_i$, $i = 1, 2$ and A is the first positive solution of the nonlinear equation

$$A\{\alpha_2 \coth[(1 - \xi)/\alpha_2] + \alpha_1 \coth[(1 - \xi)/\alpha_1]\} = \exp(-A).$$

We take the interface $\xi = 1/3$ and chouse N so that the interface is a grid node and $\alpha_1 = 0.1$, $\alpha_2 = 1$. Results are presented in Table 2. Again the second order of the difference scheme and fourth order for the extrapolated solution are confirmed.

Example 3 *System of two equation with linear local own sources:*

$$\begin{aligned} u''(x) - u(x) &= \delta(x - 0)[2u(x) + v(x)], \ x \in (-1,1), \\ v''(x) - v(x) &= \delta(x - 0)[u(x) + 2v(x)], \ x \in (-1,1). \end{aligned}$$

TABLE 2. Grid refinement analysis for **Example 2**: $\xi = 1/3$ and $\alpha_1 = 0.1$, $\alpha_2 = 1$

N	$\|E_N\|$	$\|E_{2N}\|$	**Ratio**	$\|E_N\|_{rich}$	**Ratio**
48	2.0344e-04	5.1039e-05	3.99	2.4226e-07	15.65
96	5.1039e-05	1.2788e-05	3.99	1.5245e-08	15.89
192	1.2788e-05	3.1978e-06	4.00	9.5569e-10	15.95
384	3.1978e-06	7.9949e-07	4.00	5.9733e-11	16.00

The exact solution is:

$$u(x) = v(x) = \begin{cases} c_{1l}\exp(x) + c_{2l}\exp(-x), & \text{if} \quad -1 \le x \le 0, \\ c_{1r}\exp(x) + c_{2r}\exp(-x), & \text{if} \quad 0 \le x \le 1, \end{cases}$$

where

$$c_{1l} = [2\exp(1) - 6\exp(3)/(3\exp(2)+1)], \; c_{2l} = 6\exp(3)/(3\exp(2)+1)$$

and $c_{1r} = c_{2l}$, $c_{2r} = c_{1l}$. The boundary conditions are obtained by the exact solution.

In Table 3 are presented the results with Richardson extrapolation on two consecutive grids. The results show fourth order of the method.

TABLE 3. Grid refinement analysis for **Example 3** and extrapolation on two consecutive grids

N	$\|E_N\|$	$\|E_{2N}\|$	**Ratio**	$\|E_N\|_{rich}$	**Ratio**
40	1.7018e-04	4.2555e-05	4.00	1.2077e-08	15.98
80	4.2555e-05	1.0639e-05	4.00	7.5543e-10	15.99
160	1.0639e-05	2.6599e-06	4.00	4.7477e-11	15.91
320	2.6599e-06	6.6496e-07	4.00	3.4997e-12	13.96

CONCLUSIONS

In this contribution we have studied the Richardson extrapolation for a system of reaction-diffusion equations with localized nonlinearities. We first consider one reaction-diffusion equation with single reaction: linear or nonlinear. In this case we first described the algorithm of expansion with respect to the mesh step h of the discrete solution. Then, a rigorous rate of convergence analysis is given without proof of the theorems. More detailed analysis, as well as numerical experiments, is carried to such problems in our forthcoming papers.

ACKNOWLEDGMENTS

This research is supported by the Bulgarian National Fund of Science under Project VU-MI-106/2005.

REFERENCES

1. R. O. Abasian and G. F. Carey, *Comm. in N.M. in Engn.* **13**, 533–540 (1997).
2. K. Bimpong-Bota, A. Nizan, P. Ortoleva, and J. Ross, *J. Chem. Phys.* **66**, 3650–3683 (1970).
3. C. Brezinski, *Numer. Math.* **35**, 175–187 (1980).
4. C. Brezinski and M. Redivo-Zaglia, *Extrapolation Methods - Theory and Practice*, Studies in Computational Mathematics, No. 2, North-Holland, Amsterdam 1990.
5. C.Y. Chan, and P.C. Kong, *Quart. Appl. Math.* **55**, 51–56 (1997).
6. B. Jovanovic, J. Kandilarov, and L. Vulkov, *Lect. Notes Comp. Science*, **1988**, 431–438 (2000).
7. P. M. Lima and M. M. Graca, *J. Comput. Appl. Math.* **61**, 139–164 (1995).
8. J. Kandilarov and L. Vulkov, *Appl. Num. Math.* **57**(5-7), 486–497 (2007).
9. J. Kandilarov and L. Vulkov, *Numer. Algor.* **36**, 285–307 (2004).
10. J. Kandilarov, *Kragujevac Journal of Mathematics* **30**, 151–170 (2007).
11. G. I. Marchuk and V. V. Shaidurov, *Difference Methods and Their Extrapolations*, Springer, New York, 1983.
12. A. P. Pierce and H. Rabitz, *Surface Science* **202**, 1–31 (1988).
13. E. O'Riordan, J. Stynes, and M. Stynes, "An iterative numerical algorithm for a coupled system of singularly perturbed convection diffusion two point boundary value problems," *Lect. Notes in Comp. Science.* (in press)

A Two-grid Algorithm for Implementation of Fully Conservative Difference Schemes of the Gas Dynamics Equations

M. N. Koleva

Dept. of Mathematical Analysis, FNSE, University of Rousse, 8 Studentska str., 7017 Rousse, Bulgaria

Abstract. This study is proposing a two-grid algorithm for solving a fully conservative difference schemes (FCDS) for gas dynamics equations. In this two-level scheme, the fully nonlinear problem is solved on a coarse grid with space mesh step H. The nonlinearities are expanded about the coarse grid solution and an appropriate interpolation operator is used to provide values of the coarse grid solution on the fine grid. The resulting linear system is solved on a fine mesh with step size h. Applications in the computation of shock and rarefaction waves model problems illustrate the method's effectiveness. It is shown numerically, that the coarse grid can be much coarser than the fine grid and achieve optimal approximation as long as the mesh sizes satisfy $H = \mathcal{O}(\sqrt{h})$.

Keywords: Gas dynamics, fully conservative difference schemes, two-grid method, convergence.
PACS: 44.05.+e, 44.20.+b, 47.10.A, 47.11.-j, 47.11.Bc

INTRODUCTION

The numerical modeling of gas dynamics flows indicates that one of the important properties of the difference schemes is their full conservatism [5]. The difference scheme of the gas dynamics equations is called **fully conservative**, if with the help of equivalent transformations (corresponding to the continuous case) one can obtain from it the complete set of independent discrete conservation laws (CL) and balance relations, approximating the differential problem and all its conservation laws forms. In this sense such difference approximations were constructed and applied in physics and engineering problems by A. Samarskii, Y. Popov and collaborators [5]. Results, treating the construction and convergence of fully conservative schemes for the equations of the elastodynamics could be found in [6, 7].

The application of the FCDS requires solving large systems of nonlinear algebraic equations. In this paper we develop a two-grid scheme for solution of the FCDS equations of the 1D gas dynamics. This method was first constructed for semilinear elliptic equations by O. Axelsson [2] and J. Xu [8]. The two-grid method was next implemented in many nonlinear steady and nonsteady boundary value problems [1, 4]. The main idea of the two-grid method is to solve a complicated nonlinear problem on a coarse grid (with mesh step size H) and then solve an easier problem (linear) on the fine grid (with mesh step size h, $h \ll H$).

In Section 2 is discussed the conservative laws form of 1D gas dynamic equations. Notations are established and Samarskii-Popov FCDS is presented in Section 3. The two-grid algorithm is described in the next section. Numerical results are reported in

CP1067, *Applications of Mathematics in Engineering and Economics '34—AMEE '08*, edited by M. D. Todorov
© 2008 American Institute of Physics 978-0-7354-0598-1/08/$23.00

Section 4. Comments, conclusions and future work are provided in Section 5.

DIVERGENT AND NONDIVERGENT FORMS OF THE 1D GAS DYNAMICS EQUATIONS

The equations of the gas motion can be written in the Euler variables (x,t) or in the Lagrange variables (s,t), where x is the particle coordinate, s is the initial coordinate of the particle and ρ is the density,

$$s = \int_0^x \rho(\xi,0)d\xi \;\Rightarrow\; \frac{1}{\rho} = \frac{\partial x}{\partial s} \quad \text{(CL of the mass)}, \tag{1}$$

i.e., the mass of the volume, $0 \le \xi \le x$. The gas dynamics system of equations in Lagrange coordinates (s,t) is as follows:

$$\frac{\partial v}{\partial t} = -\frac{\partial p}{\partial s} \quad \text{(CL of the momentum)}, \tag{2}$$

$$\frac{\partial}{\partial t}\left(\varepsilon + \frac{v^2}{2}\right) = -\frac{\partial}{\partial s}(pv) - \frac{\partial w}{\partial s} \quad \text{(CL of the energy)}, \tag{3}$$

$$p = P(\rho,T), \quad \varepsilon = \varepsilon(\rho,T) \quad \text{(state equations)}, \tag{4}$$

where $v = dx/dt$ is the velocity, w – heat flow, p – pressure, and ε – internal energy.
Differentiating (1) results in:

$$\frac{\partial}{\partial t}\left(\frac{1}{\rho}\right) = \frac{\partial}{\partial t}\left(\frac{\partial x}{\partial s}\right) = \frac{\partial}{\partial s}\left(\frac{\partial x}{\partial t}\right) = \frac{\partial v}{\partial s}. \tag{5}$$

In order to solve the above system of equations, we write the expression of the heat flow

$$w = -æ(\rho,T)\frac{\partial T}{\partial s}, \tag{6}$$

where $æ = æ(\rho,T)$ is the heat conduction coefficient. The functions $P(\rho,T), \varepsilon(\rho,T)$ and $æ(\rho,T)$ are given. For the perfect gas state the equations are: $p = R\rho T$, $\varepsilon = \varepsilon(T)$. For example $\varepsilon = c_v T$, where R and c_v are constants, $R/c_v = \gamma - 1$, γ is a constant, such that $\varepsilon = p/[(\gamma-1)\rho]$. We consider two extremal cases.
a) *Adiabatic flow*, when $w = 0$, i.e., the heat-conduction can be ignored, setting $æ = 0$. Rewriting the equation (2), (3) and (5) we get:

$$\frac{\partial v}{\partial t} = -\frac{\partial p}{\partial s}, \quad \frac{\partial}{\partial t}\left(\frac{1}{\rho}\right) = \frac{\partial v}{\partial s}, \quad p = p(\eta), \tag{7}$$

$$\frac{\partial}{\partial t}\left(\varepsilon + \frac{v^2}{2}\right) = -\frac{\partial}{\partial s}(pv). \tag{8}$$

So, we have four equations and four unknowns v, p, ρ, ε. Instead of the density ρ we can use the specific volume $\eta = 1/\rho$. Then we have

$$\frac{\partial \eta}{\partial t} = \frac{\partial v}{\partial s}. \tag{9}$$

Thus, equation (8) can be replaced with

$$\frac{\partial \varepsilon}{\partial t} = -p \frac{\partial v}{\partial s} \quad \text{or} \quad \frac{\partial \varepsilon}{\partial t} = -p \frac{\partial \eta}{\partial t}. \tag{10}$$

b) *Isoentropic flow* of compressible gas, when the temperature $T = \text{const}$ and there is no energy equation (3). The condition $T = \text{const}$ corresponds to the case $æ \to \infty$. Then the gas dynamic system of equations takes the form (c is a isotherm velocity)

$$\frac{\partial v}{\partial t} = -\frac{\partial p}{\partial s}; \quad \frac{\partial}{\partial t}\left(\frac{1}{\rho}\right) = \frac{\partial v}{\partial s} \quad \text{or} \quad \frac{\partial \eta}{\partial t} = \frac{\partial v}{\partial s}, \quad p\eta = c^2. \tag{11}$$

Equation (8), (9) are subjected to the initial and boundary conditions

$$v(x,0) = v_0(x), \quad \rho(x,0) = \rho_0(x), \quad p(x,0) = p_0(x),$$
$$p(0,t) = p_e(t), \quad p(1,t) = p_r(t) \quad \text{or} \quad v(0,t) = v_e(t), \quad p(1,t) = p_r(t).$$

FULLY CONSERVATIVE SCHEMES

In the domain $(0,1) \times (0,T)$, uniform mesh $\overline{\omega}_{h\tau} = \overline{\omega}_h \times \overline{\omega}_\tau$ is defined, where

$$\overline{\omega}_h = \{s_i = ih, \ i = 0,1,\ldots,N, \ Nh = 1\}, \overline{\omega}_\tau = \{t_j = j\tau, \ j = 0,1,2,,\ldots,J, \ J\tau = T\}.$$

Further, the following notations for the mesh function y (defined on $\overline{\omega}_{h\tau}$) will be used

$$y(s_i,t_j) = y_i^j = y, \quad y(s_i \pm h) = y_{i\pm1}^j = y(\pm1), \quad y(s_i,t_{j+1}) = y_i^{j+1} = \widehat{y},$$
$$y_t = (\widehat{y} - y)/\tau, \quad y_s = (y(+1) - y)/h, \quad y_{\bar{s}} = y_s(-1),$$
$$y_{\bar{s}} = \frac{y-y(-1)}{h}, \quad y^{(\sigma)} = \sigma\widehat{y} + (1-\sigma)y, \quad 0 \le \sigma \le 1.$$

We will compute the function v in the integer nodes $s = s_i$ of the mesh ω_h, and η, p in the semi-integer ones $s = s_{i+1/2} = (i+1/2)h, \ i = 0,1,\ldots,N-1$. The schemes, which satisfy not only the conservation laws of mass, momentum and the energy but also the detail balance of the energy-kinetic and internal ones was first proposed by Samarskii & Popov [5]: $v_t = -p_{\bar{s}}^{(\sigma)}$, $\eta_t = v_s^{(0.5)}$, $\varepsilon_t = p^{(0.5)}v_s^{(0.5)}$. Incorporating pseudoviscosity is not difficult during the process of modeling FCDS. It is sufficient to replace the pressure p by the expression $g = p + w$:

$$v_t = -g_{\bar{s}}^{(\sigma)}, \quad \eta_t = v_s^{(0.5)}, \quad \varepsilon_t = -g^{(\sigma)}v_s^{(0.5)}, \quad g = p + w. \tag{12}$$

In the conditions of ideal gas and linear viscosity we have

$$p\eta = (\gamma - 1)\varepsilon, \quad w = -\frac{\mu}{\eta}v_s. \tag{13}$$

To these equations the boundary conditions at $i = 0$ and $i = N$ must be added. If, for example, the pressure p_0^j, p_N^j is given, then from the motion equations for v_i^j at $i = 0$, $i = N$

$$\frac{v_0^{j+1} - v_0^j}{\tau} = \frac{(p_{1/2} - p_0)^{(\sigma)}}{0.5h}, \quad \frac{v_N^{j+1} - v_N^j}{\tau} = -\frac{(p_N - p_{N-1/2})^{(\sigma)}}{0.5h}, \qquad (14)$$

one finds v_0^{j+1} and v_N^{j+1}. The rest functions η, ε and g can be determined only in the internal nodes $s_{1/2}, s_{3/2}, \ldots, s_{N-1/2}$.

In order to compute the values of $v^{j+1}, g^{j+1}, \eta^{j+1}, \varepsilon^{j+1}$ on the new time level at $\sigma \neq 0$, we obtain a system of nonlinear algebraic equations. For the solution, Newton's method is used [5]. Equations (12) and (13) can be written in the following form:

$$\hat{v} + \sigma \tau \widehat{g_{\bar{s}}} = v - (1 - \sigma) \tau g_{\bar{s}}, \quad \hat{\eta} - 0.5 \tau \hat{v}_s = \eta + 0.5 \tau v_s, \qquad (15)$$
$$\hat{\varepsilon} + \sigma \hat{g}(\hat{\eta} - \eta) + (1 - \sigma) g \hat{\eta} = \varepsilon + (1 - \sigma) g \eta, \quad \hat{g} \hat{\eta} - \hat{\varepsilon}(\gamma - 1) + \mu \hat{v}_s = 0. \qquad (16)$$

Convergence analysis of the Newton method as well as FCDS for the gas dynamics equations and for the nonlinear viscoelasticity equations, is presented in the papers [5] and [6], respectively. The convergence rate is $\mathcal{O}(h^2 + \tau^2)$ for $\sigma = 0.5$ and the scheme is unconditionally stable for $\sigma \geq 0.5$.

TWO-GRID ALGORITHM

Now, on $(0, 1)$ we define two coarse meshes

$$\overline{\Psi}_H = \{s_i = iH, \ i = 0, 1, \ldots, N_c, \ N_c H = 1\},$$
$$\overline{\Psi}_{H/2} = \{s_i = (i + 0.5)H, \ i = 0, 1, \ldots, N_c - 1, \ (N_c - 0.5)H = 1 - 0.5H\},$$

and two fine meshes $\overline{\Psi}_h$ and $\overline{\Psi}_{h/2}$, $h < H$ with N_f and $N_f - 1$ number of grid nodes, respectively.

Step 1 We solve the system of nonlinear equations (15), (16) with appropriate chosen initial and boundary conditions (for example (14)) on the coarse mesh with step size H. For the solution of the unknown variable v we use $\overline{\Psi}_H$, while for the rest unknowns: g, η and ε we use the mesh $\overline{\Psi}_{H/2}$. Thus, we obtain $2N_c - 2$ nonlinear and $2N_c - 1$ linear equations. We solve them by some second order iteration method (for example, Newton's method). The solutions from this step we denote by v_H, g_H, η_H and ε_H.

Step 2 On the fine mesh, with step size h, we seek the corrections: e_v, e_g, e_η and e_ε, such that $v_h = v_H + e_v, g_h = g_H + e_v, \eta_h = \eta_H + e_\eta$ and $\varepsilon_h = \varepsilon_H + e_\varepsilon$ to satisfy the FCDS (15), (16) and the corresponding boundary conditions. In fact, we use the interpolated values of the solution on coarse mesh, in order to obtain the solution on fine mesh. Therefore, it is comfortable to choose the fine mesh, such that all coarse nodes coincides with some of the fine grid nodes. With v_h, g_h, η_h and ε_h are denoted the solutions v, g, η and ε respectively, on the fine mesh. As before for e_v we use the mesh $\overline{\Psi}_h$ and $e_g, e_\eta, e_\varepsilon$ are computed on the half fine mesh $\overline{\Psi}_{h/2}$. For example, for $i = 0, \ldots, N_f$, from the first equation of (15) (written in details), we have

$$v_{h_i}^{j+1} + \frac{\tau\sigma}{h}(g_{h_{i+1/2}}^{j+1} + g_{h_{i-1/2}}^{j+1}) = v_{h_i}^j - \frac{\tau(1-\sigma)}{h}(g_{h_{i+1/2}}^j + g_{h_{i-1/2}}^j). \quad (17)$$

Next, substituting $v_h = v_H + e_v$ and $g_h = g_H + e_v$ into (17) we obtain

$$e_{v_i}^{j+1} + \frac{\tau\sigma}{h}(e_{g_{i+1/2}}^{j+1} + e_{g_{i-1/2}}^{j+1}) = v_{h_i}^j - \frac{\tau(1-\sigma)}{h}(g_{h_{i+1/2}}^j + g_{h_{i-1/2}}^j)$$
$$- v_{H_i}^{j+1} - \frac{\tau\sigma}{h}(g_{H_{i+1/2}}^{j+1} + g_{H_{i-1/2}}^{j+1}). \quad (18)$$

Similarly, for the second equation of (15) we have

$$e_{\eta_{i+1/2}}^{j+1} - \frac{0.5\tau}{h}(e_{v_{i+1}}^{j+1} - e_{v_i}^{j+1}) = \eta_{h_{i+1/2}}^j + \frac{0.5\tau}{h}(v_{h_{i+1}}^j - v_{h_i}^j)$$
$$- \eta_{H_{i+1/2}}^{j+1} + \frac{0.5\tau}{h}(v_{H_{i+1}}^{j+1} - v_{H_i}^{j+1}). \quad (19)$$

The same procedure we apply to the equations (16). For the nonlinearity $\widehat{g}_h\widehat{\eta}_h = g_{h_{i+1/2}}^{j+1}\eta_{h_{i+1/2}}^{j+1}$ we use Taylor expansion of the fine mesh solution around the coarse grid solution, namely $\widehat{g}_h\widehat{\eta}_h \simeq \widehat{g}_H\widehat{\eta}_H + \widehat{e}_\eta\widehat{g}_H + \widehat{e}_g\widehat{\eta}_H$. Thus, the last two equations of the scheme, corresponding to equations in (16), are

$$\widehat{e}_\varepsilon + \sigma\widehat{e}_\eta\widehat{g}_H + \sigma\widehat{e}_g(\widehat{\eta}_H - \eta_h) + (1-\sigma)g_h\widehat{e}_\eta = \sigma\widehat{g}_H(\eta_h - \widehat{\eta}_H) - \widehat{\varepsilon}_H$$
$$+ (1-\sigma)g_h(\varepsilon_h\eta_h - \widehat{\eta}_H), \quad (20)$$

$$\widehat{g}_H\widehat{e}_\eta + \widehat{\eta}_H\widehat{e}_g - (\gamma-1)\widehat{e}_\varepsilon + \mu\widehat{e}_{v_s} = -\widehat{g}_H\widehat{\eta}_H + (\gamma-1)\widehat{\varepsilon}_H - \mu\widehat{v}_{h_s}. \quad (21)$$

Note, that $\widehat{v}_H, \widehat{g}_H, \widehat{\eta}_H$ and $\widehat{\varepsilon}_H$ are already known from step 1 and equations (18), (19), (20) and (21) are **linear**. The convergence rate of the scheme (18), (19), (20) and (21) is $\mathscr{O}(H^4 + h^2 + \tau)$ for $\sigma \neq 0.5$ and $\mathscr{O}(H^4 + h^2 + \tau^2)$ for $\sigma = 0.5$. If we chose $h = H^2$, the convergence rate is $\mathscr{O}(H^4 + \tau^2)$, see [8].

Remark 1 *The coefficient matrices of the linear systems of algebraic equations, which arise at each step of the Newton iteration process (in step 1) and in step 2 , are tri-diagonal, M-matrix (even for coarse meshes) and this "good" property is a typical for FCDS of Samarskii-Popov.*

COMPUTATIONAL RESULTS

In this section, we consider the results from a set of computational examples, associated with the two-grid method (18), (19), (20) and (21). Our computational experiments are centered around the investigation of three main areas of interest, associated with gas dynamics simulations:

- Verifying the order of convergence;
- Verifying that there is dramatic decreasing of computational cost;

• Observing various dependencies upon modeling parameter of viscosity: μ.

In order to illustrate the efficiency of the two-grid linearization, we examine the following test problem

$$v_t = -g_s, \quad \eta_t = v_s, \quad \eta g = c^2 - \mu v_s, \quad w = g - p, \quad p = \frac{c^2}{\eta},$$

$$v(0,t) = -D\left(\frac{\eta_1 + \eta_0 E(0,t)}{1+E(0,t)} - \eta_0\right), v(1,t) = -D\left(\frac{\eta_1 + \eta_0 E(1,t)}{1+E(1,t)} - \eta_0\right),$$

$$\eta^0(s) = \frac{\eta_1 + \eta_0 E(s,0)}{1+E(s,0)}, \quad v^0(s) = -D(\eta^0(s) - \eta_0), \quad g^0(s) = \frac{D^2(\eta_0 - \eta^0(s))}{1+E(s,0)} + c^2,$$

where $E(s,t) = e^{\frac{D}{\mu}(s-Dt)(\eta_0-\eta_1)}$, $s \in (0,1)$, $0 < t < \infty$, c is a constant, μ is a viscosity coefficient and D is a mass velocity. The exact solutions are

$$\eta(s,t) = \frac{\eta_1 + \eta_0 E(s,t)}{1+E(s,t)}, \qquad v(s,t) = -D(\eta - \eta_0),$$

$$g(s,t) = \frac{D^2(\eta_0 - \eta)}{1+E(s,t)} + c^2, \qquad p(s,t) = \frac{c^2}{\eta}, \quad w(s,t) = g(s,t) - p(s,t),$$

with $\eta_1 = \frac{c^2}{D^2 \eta_0}$. Let $D = 1$, $\eta_0 = 1$, $c = 0.5$.

Example 1. The computations are performed with two-grid method for $\sigma = 0.5$, $\mu = 0.05$, $N_c = 321$, $N_f = 961$ ($h = \frac{H}{3}$). In Table 1 we give the discrete max (denoted with E_∞) and L_2 (denoted with E_2) norms at $t = 0.3$ and corresponding convergence rate ($CR_{2,\infty} = \log_2[E_{2,\infty}^{2\tau}/E_{2,\infty}^{\tau}]$) with respect to the time variable. The ratios $\frac{\tau}{H^2}$ and $\frac{\tau}{h}$ are not fixed: they decrease each time, when τ is divided in two. That's way the convergence rate is slightly less than expected. Nevertheless, it is clear that the accuracy is $\mathscr{O}(\tau^2)$.

Example 2. Here we check the convergence rate with respect to the space variable, using two-grid scheme, $\sigma = 0.5$, $\mu = 0.05$ $\tau = H^2$ and $h = H^2$. Undoubtedly, after the first step of algorithm (Newton method), we have the estimate $O(\tau^2 + h^2)$ or $O(\tau + h^2)$ depending on σ. The convergence rate after the second step of the algorithm is also verified. Table 2 shows how the max and L_2 errors at $t = 0.3$ shrink with h, when $H^2 \simeq h$ and convergence rate $CR_{2,\infty} = \log_2[E_{2,\infty}^{2N_c}/E_{2,\infty}^{N_c}]$. Also, we give the CPU time, computing with two-grid algorithm and alternatively, computing only on fine mesh and solving nonlinear equations.

TABLE 1. Errors in different norms and convergence rate with respect to the time variable, $\sigma = 0.5$

τ	E_∞	CR_∞	E_2	CR_2	$\frac{\tau}{H^2}$	$\frac{\tau}{h}$
0.1	3.858408e-2		1.291521e-2		10240	96
0.05	1.017025e-2	1.9237	3.337671e-3	1.9522	5120	48
0.025	2.416144e-3	2.0736	8.170605e-4	2.0303	2560	24
0.0125	5.830566e-4	2.0510	1.985212e-4	2.0411	1280	12
0.00625	1.485629e-4	1.9726	5.055559e-5	1.9734	640	6
0.003125	3.870847e-5	1.9404	1.304286e-5	1.9546	320	3

267

TABLE 2. Errors in different norms, convergence rate and CPU time, $\tau = h = H^2$

H	h	E_∞	CR_∞	E_2	CR_2	CPU coarse-fine mesh	CPU fine mesh
2^{-2}	H^2	4.520839e-2		2.473609e-2		0.0165	0.0320
2^{-3}	H^2	2.325069e-3	4.2812	1.160678e-3	4.4136	0.1560	18.3240
2^{-4}	H^2	1.178577e-4	4.3022	4.739397e-5	4.6141	0.2030	42.8432
2^{-5}	H^2	7.070605e-6	4.0591	2.626465e-6	4.1735	0.8534	1 292.5072
2^{-6}	H^2	4.358908e-7	4.0198	1.579675e-7	4.0554	1 112.8832	754 489.7542

Remark 2 *The solutions η, g, p, w are computed in half nodes of coarse and fine meshes. At each time level we need the solution from previous time step in coarse (also in half coarse) nodes in order to start Newton iterations, therefore after executing the second step of algorithm, we have to pick out (or interpolate) only the solution in coarse grid points.*

Example 3. The two-grid method allows one to iterate on a very coarse grid and still get good approximations by executing one more iteration on the fine grid. On Figure 1 is displayed the solution of the test problem with $\mu = 0.05$, $N_c = 5$, $N_f = 37$, $\sigma = 1$, $\tau = 0.01$ at $t = 0.5$.

If we decrease the viscosity coefficient, the solution become steepen and we have to refine coarse mesh ($N_c = 5$ is quite coarse for this case), see Figure 2. For very small μ this method is not effective. We need special grids, as a base of the two-grid algorithm: for example, Shishkin meshes. This will be a subject of our future work.

CONCLUSIONS

A novel application of the two-grid method of Axelsson-Xu for development the gas dynamics FCDS is examined. The fundamental idea is that one can solve the nonlinear system of algebraic difference equations by applying the Newton-like iteration on a coarse grid, then execute one fine grid iteration, and add a new correction step on the coarse mesh at the end of the last iteration. The approximation space can be very coarse ($H \gg h$), and we can still maintain a good accuracy (except the case for very small viscosity μ). The key feature of the two-grid method was confirmed: it allows one to execute all the nonlinear iterations on a system associated with a coarse spatial grig, without sacrificing the order of accuracy of the fine grid solution.

The 1D set of PDEs considered in the present paper are *just a model problem* and even if there is a big difference between the 1D and the multidimensional problem, this work is a base of our future investigations. Extension for 2D gas dynamics equations and theoretical analysis are currently being studied.

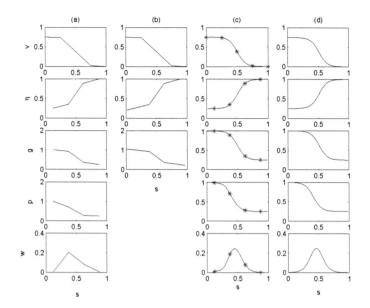

FIGURE 1. Two-grid method, $N_c = 5$, $N_f = 37$, $\tau = 0.01$, $\mu = 0.05$, $\sigma = 1$, $t = 0.5$, (a) Numerical solutions on coarse mesh, Newton method; (b) Interpolations on fine grid; (c) Numerical solutions on fine mesh, Step 2; we star the coarse grid points; (d) Exact solutions

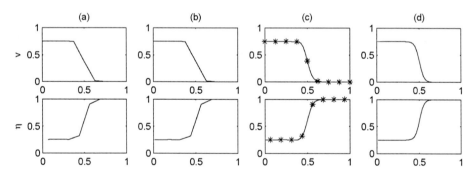

FIGURE 2. Two-grid method, $N_c = 9$, $N_f = 41$, $\tau = 0.01$, $\mu = 0.02$, $\sigma = 1$, $t = 0.5$, (a) Numerical solutions on coarse mesh, Newton method; (b) Interpolations on fine grid; (c) Numerical solutions on fine mesh, Step 2; we star the coarse grid points; (d) Exact solutions

ACKNOWLEDGMENTS

This research is supported by the Bulgarian National Fund of Science under Project VU-MI-106/2005.

REFERENCES

1. H. Abboud, V. Giranlt, and T. Sayah, *C. R. Acad. Sci. Paris, Ser I* **341**, 451–456 (2005).
2. O. Axelsson, *Applications of Math.* **38**(4-5), 249–265 (1983).
3. C. Dawson, M. Wheeler, and C. Woodward, *SIAM J. Numer. Anal.* **35**(2), 435–452 (1998).
4. M. Mu and J. Xu, *SIAM J. Numer Anal.* **45**, 1801–1813 (2007).
5. A. Samarskii and Y. Popov, *Difference Methods for Solution of Problems of Gas Dynamics*, Nauka, Moscow, 1980, 1st edn., 1992, 2nd edn. (in Russian)
6. L. Vulkov, *Diff. Uravnenia* **27**(1), 1123–1132 (1991). (in Russian).
7. L. Vulkov, *J. of Comp. Math. Math. Phys.* **31**(9), 1392–1401 (1991). (in Russian).
8. J. Xu, *SIAM J. Sci. Comput.* **15**, 231–237 (1994).

Validation of a Three-Dimensional Model of the Ocean Circulation

L.Monier, F.Brossier and F.Razafimahery

INSA 20 avenue des Buttes de Coesmes CS14315, 35043 Rennes, France

Abstract.. This paper is devoted to a numerical model of the oceanic circulation. In order to obtain the variability of the currents with respect to time and depth, we have to use 3D Navier-Stokes equations. The horizontal gradient of pressure appears as an unknown term in these equations. It is directly related to the gradient of the sea surface topography and obtained by solving a 2D model governed by shallow water equations. Numerical experiments are carried out in a parallelepiped canal located in the south hemisphere. First experiment proves coherence between the 2D and 3D models. Then, we test the influence of the Coriolis stress on a southward flow. Finally, we compute different vertical profiles of velocity depending on virtual viscosity used in order to model turbulence.

Keywords: Oceanic currents, three-dimensional model, Navier-Stokes, shallow water.
PACS: 95.75.Pq

INTRODUCTION

In this paper we develop a numerical simulation of the variability of the oceanic currents in a parallelepiped canal. The domain is characterized by great differences between horizontal and vertical scales, which oblige to use dimensionless equations. In order to obtain the variations of the currents with respect to time and depth, we have to use a three dimensional and time dependent model. The sea surface topography, related to the surface pressure, is unknown in our problem. For this reason, we have connected a shallow-water model with the three dimensional model. Physical variables computed by the two dimensional model are the velocity of the horizontal flow integrated with respect to depth in the surface layer and the topography of the sea surface. These variables are controlled by the shallow water equations. Variables computed by the three dimensional model are the three components (u,v,w) of the current velocity. The vector (u,v,w) obeys to the Navier-Stokes equations. The sea level topography present in these equations is given by the shallow-water model. The model is carried out in a rectangular canal of longitudinal extension L=500 km, latitudinal extension $3*L$=1500 km and vertical extension H=200 m (thickness of the surface layer). The domain is located in the south hemisphere and in the tropical zone. Characteristic time of the simulation is T=500*1000 seconds giving a characteristic velocity (L/T) equal to 1 m/s.

In the first section, we set the equations of the shallow-water model. In the second section we set the equations of the 3D model. The third section is devoted to numerical validation of the models.

CP1067, *Applications of Mathematics in Engineering and Economics '34—AMEE '08*, edited by M. D. Todorov

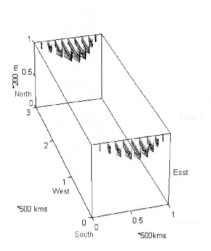

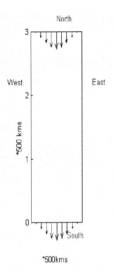

FIGURE 1. Three dimensional geometry   **FIGURE 2.** Two dimensional geometry

THE TWO-DIMENSIONAL MODEL: MODEL 1

Dimensionless Equations

Variables computed by the model are the sea surface topography Z and the horizontal flow UM obtained by integrating the horizontal velocity U with respect to the vertical coordinate z in the surface layer. So the model is time dependent and two dimensional in space. The equations satisfied by UM and Z are the shallow water equations:

$$
\begin{cases}
\dfrac{\partial UM}{\partial t} + v\Delta UM - \nabla.(UM \otimes U) = cor \wedge UM - g\,Z\,\nabla Z + F \\
\dfrac{\partial Z}{\partial t} + \nabla.UM = 0
\end{cases}
\tag{1}
$$

Symbol U denotes the mean velocity in the surface layer, v is the coefficient of eddy viscosity, cor is the Coriolis parameter and g - the gravity coefficient. The body force F represents the forcing of the model. Currents are driven by the wind-stress acting on the surface and braked by the drag stress on the bottom.

We now introduce dimensionless variables:

$$\bar{x} = \frac{x}{L} \;,\; \bar{y} = \frac{y}{L} \;,\; \bar{z} = \frac{z}{H} \;,\; \bar{U} = \frac{U}{U_0} \;,\; \bar{t} = \frac{t}{T} \quad \text{where} \quad U_0 = \frac{L}{T}\,.$$

We set $\quad \bar{v} = \dfrac{v}{LU_0} \;,\; \bar{g} = \dfrac{H}{U_0^2}g \;,\; \overline{cor} = T\,cor \;,\; \bar{F} = \dfrac{T^2}{HL}F\,.$

Equations (1) become:

$$\begin{cases} \dfrac{\partial UM}{\partial t}+v\,\Delta UM-\nabla.(UM\otimes U)=cor\wedge UM-g\,Z\nabla Z+F \\ \dfrac{\partial Z}{\partial t}+\nabla.UM=0 \end{cases} \qquad (2)$$

In order to obtain more readable equations, the bars have been omitted in (2).

Boundary Conditions

We have chosen Dirichlet boundary conditions for the velocity U. On the eastern and western coasts, we assume the adherence condition $U=0$. The velocity is given on the northern and southern boundaries. We impose a southward current with sinusoidal longitudinal dependence, as plotted on Figure 2. Outflow is equal to inflow to avoid storage of water in the canal [7].

THE THREE-DIMENSIONAL MODEL: MODEL 2

Dimensionless Equations

Variables computed by the model are the three components of the velocity $V=(u,v,w)$. The velocity V obeys to the complete Navier-Stokes equations:

$$\begin{cases} \dfrac{\partial u}{\partial t}+(V.\nabla)u-cor\,v+\dfrac{1}{\rho}\dfrac{\partial p}{\partial x}-v_h\Delta_h u-\dfrac{\partial}{\partial z}v_v\dfrac{\partial u}{\partial z}=0 \\ \dfrac{\partial v}{\partial t}+(V.\nabla)v+cor\,u+\dfrac{1}{\rho}\dfrac{\partial p}{\partial y}-v_h\Delta_h v-\dfrac{\partial}{\partial z}v_v\dfrac{\partial v}{\partial z}=0 \\ \dfrac{\partial w}{\partial t}+(V.\nabla)w+\dfrac{1}{\rho}\dfrac{\partial p}{\partial z}-v_h\,\Delta_h w-\dfrac{\partial}{\partial z}v_v\dfrac{\partial w}{\partial z}=g \\ \nabla u=0 \end{cases} \qquad (3)$$

where v_h and v_v denote the eddy viscosity coefficients and p denotes the pressure.

Eddy viscosity is introduced in order to model small scale turbulent movements in an horizontal or vertical direction. The simplest model consists in choosing constant values for these two coefficients [8].

We set $\bar{x}=\dfrac{x}{L}$, $\bar{y}=\dfrac{y}{L}$, $\bar{z}=\dfrac{z}{H}$, $\bar{U}=\dfrac{U}{U_0}$, $\bar{t}=\dfrac{t}{T}$, $\varepsilon=\dfrac{H}{L}$. $U_0=\dfrac{L}{T}$ is the scaling of the horizontal velocity , $w_0=\varepsilon U_0$ is the scaling of the vertical velocity , and $p_0=\rho U_0^2$ the scaling of the pressure. Then we set : $(\bar{u},\bar{v})=\dfrac{(u,v)}{U_0}$, $\bar{w}=\dfrac{w}{w_0}$, $\bar{p}=\dfrac{p}{p_0}$, $\bar{v_h}=\dfrac{v_h}{LU_0}$,

$\bar{v_v}=\dfrac{v_v}{\varepsilon H U_0}$. System (3) becomes:

$$\begin{cases} \dfrac{U_0}{T}\dfrac{\partial u}{\partial t}+\dfrac{U_0^2}{L}u\dfrac{\partial u}{\partial x}+\dfrac{U_0^2}{L}v\dfrac{\partial u}{\partial y}+\varepsilon\dfrac{U_0^2}{H}w\dfrac{\partial u}{\partial z}-U_0 cor\,v+\dfrac{P_0}{\rho L}\dfrac{\partial p}{\partial x}-\dfrac{U_0^2}{L}v_h\Delta_h u-\dfrac{U_0^2}{L}\dfrac{\partial}{\partial z}v_v\dfrac{\partial u}{\partial z}=0 \\[2mm] \dfrac{U_0}{T}\dfrac{\partial v}{\partial t}+\dfrac{U_0^2}{L}u\dfrac{\partial v}{\partial x}+\dfrac{U_0^2}{L}v\dfrac{\partial v}{\partial y}+\varepsilon\dfrac{U_0^2}{H}w\dfrac{\partial v}{\partial z}+U_0 cor\,u+\dfrac{P_0}{\rho L}\dfrac{\partial p}{\partial y}-\dfrac{U_0^2}{L}v_h\Delta_h v-\dfrac{U_0^2}{L}\dfrac{\partial}{\partial z}v_v\dfrac{\partial v}{\partial z}=0 \\[2mm] \varepsilon\dfrac{U_0}{T}\dfrac{\partial w}{\partial t}+\varepsilon\dfrac{U_0^2}{L}u\dfrac{\partial w}{\partial x}+\varepsilon\dfrac{U_0^2}{L}v\dfrac{\partial w}{\partial y}+\varepsilon^2\dfrac{U_0^2}{H}w\dfrac{\partial w}{\partial z}+\dfrac{P_0}{\rho H}\dfrac{\partial p}{\partial z}-\varepsilon\dfrac{U_0^2}{L}v_h\Delta_h w-\varepsilon\dfrac{U_0^2}{L}\dfrac{\partial}{\partial z}v_v\dfrac{\partial w}{\partial z}=g \\[2mm] \dfrac{U_0}{L}\dfrac{\partial u}{\partial x}+\dfrac{U_0}{L}\dfrac{\partial v}{\partial y}+\dfrac{w_0}{H}\dfrac{\partial w}{\partial z}=0 \end{cases} \tag{4}$$

As previously, the bars have been omitted in (4) in order to get more readable equations.

Using $U_0=\dfrac{L}{T}$, equations governing u and v can be written as follows :

$$\begin{cases} \dfrac{\partial u}{\partial t}+u\dfrac{\partial u}{\partial x}+v\dfrac{\partial u}{\partial y}+w\dfrac{\partial u}{\partial z}-T cor\,v+\dfrac{\partial p}{\partial x}-v_h\Delta_h u-\dfrac{\partial}{\partial z}v_v\dfrac{\partial u}{\partial z}=0 \\[2mm] \dfrac{\partial v}{\partial t}+u\dfrac{\partial v}{\partial x}+v\dfrac{\partial v}{\partial y}+w\dfrac{\partial v}{\partial z}+T cor\,u+\dfrac{\partial p}{\partial y}-v_h\Delta_h v-\dfrac{\partial}{\partial z}v_v\dfrac{\partial v}{\partial z}=0 \end{cases} \tag{5}$$

The quantity ε measures the ratio between vertical and horizontal scaling. So, it is a small parameter and the equation governing w can be simplified into:

$$\dfrac{\partial p}{\partial z}=\dfrac{H}{L^2}T^2 g. \tag{6}$$

The continuity equation remains unchanged:

$$\nabla V=0. \tag{7}$$

We assume that $w=0$ on the bottom of the domain. Then by integrating equation (7) with respect to z we can compute w depending on u and v.

If we neglect the variations of atmospheric pressure at the sea surface, by integrating equation (6) with respect to z we obtain the following relation between the horizontal gradient of pressure and the gradient of the sea surface topography Z:

$$\nabla Z=-\dfrac{L^2}{gHT^2}\nabla P.$$

So, at each step time, we compute Z by using Model 1 and then we can compute the velocity (u,v,w) by using Model 2.

Boundary Conditions

On the northern and southern boundaries, we have chosen Dirichlet conditions plotted on Figure 1. Notice that these conditions have to be coherent with conditions of Model 1: $\int_0^1 (u,v)\,dz=UM$ on these two boundaries.

On the eastern and western coasts, and on the bottom we assume that $V=0$. The flow is driven by the wind [6]. So at the sea surface we have the condition: $(\dfrac{\partial u}{\partial n},\dfrac{\partial v}{\partial n})=(F_x,F_y)$ where (F_x,F_y) denotes the wind stress.

NUMERICAL EXPERIMENTS

Numerical resolution is performed by using the toolbox FEMLAB. It allows an automatic mesh of any domain, a simple resolution of partial differential equations by finite-elements methods [9] and a graphic representation of interpolated results. The purpose of the first experiment is to prove the coherence of the two models. The purpose of the second experiment is to study the impact of the Coriolis stress. In the third experiment, we are interested by the impact of the vertical viscosity.

Characteristics of the First Experiment

We neglect the wind stress and the Coriolis force. The two models are forced by the flow imposed at the northern and southern boundaries (Figure 1a). We compute the flux $\int_0^1 (u,v)\,dz$ using (u,v) given by Model 2 and compare it to UM computed by Model 1. The results have been plotted on Figures 3 and 4. Arrows represent the velocity direction and colors the intensity of the velocity: red colors correspond to strong values and blue values to small values. There is almost no difference between the two circulations. That proves the coherence between the two models.

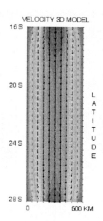

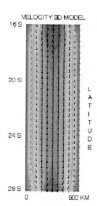

FIGURE 3. The flux UM given by Model 1   **FIGURE 4.** The flux $\int_0^1 (u,v)\,dz$ given by Model 2

Characteristics of the Second Experiment

Boundaries conditions are the same as previously but now we introduce the Coriolis stress. On Figure 5 we have plotted the circulation computed by the three-dimensional model. We observe an intensification of the circulation near the west coast. It is a classical result since our domain is located in the southern hemisphere. The flow remains southward because of the adherence condition on the western boundary; then a westward deviation of the flow is forbidden. The Coriolis stress acts on the whole domain. So this phenomenon is present at every depth.

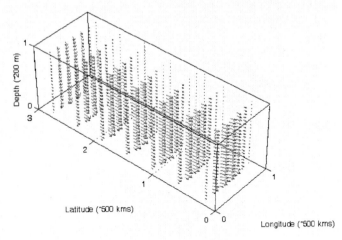

FIGURE 5. Velocity given by Model 2

Characteristics of the Third Experiment

Coefficients of eddy viscosity v_h and v_v represent turbulent movements [3]. The simplest model corresponds to constant values for these two coefficients. For horizontal viscosity, we choose $v_h=10^4 m^2/s$. The admissible values for v_v can be chosen between $10^{-4} m^2/s$ and $10^{-1} m^2/s$ [4,8]. We proceed to three experiments using three different constant values for the vertical viscosity: $v_v=10^{-4} m^2/s$, $v_v=10^{-2} m^2/s$, $v_v=10^{-1} m^2/s$. We have plotted the magnitude of the velocity at a constant latitude ($y=1.5$ which is the middle of the domain). Figures 6, 7, 8 represent a cross section in depth and longitude. As in the previous section, we observe an intensification of the velocity on the west side of the domain connected to the Coriolis stress. The influence of the vertical eddy viscosity is obvious on the three currents profiles. Indeed with strong values of the viscosity, the adherence condition causes a weakening of the circulation at the bottom (Figure 6). With lower values (Figure 7), we observe stronger values near the bottom. On Figure 8, the conflict between large values of the velocity and the null adherence condition at the bottom involves numerical instabilities.

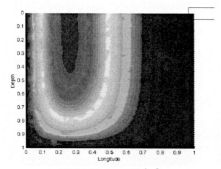

FIGURE 6. $v_v = 10^{-4} \text{m}^2/\text{s}$

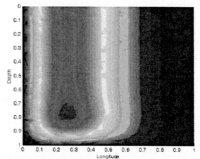

FIGURE 7. $v_v = 10^{-2} \text{m}^2/\text{s}$

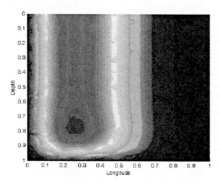

FIGURE 8. $v_v = 10^{-4} \text{m}^2/\text{s}$

CONCLUSION

Connection between the two models gives coherent results. Indeed, we observe the Coriolis deviation according to the theory. We have proceeded to other experiments giving the Ekman spiral and Ekman pumping. The impact of vertical eddy viscosity on vertical currents profiles is showed by the third numerical experiment. It will be interesting to test more sophisticated models for turbulence [1]. Then, this study shall be improved by the development of a numerical model including transport of particles and salinity.

REFERENCES

1. A.C. Bennis, M. Gomez-Marmol, R. Lewandowsky, T. Chacon-Rebollo, and F. Brossier, "A Comparison of Three Turbulence Models with an Application to the West Pacific Warm Pool", math-ph/0701059, 2007.
2. J. Cousteix, *Turbulence et Couche Limite*, Cepadues, 1989.
3. M. Lesieur, *Turbulence in Fluids*, Kluwer, 1997.

4. G. Madec, P. Delecluse, M. Imbard, and C. Levy , "O.p.a. version 8.0. Ocean General Circulation Model, reference manual," Technical Report, 1997.
5. B. Mohammadi and O. Pironneau, *Analysis of the Turbulence Model*, Wiley and Masson, 1994.
6. L. Monier, F. Brossier, and F. Razafimahery, "Dynamic of the Currents in the Mozambique Canal," in *Applications of Mathematics in Engineering and Economics*, edited by M.Marinov et al, Sozopol , Bulgaria, 2004.
7. L. Monier, F. Brossier, and F. Razafimahery, "Variabilities of the Currents and of the Sea Level in the Mozambique Canal," in *International Conference Fluxes and Structures*, Moscow, Russia, 2005.
8. R.C. Pacanowski and S.G.H. Philander, *J. Phys. Oceanogr* **11**, 1981.
9. O. Pironneau, *Méthode des Eléments Finis pour les Fluides*, Masson, 1988.

Procedure for Deconvolution of Time-Dependent Signals with Sinusoidal Shape

M.D. Todorov*, R. Djulgerova†, V. Mihailov† and J. Koperski**

*Faculty of Applied Mathematics and Informatics, Technical University of Sofia,
1000 Sofia, Bulgaria
†Institute of Solid State Physics, Bulgarian Academy of Sciences, 1784 Sofia, Bulgaria
**Institute of Physics, Jagellonian University, 30-059 Krakow, Poland

Abstract. The present paper deals with development of the deconvolution procedure of two time-dependent signals with sinusoidal shape in order to extract the genuine optogalvanic signal from the registered one in hollow cathode glow discharge.

Keywords: Deconvolution procedure, optogalvanic signals, signal waveform.
PACS: 52.70.- m, 02.60Cb

INTRODUCTION

The optogalvanic effect is a change of the plasma conductivity under radiation with resonant light [1]. It is a result of the redistribution of the energetic levels' populations, hence – of the plasma ionization processes changed effectiveness. The optogalvanic spectroscopy is widely applied in spectra identification, light sources calibration and stabilization, atom constants estimation, etc.

Dynamic optogalvanic signals (DOGS) expressing the plasma time reaction after radiation with short (~ns) laser pulses are much more interesting than the stationary ones, because they are characterized not only by amplitude and sign, but also by time shape, usually of sinusoidal type and various peculiarities. They are mainly applied in extraction of various atom constants and plasma parameters by fitting the modeled DOGS, mainly by a kinetic method, with the registered DOGS [2, 3]. Because of the fact that the registered DOGS is a convolution of plasma response resonant signal (the true signal) and apparatus signal (containing all disturbing factors such as time-constant of the registering circuit and measurement devices, discharge electric characteristics, optogalvanic signals resulting from the highly excited energetic levels photoionization and possibly from cathode walls' photo effect, etc.) it is necessary to remove the second one from the registered signal. We suggest that the non-resonant DOGS, registered as a time-dependent signal with sinusoidal shape under radiation of plasma with non-resonant light frequency should be used as apparatus signal – a complex expression of all factors mentioned above, instead of the apparatus signal with exponential function shape used at present. Thus the genuine DOGS might be defined much more precisely by its extraction through deconvolution from the registered DOGS.

CP1067, *Applications of Mathematics in Engineering and Economics '34—AMEE '08*, edited by M. D. Todorov
© 2008 American Institute of Physics 978-0-7354-0598-1/08/$23.00

DECONVOLUTION PROCEDURE

The deconvolution of the pure OGS, referred to as $y(\tau)$, from the registered OGS actually means to solve the following integral equation

$$\int_0^t K(t-\tau)d\tau = f(t) \qquad (1)$$

with respect to the unknown function $y(\tau)$. In our case the function $K(\zeta)$ is the apparatus OGS, $f(t)$ – the registered OGS and the argument t is the time. The functions $K(\zeta)$ and $f(t)$ are supposed to be smooth and $K(t-t) = \alpha > 0$, $\alpha = $ const. Equation (1) is a Volterra convolution equation of first kind and therefore its solving appears to be an inverse incorrect in sense of Hadamard problem [4]. A specious and smart way to solve it is mapping by, e.g., Laplace or Fourier transform having in mind that the left hand side is an operator with delayed argument. A shortcoming to accomplish the above is the strong non-linear untidy behaviour of both the apparatus OGS and the registered OGS, for which only experimental data are known, i.e., they are not analytically presented. For this reason, the usual treatment of equation (1) by either integral transformations or regularization method comes upon hard obstacle. In order to overcome these shortcomings we prefer to solve the above problem numerically direct in the physical space (not in transformed parametric one) assuming the experimental data available as smooth. After that we differentiate equation (1) by t and obtain the following equivalent regular Volterra equation of second kind, for which well posed numerical methods are developed [5, 6]:

$$y(t) + \int_0^t \kappa(t-\tau)d\tau = \varphi(t). \qquad (2)$$

Here we note $\varphi(t) \equiv y(t)/K_t(t-t)$ and $\kappa(t-\tau) \equiv K_t(t-t)/K(t-t)$. Further we introduce an uniform set, e.g., $t_i = ih$, $h = T/N$, $i = 0,...,N$ on the time interval $[0,T]$, $0 \le t \le T$, and replace the integral in equation (2) with an appropriate quadrature formula by using, for example, the trapezium or Simpson rule. On a discrete level equation (2) possesses an upper triangle matrix what means we should simply implement the Gaussian back substitution, i.e., simply perform the following recurrent relationship:

$$y(t_i) = \left[\varphi(t_i) - \frac{h}{2}\kappa(t_i - \tau_0)y(\tau_0) - h\sum_{j=1}^{i-1} \kappa(t_i - \tau_j)y(\tau_j) \right]$$

$$\times \left[1 + \frac{h}{2}\kappa(t_0) \right]^{-1}, \qquad i = 1,...,N \qquad (3)$$

$$y(t_0) = \varphi(0).$$

The derivatives $K_t(t_i - \tau_j)$ in (2) are approximated by central finite differences of second order of accuracy. Thus, knowing $y(t_j)$, $j = 1,...,i-1$ (pure OGS) we compute

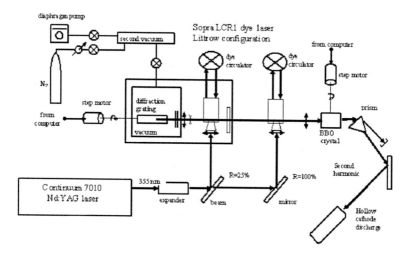

FIGURE 1. Experimental set-up for dynamic optogalvanic signal registration

its value at the next node t_i. Let us denote that following the procedure described above we filter the solution as late as obtaining the solution. Namely, we smooth the obtained solution from the noises replacing with time averaging data by using the well known integral formula $\bar{y}(t) = \frac{1}{t}\int_0^t y(\tau)d\tau$.

SOME RESULTS AND DISCUSSION

The experimental set-up for dynamic optogalvanic signals registration is shown in Figure 1. The hollow cathode has the shape of a cylinder with bottom. Ne at 3.5Torr pressure is employed as working gas. The DOGDs, recorded by oscilloscope and processed by a computer, are obtained when the incident pulsed (5ns, 10Hz) laser beam passes along the cathode axis and illuminates both the hollow cathode plasma and the cathode bottom. Non-resonant DOGSs are recorded at laser wavelengths far enough from Ne and Mn optical transitions.

The response of the hollow cathode plasma to the resonant laser light (the registered DOGS) corresponding to Ne I 540.05nm spectral line and to the non-resonant laser light (the apparatus DOGS) is shown in Figure 2 for Ne/Mn hollow cathode lamps.

The first components of the registered DOGSs are positive, which is equivalent to the plasma conductivity improvement. The reason for the resonant DOGS arising is the change in the population of the two involved excited levels (1s4 and 2p1) as a result of absorption of resonant laser light. The non-resonant DOGS is due mainly to the photoionization from highly excited Ne levels, well populated in the hollow cathode

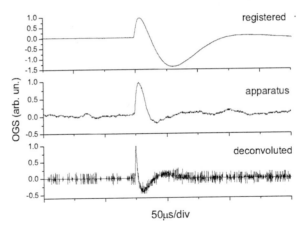

FIGURE 2. Registered, apparatus and deconvoluted dynamic optogalvanic signals for (1s4-2p1) optical transition corresponding to Ne I 540.05nm, at 4.4mA discharge current

plasma. At the same figure the obtained deconvoluted (true) DOGS is demonstrated. Its shape is quite similar to the registered and apparatus DOGSs but its time width is significantly narrower.

The review of the computed results (see Figure 2) clearly shows that the spatial mean-line, which can be easily obtained knowing the noisy dispersion agrees very well with the experimental measurements. The above described algorithm is very convenient for a fast obtaining a consistent forecast about deconvoluted signals and a comparing with experimental data because both

- The matrix of the discretized integral equation is triangle as a rule and therefore its inversion requires much less computational resources compared to, for example, the Gaussian inversion procedure;
- We treat the equation direct in the physical plane without a preliminary clumsy transforming to a parametric space and using hard regularization methods.

CONCLUSION

A quick and easy deconvolution procedure has been developed and applied in extraction of the true (pure) dynamic optogalvanic signal from the registered one in hollow cathode glow discharge. It is clear, that using this procedure and taking the non-resonant opto-galvanic signal as apparatus signal, the fitting would be more correct for more precise values of the parameters extracted from the kinetic modeling of the dynamic optogal-vanic signals.

ACKNOWLEDGMENTS

Support for this work is gratefully acknowledged by the joint research project between the Institute of Solid State Physics of Bulgarian Academy of Sciences (Sofia, Bulgaria), the Institute of Physics of Jagiellonian University (Krakow, Poland), and the Institute of Physics of Academy of Sciences and Arts in Serbia based on the scientific cooperation between our Academies.

REFERENCES

1. B. Barbieri, N. Beverini., and A. Sasso, *Review of Modern Physics* **62**, 603–644 (1990).
2. X. L. Han, M. C. Su, C. Haridass, and P. Misra, *Journal of Molecular Structure* **155**, 695–696 (2004).
3. S. Mahmood, M. Anwar-ul-Haq, M. Riaz, and M.A. Baig, *European Physical Journal D* **36**, 1–9 (2005).
4. N. Tikhonov, and V. I. Arsenin, *Solution of Ill-posed Problems*, John Wiley & Sons, New Jersey, 1977.
5. J. L. Walsh, L. M. Delver, *Numerical Solution of Integral Equations*, Oxford University Press, 1974.
6. C. T. H. Baker, *The Numerical Treatment of Integral Equation*, Clarendon Press, Oxford, 1977, Chapter 4.

Two-grid Interpolation Algorithms for Difference Schemes of Exponential Type for Semilinear Diffusion Convection-Dominated Equations

L.G. Vulkov* and A.I. Zadorin†

*Dept. of Numerical Methods and Statistics, FNSE, University of Rousse, 8 Studentska str.
7017 Rousse, Bulgaria
†Institute of Mathematics, RAS, Siberian branch, Omsk 644099, Russia

Abstract. In this paper we propose two-grid algorithms for implementation of the A.M.Il'in's scheme to diffusion convection-dominated equations. To find the solution from nonlinear algebraic systems we investigate Newton and Picard iterative methods. We offer to use the difference scheme on a coarse mesh and, using uniform interpolation, taking into account the boundary layers, to find suitable initial iteration for an iterative method on a fine mesh. We estimate the accuracy of the proposed algorithms and we count the number of the arithmetic operations. Numerical experiments illustrate the efficiency of these algorithms.

Keywords: Semilinear convection diffusion, singular perturbation, difference scheme, iterative method, two-grid method.
PACS: 01.30.Cc, 44.05.+e, 41.20.Cv

INTRODUCTION

It is shown theoretically and experimentally that classical finite difference schemes on non-adaptive meshes have a cell Reynolds number limitation when they are applied to convection-dominated equations [2, 4, 6]. For small values of the perturbation parameter, these techniques lead to spurious solutions when central differences for the advection terms are employed; on the other hand, first-order upwind methods introduce artificial diffusion that thickens the boundary layers. In order to avoid these difficulties, exponentially-fitting techniques are frequently used [2, 4, 6]. Another approach is based on the generation of layer-adapted meshes that allow resolution of the structure of the layer [3, 4, 6].

In the case of nonlinear boundary value problems, the order of ε-uniform convergence rate of the two classes special methods does not exceed 2 for elliptic reaction-diffusion equations and is not higher than 1 for convection-diffusion equations (e.g., see [3, 4, 6]) and the bibliography therein. There are a large number of papers where for 1D reaction-diffusion problems, high-order difference schemes are constructed, see [5, 7] and the references therein. In contrast to the reaction-diffusion case there are only a few results for diffusion-convection problems with second order of approximations [6].

The defect correction and the Richardson methods have been used to increase the accuracy of grid solutions for singularly perturbed boundary value problems. Note that in [7] have been considered the nonlinear case.

In this paper we shall employ in singular perturbed problems of the two-grid finite

CP1067, *Applications of Mathematics in Engineering and Economics '34—AMEE '08*, edited by M. D. Todorov
© 2008 American Institute of Physics 978-0-7354-0598-1/08/$23.00

element method that was originally proposed by Axelsson [1] and Xu [8] independently each other. The idea in [1, 8] is basically to use a coarse grid to produce a rough approximation of the solution, and then to use it as the initial guess for one Newton iteration on the fine grid. Hence, an obvious advantage in comparison with the above methods is the solution of nonlinear systems of algebraic equations with **small** number of equations.

Let us consider the boundary value problem

$$-\varepsilon u'' - a(x)u' + f(x,u) = 0, \ x \in \Omega \equiv (0,1),$$
$$u(0) = A, \ u(1) = B, \tag{1}$$

where A, B are given constants, ε is a parameter in $(0,1]$,

$$a(x) \geq \alpha > 0, \ a \in C^2(\overline{\Omega}). \tag{2}$$

For the function f we will assume that it is twice continuously differentiable with respect to x, three time differentiable with respect to u and

$$f'_u(x,u) \geq 0 \text{ on } \Omega \times R. \tag{3}$$

The goal of the present paper is to construct and study theoretically and numerically two-grid interpolation algorithms for implementation of the classical Il'in difference scheme [2] to problem (1)-(3). Some preliminary analysis is given in Section 1. In Sections 3,4 two-grid method is applied to decrease the number of calculations, when Newton and Picard methods respectively, are used to resolve the nonlinear system of difference equations. In Section 5, the problem (1)-(3) is discretized on a fine grid with mesh-size h and linearized around the interpolant u_H (see [10]) computed on a coarse grid. This strategy is motivated by the fact (Theorems 2, 3) that the global error of the two-grid interpolation algorithm is of order h, the same as would have been obtained if the non-linear problem had been solved directly on the fine grid. The coarse mesh can be quite coarse (see the experiments in Section 6) and still maintain an optimal approximation.

Our error estimates are in the maximum norm, which is sufficiently strong to capture layers and hence seems most appropriately for singularly perturbed problems. Note, that the error estimates in the original papers [1, 8] and the later ones, are given in energy Sobolev discrete norms.

PRELIMINARY ANALYSIS

At first, we get the estimate for the solution of problem (1)-(3):

$$\|u\|_\infty \leq l = \alpha^{-1}\|f(x,0)\|_\infty + \max\{|A|,|B|\}. \tag{4}$$

Let us introduce the linear operator

$$Lz(x) = -\varepsilon z''(x) - a(x)z'(x) + b(x)z(x), \ b(x) = \frac{f(x,u(x)) - f(x,0)}{u(x)} \geq 0,$$

where $z(x) \in C^2(\overline{\Omega})$. It is known, [4, 6], that maximum principle is valid for the operator L:

$$z(0) \geq 0, \ z(1) \geq 0, \ Lz(x) \geq 0, \ x \in [0,1] \rightarrow z(x) \geq 0, \ x \in [0,1].$$

Define

$$z(x) = \alpha^{-1} \|f(x,0)\|_\infty (1-x) + \max\{|A|,|B|\} \pm u(x).$$

It is clear that the requirements of the maximum principles are fulfilled, see [4, 6]. Therefore $z(x) \geq 0$, $x \in [0,1]$, which implies (4).

Let $u(x)$ be the solution of (1)-(3) and write

$$u(x) = V(x) + p(x),$$

where $V(x) = \rho \exp(-a_0 \varepsilon^{-1} x)$, $\rho = -\varepsilon u'(0)/a_0$, $a_0 = a(0)$. Constant ρ is chosen so that $p'(0) = 0$. The next lemma holds.

Lemma 1 *There exists a constant $C_1 > 0$ independent of ε such that*

$$|p'(x)| \leq C_1, \ x \in [0,1].$$

Therefore the solution of problem (1)-(3) has an exponential boundary layer near to the boundary $x = 0$. To obtain reliable numerical approximations of layer solutions in an efficient way, we will use exponentially-fitted difference schemes. Let introduce the uniform mesh:

$$\Omega_h = \{x_i = ih, \ i = 0,1,\dots,N, \ x_0 = 0, x_N = 1\}.$$

We discretize (1) using the Il'in's difference scheme [2], taking into account boundary layer component $V(x)$ in the solution $u(x)$:

$$L_i^h u^h = -\varepsilon_i^h \lambda_{xx}^h u_i^h - a_i \lambda_x^h u_i^h + f(x_i, u_i^h) = 0, \ i = 1, 2, \dots, N-1,$$

$$u_0^h = A, \ u_n^h = B,$$

(5)

where

$$x_i \in \Omega_h, \ a_i = a(x_i), \ \varepsilon_i^h = \frac{a_i h}{2} \coth \frac{a_i h}{2\varepsilon},$$

$$\lambda_x^h u_i^h = \frac{u_{i+1}^h - u_{i-1}^h}{2h}, \ \lambda_{xx}^h u_i^h = \frac{u_{i+1}^h - 2u_i^h + u_{i-1}^h}{h^2}.$$

Theorem 1 *There exists constant C_0 independent of ε and h, such that*

$$\|u^h - [u]_{\Omega_h}\|_{\infty,h} \leq C_0 h,$$

(6)

were $[u]_{\Omega_h}$ means projection of function $u(x)$ on grid Ω_h.

For the proof, see [9].

Our main object is the investigation of the two-grid method for numerical solution of the problem (1)-(3) with using of the difference scheme (5). We take into account, that the solution of problem (1)-(3) contains a boundary layer component and offer to use the exponential interpolation proposed in [10] to keep the accuracy of the two-grid method. To solve the system of nonlinear algebraic equations (5), we investigate Newton's and Picard's iterative methods.

NEWTON'S METHOD

To compute the solution of scheme (5) we consider the Newton method:

$$-\varepsilon_i^h \lambda_{xx}^h u_i^{m+1} - a_i \lambda_x^h u_i^{m+1} + f(x_i, u_i^m) + f_u'(x_i, u_i^m)(u_i^{m+1} - u_i^m) = 0,$$
$$0 < i < N, \quad u_0^{m+1} = A, \quad u_N^{m+1} = B, \quad m = 0, 1, 2, \ldots . \tag{7}$$

Let u^0 be an initial guess such that $\max_i |u_i^0 - u_i^h| \leq \delta$, $\delta > 0$ is a given constant. First we study the convergence of the method (7). Letting $z_i^m = u_i^m - u_i^h$ we have from (5), (7)

$$-\varepsilon_i^h \lambda_{xx}^h z_i^{m+1} - a_i \lambda_x^h z_i^{m+1} + f_u'(x_i, u_i^m) z_i^{m+1} = f_{uu}''(x_i, r_i^m)(u_i^m - s_i^m) z_i^m,$$
$$z_0^{m+1} = 0, \quad z_N^{m+1} = 0, \tag{8}$$

where $r_i^m, s_i^m \in [u_i^m, u_i^h]$. Define the constant

$$\theta = \max_{x \in \overline{\Omega}, |\xi| \leq l + \delta} |f_{uu}''(x, \xi)|.$$

The application of the maximum principle to problem (8) implies

$$\|u^{m+1} - u^h\|_{\infty, h} \leq \alpha^{-1} \theta \|u^m - u^h\|_{\infty, h}^2. \tag{9}$$

It follows from (9), that the Newton method converges, if

$$\alpha^{-1} \theta \delta < 1. \tag{10}$$

Also, (9) implies

$$\|u^m - u^h\|_{\infty, h} \leq \alpha \theta^{-1} (\alpha^{-1} \theta \delta)^{2^m}, \quad m \geq 0. \tag{11}$$

Further, we will find a lower bound for the necessary number m_h of iterations in order the estimate to be fulfilled: $\max_i |u_i^{m_h} - u_i^h| \leq h$. Taking into account (11), we get:

$$m_h \geq \log_2 \frac{\ln(\alpha^{-1} \theta h)}{\ln(\alpha^{-1} \theta \delta)}. \tag{12}$$

Now we will count the number of the arithmetical operations N_h for m_h iterations. Suppose that on each iteration of the method (7) we need dN operations, where $d > 0$ is a constant. Then, for m_h iterations we need

$$N_h \approx dN \log_2 \frac{\ln(\alpha^{-1} \theta h)}{\ln(\alpha^{-1} \theta \delta)}.$$

Following the idea of the to two-grid method, we introduce the coarse grid Ω_H with n uniform intervals. Let u^H be the solution of Il'in's scheme on the grid Ω_H:

$$-\varepsilon_i^H \lambda_{xx}^H u_i^H - a_i \lambda_x^H u_i^H + f(x_i, u_i^H) = 0, \quad 0 < i < N, \quad u_0^H = A, \quad u_n^H = B. \tag{13}$$

Then, in view of (6), we have $||u^H - [u]_{\Omega_H}||_{\infty,H} \leq C_0 H$. To compute the solution u^H, we use the Newton method and let u^m is the approximate solution on the m-th iteration.

We will count the number of iterations to achieve $\max_i |u_i^m - u_i^H| \leq H$. If m_H is the necessary number of iterations, then analogously as in (12), we have

$$m_H \geq \log_2 \frac{\ln(\alpha^{-1}\theta H)}{\ln(\alpha^{-1}\theta\delta)}.$$

Now we have to interpolate the approximate mesh solution u^m from nodes of a coarse grid Ω_H to nodes of the fine grid Ω_h. To do this, we use the exponential interpolation in [10] of accuracy $O(H)$, uniformly with respect to the parameter ε, and we construct the function $u_H^I(x)$, $x \in [0,1]$ with the property $||u - u_H^I||_\infty \leq C_1 H$. So, using iterations on a coarse grid and exponential interpolation, we will get the initial guess $[u_H^I]_{\Omega_h}$ for the method (7) on a fine grid with accuracy $O(H)$. Then we continue the iterations (7) to find u^h with accuracy $O(h)$. Let us count the number of arithmetical operations for this two-grid method:

$$N_{hH} \approx d n \log_2 \frac{\ln(\alpha^{-1}\theta H)}{\ln(\alpha^{-1}\theta\delta)} + d N \log_2 \frac{\ln(\alpha^{-1}\theta h)}{\ln(\alpha^{-1}\theta H)} + I_H,$$

where I_H is the number of operations for an interpolation.

Suppose, that $H \ll \alpha^{-1}\theta$ and let us account the economy of operations:

$$N_h - N_{hH} \approx d(N-n) \log_2 \frac{\ln(H)}{\ln(\alpha^{-1}\theta\delta)} - I_H.$$

Consider the case $h = H^2$. Then (9) shows that we have to perform *only one iteration* (7) on a fine grid to find the solution of the scheme (5) with accuracy $O(h)$. In the case $h = H^4$ we *need two iterations* (7) to find u^h with accuracy $O(h)$.

PICARD'S METHOD

Suppose, that instead of condition (3) we have the more stronger condition:

$$\beta \geq f_u'(u,x) \geq \gamma > 0 \text{ on } \Omega \times R.$$

To find the solution of scheme (5) we consider the Picard iterative method:

$$\varepsilon_i^h \lambda_{xx}^h u_i^{m+1} + a_i \lambda_x^h u_i^{m+1} - \beta u_i^{m+1} = f(x_i, u_i^m) - \beta u_i^m, \ 0 < i < N,$$

$$u_0^{m+1} = A, \ u_N^{m+1} = B.$$

Using maximum principle, we can prove, that

$$||u^{m+1} - u^h||_{\infty,h} \leq \left(1 - \frac{\gamma}{\beta}\right) ||u^m - u^h||_{\infty,h}.$$

Now, similarly as the Newton method one obtains:

$$N_h \approx \frac{dN \ln \frac{h}{\delta}}{\ln(1 - \frac{\gamma}{\beta})}, \quad N_{hH} \approx \frac{dn \ln \frac{H}{\delta}}{\ln(1 - \frac{\gamma}{\beta})} + \frac{dN \ln \frac{h}{H}}{\ln(1 - \frac{\gamma}{\beta})} + I_H.$$

Therefore, an economy of operations can be realized, if we use the two-grid Picard method:

$$N_h - N_{hH} \approx \frac{d(N - n)}{\ln(1 - \frac{\gamma}{\beta})} \ln \frac{H}{\delta} - I_H.$$

Remark As it is known, the Newton method requires an initial guess, closed to the exact solution. According to (10), it corresponds to inequality $\alpha^{-1}\theta\|u^0 - u^H\|_{\infty,H} < 1$. So, first iterations on a coarse grid we can perform, using Picard method to achieve the inequality $\alpha^{-1}\theta\|u^m - u^H\|_{\infty,H} < 1$ and then we may use Newton iterations.

TWO-GRID ALGORITHMS OF HIGH ACCURACY

The results obtained in the previous sections can be used for formulation of high order accuracy algorithms

Algorithm 1

1. Solve the nonlinear difference scheme (13) and let u^H be the solution of the difference scheme. We seek u^H with accuracy $O(H)$ using Newton or Picard method.

2. Interpolate the mesh function u^H, using exponential interpolation [10], $u^I_H(x) = \text{Int}(u^H, x)$.

3. Let $h = H^2$, $y^I_H = [u^I_H]_{\Omega_h}$. Solve the linear problem on the fine mesh Ω_h:

$$-\varepsilon^h_i \lambda^h_{xx} u^h_i - a_i \lambda^h_x u^h_i + f(x_i, (y^I_H)_i) + f'_u(x_i, (y^I_H)_i)(u^h_i - (y^I_H)_i) = 0,$$

$$0 < i < N, \quad u^h_0 = A, \quad u^h_N = B.$$

4. Perform interpolation [10] from nodes of the fine grid: $u^I_h(x) = \text{Int}(u^h, x)$.
Using results of Sections 3, 4 one can prove the following theorem.

Theorem 2 *There exists a constant C independent of ε and h such that*

$$\|u^I_h - u\|_\infty \leq CH^2.$$

Algorithm 2

1. Solve the nonlinear difference scheme (13) and let u^H be the solution. We seek u^H with accuracy $O(H)$ using Newton's or Picard's method.

2. Interpolate the mesh function u^H, using exponential interpolation, $u^I_H(x) = \text{Int}(u^H, x)$, see [10].

3. Let $h = H^4$, $y^I_H = [u^I_H]_{\Omega_h}$.

4. Solve the linear problem on the fine mesh Ω_h:

$$-\varepsilon^h_i \lambda^h_{xx} u^{hH}_i - a_i \lambda^h_x u^{hH}_i + f(x_i, (y^I_H)_i) + f'_u(x_i, (y^I_H)_i)(u^{hH}_i - (y^I_H)_i) = 0,$$

289

$$0 < i < N, \quad u_0^{hH} = A, \quad u_N^{hH} = B.$$

5. Let $u_{hH}^I = \text{Int}(u^h, x)$.

Again, using results from Sections 3,4 and ideas from [8] one can prove the following assertion.

Theorem 3 *There is constant C, independent of ε and H, such that*

$$||u_{hH}^I - u||_\infty \le CH^4.$$

NUMERICAL EXPERIMENTS

In order to verify the theoretical results we performed several numerical experiments on the test example from [9]:

$$\varepsilon u'' - u' = -\exp(-u), \quad u\left(\frac{1}{2}\right) = \ln\frac{1}{2}, \quad u(1) = 1.$$

For "exact" solution we take the asymptotic approximate solution

$$u_\varepsilon(x) = \exp\left(\frac{x-1}{\varepsilon}\right) + \ln x, \quad u(x) = u_\varepsilon(x) + O(\varepsilon).$$

The results of Step 1, Step2 are shown in Tables 1, 2. The convergence rate is $O(\varepsilon + H)$ at Step 1 and $O(\varepsilon + h + H^2)$ at Step 2.

TABLE 1. Error (E^H) and convergence rate (CR) in max discrete norm, $\varepsilon = h = H^2$

H	step 1		step 2	
	E^H	CR	E^H	CR
0.05	1.620904e-2		2.073142e-3	
0.025	8.389118e-3	*0.9502*	5.472845e-4	*1.9215*
0.0125	4.263923e-3	*0.9763*	1.406949e-4	*1.9597*
0.00625	2.149091e-3	*0.9885*	3.567948e-5	*1.9794*
0.003125	1.078802e-3	*0.9943*	8.984669e-6	*1.9896*

TABLE 2. Max discrete error and **uniform** convergence, $H = 1/6$, $h = H^2$, $\varepsilon < h$

ε	step 1	step 2
2^{-5}	4.508419e-2	1.966952e-2
2^{-6}	4.417110e-2	1.242488e-2
2^{-7}	**4.416666e-2**	1.030566e-2
2^{-8}	4.416666e-2	1.005556e-2
2^{-9}	4.416666e-2	**1.005208e-2**
2^{-10}	4.416666e-2	1.005208e-2
2^{-11}	4.416666e-2	1.005208e-2

TABLE 3. Error (E^H) and convergence rate (CR) in max discrete norm, $\varepsilon = h = H^3$

H	E^H	CR
0.1	1.132694e-3	
0.05	9.353906e-5	*3.5980*
0.025	1.107057e-5	*3.0788*
0.0125	1.173113e-6	*3.2383*

In Table 2, because the numerical methods are uniform convergent with respect to ε, the errors stabilizes in two steps, see black numbers.

The results of Step 3 are produced in Table 3. The convergence rate is $O(\varepsilon + h + H^3)$

On Table 4 are computed all Steps 1, 2, 3, 4. We suppose that, the convergence rate at Step 4 is $O(\varepsilon + |\log h|^{\frac{1}{2}} H^6)$. We also recorded CPU times for the parameters ε, H, see Table 4, where by 'cpu f' and 'cpu f-c' are denoted CPU time in fine mesh and on the two-grid method, respectively.

TABLE 4. Max error for different relations between $H = 0.0125$ and h, **uniform** convergence

h	ε	step 1	step 2	step 3	step 4	cpu f	cpu f-c
	10^{-5}	**3.592649e-2**	1.695773e-4	1.539661e-4	7.259820e-5	430.2653	14.0617
H^6	10^{-6}	3.592649e-2	**1.697145e-4**	**1.548449e-4**	**7.172903e-5**	485.2344	15.8521
	10^{-7}	3.592649e-2	1.697145e-4	1.548449e-4	7.172903e-5	492.5612	16.8012
H^4	10^{-6}	3.592649e-2	4.136344e-4	1.175643e-4	3.174635e-4	231.7240	12.7513
H^2	10^{-6}	3.592649e-2	6.649135e-3	6.432556e-3	6.596393e-3	98.3561	10.6406
$H/4$	10^{-6}	3.592649e-2	8.407888e-3	8.232647e-3	8.371683e-3	28.5156	9.9687
$H/2$	10^{-6}	3.592649e-2	1.740000e-2	1.733059e-2	1.739343e-3	28.4531	9.0594

CONCLUSIONS

In summary, for the model nonlinear convection-diffusion problem (1)-(3) we implemented *the two-grid method of Axelsson and Xu*. For numerical solution of (1) we used the classical Il'in's scheme on uniform meshes. The method applied in this paper is also applicable for coupled system of singularly perturbed nonlinear convection-diffusion problems. Nevertheless, the justification of the method requires another new techniques, which is out of the scope of the present paper.

ACKNOWLEDGMENTS

This research is supported by the Bulgarian National Fund of Science under Project VU-MI-106/2005.

REFERENCES

1. O. Axelsson, *Appl. of Mathematics* **38**(4-5), 249–265 (1993).
2. A. Il'in, *Mat. Zametki* **6**, 237–248 (1969). (in Russian)
3. T. Linss, *Layer-adapted meshes for convection-diffusion problems*, PhD thesis, Habilitation, TU Dresden, 2007.
4. J. Miller, E. O'Riordan, and G. Shishkin, *Fitted Numerical Methods for Singular Perturbed Problems*, World Scientific, Singapore (1996).
5. I. Radeka and D. Herceg, *Novi Sad. J. Math.* **33**(2), 139–161 (2003).
6. H. Roos, M. Stynes, and L. Tobiska, *Numerical Methods for Singularly Perturbed Differential Equations, Convection-Diffusion and Flow problems*, Springer-Berlin, 2008.
7. G. Shishkin and L. Shishkina, *Diff. Eqns* **41**(7), 1030–1039 (2005).
8. J. Xu, *SIAM J. Sci. Comput.* **15**(1), 231–237 (1994).
9. A. Zadorin, *Num. Methods in Cont. Mech.* **17**(6), 35–44 (1986). (in Russian)
10. A. Zadorin, *Sib. J. of Numer Math.* **10**(3), 267–275 (2007). (in Russian)

Efficient Modified Filon-Type Quadrature for Highly Oscillatory Bessel Transformations

S. Xiang

Dept. of Applied Mathematics and Software, Central South University, Changsha, Hunan 410083, China

Abstract. In this paper, we consider efficient modified Filon-type method for the integration of systems containing Bessel function and gives error analysis for these quadratures. Preliminary numerical results show the effectiveness and accuracy of the quadrature for large arguments of integral systems.

Keywords: Oscillatory integral, Bessel function, error bound, Filon-type method, numerical quadrature.
PACS: 02.30.Uu, 02.60.Jh

INTRODUCTION

The occurrence of Bessel functions in the work of Fourier, the study of asymptotics related to these functions by Airy, Stokes, Lipschitz, and Riemann, took well over 100 years ago. The integration of systems involving Bessel function

$$\int_a^b f(x)J_m(rx)dx, \qquad \int_a^{+\infty} f(x)J_m(rx)dx \tag{1}$$

is a central point in many practical problems in physics, chemistry and engineering, where $J_m(rx)$ is the Bessel function of the first kind of order m, m and r are arbitrary real numbers. In most of the cases, these transformations cannot be done analytically and one has to rely on numerical methods. For large r, the integrands become highly oscillatory and thereby present serious difficulties in obtaining numerical convergence of the integrations.

Levin [13] presented a collocation method for $\int_a^b f(x)J_m(rx)dx$ with an error bound $O(r^{-2.5})$ [14, 25] under the assumption that $0 \notin [a,b]$. The collocation method is applicable to a wide class of oscillatory integrals with weight functions satisfying certain differential conditions.

Levin's collocation method [13]: Given a general class of highly oscillatory integrals of the form

$$I[f] = \int_a^b f(x)J_m(rx)dx \equiv \int_a^b F(x)^T W(rx)dx, \tag{2}$$

CP1067, *Applications of Mathematics in Engineering and Economics '34—AMEE '08*, edited by M. D. Todorov
© 2008 American Institute of Physics 978-0-7354-0598-1/08/$23.00

where $F(x) = [0, f(x)]^T$ and $W(rx) = [J_{v-1}(rx), J_m(rx)]^T$. Select linearly independent basis 2-vector functions $\{U_k(x)\}_{k=1}^m$ and let $P(x) = \sum_{k=1}^m a_k U_k(x)$ satisfy

$$P'(c_j) + A^T(r, c_j)P(c_j) = F(c_j), \quad j = 1, 2, \ldots, v, \tag{3}$$

at nodes $a = c_1 < c_2 < \cdots < c_m = b$ and calculate the approximation

$$Q^L[f] = \int_a^b [P'(x) + A^T(r, x)P(x)]^T W(rx)dx = P(b)^T W(rb) - P(a)^T W(ra), \tag{4}$$

where

$$A(r, x) = \begin{pmatrix} \frac{v-1}{x} & -r \\ r & -\frac{v}{x} \end{pmatrix}.$$

The error $I[f] - Q^L[f] \sim (r^{-2.5})$ [25].

Levin's collocation method can be developed to higher order Levin-type method for m-vector functions [16, 26]: Let s be some positive integer and let $\{m_k\}_1^v$ be a set of multiplicities associated with the node points $a = c_1 < \cdots < c_v = b$ such that $m_1, m_v \geq s$, and $\{\psi_k(x)\}_{k=0}^n$ be a set of linearly independent basis functions. Let $P(x) = \left[\sum_{k=0}^n a_k^{(1)} \psi_k(x), \ldots, \sum_{k=0}^n a_k^{(m)} \psi_k(x))\right]^T$ satisfy

$$P'(c_j) + A^T(r, c_j)P(c_j) = F(c_j), \quad j = 1, 2, \ldots, v, \tag{5}$$

and

$$[P'(x) + A^T(r, x)P(x)]_{x=c_j}^{(k)} = F^{(k)}(c_j), \quad k = 1, 2, \ldots, m_j - 1, \quad j = 1, 2, \ldots, v. \tag{6}$$

The Levin-type method is defined by

$$Q_s^L[F] = \int_a^b [P'(x) + A^T(r, x)P(x)]^T W(rx)dx = P(b)^T W(rb) - P(a)^T W(ra). \tag{7}$$

Evans and Webster [4], Chung, Evans and Webster [2] and Evans and Chung [5] introduced **a generalized quadrature rule**, which generates quadrature rules for more general irregular oscillatory weight functions $w(rx)$ based on Lagrange's identity

$$z\mathscr{L}w - w\mathscr{M}z = \{Z[w(rx), z(x)]\}_x' \tag{8}$$

for some function $Z[w(rx), z(x)]$, where $\mathscr{L}$ is a differential operator such that $\mathscr{L}w = 0$ and $\mathscr{M}$ is the adjoint operator of $\mathscr{L}$. For Bessel transformation $\int_a^b f(x)J_m[rq(x)]dx$ with $q(x) \neq 0$ and $q'(x) \neq 0$ for all $x \in [a, b]$, the generalized quadrature rule is efficient with an error bound $O(r^{-2.5})$ [28].

However, both the Levin-type method and generalized quadrature rule fail to compute $\int_0^1 f(x)J_m(rx)dx$ and $\int_0^{+\infty} f(x)J_m(rx)dx$. There are many methods developed for approximation of Bessel transformation over a finite or infinite interval including the turning point 0 [7, 17, 18, 19, 20, 22]. Those methods are efficient for highly oscillatory Bessel function $J_m(rx)$ of integer order m satisfying $0 \leq m \leq 10$.

Filon-type methods for generalized Fourier transformation $\int_a^b f(x)e^{i\omega g(x)}dx$ have been extensively studied [3, 6, 8, 9, 10, 11, 12, 26]. But for other special functions, by and large, this is an uncharted territory [11]. In this paper, we consider modified Filon-type methods for approximation of $\int_0^1 f(x)J_m(rx)dx$ and $\int_0^{+\infty} f(x)J_m(rx)dx$ for Bessel function of arbitrary order m. Without loss of generality, in this paper we assume $r \geq 1$.

MODIFIED FILON-TYPE METHOD

Filon-type method was first introduced by Iserles and Nørsett [9, 10] for Fourier transformation

$$\int_a^b f(x)e^{i\omega g(x)}dx.$$

The modified Filon-type method together with the asymptotic method for Bessel transformation

$$\int_a^b f(x)J_m(rx)dx$$

is defined as follows: Let $p(x) = \sum_{k=0}^n a_k x^k$ be a polynomial. Assume that $p(x)e^{-x}$ is the solution of the system of equations

$$p(c_k)e^{-c_k} = f(c_k), \quad [p(x)e^{-x}]^{(j)}|_{c_k} = f^{(j)}(c_k), \ j = 1, \ldots, m_k - 1, \qquad (9)$$

for every integer $1 \leq k \leq v$. Then $I[f]$ can be approximated by $I[p(x)e^{-x}]$. We rewrite $I[p(x)e^{-x}]$ in the form of

$$I[p(x)e^{-x}] = \int_0^{+\infty} p(x)e^{-x}J_m(rx)dx - \int_1^{+\infty} p(x)e^{-x}J_m(rx)dx$$

$$= \sum_{k=0}^n a_k I[k,m,r,+\infty] - \int_1^{+\infty} p(x)e^{-x}J_m(rx)dx, \quad (10)$$

where $I[k,m,r,+\infty] = \int_0^{+\infty} x^k e^{-x}J_m(rx)dx$ can be computed explicitly by Gamma functions and Gauss hypergeometric function denoted by $\text{hypergeom}([a,b],[c],z)$ ([23],

295

p.385) as follows

$$I[k,m,r,+\infty]$$
$$= \frac{r^m \Gamma(k+m+1)}{2^m \Gamma(m+1)} \text{hypergeom}\left(\left[\frac{k+m+1}{2}, \frac{k+m+2}{2}\right], [m+1], -r^2\right). \quad (11)$$

Expression (11) can be represented for $r \geq 1$ by the transformation in [1], p.559 as

$$I[k,m,r,+\infty] = \frac{2^k}{r^{k+2}} \left\{ \frac{\Gamma\left(\frac{1+k+m}{2}\right) r}{\Gamma\left(\frac{1+m-k}{2}\right)} \text{hypergeom}\left(\left[\frac{1+k+m}{2}, \frac{1-m+k}{2}\right], \left[\frac{1}{2}\right], -\frac{1}{r^2}\right) \right.$$
$$\left. - \frac{\Gamma\left(\frac{2+k+m}{2}\right)}{\Gamma\left(\frac{m-k}{2}\right)} \text{hypergeom}\left(\left[\frac{2+k+m}{2}, \frac{2-m+k}{2}\right], \left[\frac{3}{2}\right], -\frac{1}{r^2}\right) \right\}. \quad (12)$$

The Gauss hypergeometric function

$$\text{hypergeom}([a,b],[c],-r^2) \quad \text{or} \quad \text{hypergeom}\left([a,b],[c],-\frac{1}{r^2}\right)$$

can be efficiently approximated by a truncated series of its asymptotic expansion [21], p.73

$$\text{hypergeom}([a,b],[c],-r^2) = \sum_{k=0}^{\infty} \frac{(-1)^k \Gamma(a+k)\Gamma(b+k)r^{2k}}{\Gamma(a)\Gamma(b)k!}, \quad r \leq 1$$

$$\text{hypergeom}\left([a,b],[c],-\frac{1}{r^2}\right) = \sum_{k=0}^{\infty} \frac{(-1)^k \Gamma(a+k)\Gamma(b+k)}{\Gamma(a)\Gamma(b)r^{2k}k!}, \quad r > 1.$$

Remark 1 *The Gauss hypergeometric function* $\text{hypergeom}([a,b],[c],-r^2)$ *can also be efficiently computed by software MAPLE and MATLAB for all values of r.*

The second term in the right side of the second identity of (10) can be evaluated by the following asymptotic method $Q_s^A[p(x)e^{-x}]$, which was introduced by Xiang [27] and by Olver [15] for Fourier transformation $\int_1^{+\infty} f(x)e^{i\omega g(x)}dx$ and Airy transformation $\int_a^{+\infty} f(x)Ai(-rx)dx$ $(a > 0)$, and can be considered as Levin-type method Q_s^L with nodes $\{1,+\infty\}$ and multiples $\{s,s\}$: Let

$$F_1(x) = [0, f(x)]^T, \quad F_{k+1}(x) = [B^T(r,x)F_k(x)]', k = 1,2,\ldots,$$

where $B(r,x)$ is defined by

$$B(r,x) = \left[\frac{A(r,x)}{r}\right]^{-1} = \begin{pmatrix} \frac{mrx}{m^2-m-r^2x^2} & -\frac{r^2x^2}{m^2-m-r^2x^2} \\ \frac{r^2x^2}{m^2-m-r^2x^2} & -\frac{(m-1)rx}{m^2-m-r^2x^2} \end{pmatrix}. \quad (13)$$

The asymptotic quadrature

$$Q_s^A[p(x)e^{-x}] = \sum_{k=1}^{s} \frac{(-1)^k}{r^k} F_k(1)^T B(r,1) W(r) \tag{14}$$

is efficient for large r with an error bound $O\left(\frac{1}{r^{s+3/2}}\right)$.

Theorem 1 *Let $p(x)$ be a polynomial. Then the error for computing $\int\limits_{1}^{+\infty} p(x)e^{-x}J_m(rx)dx$ by asymptotic method $Q_s^A[p(x)e^{-x}] = \sum\limits_{k=1}^{s} \frac{(-1)^k}{r^k} P_k(1)^T B(r,1) W(r)$ is*

$$\int\limits_{1}^{+\infty} p(x)e^{-x}J_m(rx)dx - Q_s^A[p(x)e^{-x}] = O\left(\frac{1}{r^{s+3/2}}\right), \quad r \gg 1, \tag{15}$$

where $P_1(x) = [0, p(x)e^{-x}]^T$ and $P_{k+1}(x) = [B^T(r,x)P_k(x)]'$, $k = 1,2,\ldots$.

Theorem 2 *Let s be some positive integer and let $\{m_k\}_1^v$ be a set of multiplicities associated with the node points $0 = c_1 < c_2 < \cdots < c_v = 1$ such that $m_1, m_v \geq s$. Suppose that $p(x) = \sum\limits_{k=0}^{n} a_k x^k$ with $n = \sum\limits_{k=1}^{v} m_k - 1$ and $p(x)e^{-x}$ is the solution of (9). Define*

$$Q_s^F[f] = \sum_{k=0}^{n} a_k I[k,m,r,+\infty] - \sum_{k=1}^{s} \frac{(-1)^k}{r^k} F_k(1)^T B(r,1) W(r). \tag{16}$$

Then

$$I[f] - Q_s^F[f] = O\left(r^{-s-1}\right), \quad r \gg 1.$$

Example 1 Let us consider quadrature (16) for $I[e^x] = \int\limits_{0}^{1} e^x J_{4\pi}(rx)dx$ at nodes $c_k = \frac{1+\cos[\frac{(k-1)\pi}{9}]}{2}$ $(k = 1,\ldots,10)$ with $m_1 = \cdots = m_{10} = 1$ (Figure 1), and at nodes $c_1 = 0$ and $c_2 = 1$ with $m_1 = m_2 = 2$ (Table 1) respectively. Here we consider the asymptotic method $Q_s^A[p(x)e^{-x}]$ $(s = 1, 2)$ for the first case.

In (10), from the special form for computing $\int\limits_{1}^{+\infty} p(x)e^{-x}J_m(rx)dx$, an alternative approach of the integral is by choosing a large b, for example, $b = 30$, such that it can be efficiently approximated by $\int\limits_{1}^{b} p(x)e^{-x}J_m(rx)dx$ which can be computed by Levin-type method [16, 27]. For a fixed r, the error can be controlled.

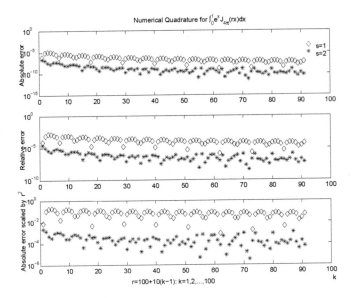

FIGURE 1. Quadrature (16) combined with $Q_s^A[p(x)e^{-x}]$ ($s = 1, 2$) for $\int_0^1 e^x J_{4\pi}(rx)dx$ at nodes $c_k = \frac{1+\cos\frac{(k-1)\pi}{9}}{2}$ ($k = 1, ..., 10$) with $m_1 = \cdots = m_{10} = 1$, r from 100 to 1000

In the following we consider an application about the above modified Filon-type method to $I_1[f] := \int_0^{+\infty} f(x)J_m(rx)dx$.

Theorem 3 *Let* $p(x) = \sum_{k=0}^{n} a_k x^k$ *be a polynomial and* $p(x)e^{-x}$ *satisfy (9) in* $[0,1]$. *Suppose that* $f(x)$ *and all its derivatives are bounded in* $[0,+\infty)$. *Set*

$$\widetilde{Q}_s^F[f] = \int_0^{+\infty} p(x)e^{-x}J_m(rx)dx = \sum_{k=0}^{n} a_k I[k,m,r,+\infty]. \qquad (17)$$

Then

$$I_1[f] - \widetilde{Q}_s^F[f] = O\left(\frac{1}{r^{s+1}}\right), \quad r \gg 1. \qquad (18)$$

Example 2 Let us consider the modified Filon-type method (17) for $\int_0^{+\infty} \cos(x)J_{4\pi}(rx)dx$ at nodes $c_1 = 0$ and $c_2 = 1$ with $m_1 = m_2 = 2$ (Figure 2).

Remark 2 *For large values of* r, *the integral* $\int_0^{+\infty} f(x)J_m(rx)dx$ *is mainly determined by a small neighborhood of the turning point* 0. *The choices of the initial interval* $[0,a]$

298

TABLE 1. Numerical Quadrature (16) combined with asymptotic method Q_2^A for $\int_0^1 e^x J_{4\pi}(rx)dx$ at nodes $c_k = \frac{1+\cos[\frac{(k-1)\pi}{9}]}{2}$ $(k=1,...,10)$ with $m_1 = \cdots = m_{10} = 1$ (denoted by Q_1), and at $c_1 = 0$, $c_2 = 1$ with $m_1 = m_2 = 2$ (denoted by Q_2) respectively

r	10^3	10^4
Exact	$9.7266793794774e-4$	$1.0224502528770e-4$
Q_1	$9.7266799692112e-4$	$1.0244503532900e-4$
Q_2	$9.7242329640788e-4$	$1.0224477337685e-4$

r	10^5	10^6
Exact	$1.0069178988621e-5$	$9.9807081843987e-7$
Q_1	$1.0069179989289e-5$	$9.9807091842114e-7$
Q_2	$1.0069178736594e-5$	$9.9807081816837e-7$

TABLE 2. Numerical Quadrature (10) at $c_1 = 0$ and $c_2 = 1$ with $m_1 = m_2 = 2$ for $\int_0^1 e^x J_{4\pi}(rx)dx$, where $\int_1^{+\infty} p(x)e^{-x}J_m(rx)dx$ is approximated by $\int_1^{30} p(x)e^{-x}J_m(rx)dx$, which is evaluated by Levin-type method [16, 27] with nodes $\{1, 30\}$ and multiples $\{2, 2\}$

r	10^3	10^4	10^5	10^6
Exact	$9.7266793795e-4$	$1.0224502529e-4$	$1.0069178989e-5$	$9.9807081844e-7$
Quadrature	$9.7242328926e-4$	$1.0224477341e-4$	$1.0069178738e-5$	$9.9807081842e-7$

in which $f(x)$ is approximated by $p(x)e^{-x}$ will lead to different numerical results (see Figure 3). Generally speaking, for a lower interpolation polynomial with degree $n < 10$, the interval $[0, 0.1]$ is a good choice. For $n = 10$, the interval $[0, 1]$ is a good choice. For large n, the initial interval can be chosen as $[0, n/10]$.

ACKNOWLEDGMENTS

This work is supported partly by NSF of China (No 10771218) and partly by Program for New Century Excellent Talents in University, State Education Ministry, China.

The author is grateful to the fruitful discussion on the 34th Conference "Applications of Mathematics in Engineering and Economics" held at Sozopol, Bulgaria, and to the referee's helpful comments for improvement of this paper.

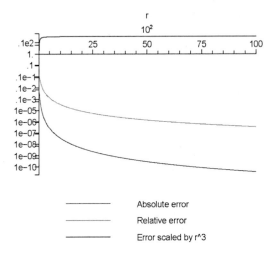

FIGURE 2. Absolute error (bottom figure), relative error (middle figure) and the absolute error scaled by r^3 (top figure) by the modified Filon-type method $Q_2^F[\cos(x)]$ (17) at nodes $c_1 = 0$ and $c_2 = 1$ with $m_1 = m_2 = 2$ for $\int\limits_0^{+\infty} \cos(x)J_{4\pi}(rx)dx$: r from 10^2 to 10^4

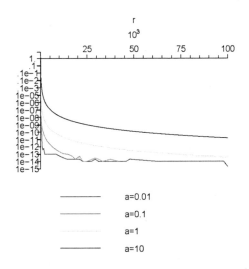

FIGURE 3. Absolute error by modified Filon-type method $Q_2^F[\cos(x)]$ (17) for $\int\limits_0^{+\infty} \cos(x)J_{4\pi}(rx)dx$ with nodes $\{0, 1\}$ and multiples $\{2, 2\}$: $a = 0.01, 0.1, 1, 10$ corresponding to the bottom, third, second and top respectively, and r from 2 to 10^5

REFERENCES

1. M. Abramowitz, I. A. Stegun, *Handbook of Mathematical Functions: with Formulas, Graphs, and Mathematical Tables*, Dover Publications, Inc., New York, 1965.
2. K. C. Chung, G. A. Evans, and J. R. Webster, *Appl. Numer. Math.* **34**, 85–93 (2000).
3. P. I. Davis, P. Rabinowitz, *Methods of Numerical Integral Integration*, Academic Press, Orlando, 1984.
4. G. A. Evans, J. R. Webster, *Appl. Numer. Math.* **23**, 205–218 (1997).
5. G. A. Evans, K. C. Chung, *J. Complexity* **19**, 272–285 (2003).
6. L. N. G. Filon, *Proc. Royal Soc. Edinburgh.* **49**, 38–47 (1928).
7. Q. Haider, L. C. Liu, *J. Phys. A: Math. Gen.* **25**, 6755–6760 (1992).
8. D. Huybrechs, S. Vandewalle, *SIAM J. Numer. Anal.* **44**, 1026–1048 (2006).
9. A. Iserles, S. P. Nørsett, *BIT* **44**, 755–772 (2004).
10. A. Iserles, S. P. Nørsett, *Proc. Royal Soc. A* **461**, 1383–1399 (2005).
11. A. Iserles, S. .P. Nørsett, S. Olver, "Highly oscillatory quadrature: The story so far," in *Proc. of ENumath, Santiago de Compostela*, edited by A. Bermudez de Castro et al, Springer-Verlag, 2006, pp. 97–118.
12. D. Levin, *Math. Comp.* **38**, 531–538 (1982).
13. D. Levin, *J. Comput. Appl. Math.* **67**, 95–101 (1996).
14. D. Levin, *J. Comput. Appl. Math.* **78**, 131–138 (1997).
15. S. Olver, *Numerical approximation of highly oscillatories, Smith-knight Essasy*, University of Cambridge, 2006.
16. S. Olver, *BIT* **47**, 637–655 (2007).
17. R. Piessens, *Comput. Phys. Commun.* **25**, 289–295 (1982).
18. R. Piessens, M. Branders, *BIT* **23**, 370–381 (1983).
19. R. Piessens, F. Poleunis, *BIT* **11**, 317–327 (1971).
20. M. Puoskari, *J. Comput. Phys.* **75**, 334–344 (1988).
21. E. D. Rainville, *Special Functions*, McGraw-Hill Book Company, New York, 1960.
22. B. Sommer, J. G. Zabolitzky, *Comput. Phys. Commun.* **16**, 383–387 (1979).
23. G. N. Watson, *A Treaties on the Theory of Bessel Functions*, Cambridge University Press, Cambridge, 1952.
24. S. Xiang, *J. Comput. Appl. Math.* **177**, 231–139 (2005).
25. S. Xiang, *BIT* **47**, 469–482 (2007).
26. S. Xiang, *Numer. Math.* **105**, 633–658 (2007).
27. S. Xiang, W. Gui, P. Mo, *Appl. Numer. Math.* **58**, 1247–1261 (2008).
28. S. Xiang, W. Gui, *Appl. Math. Comp.* **197**, 60–75 (2008).

ANALYSIS AND APPLICATIONS

Comparison of the Two-grid Method on Different Meshes for Singularly Perturbed Semilinear Problems

I. T. Angelova and L. G. Vulkov

University of Rousse, Dept. of Mathematics, 8 Studentska str., 7017 Rousse, Bulgaria

Abstract. A two-grid algorithm for FEM solution of nonlinear reaction diffusion equations was constructed. The algorithm involves solving one small, nonlinear problem on the coarse-grid system, one-linear problem on the fine-grid system. Three meshes (Shishkin's, Bakhvalov's and Vulanovic's) are examined. All of the approximations are uniformly convergent with respect to the small parameter. We prove that when the mesh step parameters are chosen appropriately, the error in the two-grid algorithm is comparable to the error in solving the fully nonlinear problem on the fine mesh. We also provide an analysis of numerical experiments.

Keywords: Reaction-diffusion, singularly perturbation, difference schemes, finite element, two-grid method.
PACS: 01.30.cc, 44.05.+e, 41.20.Cv

INTRODUCTION

We consider the problem

$$-\varepsilon^2 u'' + f(x,u) = 0, \; x \in \Omega \equiv (0,1), \tag{1}$$
$$u(0) = u(1) = 0,$$

where $0 < \varepsilon \ll 1$, is a small perturbation parameter, assuming that the following conditions are fulfilled:

$$f \in C^2(\Omega \times R), \tag{2}$$

$$f_u(x,u) \geq c_0^2 > 0, \; (x,u) \in (\Omega \times R). \tag{3}$$

The condition (3) is the standard stability condition, which implies that both (1) and the reduced problem $f(x,u) = 0$ have unique smooth solutions u_ε and u_0, respectively.

It is shown theoretically and experimentally in [3] that there is no finite difference scheme (or finite element approximation) of (1) on standard meshes whose solution can be guaranteed to converge to the solution u in the maximum norm, uniformly with respect to the perturbation parameter ε.

Nowadays, two basic types of non-equidistant (layer adapted) meshes, Bakhvalov's and Shishkin's are used for solving singularly perturbed problems [2, 4, 8, 11]. An explicit mesh construction method to solve a singularly perturbed problem of type (1) was used first by Bakhvalov [2], where he obtained the special discretization mesh $w_h = \{x_i = \lambda(i/N) : i = 0, 1, \ldots N\}, h = 1/N$, where by h is the mesh generating function

CP1067, *Applications of Mathematics in Engineering and Economics '34—AMEE '08*, edited by M. D. Todorov
© 2008 American Institute of Physics 978-0-7354-0598-1/08/$23.00

that consists of three parts: λ_1, λ_2 and λ_3. Functions λ_1 and λ_3 generate mesh points in the boundary layers in the neighborhood of $x = 0$ and $x = 1$ respectively. Function λ_2 generates mesh points outside the boundary layers and it is a tangent line to both λ_1 and λ_3, and $\lambda_2(0.5) = 0.5$.

A much simpler mesh was constructed by Shishkin, see [8], but many difference schemes applied to Bakhalov's mesh show better results.

In order to simplify Bakhvalov's mesh, but also to increase the density of mesh points in the boundary layers, according to Shishkin's mesh, Vulanovic modified the former mesh generating functions [11].

There is a wide range of publications that deal with layer adapted meshes. In the thesis [7] a rather complete overview is given.

The construction of high-order approximations to singularly perturbed problems attracted a lot of attention in the recent years. But such constructions often lead to extension of the stencil or to discretizations that are not inverse monotone [7]. Another way to increase the accuracy of the numerical solution to singularly problems is the use of Richardson extrapolation.

The main objective of this paper is to present a two-grid algorithm using a standard FEM approximation on different adaptive meshes for the boundary value problem (1). The two-grid method used for high-order and in the same time expensive computation was first introduced by Axellson [1] and Xu [12] independently of each other. Later on, it was further investigated by many other authors, for instance, for elliptic and parabolic, Stokes-Darcy equations, see [9] and the references therein.

The rest of the paper is organized as follows: In Section 2 we describe the three meshes. In Section 3 we present the discrete problem. In Section 4 we describe the two-grid algorithm (TGA) and provide error estimates for the difference scheme discretization of the TGA, Theorem 2, Section 4 includes numerical results that illustrate the theoretical estimates. Finally, conclusions and directions for future work are also presented. Although our result will be presented in a model, one-dimensional situation, the algorithm possess a wider generality. There are many interesting and relevant boundary value problems of the form (1), for which the condition (3) is not satisfied, see [6], and the computational techniques still work well.

THE MESHES

For a given integer N, we introduce on $\Omega = [0,1]$ the mesh

$$w_H = \{0 = X_0 < X_1 < \ldots < X_{N-1} < X_N = 1\},$$

with $X_i = \lambda(i/N)$, $H_i = X_i - X_{i-1}$, $i = 0, 1, \ldots, N$. Following [2] we present all meshes by their mesh-generating functions.
Bakhvalov's generation function is given by

$$\lambda(t) = \begin{cases} \phi(t) := a\varepsilon \ln \frac{q}{q-t}, & t \in [0, \alpha], \\ \phi(\alpha) + \phi'(\alpha)(t - \alpha), & t \in [\alpha, 0.5], \\ 1 - \lambda(1-t), & t \in [0.5, 1], \end{cases}$$

TABLE 1. Points in boundary layers (%)

		8	16	32	64
	N	64	256	1024	4096
	n				
S-mesh	I st	25	12.5	12.5	6.25
	II st	6.25	4.69	3.71	3.03
V-mesh	I st	50	50	43.75	40.63
	II st	40.63	40.63	40.04	40.04
B-mesh	I st	25	25	18.75	18.75
	II st	18.75	17.97	17.77	17.72

where a and q are constants, independent of ε, such that

$$q \in (0, 0.5), \ a \in \left(0, \frac{q}{\varepsilon}\right), \ ac_0^2 \geq 2. \tag{4}$$

Here α is the abscissa of the contact point of the tangent line from $(0.5, 0.5)$ to $\phi(t)$. The generated mesh will be called B-mesh.

Shishkin's mesh (S-mesh) is a piecewise equidistant and consequently much simple than the mesh above. The generating function for the mesh is

$$\lambda(t) = \begin{cases} 4\alpha t, & t \in [0, \alpha], \\ \alpha + 2(1 - 2\alpha)(t - 0.25), & t \in [\alpha, 0.5], \\ 1 - \lambda(1 - t), & t \in [0.5, 1] \end{cases}$$

with

$$\alpha = \min\{1/4, 2\gamma_0^{-1}\varepsilon \ln N\}, \ \gamma_0 = \min\{\gamma, 1\}.$$

In [4, 9], Vulanovic has shown that λ_1 need not be a logarithmic function. A class of suitable mesh generating functions was given and it includes functions of a much simpler rational form. Out of those functions we select the following one

$$\lambda(t) = \begin{cases} \mu(t) := \frac{\alpha \varepsilon t}{q - t}, & t \in [0, \alpha], \\ \mu(\alpha) + \mu'(\alpha)(t - \alpha), & t \in [\alpha, 0.5], \\ 1 - \lambda(1 - t), & t \in [0.5, 1], \end{cases}$$

with q and a satisfying the conditions (4) and the parameter α has the same meaning as of Bakhvalov's mesh, but here it can be explicitly calculated

$$\alpha = \frac{q - \sqrt{\varepsilon a q(1 - 2q + 2\varepsilon a)}}{1 + 2\varepsilon a}.$$

The mesh will be called V-mesh.

In order to emphasize the difference between meshes we present Table 1, where the percentage of the number of mesh points in the boundary layers, i.e., in $[0, \varepsilon] \cup [1 - \varepsilon, 1]$, is given, with $q = 0.4$, $a = 1$ and $\varepsilon = 2^{-8}$.

THE DISCRETE PROBLEM ON COARSE GRID

In order to illustrate the advantages of the two-grid method we approximate the problem (1)-(3) by the FEM second order difference scheme [4, 10]

$$-\varepsilon^2 y_{\bar{X}\hat{X},i} + + \frac{H_i}{6\hat{H}_i}f(X_{i-1},y_{i-1}) + \frac{2}{3}f(X_i,y_i) + \frac{H_{i+1}}{6\hat{H}_i}f(X_{i+1},y_{i+1}) = 0, \ i = 1,\ldots,N-1,$$

(5)

$$y_0 = 0, \ y_N = 0,$$

where, [8], $y_i \approx u(x_i)$ and

$$y_{\bar{X}\hat{X},i} = \frac{1}{\hat{H}_i}\left(\frac{y_{i+1}-y_i}{H_{i+1}} - \frac{y_i - y_{i-1}}{H_i}\right), \ \hat{H}_i = 0.5(H_i + H_{i+1}).$$

Newton's method can be implemented for solution of the system of nonlinear algebraic equations

$$-\varepsilon^2 y_{\bar{X}\hat{X},i}^{k+1} + \frac{H_i}{6\hat{H}_i}f_u'(X_{i-1},y_{i-1}^k)y_{i-1}^{k+1} + \frac{2}{3}f_u'(X_i,y_i^k)y_i^{k+1} + \frac{H_{i+1}}{6\hat{H}_i}f_u'(X_{i+1},y_{i+1}^k)y_{i+1}^{k+1}$$

$$= \frac{H_i}{6\hat{H}_i}f_u'(X_{i-1},y_{i-1}^k)y_{i-1}^k + \frac{2}{3}f_u'(X_i,y_i^k)y_i^k + \frac{H_{i+1}}{6\hat{H}_i}f_u'(X_{i+1},y_{i+1}^k)y_{i+1}^k$$

(6)

$$-\frac{H_i}{6\hat{H}_i}f(X_{i-1},y_{i-1}^k) - \frac{2}{3}f(X_i,y_i^k) - \frac{H_{i+1}}{6\hat{H}_i}f(X_i,y_{i+1}^k), \ i = 1,\ldots,N-1,$$

where k is the iteration number, $k = 0,1,2,\ldots$.

The rate of convergence of the numerical solution is second-order uniformly with respect to ε for the Bakhalov and the Vulanovic meshes, while for the Shishkin mesh – almost of second order.

The tables below present the errors

$$E_N = \|u_\varepsilon - y\|_\infty,$$

where y is the numerical solution on a mesh with N mesh steps. Also, we calculated numerical orders of convergence by the formula:

$$O_N = \frac{\ln E_N - \ln E_{2N}}{\ln 2}.$$

Numerical results are presented in Tables 2, 3, 4 which validate the theoretical ones established in Theorems 1, 2. It is interesting to discuss the computations for small ε. In our case the methods are uniform convergent, the errors stabilize for each N as $\varepsilon \to 0$. See the results in the Tables for $\varepsilon = 10^{-2}$, $\varepsilon = 10^{-3}$, $\varepsilon = 10^{-4}$. So, we will discuss the correspondent rows in the tables for $\varepsilon = 10^{-2}$. For example the maximum error at the Ist step for $N = 64$ in Table 2 (S-mesh) is $5.300e-3$, in Table 3 (V-mesh) is $1.500e-3$ and in Table 4 (B-mesh) is $8.491e-4$, while at the IIst (TGA) the corresponding errors are $1.298e-5$, $7.363e-7$, $3.313e-7$. Therefore: (1) the (TGA) significantly increase

TABLE 2. The maximum error and the numerical order of convergence for $\varepsilon = 1$, 10^{-1}, 10^{-2}, 10^{-3}, 10^{-4} for the scheme (5) and the TGA on S-mesh, $\sigma = 2$

ε	N	8	16	32	64
	n	64	256	1024	4096
1	I st	1,741E-4	3,582E-5	8,952E-6	2,238E-6
	O_N	2,2815	2,0003	2,0001	
	II st	2,262E-6	1,414E-7	8,838E-9	5,381E-10
	O_n	3,9996	4,0001	4,0378	
10^{-1}	I st	3,230E-02	7,500E-3	1,900E-3	4.682E-4
	O_N	1,4602	1,9809	2,0209	
	II st	8,295E-4	4,450E-5	2,844E-6	1,763E-7
	O_n	4,2204	3,9679	4,0117	
10^{-2}	I st	7,460E-2	3,920E-2	1,480E-2	5,300E-3
	O_N	0,9283	1,4053	1,4815	
	II st	7,000E-3	1,100E-3	1,299E-4	1,298E-5
	O_n	2,6699	3,0823	3,3234	
10^{-3}	I st	7,460E-2	3,920E-2	1,480E-2	5,300E-3
	O_N	0,9283	1,4053	1,4815	
	II st	7,000E-3	1,100E-3	1,299E-4	1,298E-5
	O_n	2,6699	3,0821	3,3235	
10^{-4}	I st	7,460E-2	3,920E-2	1,480E-2	5,300E-3
	O_N	0,9283	1,4053	1,4815	
	II st	7,000E-3	1,100E-3	1,299E-4	1,298E-5
	O_n	2,6699	3,0821	3,3234	

the accuracy and the experiments confirm Theorems 1, 2. (2) It is known (see [4]) that the most accurate is the B-mesh and now for the TGA the situation is similar.

Theorem 1 *Assume that (2), (3) hold. Further*

(i) Let y be the solution of (5) on Bakhalov's or Vulanovic's meshes. Then there exist a constant C independent of ε and N such that, (see [4, 6, 8]),

$$\|u - y\|_\infty \leq CN^{-2}. \tag{7}$$

(ii) Let y be the solution of (5) on Shishkin's mesh. Now the convergence rate is almost of second order, [3],

$$\|u - y\|_\infty \leq CN^{-2} \ln^2 N. \tag{8}$$

The constant C is independent of ε and N and different for the three meshes.

TWO-GRID ALGORITHM

In this section we propose a simple algorithm based on two grids [1, 9, 12]. For this we introduce the fine grids

$$w_h = \{0 = x_0 < x_1 < \ldots < x_{N^2} = 1\}, \text{ with } x_i = \lambda(i/N^2), h_i = x_i - x_{i-1}, i = 1, \ldots, N^2.$$

Our algorithm reads as follows:

TABLE 3. The maximum error and the numerical order of convergence for $\varepsilon = 1$, 10^{-1}, 10^{-2}, 10^{-3}, 10^{-4} for the scheme (5) and the TGA on V-mesh, $a = 1$, $q = 0.4$

ε	N	8	16	32	64
	n	64	256	1024	4096
1	I st	1,741E-4	3,582E-5	8,952E-6	2,238E-6
	O_N	2,2815	2,0003	2,0001	
	II st	2,262E-6	1,414E-7	8,838E-9	5,381E-10
	O_n	3,9996	4,0001	4,0378	
10^{-1}	I st	2,790E-2	8,000E-3	2,000E-3	5,177E-4
	O_N	1,8022	2,0000	1,9498	
	II st	6,929E-4	4,647E-5	2,942E-6	1,873E-7
	O_n	3,8985	3,9812	3,9738	
10^{-2}	I st	1,368E-1	2,090E-2	5,900E-3	1,500E-3
	O_N	2,7105	1,8247	1,9758	
	II st	4,600E-03	2,051E-4	1,246E-5	7,363E-7
	O_n	4,4873	4,0404	4,0813	
10^{-3}	I st	1,488E-1	2,210E-2	5,900E-3	1,500E-3
	O_N	2,7513	1,9053	1,9758	
	II st	5,400E-3	2,165E-4	1,248E-5	7,363E-7
	O_n	4,6408	4,1169	4,0827	
10^{-4}	I st	1,493E-1	2,220E-2	5,900E-3	1,500E-3
	O_N	2,7496	1,9118	1,9758	
	II st	5,400E-3	2,170E-4	1,248E-5	7,363E-7
	O_n	4,6369	4,1207	4,0827	

1. Solve the discrete problem (5) on the coarse grid w_H and then we do a linear interpolation to obtain the function $u_H^I(x)$ defined on the domain $[0,1]$.

2. Solve the linear discrete problem

$$-\varepsilon^2 y_{\bar{x}\hat{x},i} + \tfrac{h_i}{6\hat{h}_i} f'_u(x_{i-1}, u_H^I(x_{i-1}))y_{i-1} + \tfrac{2}{3} f'_u(x_i, u_H^I(x_i))y_i + \tfrac{h_{i+1}}{6\hat{h}_i} f'_u(x_{i+1}, u_H^I(x_{i+1}))y_{i+1}$$

$$= \tfrac{h_i}{6\hat{h}_i} f'_u(x_{i-1}, u_H^I(x_{i-1}))y_{i-1} + \tfrac{2}{3} f'_u(x_i, u_H^I(x_i))y_i + \tfrac{h_{i+1}}{6\hat{h}_i} f'_u(x_{i+1}, u_H^I(x_{i+1}))y_{i+1} \quad (9)$$

$$-\tfrac{h_i}{6\hat{h}_i} f(x_{i-1}, u_H^I(x_{i-1})) - \tfrac{2}{3} f(x_i, u_H^I(x_i)) - \tfrac{h_{i+1}}{6\hat{h}_i} f(x_i, u_H^I(x_{i+1})), \; i = 1, \ldots, N^2 - 1,$$

$$y_0 = 0, \; y_{N^2} = 0.$$

to find the fine mesh numerical solution y^h.

3. Interpolate y^h to obtain $u_{h,H}^I(x)$, $x \in \overline{\Omega}$.

The next theorem gives the main theoretical results of the present paper.

Theorem 2 *Let the assumption in Theorem 1 hold. If $u \in C^2(\overline{\Omega})$ then:*

(i) In the case of Bakhalov's or Vulanovic's meshes we have

$$\|u_{h,H}^I - u\|_{\infty,H} \leq CN^{-4}. \tag{10}$$

(ii) For Shishkin's mesh the estimate holds

$$\|u_{h,H}^I - u\|_{\infty,H} \leq CN^{-4} \ln^4 N. \tag{11}$$

TABLE 4. The maximum error and the numerical order of convergence for $\varepsilon = 1, 10^{-1}, 10^{-2}, 10^{-3}, 10^{-4}$ for the scheme (5) and the TGA on B-mesh, $a = 4$, $q = 0.4$

ε		8	16	32	64	256	1024
	n	64	256	1024	4096	-	-
1	I st	1,741E-4	3,582E-5	8,952E-6	2,238E-6	1,399E-7	8,7419e-9
	O_N	2,2815	2,0003	2,0001	2,0000		
	CPU	0.0469	0.0625	0.0781	**0.1563**	**1.2344**	**34.7813**
	II st	2,262E-6	1,414E-7	8,838E-9	5,381E-10		
	O_n	3,9996	4,0001	4,0378			
	CPU	**0.1094**	**0.2031**	**0.6250**	9.4531		
10^{-1}	I st	3,230E-2	7,500E-3	1,900E-3	4,682E-4	2,925E-5	1,8281e-6
	O_N	2,1066	1,9809	2,0209	2,0005		
	CPU	0.0469	0.0625	0.0781	**0.1406**	**1.2344**	**34.9688**
	II st	8,295E-4	4,450E-5	2,844E-6	1,763E-7		
	O_n	4,2204	3,9679	4,0117			
	CPU	**0.0938**	**0.1875**	**0.6094**	8.8906		
10^{-2}	I st	5,810E-2	1,380E-2	3,400E-3	8,491E-4	5,309E-5	3,3181e-6
	O_N	2,0739	2,0211	2,0016	1,9994		
	CPU	0.0313	0.0469	0.0938	**0.1406**	**1.2656**	**34.7500**
	II st	1,700E-3	8,720E-5	5,323E-6	3,317E-7		
	O_n	4,2850	4,0340	4,0044			
	CPU	**0.1094**	**0.1719**	**0.5781**	7.7344		
10^{-3}	I st	5,810E-2	1,380E-2	3,400E-3	8,491E-4	5,309E-5	3,3181e-6
	O_N	2,0739	2,0211	2,0016	1,9994		
	CPU	0.0469	0.0625	0.0781	**0.1563**	**1.2500**	**34.3750**
	II st	1,700E-3	8,720E-5	5,323E-6	3,317E-7		
	O_n	4,2850	4,0340	4,0044			
	CPU	**0.0781**	**0.2188**	**0.5625**	7.0938		
10^{-4}	I st	5,810E-2	1,380E-2	3,400E-3	8,491E-4	5,309E-5	3,3181e-6
	O_N	2,0739	2,0211	2,0016	1,9994		
	CPU	0.0581	0.0781	0.0938	**0.1494**	**1.2500**	**34.8594**
	II st	1,700E-3	8,720E-5	5,323E-6	3,317E-7		
	O_n	4,2850	4,0340	4,0044			
	CPU	**0.1094**	**0.2031**	**0.5625**	6.9844		

The constant C is independent of ε and N and different for the three meshes.

NUMERICAL RESULTS

In this section we discuss numerical results from a set of computational experiments associated with the TGA. We consider the test problem [11]

$$-\varepsilon^2 u'' + \frac{u-1}{2-u} + f(x) = 0, \; u(0) = u(0) = 0,$$

311

where $f(x)$ is chosen so that the exact solution is

$$u_\varepsilon(x) = 1 - \frac{\exp(-x/\varepsilon) + \exp(-(1-x)/\varepsilon)}{1 + \exp(-1/\varepsilon)}. \tag{12}$$

Finally, the CPU time is given in Table 4. For example, for $\varepsilon = 1$, one must compare the value 0.1094 with 0.1563, 0.2031 with 1.2344, 0.6250 with 34.7813. The computational cost of the TGA is significant and increases with N.

CONCLUSIONS

A novel implementation of the Axelsson-Xu two-grid method of for numerical solution of nonlinear reaction-diffusion problems is proposed. One solves the nonlinear system of algebraic equations by applying a Newton iteration on a coarse grid, then executes one fine grid iteration, and adds a new correction step on the coarse mesh at the and of the last iteration. Three layer-adapted meshes are examined. All of the approximations are uniformly convergent with respect to the small parameter. For appropriately chosen step parameters $(n = N^2)$, the error in the TGA is in the same order as the error in solving the fully nonlinear problem on the fine mesh. In the same time there is a significant decrease in computational cost using the TGA. Numerical and theoretical extensions of the TGA for 2D nonlinear reaction-diffusion equations are currently studied.

ACKNOWLEDGMENTS

This research is supported by the Bulgarian National Fund of Science under Project VU-MI-106/2005.

REFERENCES

1. O. Axelsson, *Applications of Math.* **15**, 249–265 (1993).
2. N. S. Bakhvalov, *Zh. Vychisl. Math. Fiz.* **9**, 841–859 (1969).
3. P.A. Farrell, J.J.H. Miller, E. O'Riordan, and G.I. Shishkin, *Math. of Comp.* **67**, 603–617 (1999).
4. D. Herceg, K. Surla, I.Radeka, and H. Malicic, *Novi Sad J. Math.* **31**, 93–101 (2001).
5. J. Jin, S. Shu, and J. Xu, *Math. of Comp.* **75**, 1617–1626 (2006).
6. N. Kopteva, and M. Stynes, *Comput. Math. Appl.* **51**, 857–864 (2006).
7. T. Linss, "Layer-adapted meshex for convection-diffusion problems," PhD thesis, Habilitation, TU Dresden, 2007.
8. J.H. Miller, E. O'Riordan, and G. Shishkin, *Fitted Numerical Methods for Singular Perturbation Problems. Error Estimates in the Maximum Norm for Linear Problems in One and Two Dimensions,* World Scientific, Singapore, 1996.
9. M. Mu and J. Xu, *SIAM J. Numer. Anal.* **45**, 1801–1813 (2007).
10. G. Sun and M. Stynes, *Numer. Mdth.* **7a**, 487–500 (1995).
11. R. Vulanovic, and D. Herceg, *J. Comput. Math.* **11**, 162–171 (1993).
12. J. Xu, *SIAM, J. Sci. Comput.* **38**, 231–237 (1994).

Quantity Stickiness versus Stackelberg Leadership

F. A. Ferreira

ESEIG, Instituto Politécnico do Porto, Rua D. Sancho I, 981, 4480-876 Vila do Conde, Portugal

Abstract. We study the endogenous Stackelberg relations in a dynamic market. We analyze a twice-repeated duopoly where, in the beginning, each firm chooses either a quantity-sticky production mode or a quantity-flexible production mode. The size of the market becomes observable after the first period. In the second period, a firm can adjust its quantity if, and only if, it has adopted the flexible mode. Hence, if one firm chooses the sticky mode whilst the other chooses the flexible mode, then they respectively play the roles of a Stackelberg leader and a Stackelberg follower in the second marketing period. We compute the supply quantities at equilibrium and the corresponding expected profits of the firms. We also analyze the effect of the slope parameter of the demand curve on the expected supply quantities and on the profits.

Keywords: Game Theory, Cournot model, entry, licensing.
PACS: 89.65.Gh

INTRODUCTION

Stackelberg leader-follower relations have most often been modelled in association with the chronological order of moves. Namely, there are a first mover (leader) and a second mover (follower). In spite of such a supposedly dynamic setting, it has been common to overlook what happens during the period between these two moves, by assuming a static market which clears only once, after the second mover's move. This builds certain biases into the analysis of firms' strategic incentives either to lead or to follow, which are the contributing forces to endogenous Stackelberg outcomes.

In the earlier literature, endogenous leader/follower has been imbedded most often in the context of a timing game played by oligopolists. Hamilton and Slutsky [2] construct an 'extended game' framework, in which each firm faces the choice of production timing. A fair number of theoretical explanations have been attempted with regard to firms' incentives for Stackelberg behaviour, especially a follower's incentive to wait. Robson [12] imposes costs associated with an early action. Albaek [1] takes into account cost uncertainty. The effect of *a priori* informational heterogeneity between firms, broadly defined, have been discussed in several studies, including Mailath [7] and Normann [8]. On the other hand, when the oligopolists are *a priori* equally uncertain about the market demand, as in Spencer and Brander [16], Sadanand and Sadanand [13] and Maggi [6], earlier production can utilize less information in exchange for the strategic advantage of commitment, whereas later production does the converse. Hirokawa and Sasaki [5] employed a similar framework to Hamilton and Slutsky's 'extended game', except that the static market is replaced with an explicitly two-period market. This can be carried out in two alternative ways. One is to interpret firms' moves as their entry timing. This means that, if there is only one 'first mover', then it becomes a monopolist in its first production period. Hirokawa and Sasaki [4] adopts this interpretation. The other alternative is to assume that, at the beginning of the game, firms are already operating in the market. That is, even a 'second mover' is producing in the first marketing period as well as in the second marketing period. This is the promise we adopt in this paper (as Hirokawa and Sasaki [5] did).

In Hirokawa and Sasaki's model, firms have to choose between a quantity-flexible production mode and a quantity-sticky production mode, the latter implying that the firm's production quantity be unchanged between the two periods.[1] Quantity stickiness entails two effects. On one hand, it serves as a device for quantity commitment throughout the two periods. This is a positive, deterministic effect. On the other hand, it hinders the firm's flexibility in adjusting its supply quantity once demand uncertainty resolves in the second period. This is a negative, stochastic effect. Under some conditions, the tradeoff between these two effects can give rise to *a posteriori* asymmetric Stackelberg-like behaviour, even if the firms are *a priori* identical. Firms' payoffs consists not only of the Stackelberg leader's and follower's profits in the second marketing period, but also of the pre-Stackelberg profits in the first period.

In this paper we consider a more general inverse demand function in the model of Hirokawa and Sasaki [5], and we compute the outputs at equilibrium.

THE MODEL AND THE EQUILIBRIUM

Two *a priori* identical firms operate in the same industry. The game lasts through two marketing periods. At the beginning of the game, each firm chooses between a quantity-flexible production mode and a quantity-sticky production mode. If a firm chooses the stick mode, then the firm must supply the same quantity in both periods. This restriction dos not apply if the firm chooses the flexible production mode in the beginning. We assume that quantity stickiness becomes mutually observable before each firm sets its supply quantity.

For simplicity, the two firms are assumed to sell homogeneous products (i.e., perfect substitutes). The inverse demand function in each marketing period is $P = A - bQ$, where $Q = q_1 + q_2$ is the sum of the two firms' outputs. The demand intercept A is *ex ante* stochastic, of which the prior cumulative distribution $F(a)$ is commonly known to the two firms, with finite mean $E(A) > 0$ and variance $V(A)$. This intercept stays unchanged throughout the two marketing periods. The parameter $b \geq 1$ is the slope parameter of the demand curve. We consider prices net of marginal costs. The discount factor is δ, where $0 < \delta \leq 1$. To ensure individual rationality (non-negative quantities and prices), we assume

$$\inf(A) \geq \frac{4 + 3\delta}{2(3 + 2\delta)} E(A). \tag{1}$$

---

[1] Quantity stickiness can arise from various sources, including technology, precommitted capacity, advance production and inventory investment, binding contracts, and so on.

The intercept A is unobservable to the firms until the price and quantities are realized at the end of the first period. Then, once the state of demand has been observed, a firm can make use of this information to optimize its second-period supply quantity if, and only if, the firm has chosen the flexible mode in the beginning. On the other hand, if only one of the firms has selected the sticky mode whilst the other has selected the flexible mode, then the stickiness entitles the firm to Stackelberg leadership in the second marketing period. Therefore, at the beginning of the game, firms face the tradeoff between the strategic advantage of commitment and the adjustability to the demand realization.

The profit maximization problems for each firm F_i, with $i \in \{1,2\}$, are as follows, depending upon the two firms' commitment decisions (see Figure 1): $q_i^{I|X}$, $q_i^{II|X}(A)$ and $q_i^{L|X}$ denote an uncommitted (quantity-flexible) firm's quantity in the first and the second marketing stages, and a committed (quantity-sticky) firm's quantity throughout the game, respectively, where X indicates the opponent firm's quantity stickiness: $X = F$ if the opponent is uncommitted, or $X = L$ if committed. Note that only an uncommitted firm's second-period quantity can be made contingent upon the state A.

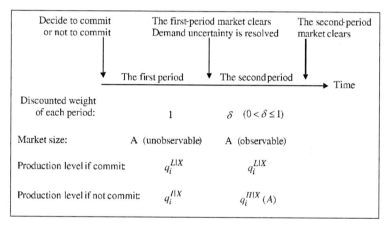

FIGURE 1. Time structure of the game

[a] If both firms commit:

$$\max_{q_i^{L|L}} E\left[\left(A - bq_1^{L|L} - bq_2^{L|L}\right) q_i^{L|L}\right], \quad i = 1,2.$$

[b] If firm F_i commits whilst the other firm F_j does not, then the game is solved backwards. In the second marketing stage, firm F_j solves:

$$\max_{q_j^{II|L}(A)} \left[\left(A - bq_i^{L|F} - bq_j^{II|L}(A)\right) q_j^{II|L}(A)\right].$$

Let $q_j^{II}(A, q_i^{L|F})$ denote the solution for this maximization. Back in the first stage, the two firms solve, respectively:

$$\max_{q_i^{L|F}} E\left[\left(A - bq_i^{L|F} - bq_j^{I|L} \right) q_i^{L|F} + \delta \left(A - bq_i^{L|F} - bq_j^{II}(A, q_i^{L|F}) \right) q_i^{L|F} \right],$$

$$\max_{q_j^{I|L}} E\left[\left(A - bq_i^{L|F} - bq_j^{I|L} \right) q_j^{I|L} \right].$$

[c] If neither firm commits, in the second stage:

$$\max_{q_i^{II|F}(A)} \left[\left(A - bq_1^{II|F}(A) - bq_2^{II|F}(A) \right) q_i^{II|F}(A) \right],$$

with $i \in \{1,2\}$; and back in the first marketing stage:

$$\max_{q_i^{I|F}} E\left[\left(A - bq_1^{I|F} - bq_2^{I|F} \right) q_i^{I|F} \right], \quad i \in \{1,2\}.$$

The game specified above can be summarized into the following payoff matrix:

TABLE 1. Payoff matrix

	Opponent	
Firm	Commit	Not commit
Commit (quantity-sticky)	$\pi^{L|L}$	$\pi^{L|F}$
Not commit (quantity-flexible)	$\pi^{F|L}$	$\pi^{F|F}$

Lemma 1

- *In Cournot-Nash outcomes (i.e., outcomes [a] and [c]), each firm's expected supply quantities are given by*

$$q_i^{L|L} = q_i^{I|F} = \frac{E(A)}{3b}, \quad q_i^{II|F}(A) = \frac{A}{3b},$$

with $i \in \{1,2\}$.
- *In a Stackelberg-like outcome, the committed firm's supply quantity is given by*

$$q_i^{L|F} = \frac{1+\delta}{b(3+2\delta)} E(A),$$

whilst the uncommitted firm's supply quantities are given by

$$q_i^{I|L} = \frac{2+\delta}{2b(3+2\delta)} E(A),$$

$$q_i^{II|L}(A) = \frac{A}{2b} - \frac{1+\delta}{2b(3+2\delta)}E(A),$$

with $i \in \{1,2\}$.

Proof: The results follow by solving optimization problems [a]-[c], assuming that positivity constraints are unbinding. ∎

In the next theorem, we present the firms' equilibrium choices between flexible and sticky modes.

Theorem 1 *Firms' equilibrium choices between flexible and sticky modes yield those three outcomes specified previously, under the following parametric conditions.*

$\{1\}$ *(Commit, Commit) is the dominant strategy equilibrium if, and only if,*

$$\frac{V(A)}{E(A)^2} < \min \left\{ \frac{9(1-b)(2+\delta)^2 + \delta(1+\delta)(12+7\delta)}{9\delta(3+2\delta)}, \right.$$

$$\left. \frac{2(1-b)(3+2\delta)^2 + \delta(1+\delta)(3+\delta)}{2\delta(3+2\delta)^2} \right\}.$$

$\{2\}$ *(Commit, Not commit) is a pure strategy subgame perfect equilibrium if, and only if,*

$$\frac{9(1-b)(2+\delta)^2 + \delta(1+\delta)(12+7\delta)}{9\delta(3+2\delta)} \leq \frac{V(A)}{E(A)^2}$$

$$\leq \frac{2(1-b)(3+2\delta)^2 + \delta(1+\delta)(3+\delta)}{2\delta(3+2\delta)^2}.$$

$\{3\}$ *(Not commit, Not commit) is the dominant strategy equilibrium if, and only if,*

$$\frac{V(A)}{E(A)^2} > \max \left\{ \frac{9(1-b)(2+\delta)^2 + \delta(1+\delta)(12+7\delta)}{9\delta(3+2\delta)}, \right.$$

$$\left. \frac{2(1-b)(3+2\delta)^2 + \delta(1+\delta)(3+\delta)}{2\delta(3+2\delta)^2} \right\}.$$

$\{1,3\}$ *Both (Commit, Commit) and (Not commit, Not commit) are pure strategy subgame perfect equilibria if, and only if,*

$$\frac{2(1-b)(3+2\delta)^2 + \delta(1+\delta)(3+\delta)}{2\delta(3+2\delta)^2} \leq \frac{V(A)}{E(A)^2}$$

$$\leq \frac{9(1-b)(2+\delta)^2 + \delta(1+\delta)(12+7\delta)}{9\delta(3+2\delta)}.$$

Proof:

{1} (Commit, Commit) is the dominant strategy equilibrium if, and only if,

$$\bar{\pi}^{L|L} > \bar{\pi}^{F|L} \text{ and } \bar{\pi}^{L|F} > \bar{\pi}^{F|F},$$

which is equivalent to

$$\frac{V(A)}{E(A)^2} < \min\left\{ \frac{9(1-b)(2+\delta)^2 + \delta(1+\delta)(12+7\delta)}{9\delta(3+2\delta)}, \right.$$

$$\left. \frac{2(1-b)(3+2\delta)^2 + \delta(1+\delta)(3+\delta)}{2\delta(3+2\delta)^2} \right\}.$$

{2} (Commit, Not commit) is a pure strategy subgame perfect equilibrium if, and only if,

$$\bar{\pi}^{L|L} \leq \bar{\pi}^{F|L} \text{ and } \bar{\pi}^{F|F} \leq \bar{\pi}^{L|F},$$

which is equivalent to

$$\frac{9(1-b)(2+\delta)^2 + \delta(1+\delta)(12+7\delta)}{9\delta(3+2\delta)} \leq \frac{V(A)}{E(A)^2}$$

$$\leq \frac{2(1-b)(3+2\delta)^2 + \delta(1+\delta)(3+\delta)}{2\delta(3+2\delta)^2}.$$

{3} (Not commit, Not commit) is the dominant strategy equilibrium if, and only if,

$$\bar{\pi}^{F|L} > \bar{\pi}^{L|L} \text{ and } \bar{\pi}^{F|F} > \bar{\pi}^{L|F},$$

which is equivalent to

$$\frac{V(A)}{E(A)^2} > \max\left\{ \frac{9(1-b)(2+\delta)^2 + \delta(1+\delta)(12+7\delta)}{9\delta(3+2\delta)}, \right.$$

$$\left. \frac{2(1-b)(3+2\delta)^2 + \delta(1+\delta)(3+\delta)}{2\delta(3+2\delta)^2} \right\}.$$

{1,3} Both (Commit, Commit) and (Not commit, Not commit) are pure strategy subgame perfect equilibrium if, and only if,

$$\bar{\pi}^{L|L} \geq \bar{\pi}^{F|L} \text{ and } \bar{\pi}^{F|F} \geq \bar{\pi}^{L|F},$$

which is equivalent to

$$\frac{2(1-b)(3+2\delta)^2 + \delta(1+\delta)(3+\delta)}{2\delta(3+2\delta)^2} \leq \frac{V(A)}{E(A)^2}$$

$$\leq \frac{9(1-b)(2+\delta)^2 + \delta(1+\delta)(12+7\delta)}{9\delta(3+2\delta)}. \blacksquare$$

From Lemma 1, we get the following corollaries:

Corollary 1

- *In Cournot-Nash outcomes, each firm's expected supply quantities do not depend upon the discount factor δ.*
- *In a Stackelberg-like outcome, the committed firm's supply quantity increases in δ, whilst the uncommitted firm's supply quantities decrease in δ.*

Corollary 2 *Each firm's expected supply quantities decrease in the slope parameter b of the demand curve.*

Corollary 3 *The expected profits of the firms are as follows:*

$$
\begin{aligned}
\overline{\pi}^{L|L} &= (1+\delta)\frac{E(A)^2}{9b}, \\
\overline{\pi}^{L|F} &= \frac{2+\delta}{2b}\left(\frac{1+\delta}{3+2\delta}E(A)\right)^2, \\
\overline{\pi}^{F|L} &= \left(1+\frac{\delta}{b}\right)\left(\frac{2+\delta}{2(3+2\delta)}E(A)\right)^2 + \frac{\delta}{4b}V(A), \\
\overline{\pi}^{F|F} &= \left(1+\frac{\delta}{b}\right)\left(\frac{E(A)}{3}\right)^2 + \frac{\delta}{9b}V(A)
\end{aligned}
$$

when, and only when, condition (1) is met.

CONCLUSIONS

We have analyzed endogenous Stackelberg-like behavior in light of quantity stickiness in an explicitly dynamic market. We computed the supply quantities and the corresponding expected profits of the firms. We proved that a Stackelberg-like outcome can endogenously arise when the market is dynamic (i.e., lasts multiple periods and clears every period) and each firm has a choice between a quantity-flexible production mode and a quantity-sticky production mode. We also saw that each firm's expected supply quantities and the corresponding firms' expected profits decrease in the slope parameter b of the demand curve.

ACKNOWLEDGMENTS

We thank ESEIG - Instituto Politécnico do Porto, Centro de Matemática da Universidade do Porto and the Programs POCTI and POSI by FCT and Ministério da Ciência, Tecnologia e do Ensino Superior for their financial support.

REFERENCES

1. S. Albaek, *Journal of Industrial Economics* **38**, 335–347 (1990).
2. J.H. Hamilton and S.M. Slutsky, *Games and Economic Behavior* **2**, 29–46 (1990).
3. M. Hirokawa and D. Sasaki, "Reexamination of Stackelberg leadership: fixed supply contracts under demand uncertainty," *Research paper 600, Department of Economics, University of Melbourne, and working paper 61, Faculty of Economics, Hosei University* (1998).
4. M. Hirokawa and D. Sasaki, "Endogenously asynchronous entries into an uncertain industry," *Research paper 608, Department of Economics, University of Melbourne* (1998).
5. M. Hirokawa and D. Sasaki, *Bulletin of Economic Research* **53**, 19–34 (2001).
6. G. Maggi, *RAND Journal of Economics* **27**, 641–659 (1996).
7. G. Mailath, *Journal of Economic Theory* **59**, 169–182 (1993).
8. H.T. Normann, *Journal of Economics* (Zeitschrift für Nationalökonomie) **66**, 177–187 (1997).
9. D. Pal, *Journal of Economic Theory* **55**, 441–448 (1991).
10. D. Pal, *Games and Economic Behavior* **12**, 81–94 (1996).
11. S. Poddar and D. Sasaki, *European Journal of Political Economy* **18**, 579–595 (2002).
12. A.J. Robson, *International Economic Review* **31**, 263–274 (1990).
13. A. Sadanand and V. Sadanand, *Journal of Economic Theory* **70**, 516–530 (1996).
14. G. Saloner, *Journal of Economic Theory* **42**, 183–187 (1987).
15. L. Schoonbeek, *Journal of Economics* (Zeitschrift für Nationalökonomie) **66**, 271–282 (1997).
16. B.J. Spencer and J.A. Brander, *European Economic Review* **36**, 1601–1626 (1992).

Welfare Effects of Entry into International Markets with Licensing

F. A. Ferreira and Fl. Ferreira

ESEIG, Instituto Politécnico do Porto, Rua D. Sancho I, 981, 4480-876 Vila do Conde, Portugal

Abstract. We study the effects of entry of a foreign firm on domestic welfare in the presence of licensing, when the entrant is technologically inferior to the incumbent. We show that foreign entry increases domestic welfare for intermediate (respectively, sufficiently large) technological differences between the firms under licensing with fixed fee (respectively, output royalty).

Keywords: Game theory, Cournot model, entry, licensing.
PACS: 89.65.Gh

INTRODUCTION

The study of the effect of entry on social welfare has been addressed in Klemperer [4], Cordella [2] and Collie [1]. In a closed economy, Klemperer [4] shows that entry reduces social welfare if cost of the entrant is sufficiently higher than that of the incumbent[1]. Collie [1] examines this issue in an open economy and shows that entry of a foreign firm reduces domestic welfare unless the cost of the foreign firm is sufficiently lower than that of the incumbent. Cordella [2] also considers this issue in an open economy, showing the effects of the number of firms. Technological difference is an important reason for cost differences between firms, which may encourage them to share their technological information through licensing. Fauli-Oller and Sandonis[2] [3] show that higher welfare under entry is more likely in the presence of licensing by the technologically efficient incumbent compared with no licensing. Their results suggest that entry always increases welfare if there is licensing with output royalty but licensing with fixed fee only increases the likelihood of higher welfare under entry rather than eliminating the possibility of lower welfare under entry. While they have considered the situation of a closed economy, Mukherjee and Mukherjee [7] show the welfare implications of entry in the presence of technology licensing in an open economy. If either the entrant or the incumbent has a relatively superior technology[3], it creates the possibility of technology licensing. Mukherjee and Mukherjee [7] show that if there is licensing with upfront fixed fee, entry

---

[1] Lahiri and Ono [5] have considered welfare implications of cost reduction in a technologically ineffi-cient firm. Although they did not consider welfare implications of entry directly, their analysis provides conclusions similar to Klemperer [4].

[2] Fauli-Oller and Sandonis [3] have examined welfare implications of horizontal merger and technology licensing rather than addressing welfare implications of entry. However, their analysis of horizontal merger can be viewed as the opposite situation of entry.

[3] In our analysis, technology is defined by the marginal cost of production. Lower marginal cost implies better technology.

CP1067, *Applications of Mathematics in Engineering and Economics '34—AMEE '08*, edited by M. D. Todorov
© 2008 American Institute of Physics 978-0-7354-0598-1/08/$23.00

of a foreign firm not only increases domestic welfare when the foreign firm is sufficiently technologically superior to the domestic firm[4], it also increases domestic welfare if the foreign firm's technological inferiority is neither very small nor very large. However, if there is licensing with output royalty, foreign entry increases domestic welfare when the foreign firm is either sufficiently technologically superior or sufficiently technologically inferior to the domestic firm. So, the presence of technology licensing significantly affects the result of Collie [1], which considers the welfare effect of foreign entry without licensing. Mukherjee and Mukherjee [7]'s results also differ from the case of domestic entry with licensing that follows from Faulí-Oller and Sandonis [3]. Hence, their results have important implications for competition policies and show that the policymakers need to be concerned about the technological efficiency of the foreign firm and the type of licensing contract (i.e., fixed-fee or royalty licensing) available to the firms. In this paper, we study the same problem, in the case where the entrant is technologically inferior to the incumbent, by considering a more general inverse market demand function.

THE MODEL AND THE RESULTS

Consider a country, called the domestic country, with the inverse market demand function

$$P = a - bq,$$

where P is the price, q the quantity in the market, $a > 0$ the demand intercept and $b \geq 1$ the slope parameter from the demand curve. We assume that there is a monopolist in the domestic country, called incumbent. To study the implications of entry, we will consider the following two situations in our analysis: (i) where the incumbent is a monopolist in the domestic country; and (ii) where a foreign firm, called entrant, enters the market and competes with the incumbent. We suppose that the entrant is technologically inferior to the incumbent.

The Case of a Monopoly

Let us first consider the situation where the incumbent is a monopolist in the domestic country. We assume that the incumbent can produce a good with the constant marginal production cost c_1. For simplicity, we assume that there is no other production cost. The incumbent maximizes the following objective function to maximize its profit: $\max_q (a - bq - c_1)q$. Optimal output of the incumbent is $q^* = (a - c_1)/2b$ and its profit and consumer surplus are, respectively, $(a - c_1)^2/4b$ and $(a - c_1)^2/8$. Therefore, in the monopoly case, welfare W^m of the domestic country, which is the summation of

---

[4] If the foreign firm is technologically superior to the domestic firm, fixed-fee licensing increases the range of cost differences for which entry increases domestic welfare compared with the situation without licensing.

consumer surplus and profit of the incumbent, is given by

$$W^m = \frac{(2+b)(a-c_1)^2}{8b}.$$

Entry without Licensing

To show the implications of licensing, let us first consider entry of a foreign firm without licensing. Assume that there is a foreign firm, called entrant, who can produce the good at the constant marginal production cost c_2 that can be greater than, less than or equal to the marginal cost c_1 of the domestic firm. We assume that this difference in the production cost is due to the technological differences between the firms. We also assume that there is no other production cost of the entrant, namely we assume that there is no transportation costs and/or tariff. In our stylized framework, we assume that the entrant exports its product to the domestic country and the firms (the incumbent and the entrant) compete like Cournot duopolists with homogeneous products. We assume that, in the case of entry, both firms always produce positive outputs, i.e., $\max\{0, 2c_1 - a\} < c_2 < (a+c_1)/2$. The incumbent and the entrant choose their outputs to maximize, respectively, the following expressions:

$$\max_{q_1}(a - bq_1 - bq_2 - c_1)q_1 \quad \text{and} \quad \max_{q_2}(a - bq_1 - bq_2 - c_2)q_2,$$

where q_1 and q_2 are the outputs of the incumbent and the entrant, respectively. Optimal outputs of the incumbent and the entrant are, respectively, $(a - 2c_1 + c_2)/3b$ and $(a - 2c_2 + c_1)/3b$. Profits of the incumbent, the entrant and consumer surplus are $(a - 2c_1 + c_2)^2/9b$, $(a - 2c_2 + c_1)^2/9b$ and $(2a - c_1 - c_2)^2/18$, respectively. So, in the case of entry without licensing, domestic welfare W_{nl}^e is given by

$$W_{nl}^e = \frac{(a - 2c_1 + c_2)^2}{9b} + \frac{(2a - c_1 - c_2)^2}{18}. \tag{1}$$

Proposition 1 *Assume that there is no possibility of technology licensing between the firms.*

- *For $b > 2$, entry always increases domestic welfare.*
- *For $1 \le b \le 2$, if $c_1 > \frac{(10 - 7b)a}{14 - 5b}$ and $c_2 < \frac{(7b - 10)a + (14 - 5b)c_1}{2(b+2)}$, entry increases domestic welfare. It reduces welfare otherwise.*

Proof: The result follows, since $W_{nl}^e > W^m$ is equivalent to

$$c_2 < \min\left\{\frac{(7b - 10)a + (14 - 5b)c_1}{2(b+2)}, \frac{a+c_1}{2}\right\} = \begin{cases} \frac{a+c_1}{2}, & \text{if } b > 2 \\ \frac{(7b-10)a+(14-5b)c_1}{2(b+2)}, & \text{if } 1 \le b \le 2 \end{cases}$$

and, by assumption, $c_2 < (a+c_1)/2$. ∎

Entry with Licensing

Now, we are going to analyse the case of the entry under licensing. We consider two important types of licensing contracts (see, for example, Wang [9]): (i) fixed-fee licensing, where the licenser charges an upfront fixed fee for its technology; and (ii) licensing with output royalty, where the licenser charges royalty per unit of output[5].

Recall that we are considering the situation where the entrant (foreign firm) is technologically inferior to the incumbent, i.e., $c_1 < c_2$. Think of the incumbent as a technology leader, who has patented several technologies. One of those may expire and induce entry of a firm. Alternatively, we may assume that the entrant has developed a new technology through its own R&D but, due to its lower R&D capability, it can generate a technology that is inferior to that of the incumbent[6]. This situation fits well if we consider the domestic country as a developed country and the entrant is from a developing country with a relatively inferior production technology.

We consider the following game under entry. At stage 1, the technologically superior incumbent decides whether to offer a take-it-or-leave-it licensing contract to the entrant and the entrant accepts the licensing contract, if it is not worse off under licensing compared with no licensing[7]. At stage 2, the firms compete à la Cournot.

Since the entrant is technologically inefficient, we need $c_2 < (a+c_1)/2$ to ensure that the entrant always produces positive output.

Fixed-fee Licensing

We have seen above that profits of the incumbent and the entrant under no licensing are, respectively, $(a - 2c_1 + c_2)^2/9b$ and $(a - 2c_2 + c_1)^2/9b$.

Now, consider the situation under licensing. If licensing occurs, both firms produce with c_1 since the incumbent charges an upfront fixed fee for its technology. Profits of the incumbent and the entrant are, respectively, $(a - c_1)^2/9b + F$ and $(a - c_1)^2/9b - F$, where F is the optimal licensing fee charged by the incumbent.

So, licensing is profitable, if the following two conditions are satisfied for the incumbent and the entrant, respectively (with at least one strict inequality):

$$\frac{(a-c_1)^2}{9b} + F \geq \frac{(a - 2c_1 + c_2)^2}{9b}, \tag{2}$$

$$\frac{(a-c_1)^2}{9b} - F \geq \frac{(a - 2c_2 + c_1)^2}{9b}. \tag{3}$$

---

[5] Licensing with fixed fee (output royalty) can be the optimal licensing contract under costless imitation (no imitation) even if we allow the licenser to use a two-part tariff under licensing (Rockett, [8]).

[6] Mills and Smith [6] provide strategic reasons for using different technologies by different firms.

[7] Under no entry, the incumbent does not license its technology since the firms produce homogeneous products and the incumbent is a monopolist.

Since the incumbent gives a take-it-or-leave-it offer to the entrant, the fixed fee makes the entrant indifferent between licensing and no licensing, i.e.,

$$F = \frac{(a-c_1)^2}{9b} - \frac{(a-2c_2+c_1)^2}{9b}.$$

So, we get the following result.

Lemma 1 *Licensing occurs if, and only if,*

$$\frac{2(a-c_1)^2}{9b} > \frac{(a-2c_2+c_1)^2 + (a-2c_1+c_2)^2}{9b}, \tag{4}$$

which is satisfied for $c_2 < (2a+3c_1)/5$, *where* $(2a+3c_1)/5 < (a+c_1)/2$.

Under fixed-fee licensing, the profit of the incumbent and consumer surplus are, respectively, $(2(a-c_1)^2 - (a-2c_2+c_1)^2)/9b$ and $2(a-c_1)^2/9$, if $c_2 < (2a+3c_1)/5$. However, the profit of the incumbent and consumer surplus are, respectively, $(a-2c_1+c_2)^2/9b$ and $(2a-c_1-c_2)^2/18b$ for $(2a+3c_1)/5 < c_2 < (a+c_1)/2$. So, domestic welfare $W_{l,f}^e$ under fixed-fee licensing is given by

$$W_{l,f}^e = \frac{(2b+2)(a-c_1)^2 - (a+c_1-2c_2)^2}{9b}.$$

Proposition 2 *Under fixed-fee licensing, entry increases domestic welfare for*

$$c_2 \in \left(c_1 + \frac{(4-\sqrt{14b-4})(a-c_1)}{8}, \frac{2a+3c_1}{5} \right).$$

Proof: Fixed fee licensing occurs for $c_2 < (2a+3c_1)/5$. Furthermore, we have that $W_{l,f}^e > W^m$ is equivalent to

$$c_1 + \frac{(4-\sqrt{14b-4})(a-c_1)}{8} < c_2 < c_1 + \frac{(4+\sqrt{14b-4})(a-c_1)}{8}.$$

Noting that

$$\frac{2a+3c_1}{5} < c_1 + \frac{(4-\sqrt{14b-4})(a-c_1)}{8},$$

we get the result. ∎

Licensing with Output Royalty

Now, consider licensing with per-unit output royalty, where the incumbent charges a per-unit output royalty for its technology. In that case of licensing, the effective marginal cost of the entrant is $c_1 + r$, where r is the optimal per-unit output royalty. The optimal outputs of the incumbent and the entrant are, respectively, $(a-c_1+r)/3b$ and

$(a-c_1-2r)/3b$. So, their profits are, respectively, $((a-c_1+r)^2+3r(a-c_1-2r))/9b$ and $(a-c_1-2r)^2/9b$. The incumbent maximizes the following expression to determine the optimal royalty rate:

$$\max_r \frac{(a-c_1+r)^2+3r(a-c_1-2r)}{9b} \qquad (5)$$

subject to the constraint $r \le c_2 - c_1$.[8] Maximizing (5) and ignoring the constraint $r \le c_2 - c_1$, the optimal royalty rate is $(a-c_1)/2$. Since $(a-c_1)/2$ is greater than $c_2 - c_1$, for all $c_2 < (a+c_1)/2$, the optimal per-unit output royalty becomes $c_2 - c_1$.

Optimal outputs and therefore profit of the entrant and consumer surplus are the same under licensing and no licensing, but profit of the incumbent increases by the royalty income, which is $(c_2-c_1)(a-2c_2+c_1)/3b$. This immediately implies that licensing with per-unit output royalty occurs for all values of $c_2 \in (c_1,(a+c_1)/2)$. So, domestic welfare $W_{l,r}^e$ under royalty licensing is given by

$$W_{l,r}^e = \frac{(a-2c_1+c_2)^2+3(c_2-c_1)(a-2c_2+c_1)}{9b} + \frac{(2a-c_1-c_2)^2}{18}.$$

Proposition 3 *In the case of licensing with per-unit output royalty,*

- *if $1 \le b < 10$, entry increases domestic welfare for $c_2 \in \left(\frac{(10-7b)a+5(b+2)c_1}{2(10-b)}, \frac{a+c_1}{2} \right)$;*
- *if $b \ge 10$, entry increases domestic welfare for $c_2 < \frac{a+c_1}{2}$.*

Proof: Licensing with output royalty occurs for all $c_2 \in \left(c_1, \frac{a+c_1}{2} \right)$. Furthermore, we have that $W_{l,r}^e > W^m$ is equivalent to

$$\frac{(10-7b)a+5(b+2)c_1}{2(10-b)} < c_2 < \frac{a+c_1}{2}$$

if $1 \le b < 10$, and it is equivalent to $c_2 < \frac{a+c_1}{2}$, if $b \ge 10$, which proves the result. ∎

CONCLUSIONS

We studied the effects of foreign entry on social welfare in the presence of licensing. We consider both fixed-fee licensing and royalty licensing, in the case where the foreign firm (the entrant) is technologically inferior to the domestic firm (the incumbent). We found that the welfare implications of entry depend upon the type of licensing contract and upon the values of the slope parameter of the demand curve.

---

[8] If $r > c_2 - c_1$, licensing will make the entrant worse off compared with no licensing and the licensing contract will be rejected.

ACKNOWLEDGMENTS

We thank ESEIG-Instituto Politécnico do Porto, Centro de Matemática da Universidade do Porto and the Programs POCTI and POSI by FCT and Ministério da Ciência, Tecnologia e do Ensino Superior for their financial support.

REFERENCES

1. D. Collie, *Recherches Economiques de Louvain* **62**, 191–202 (1996).
2. T. Cordella, *Recherches Economiques de Louvain* **59**, 355–363 (1993).
3. R. Faulí-Oller and J. Sandonis, *International Journal of Industrial Organization* **21**, 655–672 (2003).
4. P. Klemperer, *Journal of Industrial Economics* **37**, 159–165 (1988).
5. S. Lahiri and Y. Ono, *Economic Journal* **98**, 1199–1202 (1988).
6. D. Mills and W. Smith, *International Journal of Industrial Organization* **14**, 317–329 (1996).
7. A. Mukherjee and S. Mukherjee, *The Manchester School* **73**, 653–663 (2005).
8. K. Rockett, *International Journal of Industrial Organization* **8**, 559–574 (1990).
9. X. H. Wang, *Economics Letters* **60**, 55–62 (1998).

A Localization of the Zeros of Some Holomorphic Functions

T. Stoyanov

Economic University, Dept. of Mathematics, 177 Knyaz Boris Blvd., Varna 9002, Bulgaria

Abstract. In this article we localize the zeros of the derivatives of some holomorphic functions. Theorem 1 is generalization of the Gauss-Lucas theorem. Theorem 2 and Theorem 3 refer to real entire functions of finite genus. In Theorem 2 we examine the function $f(z) = z^n \exp(c + bz + az^2) \prod_{k=1}^{\infty}(1 - z/z_k)\exp(z/z_k)$, and for Theorem 3 this function is $f(z) = \exp(d + cz^m + bz^{m+1} + az^{m+2})\prod_{k=1}^{\infty} E_m(\frac{z}{z_k})$. Both functions are real entire. Under the conditions in Theorem 2 and Theorem 3 we state: the zeros of their derivatives are real. All results about zeros of entire functions and their derivatives have good applications in fields of physics, mechanics and chemistry. In the [5] and [6], we observe how then stability of the system depends of the zeros of some entire functions.

Keywords: Entire function, zeros of polynomials.
PACS: 02.30.-f

Definition 1 *By 'real entire function' will be meant an entire function, whose Maclaurine-series expansion has only real coefficients.*

Definition 2 $E_m(\zeta) = (1 - \zeta)\exp(\zeta + \frac{\zeta^2}{2} + ... + \frac{\zeta^m}{m})$ *is the m-th Weierstrass factor.*

Further all sequences $\{z_k\}_{k=1}^{\infty}$ will satisfy $\lim_{k\to\infty}|z_k| = \infty$. G is simply connected region in the complex plane.

Theorem 1 *Let* $0, \; z_k \in G \subseteq \mathbf{C}, k = 1, 2, ...,$ *and* $\sum_{k=1}^{\infty} 1/|z_k|$ *be convergent. If we put*

$\varphi(z) = cz^m \prod_{k=1}^{\infty}(1 - z/z_k)$, *then the zeros of* $\varphi'(z)$ *are in the convex hull* coG.

Proof: Let the zeros of $\varphi(z)$ are enumerated thus: $z_1 = z_2 = .. = z_m = 0$ and

$|z_{m+1}| \leq |z_{m+2}| \leq ... \leq |z_k| \leq ....$ Calculating $(\log\varphi(z))' = \varphi'(z)/\varphi(z) = \sum_{k=1}^{\infty} 1/(z - z_k)$, and if

CP1067, *Applications of Mathematics in Engineering and Economics '34—AMEE '08*, edited by M. D. Todorov
© 2008 American Institute of Physics 978-0-7354-0598-1/08/$23.00

z_0 is a zero of $\varphi'(z)$, then $\sum_{k=1}^{\infty} 1/(\overline{z_0} - \overline{z_k})$, i.e., $\sum_{k=1}^{\infty}(z_0 - z_k)/|z_0 - z_k|^2 = 0$. If we put

$$r = \sum_{k=1}^{\infty} 1/|z_0 - z_k|^2, \text{ then } rz_0 = \sum_{k=1}^{\infty} z_k/|z_0 - z_k|^2 = \sum_{k=1}^{n} z_k/|z_0 - z_k|^2 + \sum_{k=n+1}^{\infty} z_k/|z_0 - z_k|^2 \quad (1).$$

If $r_n = \sum_{k=1}^{n} 1/|z_0 - z_k|^2$ and $p_k = 1/\left[r_n|z_0 - z_k|^2\right]$, then by (1) we obtain

$$rz_0 = r_n \sum_{k=1}^{\infty} p_k z_k + \sum_{k=n+1}^{\infty} z_k/|z_0 - z_k|^2, \quad r_n \xrightarrow[n \to \infty]{} r. \text{ If we note } u_n = \sum_{k=1}^{n} p_k z_k,$$

$\upsilon_n = \sum_{k=n+1}^{\infty} z_k/|z_0 - z_k|^2$, then obviously $u_n \in coG$, and $\upsilon_n \xrightarrow[n \to \infty]{} 0$, thanks to the fact,

that $\sum_{k=1}^{\infty} 1/|z_k|$ is convergent (because $\upsilon_n = \sum_{k=n+1}^{\infty} z_k/|z_0 - z_k|^2$ is the tail of the series). If

we assume $z_0 \notin coG$, that is impossible because of $z_0 = (r_n/r)u_n + (1/r)\upsilon_n$, and

$u_n \xrightarrow[n \to \infty]{} z_0$ (We essentially use that $r_n \xrightarrow[n \to \infty]{} r$ and $\upsilon_n \xrightarrow[n \to \infty]{} 0$).

Corollary 1 *If $\varphi(x)$ satisfy the conditions of Theorem1 and zeros of $\varphi(x)$ belong to a line, then zeros of $\varphi'(x)$ belong to the some line.*

Theorem 2 *Let $f(z) = z^n \exp(c + bz + az^2)\prod_{k=1}^{\infty}(1 - z/z_k)\exp(z/z_k)$ be a real entire function, where n is non-negative integer, $z_k \neq 0, a, b, c \in R$ and $a \leq 0$ Then all the roots of $f'(z)$ are real.*

Proof: Let z_0 be such that $f'(z_0) = 0$ and $z_0 \notin R, z_0 = x_0 + iy_0, x_0, y_0 \in R, y_0 \neq 0$. Let us put

$$A = (\log f(z_0))' = f'(z_0)/f(z_0) = n/z_0 + b + 2az_0 + \sum_{k=1}^{\infty}\left[\frac{\overline{z_0} - \overline{z_k}}{|z_0 - z_k|^2} + 1/z_k\right]$$

$$= n\overline{z_0}/|z_0|^2 + b + 2az_0 + \sum_{k=1}^{\infty}\left[\frac{\overline{z_0} - \overline{z_k}}{|z_0 - z_k|^2} + 1/z_k\right]$$

$$= n(x_0 - iy_0)/|z_0|^2 + b + 2a(x_0 + iy_0) + \sum_{k=1}^{\infty}\left[\frac{\overline{z_0} - \overline{z_k}}{|z_0 - z_k|^2} + 1/z_k\right] = 0.$$

Then $\operatorname{Im} A = -y\left[n/|z_0|^2 - 2a + \sum_{k=1}^{\infty} 1/|z_0 - z_k|^2\right] = 0$. But according to the condition

$n/|z_0|^2 - 2a + \sum_{k=1}^{\infty} 1/|z_0 - z_k|^2 > 0$, which is impossible.

Theorem 3 *Let* $f(z) = \exp(d + cz^m + bz^{m+1} + az^{m+2})\prod_{k=1}^{\infty} E(\frac{z}{z_k})$ *be an real entire function, where m is an even integer, $z_k, a, b, c, d \in R, a \leq 0, c \geq 0$. Denote by $E_m(\zeta) = (1 - \zeta)\exp(\zeta + \frac{\zeta^2}{2} + \ldots + \frac{\zeta^m}{m})$ the m-th Weierstrass factor. Then we state: all the zeros of $f'(z)$ are real.*

Proof: Let z_0 be such that $f'(z_0) = 0$ and $z_0 \notin R, z_0 = x_0 + iy_0, x_0, y_0 \in R, y_0 \neq 0$. Let us put

$$A = (\log f(z_0))' = f'(z_0)/f(z_0)$$

$$= mcz_0^{m-1} + (m+1)bz_0^m + (m+2)az_0^{m+1} + \sum_{k=1}^{\infty}(\frac{1}{z_0 - z_k} + \frac{1}{z_k} + \ldots + \frac{z_0^{m-1}}{z_k^m})$$

$$= mcz_0^m\overline{z_0}/|z_0|^2 + (m+1)bz_0^m + (m+2)az_0^{m+1} + \sum_{k=1}^{\infty}\frac{(\frac{z_0}{z_k})^m}{z_0 - z_k}$$

$$= z_0\left\{ mc\overline{z_0}/|z_0|^2 + (m+1)b + (m+2)az_0 + \sum\frac{\overline{z_0 - z_k}}{|z_0 - z_k|^2}\right\} = 0$$

Let $B = mc\overline{z_0}/|z_0|^2 + (m+1)b + (m+2)az_0 + \sum_{k=1}^{\infty}\frac{\overline{z_0 - z_k}}{z_k^m|z_0 - z_k|^2}$.

Then $\operatorname{Im}B = -y_0\left\{ mc/|z_0|^2 - (m+2)a + \sum_{k=1}^{\infty}\frac{1}{z_k^m|z_0 - z_k|^2}\right\}$. But m is an even integer,

$a \leq 0, c \geq 0$, therefore $mc/|z_0|^2 - (m+2)a + \sum_{k=1}^{\infty}\frac{1}{z_k^m|z_0 - z_k|^2} > 0$, which is impossible since

$\operatorname{Im}B = 0$. This contradiction confirms the assertion.

Remark 1 *If all the roots z_k are non-negative, then it is not necessary m to be even. The assertion of Theorem 3 holds whatever the non-negative integer m be.*

Remark 2 *If $a \geq 0, c \leq 0, z_k \leq 0$ and m is odd, then the assertion of Theorem 3 is fulfilled again.*

REFERENCES

1. G. Polya and G. Szego, *Problems and Theorems in Analysis*, Vol.1, Nauka i Izkustvo, Sofia (1973). (Bulgarian translation)
2. M. A. Titchmarsh, *Theory of Functions*, Nauka, Moscow (1980). (in Russian)
3. T. Stoyanov, *J. of Australian Math. Society (Series A)* **68**, 165-169 (2000).
4. T. Stoyanov, *J. of Australian Math. Society (Series A)* **72**, 87-91 (2002).
5. U.I. Neimark, *The Stability of Linear Systems*, Gostehizdat, Leningrad, 1949. (in Russian)
6. M.M. Postnikov, *Stable Polynomials*, Nauka, Moscow, 1981. (in Russian)

DIFFERENTIAL EQUATIONS

Approximate Analytic Solutions of the Lotka-Volterra System

A.O. Antonova

National Aviation University, 1Kosmonavta Komarova Ave., 03058 Kyiv, Ukraine

Abstract. The equation for the orbit of the Lotka-Volterra system is approximately solved for one of two variables in terms of the other by using the Pade approximant of order (2, 2) for the function $e^x - x - 1$. The approximate formulas are found to describe the time behavior of the solutions. A comparison with numerical calculations is demonstrated.

Keywords: Lotka-Volterra system, Hamiltonian, Pade approximation, period.
PACS: 02.30.Hqh

INTRODUCTION

The Lotka-Volterra (LV) predator-prey model consists of the system of ODE's

$$\frac{dx}{dt} = x(-a + by), \qquad \frac{dy}{dt} = y(-cx + d) \tag{1}$$

where $x(t) > 0$ and $y(t) > 0$ are the sizes of the prey and predator populations at the time t; a, b, c and d are positive constants. The importance of equations (1) can be found with its applications in population dynamics, chemical dynamics, economics [1-3].

The solutions of equations (1) are periodic functions. In the xy-plane they are represented by a family of the nesting closed curves surrounding the fixed point $(d/c; \ a/b)$ [4]. The predator-prey oscillations are nonlinear and the period of oscillation increases with increasing its amplitude [5, 6, 8].

Formally the solution of equations (1) can be written in terms of quadrature [7, 8]. However this form of solution is too complicated for the practice. If amplitudes of the predator-prey oscillations are small, then the perturbation methods can be used to solve the LV system [9, 10].

In this paper, we present an improved perturbation method which is based on the Waldvogel approach to analysis of the LV system [5, 6].

APPROXIMATE LOTKA-VOLTERRA SYSTEM

We begin with a brief introduction of Waldvogel's approach [5, 6]. It is convenient to introduce the coordinates u and v by

$$x = \frac{d}{c}e^u, \quad y = \frac{a}{b}e^v.$$

CP1067, *Applications of Mathematics in Engineering and Economics '34—AMEE '08*, edited by M. D. Todorov
© 2008 American Institute of Physics 978-0-7354-0598-1/08/$23.00

Then (1) transforms into a Hamiltonian system

$$\frac{du}{dt} = a(e^v - 1), \qquad \frac{dv}{dt} = -d(e^u - 1) \tag{2}$$

with the Hamiltonian

$$H = ah(v) + dh(u) \tag{3}$$

where

$$h = e^u - u - 1 .$$

From equations (2) and (3), one can check that H remains constant along the trajectories and is defined by the initial conditions $u(0) = u_0$, $v(0) = v_0$:

$$H = ah(v_0) + dh(u_0) .$$

Waldvogel defines two increasing functions $G_1(u), G_2(v)$ according to the relationships

$$G_1^2 = 2dh(u), \quad G_2^2 = 2ah(v)$$

and introduces new coordinates ξ, η by

$$\xi = G_1(u), \quad \eta = G_2(v),$$
$$u = g_1(\xi), \quad v = g_2(\eta) \tag{4}$$

where g_1, g_2 are the inverse of G_1, G_2. equations (2) in ξ, η coordinates reads [5, 6]

$$\frac{d\xi}{dt} = \frac{\eta}{g_1'(\xi)g_2'(\eta)}, \qquad \frac{d\eta}{dt} = -\frac{\xi}{g_1'(\xi)g_2'(\eta)} \tag{5}$$

equation (3) reduces to

$$\xi^2 + \eta^2 = 2H . \tag{6}$$

Our aim is to construct a simple approximate expressions for g_1, g_2. It can be shown that the (2,2) Pade approximant of h, denoted by h_P, can be expressed as

$$h \approx h_P = \frac{u^2}{2}\left(1 - \frac{u}{6}\right)^{-2} . \tag{7}$$

The approximation (7) is superior to the 4-term Taylor series expansion of h, denoted by h_T

$$h_T = \frac{u^2}{2} + \frac{u^3}{6} + \frac{u^4}{24} .$$

Indeed, from Taylor's formula, we obtain

$$h(u) - h_T(u) = \frac{u^5}{120} + o(u^6)$$

whereas

$$h(u) - h_P(u) = -\frac{u^5}{1080} + o(u^6) .$$

The functions $h(u), h_P(u), h_T(u)$ and their derivatives are shown in Figure 1. From Figure 1 we determine the range of h where $h = h_P$:

$$0 \leq h \leq h_P^* \approx 2.5 .$$

The approximation $h = h_T$ is valid only in the range $0 \le h \le h_T^* \approx 1$.

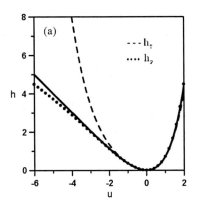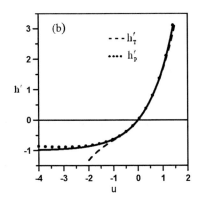

FIGURE 1. a) Functions $h(u)$ (solid line), $h_T(u)$ (dashed line) and $h_P(u)$ (dotted line). b) Derivatives $h'(u)$ (solid line), $h_T'(u)$ (dashed line) and $h_P'(u)$ (dotted line)

Using (7), we have

$$H = \frac{av^2}{2}\left(1-\frac{v}{6}\right)^{-2} + \frac{du^2}{2}\left(1-\frac{u}{6}\right)^{-2},$$

$$G_1 = \sqrt{d}\,u\left(1-\frac{u}{6}\right)^{-1}, \quad G_2 = \sqrt{a}\,v\left(1-\frac{v}{6}\right)^{-1},$$

$$g_1 = \frac{\xi}{\sqrt{d}}\left(1+\frac{\xi}{6\sqrt{d}}\right)^{-1}, \quad g_2 = \frac{\eta}{\sqrt{a}}\left(1+\frac{\eta}{6\sqrt{a}}\right)^{-1}.$$

We introduce the variables

$$U = \frac{\xi}{\sqrt{d}}, \quad V = \frac{\eta}{\sqrt{a}}$$

in place of ξ and η. Then we obtain

$$U = u\left(1-\frac{u}{6}\right)^{-1}, \quad V = v\left(1-\frac{v}{6}\right)^{-1}, \quad u = U\left(1+\frac{U}{6}\right)^{-1}, \quad v = V\left(1+\frac{V}{6}\right)^{-1}. \qquad (8)$$

Substituting g_1, g_2 into (5) gives

$$\frac{dU}{dt} = aV\left(1+\frac{U}{6}\right)^2\left(1+\frac{V}{6}\right)^2, \quad \frac{dV}{dt} = -dU\left(1+\frac{U}{6}\right)^2\left(1+\frac{V}{6}\right)^2. \qquad (9)$$

We call (9) as the approximate LV system.

The explicit equation of trajectories in the uv-plane is given by

$$v = \begin{cases} s\left(1+\dfrac{s}{6}\right)^{-1} & \text{for } v \ge 0, \\[2ex] -s(u)\left(1-\dfrac{s}{6}\right)^{-1} & \text{for } v < 0, \end{cases} \qquad s(u) = \sqrt{\frac{2H}{a} - \frac{d}{a}u^2\left(1-\frac{u}{6}\right)^{-2}}.$$

335

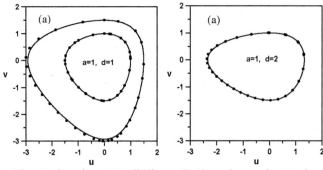

FIGURE 2. a) Exact trajectories $v(u)$ (solid lines) (Eq.(3)) and approximate trajectories (dotted lines) (Eqs. (12)) for $H = 0.72, H = 1.98$ and $a = d = 1$. b) The same for $H = 0.72$ and $a = 1, d = 2$

In the UV-plane the trajectories are a family of concentric ellipses centered on the fixed point $(0;0)$

$$aU^2 + dV^2 = 2H .\tag{10}$$

With the parameterization

$$U = U_0 \sin\psi, \quad V = V_0 \cos\psi, \quad U_0 = \sqrt{\frac{2H}{d}}, \quad V_0 = \sqrt{\frac{2H}{a}}\tag{11}$$

equations (8) take the form

$$u = U_0 \sin\psi \left(1 + \frac{U_0 \sin\psi}{6}\right)^{-1}, \quad v = V_0 \cos\psi \left(1 + \frac{V_0 \cos\psi}{6}\right)^{-1} .\tag{12}$$

Substituting $U(\psi)$ and $V(\psi)$ into (9) gives

$$\frac{d\psi}{dt} = \omega_0 F(\psi) .\tag{13}$$

Here

$$F = \left(1 + \frac{V_0 \cos\psi}{6}\right)^2 \left(1 + \frac{U_0 \sin\psi}{6}\right)^2, \quad \omega_0 = \sqrt{ad} .$$

Equation (13) shows that $\dfrac{d\psi}{dt} > 0$ for all ψ and so the trajectories of (9) spiral outwards as t increases. Letting $\psi(0) = \psi_0 = \tan^{-1}\dfrac{U(0)}{V(0)}$, we get the solution

$$\omega_0 t = \int_{\psi_0}^{\psi} F^{-1}(\xi) d\xi .\tag{14}$$

Thus $U(t)$ and $V(t)$ oscillate in phase with period

$$T = \frac{1}{\omega_0} \int_0^{2\pi} \left(1 + \frac{V_0 \cos\psi}{6}\right)^{-2} \left(1 + \frac{U_0 \sin\psi}{6}\right)^{-2} d\psi .$$

The direct numerical simulations demonstrate that the solutions of the approximate LV system (9) are close to the solutions of equations (2) if

$$\frac{H}{\min(a,d)} \le 2 .$$

Some results of the simulations are shown in Figures 2 and 3.

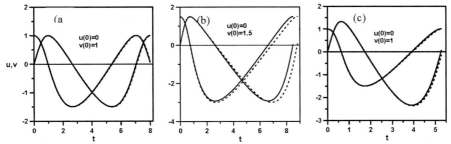

FIGURE 3. a) Exact solutions $u(t)$, $v(t)$ (solid lines) (Eqs. (2)) and approximate solutions (dotted lines) (Eqs. (12), (13)) for: a) $u(0) = 0$, $v(0) = 1$, $a = d = 1$; b) $u(0) = 0$, $v(0) = 1.5$, $a = d = 1$; and c) $u(0) = 0$, $v(0) = 1$, $a = 1$, $d = 2$

We observe that the agreement is very good.

DISCUSSION

It may be possible to evaluate the integral (14) analytically. But the expression for $t(\psi)$ is too cumbersome for use. Therefore it is useful to deduce an approximate analytical formulas for $\psi(t)$ and $U(t)$, $V(t)$. We shall carry out them below.

For our purpose here it will be necessary to assume that

$$\max(|U_0|, |V_0|) << 6 . \qquad (15)$$

The linearized problem (13) ($F^{-1}(\psi) = 1$) subject to the initial condition $\psi(0) = \psi_0$ has the solution

$$\psi = \psi_0 + \omega_0 t .$$

Hence the trajectory of the linearized problem is an ellipse with the period $2\pi / \omega_0$ in the UV -plane:

$$U_{lin} = U_0 \sin(\omega_0 t + \psi_0), \quad V_{lin} = V_0 \cos(\omega_0 t + \psi_0).$$

The harmonic oscillations in $U_{lin}(t)$, $V_{lin}(t)$ are symmetric. But the corresponding uv oscillations

$$u_{lin} = U_{lin}(t)\left(1 - \frac{U_{lin}(t)}{6}\right)^{-1}, \quad v_{lin} = V_{lin}(t)\left(1 - \frac{V_{lin}(t)}{6}\right)^{-1}$$

are non harmonic and the trajectory is asymmetric in the uv - plane.

Now we consider the effect of a small amplitude oscillation in (13). For definiteness we suppose that $k = \dfrac{a}{d} \leq 1$, $\psi_0 = 0$ and expand F^{-1} as

$$F^{-1} \approx f_0 + \varepsilon f_1 + \varepsilon^2 f_2 + \varepsilon^3 f_3, \qquad \varepsilon = V_0 / 6.$$

Using this expansion in (13) we get the solution

$$\omega_0 t = C_\psi \psi + \sum_{n=1}^{3} \varepsilon^n (A_n \cos n\psi + B_n \sin n\psi - A_n)). \tag{16}$$

Here

$$C_\psi = \pi \left(2 + 3\varepsilon^2 (1 + k^2) \right),$$

$$A_1 = \frac{k}{2} \left(4 + 6k^2 \varepsilon^2 + 3\varepsilon^2 \right), \quad A_2 = -k, \quad A_3 = \frac{k}{6} \left(3 - 2k^2 \right),$$

$$B_1 = -\frac{1}{2} \left(4 + 3k^2 \varepsilon^2 + 6\varepsilon^2 \right), \quad B_2 = -\frac{3}{4} \left(k^2 - 1 \right), \quad B_3 = \frac{1}{6} \left(-2 + 3k^2 \right).$$

Equations (9) can be solved by expanding $U(t)$ and $V(t)$ in the small parameter ε and using one of the averaging methods. We have applied the Lindstedt-Poincaré method to find the periodic solutions of equations (9). For $U(0) = 0$, $V(0) = V_0$ we have shown that $U(t)$ and $V(t)$ are given by the following:

$$U = kV_0 \sin\phi + \gamma_{U2} V_0^2 + \gamma_{U3} V_0^3, \tag{17}$$

$$V = V_0 \cos\phi + \gamma_{V2} V_0^2 + \gamma_{V3} V_0^3. \tag{18}$$

In these formulae

$$\phi = \omega t, \quad \omega = \omega_0 \left(1 - \frac{3}{2} \varepsilon^2 (1 + k^2) \right),$$

$$\gamma_{U2} = k \frac{[\sin 2\phi - k(\cos 2\phi + 2\cos\phi - 1)]}{6},$$

$$\gamma_{V2} = \frac{k \sin 2\phi + \cos 2\phi - 2k \sin\phi - 1}{6},$$

$$\gamma_{U3} = -\frac{k\left[(7 + 25k^2)\sin\phi + 12k\cos\phi - 32k(k\sin 2\phi + \cos 2\phi) - 9(k^2 - 1)\sin 3\phi + 20k\cos 3\phi \right]}{288},$$

$$\gamma_{V3} = \frac{\left[4\sin\phi - (23k^2 + 9)\cos\phi - 32k(\sin 2\phi - k\cos 2\phi) + 9(k^2 - 1)\cos 3\phi + 20k\sin 3\phi \right]}{288}.$$

The approximate model (9) may be used to obtain some important information about the properties of the LV solution. As an example consider any periodic solution of the LV system. In the phase plane the corresponding $v(u)$ contour is shown schematically in Figure 4.

This curve is divided into four segments (branches) separated by the points E, N, W and S. For definiteness we suppose that

$$v(0) = v_0 = v_{\max} > 0, \quad u(0) = 0.$$

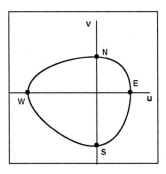

FIGURE 4. Sketch the trajectory in the phase plane

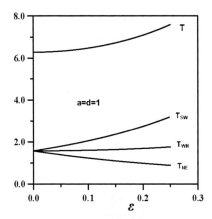

FIGURE 5. Times T_{NE}, T_{WN}, T_{SW} and T versus ε for $a = d = 1$

It then follows from equations (8) that

$$V(0) = V_0 = v_0 \left(1 - \frac{v_0}{6}\right)^{-1}, \quad U(0) = 0.$$

Therefore the coordinates of E, N, W and S are

$$E(u_{max}; 0), \quad N(0; v_{max}), \quad W(u_{min}; 0), \quad S(0; v_{min})$$

where

$$u_{max} = kV_0 \left(1 + \frac{kV_0}{6}\right)^{-1}, \quad u_{min} = -kV_0 \left(1 - \frac{kV_0}{6}\right)^{-1}, \quad v_{min} = -V_0 \left(1 - \frac{V_0}{6}\right)^{-1}.$$

Let the phase point $(u; v)$ pass through the points E, N, W, S at the time t_E, t_N, t_W, t_S. We write $T_{NE} = t_N - t_E$ ("NE traveling time"), etc. Using equation (16) with $n = 2$, we obtain the following expressions:

$$T_{NE} = \frac{T}{4} - \frac{2\left[(1+k)\varepsilon - k\varepsilon^2\right]}{\omega_0}, \quad T_{WN} = \frac{T}{4} + \frac{2\left[(1-k)\varepsilon - k\varepsilon^2\right]}{\omega_0},$$

339

$$T_{SW} = \frac{T}{4} + \frac{2\left[(1+k)\varepsilon + k\varepsilon^2\right]}{\omega_0}, \qquad T_{ES} = \frac{T}{4} - \frac{2\left[(1-k)\varepsilon + k\varepsilon^2\right]}{\omega_0}.$$

Here T is the approximate period of the nonlinear predator-prey oscillations,

$$T = \frac{2\pi}{\omega_0}\left(1 + \frac{3}{2}\varepsilon^2(1+k^2)\right).$$

The graphs of T_{NE}, $T_{WN} = T_{ES}$, T_{SW} and T for $a = d = 1$ are shown in Figure 5.

CONCLUSIONS

We have presented the useful explicit approximations to the solutions of the Lotka-Volterra predator-prey system. These formulae are valid for the significant deviations from the fixed point.

We also find the approximate explicit expression for the orbit in the phase plane and for the traveling times along its branches.

ACKNOWLEDGMENTS

It is a pleasure to thank Dr. S. Reznik for stimulating discussion and his assistance in carrying out the computations.

REFERENCES

1. J. D. Murray, *Lectures on Nonlinear Differential-Equation Models in Biology*, Clarendon Press, Oxford, 1977, pp. 135-139.
2. A. D. Bazykin, *Mathematical Biophysics of Interacting Populations*, Nauka, Moscow, 1985, pp. 23-26. (in Russian).
3. R. M. Goodwin, "A growth cycle," in *Socialism, Capitalism and Economic Growth*, edited by C. H. Feinstein, Cambridge Univ. Press, 1967, pp. 54–58.
4. D.K. Arrowsmith and C.M. Place, *Dynamical Systems. Differential Equations, Maps and Chaotic Behavior*, Chapman & Hall, London, 1992, pp. 183-185.
5. J. Waldvogel, *SIAM J.Numer. Anal.* **20**, 1264-1272 (1983).
6. J. Waldvogel, *J. Math. Anal. Appl.* **114**, 178-184 (1986).
7. R. M. Corless, G. H. Gonnet, D. E. G. Hare, D. J. Jeffrey, and D. E. Knuth, *Adv. Comput. Math.* **5**, 329-359 (1996).
8. S.-D. Shih, *Taiwanese J. Math.* **1**, 451-470 (1997).
9. R. Dutt, *Bull. Math. Biol.* 38, 459-465 (1976).
10. T. Grozdanovski and J. Shepherd, *ANZIAM Journal.* **49**, 243-257 (2007).

Global Solutions to the L^2–Critical Schrödinger–Poisson System

S. Avramska and G.P. Venkov

Faculty of Applied Mathematics and Informatics, Technical University of Sofia
1000 Sofia, Bulgaria

Abstract. The main purpose of the present paper is to study the global behavior of solutions to the L^2-critical Schrödinger–Poisson system

$$i\partial_t \psi(t,x) + \triangle \psi(t,x) = V(t,x)\psi(t,x), \quad (t,x) \in \mathbb{R}_+ \times \mathbb{R}^3,$$

$$\triangle V(t,x) = -4\pi |\psi(t,x)|^{\frac{8}{3}}, \quad \psi(0,x) = \psi_0(x).$$

More precisely, we shall establish the local existence of solutions for initial data ψ_0 in $L^2(\mathbb{R}^3)$, as well as the existence of global solutions for small initial data. Finally, we shall prove the existence of scattering operator.

Keywords: Schrödinger–Poisson system, Hartree equation, scaling symmetry, global well-posedness, wave and scattering operators.
PACS: 02.30.Em, 02.30.Jr, 02.30.Sa, 03.65.Nk

INTRODUCTION

In this paper we analyze the global properties of solutions to the Cauchy problem for the Schrödinger equation

$$i\partial_t \psi + \triangle \psi = V\psi, \quad (t,x) \in \mathbb{R}_+ \times \mathbb{R}^3, \tag{1}$$

$$\psi(0,x) = \psi_0(x), \quad x \in \mathbb{R}^3, \tag{2}$$

coupled to the Poisson equation

$$\triangle V = -4\pi |\psi|^{\frac{8}{3}}, \tag{3}$$

where the negative sign in the Poisson equation (3) corresponds to the repulsive type interaction. This Schrödinger–Poisson system governs the temporal evolution of the pure quantum mechanical state $\psi(t,x)$ associated with the particle and is closely related to the quantum transport in semiconductors, the classical modelling of electron gases and in the stationary case, to the Thomas–Fermi theory for matter in Coulomb field.

Let us note that the potential V will be a weak solution of the Poisson equation (3), which can be explicitly written as

$$V(t,x) = \int_{\mathbb{R}^3} \frac{|\psi(t,y)|^{\frac{8}{3}}}{|x-y|} dy. \tag{4}$$

CP1067, *Applications of Mathematics in Engineering and Economics '34—AMEE '08*, edited by M. D. Todorov
© 2008 American Institute of Physics 978-0-7354-0598-1/08/$23.00

Thus, substituting (4) in the righthand side of the Schrödinger equation (1), we obtain a single equation with respect to the wave function $\psi(t,x)$ of the form

$$i\partial_t \psi + \triangle \psi = (|x|^{-1} * |\psi|^{\frac{8}{3}})\psi, \quad (t,x) \in \mathbb{R}_+ \times \mathbb{R}^3, \tag{5}$$

where "$*$" denotes the usual convolution operator in $\mathbb{R}^3$. Equation (5) is known as the defocussing Schrödinger equation with nonlocal power nonlinearity of a Hartree type.

In order to obtain some initial information on the Schrödinger–Poisson system (1)–(3), let us consider the scaling symmetry of the system under the transformation

$$\psi_\lambda(t,x) = \frac{1}{\lambda^{3/2}}\psi(\frac{t}{\lambda^2}, \frac{x}{\lambda}), \quad V_\lambda(t,x) = \frac{1}{\lambda^2}V(\frac{t}{\lambda^2}, \frac{x}{\lambda}), \tag{6}$$

for $\lambda > 0$. Note that the above transformation rescales the L^p-norm

$$\|\psi_\lambda\|_{L^p(\mathbb{R}^3)} = \frac{1}{\lambda^{3(1/2-1/p)}}\|\psi\|_{L^p(\mathbb{R}^3)}. \tag{7}$$

A simple calculation shows that the scaling transformation (6) leaves the system unperturbed and, in addition, preserves the L^2-norm. In general, the scaling symmetry (6) relates (1)–(3) to the L^2-critical or mass-critical Schrödinger–Poisson systems. The name comes from the fact that the transform (6) leaves both the system and the mass (the L^2-norm of ψ) invariant. Mass is one of the basic structures used in physics, defined by

$$M(\psi(t)) = \int_{\mathbb{R}^3} |\psi(t,x)|^2 dx \tag{8}$$

and we shall prove that the mass for (1)–(3) is a conserved quantity, i.e., $M(\psi(t)) = M(\psi_0)$.

The main purpose of the present work is to consider the problems of local and global existence, well-posedness and scattering of solutions to the L^2-critical Schrödinger–Poisson system (1)–(3). As in the papers [6, 8], our results make use of mixed spaces of the type $L^q([0,T], L^r(\mathbb{R}^3))$ for Schrödinger–admissible q and r, i.e., for q and r satisfying

$$\frac{2}{q} = 3\left(\frac{1}{2} - \frac{1}{r}\right), \quad 2 \le q \le \infty. \tag{9}$$

Following the strategy developed in [1, 2, 12], we aim to establish the local well-posedness theory for (1)–(3) and to construct global solutions for sufficiently small L^2-initial data. More precisely, a function $\psi : [0,T^*) \times \mathbb{R}^3 \mapsto \mathbb{C}$, $0 < T^* \le \infty$ is a $L^2(\mathbb{R}^3)$ solution to (1)–(3) (and (5)) if $\psi \in C^0([0,T], L^2(\mathbb{R}^3)) \cap L^{14/3}([0,T], L^{14/5}(\mathbb{R}^3))$ for $0 < T < T^*$, and we have Duhamel's integral representation

$$\psi(t) = U(t)\psi_0 - i\int_0^t U(t-s)V(s)\psi(s)ds, \tag{10}$$

for any $t \in [0,T]$. Here $U(t) = e^{it\Delta}$ is the free Schrödinger evolution group defined via the Fourier transform

$$\hat{f}(\xi) = \frac{1}{(2\pi)^{3/2}} \int_{\mathbb{R}^3} e^{-ix\cdot\xi} f(x)dx, \tag{11}$$

by

$$\widehat{e^{it\Delta}f}(\xi) = e^{-it|\xi|^2} \hat{f}(\xi). \tag{12}$$

We say that ψ is a global solution if $T^* = \infty$.

The first main result of the present paper is the following

Theorem 1 *For every initial data $\psi_0 \in L^2(\mathbb{R}^3)$ there exists a unique maximal solution $\psi \in C^0([0,T^*), L^2(\mathbb{R}^3)) \cap L^{\frac{14}{3}}([0,T^*), L^{\frac{14}{5}}(\mathbb{R}^3))$ of (1)–(3). Furthermore:*

(i) $\psi \in L^q([0,T], L^r(\mathbb{R}^3))$, *for $0 < T < T^*$ and every admissible pair (q,r);*

(ii) *the mass is conserved, i.e., $M(\psi(t)) = M(\psi_0)$ for $t \in [0,T^*)$;*

(iii) *there exists a constant $\varepsilon > 0$ sufficiently small, such that if $\|\psi_0\|_{L^2(\mathbb{R}^3)} < \varepsilon$, then*
$$T^* = \infty \text{ and } \psi \in L^q(\mathbb{R}_+, L^r(\mathbb{R}^3)) \text{ for every admissible pair } (q,r);$$

(iv) *if $T^* < \infty$, then $\|\psi\|_{L^q([0,T^*), L^r(\mathbb{R}^3))} = \infty$ for every $r > 14/5$;*

(v) ψ *depends continuously on the initial data $\psi_0 \in L^2(\mathbb{R}^3)$ in the space*
$$\psi \in C^0([0,T^*), L^2(\mathbb{R}^3)) \cap L^{\frac{14}{3}}([0,T^*), L^{\frac{14}{5}}(\mathbb{R}^3)).$$

There exist an extensive literature on the scattering theory for the Schrödinger-Poisson system and for the Hartree equation [3, 4, 5], of which the existence of a wave operator is the question of crucial importance. Let $v(t) = U(t)\psi_+$ be a solution to the free Schrödinger equation

$$i\partial_t v + \Delta v = 0, \tag{13}$$

with initial data $\psi_+ \in X$ (called the asymptotic state), where $X = X_{\psi_0}$ is a suitable Banach space, depending on the initial data. That question can be formulated as follows. Does there exist a solution of (1)–(3), which behaves asymptotically as v when $t \to \infty$ in a suitable sense, depending on the choice of the space X? If that is the case, then the map $\Omega_+ : X \mapsto X$ is called the wave operator for positive times. In other words, a global strong X-solution ψ to (1)–(3) with an initial data ψ_0 scatters in X to a solution $v(t) = U(t)\psi_+$ if we have

$$\lim_{t\to\infty} \|\psi(t) - U(t)\psi_+\|_X = 0, \tag{14}$$

or equivalently (by using the unitarity of $U(t)$)

$$\lim_{t\to\infty} \|U(-t)\psi(t) - \psi_+\|_X = 0. \tag{15}$$

Suppose that for every asymptotic state $\psi_+ \in X$, there exists a unique initial data $\psi_0 \in X$, whose corresponding X-wellposed solution is global and scatters to $v(t)$ as $t \to \infty$. Then, we can define the wave operator $\Omega_+ : X \mapsto X$ in the sense of the space X by

$$\Omega_+ \psi_+ = \psi_0. \tag{16}$$

The problem of the existence of ψ for given ψ_+ is referred to as the problem of existence of the wave operator. When the wave operator Ω_+ is injective, we say that the Cauchy problem (1)–(3) is asymptotically complete in X. The same problem can be constructed for negative times, but for definiteness, hereinafter, we shall restrict our attention only to positive time.

A standard way to construct the wave operator Ω_+ consists in solving the Cauchy problem for (1)–(3) with initial data ψ_+ at $t = \infty$ in the form of the integral equation

$$\psi(t) = U(t)\psi_0 + i \int_t^\infty U(t-s)V(s)\psi(s)ds. \tag{17}$$

One usually solves (17) by a contraction method in a neighborhood of infinity in time (or in the time interval $[T,\infty)$ for T sufficiently large) and then continues that solution to all times. Thus, the problem is an immediate consequence of the global well-posedness and uses the results of Theorem 1.

With our second main result, we shall construct scattering theory in $L^2(\mathbb{R}^3)$ with small initial data. In fact, we shall prove the following

Theorem 2 *Let $\varepsilon > 0$ be sufficiently small and consider the ball $B_\varepsilon = \{\psi \in L^2(\mathbb{R}^3); \|\psi\|_{L^2} < \varepsilon\}$.*

Let $\psi \in C^0([0,T^*),L^2(\mathbb{R}^3)) \cap L^{\frac{14}{3}}([0,T^*),L^{\frac{14}{5}}(\mathbb{R}^3))$ be the unique maximal solution of (1)–(3), given by part (iii) of Theorem 1. Then the wave operators $\Omega_\pm$ exist and are asymptotically complete in B_ε. Moreover, the scattering operator $S = \Omega_+^{-1} \circ \Omega_-$ is a homeomorphism from B_ε onto itself and an isometry in the $L^2(\mathbb{R}^3)$ norm.

We shall conclude this section by giving some of the notations, used in the paper. As usual, $L^r(\mathbb{R}^n) = \{\varphi \in \mathscr{S}'; \|\varphi\|_{L^r} < \infty\}$, where $\|\varphi\|_{L^r} = (\int |\varphi(x)|^r dx)^{\frac{1}{r}}$ if $1 \le r < \infty$ and $\|\varphi\|_{L^\infty} = \text{ess.sup}\{|\varphi(x)|; x \in \mathbb{R}^n\}$ if $r = \infty$. We use r' for denoting the exponent dual to r and defined by $1/r + 1/r' = 1$. Given Lebesgue exponents q, r and a function $f(t,x)$ in $L^q(\mathbb{R},L^r(\mathbb{R}^3))$, we write $\|f\|_{L^q(\mathbb{R},L^r(\mathbb{R}^3))} = (\int \|f(t)\|_{L^r(\mathbb{R}^3)}^q dt)^{\frac{1}{q}}$.

THE LOCAL EXISTENCE RESULT

We shall start this section by collecting some preliminaries and useful results.

As we shall see, the $L_t^{\frac{14}{3}} L_x^{\frac{14}{5}}$ norm in space-time plays a fundamental role. This is better understood if we recall some of the estimates available for the corresponding linear problem. We begin by recalling the following properties of the free Schrödinger evolution group $U(t) = e^{it\triangle}$ (see, for instance [7, 8]).

Lemma 1 *Let (q,r) be an admissible pair. Then, for every $\varphi \in L^2(\mathbb{R}^3)$ the following estimate holds*

$$\|U(t)\varphi\|_{L^q(\mathbb{R},L^r(\mathbb{R}^3))} \le C_0 \|\varphi\|_{L^2(\mathbb{R}^3)}. \tag{18}$$

Moreover, for every admissible pair (θ, ρ) and $f \in L^{\theta'}([0,T], L^{\rho'}(\mathbb{R}^3))$ we have

$$\left\| \int_0^{\cdot} U(\cdot - s) f(s) ds \right\|_{L^q([0,T], L^r(\mathbb{R}^3))} \leq C \|f\|_{L^{\theta'}([0,T], L^{\rho'}(\mathbb{R}^3))}, \tag{19}$$

for $0 < T \leq \infty$. Here the constants $C_0, C > 0$ and depend only on the spatial exponents r and ρ.

The classical Strichartz estimates for the Schrödinger equation [11] are one of the main tools in the study of local and global existence, time decay and scattering both for the linear and the nonlinear equation, due to the fact that they fit the assumptions required by the contraction argument (see, for instance [3, 8, 11, 12]).

The Strichartz type estimates for the inhomogeneous Schrödinger equation

$$i\partial_t v(t,x) + \triangle v(t,x) = F(t,x), \quad v(0,x) = f(x), \tag{20}$$

in $\mathbb{R}_+ \times \mathbb{R}^3$ are given, up to the end-point, namely the pair $(q,r) = (2,6)$ by the following result due to Keel and Tao [8].

Lemma 2 *If (q,r) and $(\tilde{q}, \tilde{r})$ satisfy (9), then the solution to the Cauchy problem (20) satisfies the estimate*

$$\|v\|_{L^q([0,T^*), L^r(\mathbb{R}^3))} + \|v\|_{L^\infty([0,T^*), L^2(\mathbb{R}^3))}$$
$$\leq C(\|F\|_{L^{\tilde{q}'}([0,T^*), L^{\tilde{r}'}(\mathbb{R}^3))} + \|f\|_{L^2(\mathbb{R}^3)}). \tag{21}$$

Very important tools in our functional analysis background are the following lemma.

Lemma 3 *(Hardy-Littlewood-Sobolev Inequality) For $0 < \alpha < 3$ consider the Riesz potential*

$$I_\alpha(g)(x) = \int_{\mathbb{R}^3} \frac{g(y)}{|x-y|^{3-\alpha}} dy. \tag{22}$$

Then for any $1 < \theta < r < \infty$ and $g \in L^r(\mathbb{R}^3)$, we have

$$\|I_\alpha(g)\|_{L^\theta} \leq C \|g\|_{L^r}, \tag{23}$$

where $\frac{1}{\theta} = \frac{1}{r} - \frac{\alpha}{3}$.

For the proof of Lemma 3, see equation (31), Chapter VIII.4.2 in Stein [10].

The first important result in the present study is the following conservation law.

Lemma 4 *Let $\psi \in C^0([0,T^*), L^2(\mathbb{R}^3))$ be a solution to (1)–(3) with initial data $\psi(0) = \psi_0 \in L^2(\mathbb{R}^3)$. Then*

$$M(\psi(t)) = M(\psi_0), \tag{24}$$

for any $0 \leq t < T^*$.

Proof: We multiply (1) by $\overline{\psi}$ to get

$$i\overline{\psi}\partial_t \psi + \overline{\psi}\triangle \psi = V|\psi|^2. \tag{25}$$

We conjugate (25) and subtract the result from the above expression (note that V is real) to obtain

$$i\partial_t |\psi|^2 = (\psi \triangle \overline{\psi} - \overline{\psi} \triangle \psi) = \nabla \cdot (\psi \nabla \overline{\psi} - \overline{\psi} \nabla \psi). \tag{26}$$

By integration over $\mathbb{R}^3$ we obtain the desired result

$$\frac{d}{dt} M(\psi(t)) = \frac{d}{dt} \|\psi(t)\|^2_{L^2(\mathbb{R}^3)} = 0 \tag{27}$$

and the proof of the Lemma is completed.

The arguments of Theorem 1 rely primarily on the Strichartz estimate (21) in Lemma 2 and on the Hölder inequality.

Let us denote by

$$N(\psi) = (|x|^{-1} * |\psi|^{\frac{8}{3}}) \psi \tag{28}$$

the term in the righthand side of equation (1). Consider the form

$$V(\psi_1, \psi_2)(t, x) = |x|^{-1} * |\psi_1(t,x)|^\alpha |\psi_2(t,x)|^{\frac{8}{3} - \alpha}, \quad 0 \le \alpha \le \frac{8}{3}. \tag{29}$$

Then

$$N(\psi) = V(\psi, \psi) \psi$$

and we shall need the following estimates.

Lemma 5 *For any $3 < p < \infty$ and r_1, r_2 satisfying $\frac{1}{p} = \frac{\alpha}{r_1} + \frac{8/3 - \alpha}{r_2} - \frac{2}{3}, 0 \le \alpha \le \frac{8}{3}$ we have*

$$\|V(\psi_1, \psi_2)(t, \cdot)\|_{L^p(\mathbb{R}^3)} \le C \|\psi_1(t, \cdot)\|^\alpha_{L^{r_1}(\mathbb{R}^3)} \|\psi_2(t, \cdot)\|^{\frac{8}{3} - \alpha}_{L^{r_2}(\mathbb{R}^3)}. \tag{30}$$

Proof: It is sufficient to apply the Hardy-Littlewood-Sobolev inequality (23) and Hölder inequality to get

$$\||\cdot|^{-1} * |\psi_1(t,x)|^\alpha |\psi_2(t,x)|^{\frac{8}{3} - \alpha}\|_{L^p} \le C \|\psi_1(t, \cdot)\|^\alpha_{L^{r_1}} \|\psi_2(t, \cdot)\|^{\frac{8}{3} - \alpha}_{L^{r_2}},$$

provided

$$3 < p < \infty, \quad \frac{1}{p} = \frac{\alpha}{r_1} + \frac{8/3 - \alpha}{r_2} - \frac{2}{3}.$$

Lemma 6 *For any $\tilde{r}$ satisfying $2 \le \tilde{r} \le 6$ and for any $r_1, r_2, r_3, 2 \le r_j \le 6, j = 1, 2, 3$, satisfying*

$$\frac{1}{\tilde{r}'} = \frac{\alpha}{r_1} + \frac{8/3 - \alpha}{r_2} + \frac{1}{r_3} - \frac{2}{3}, \tag{31}$$

we have

$$\|V(\psi_1, \psi_2)(t, \cdot) \psi_3(t, \cdot)\|_{L^{\tilde{r}'}(\mathbb{R}^3)}$$
$$\le C \|\psi_1(t, \cdot)\|^\alpha_{L^{r_1}(\mathbb{R}^3)} \|\psi_2(t, \cdot)\|^{\frac{8}{3} - \alpha}_{L^{r_2}(\mathbb{R}^3)} \|\psi_3(t, \cdot)\|_{L^{r_3}(\mathbb{R}^3)}. \tag{32}$$

Proof: The Hölder inequality implies

$$\|V(\psi_1,\psi_2)(t,\cdot)\psi_3(t,\cdot)\|_{L^{\tilde{r}'}} \leq C\|V(\psi_1,\psi_2)(t,\cdot)\|_{L^p}\|\psi_3(t,\cdot)\|_{L^{r_3}},$$

where

$$\frac{1}{\tilde{r}'} = \frac{1}{p} + \frac{1}{r_3}.$$

Applying further the estimate of Lemma 5, we complete the proof of the Lemma.

The corresponding space-time estimate follows easily from the above estimate.

Lemma 7 *Let $0 < T \leq \infty$ and consider the Schrödinger-admissible pair $(q,r) = (14/3,14/5)$. Then, we have the estimate*

$$\|N(\psi)\|_{L^{\frac{14}{11}}([0,T],L^{\frac{14}{9}}(\mathbb{R}^3))} \leq C\|\psi\|_{L^{\frac{14}{3}}([0,T],L^{\frac{14}{5}}(\mathbb{R}^3))}^{\frac{11}{3}}. \tag{33}$$

Proof: Applying Hölder inequality in time to (32) in Lemma 6 we obtain

$$\|V(\psi_1,\psi_2)(t,\cdot)\psi_3(t,\cdot)\|_{L^{\tilde{q}'}([0,T],L^{\tilde{r}'})}$$

$$\leq C\|\psi_1(t,\cdot)\|_{L^{q_1}([0,T],L^{r_1})}^{\alpha}\|\psi_2(t,\cdot)\|_{L^{q_2}([0,T],L^{r_2})}^{\frac{8}{3}-\alpha}\|\psi_3(t,\cdot)\|_{L^{q_3}([0,T],L^{r_3})},$$

where $0 \leq \alpha \leq 8/3$ and

$$\frac{1}{\tilde{r}'} = \frac{\alpha}{r_1} + \frac{8/3-\alpha}{r_2} + \frac{1}{r_3} - \frac{2}{3}, \tag{34}$$

$$\frac{1}{\tilde{q}'} = \frac{\alpha}{q_1} + \frac{8/3-\alpha}{q_2} + \frac{1}{q_3}. \tag{35}$$

Now we shall choose the couples (q_j,r_j) so that

$$\frac{1}{q_j} = \frac{3}{2}\left(\frac{1}{2}-\frac{1}{r_j}\right), \quad j = 1,2,3 \tag{36}$$

and the relations (34), (35) are satisfied. Indeed, the simplest choice is

$$r = \tilde{r} = r_1 = r_2 = r_3, \quad q = \tilde{q} = q_1 = q_2 = q_3.$$

Then (34) reads as

$$\frac{1}{\tilde{r}'} = \frac{11}{3r} - \frac{2}{3}, \tag{37}$$

while (35) becomes

$$\frac{1}{\tilde{q}'} = \frac{11}{3q} \tag{38}$$

which together give the couple $(q,r) = (14/3,14/5)$ and the proof is completed.

Lemma 8 *Let $0 < T \leq \infty$ and let (q,r) be a Schrödinger-admissible pair. Then there exists a constant $C > 0$, independent of T such that*

$$\left\| \int_0^{\cdot} U(\cdot - s)[N(\psi)(s) - N(\chi)(s)]ds \right\|_{L^q([0,T],L^r)}$$

$$\leq C \left(\|\psi\|_{L^{\frac{14}{3}}([0,T],L^{\frac{14}{5}})}^{\frac{8}{3}} + \|\chi\|_{L^{\frac{14}{3}}([0,T],L^{\frac{14}{5}})}^{\frac{8}{3}} \right) \|\psi - \chi\|_{L^{\frac{14}{3}}([0,T],L^{\frac{14}{5}})}, \quad (39)$$

for every $\psi, \chi \in L^{\frac{14}{3}}([0,T], L^{\frac{14}{5}}(\mathbb{R}^3))$.

Proof: To prove the Lemma, we shall use the estimates in Lemma 1. Then the estimate (39) follows directly from (32), (33), (19) and Hölder inequality.

Proof of Theorem 1: The proof of (ii) follows from Lemma 4.

We shall prove the existence of solution to (1)–(3) by a fix point argument. Let $\psi_0 \in L^2(\mathbb{R}^3)$ with $\|\psi_0\|_{L^2(\mathbb{R}^3)} < \varepsilon$, where $\varepsilon > 0$ is sufficiently small. Let $R > 0$ and consider the ball

$$B_{2R}(T) = \{\psi \in C^0([0,T],L^2)) \cap L^{\frac{14}{3}}([0,T],L^{\frac{14}{5}}); \|\psi\|_{L^{\frac{14}{3}}([0,T],L^{\frac{14}{5}})} < 2R\},$$

endowed with the metric

$$d(\psi,\chi) = \|\psi - \chi\|_{L^{\frac{14}{3}}([0,T],L^{\frac{14}{5}})}.$$

Since the space $L^{\frac{14}{3}}([0,T],L^{\frac{14}{5}})$ is reflexive, the ball B_{2R} is weakly compact, implying B_{2R} is a complete metric space.

Consider the map $\Phi[\psi](t)$, defined by the right-hand side of the Duhamel's integral representation (10). Then, for $\psi \in B_{2R}$, using (18), (19) and (33), we can write

$$\|\psi\|_{L^{\frac{14}{3}}([0,T],L^{\frac{14}{5}})} \leq \|U(\cdot)\psi_0\|_{L^{\frac{14}{3}}([0,T],L^{\frac{14}{5}})} + C_1 \|\psi\|_{L^{\frac{14}{3}}([0,T],L^{\frac{14}{5}})}^{\frac{11}{3}}$$

$$\leq C_0 \varepsilon + C_1 \|\psi\|_{L^{\frac{14}{3}}([0,T],L^{\frac{14}{5}})}^{\frac{11}{3}}. \quad (40)$$

Now, write $R = C_0\varepsilon$ and choose $\varepsilon > 0$ so small that there exists a positive number y, satisfying $C_1 y^{\frac{11}{3}} - y + R < 0$, $0 < y \leq 2R$. For that purpose, it is sufficient to take

$$\varepsilon < \frac{1}{2C_0(2C_1)^{\frac{3}{8}}},$$

implying that $\Phi[\psi] \in C^0([0,T],L^2)) \cap L^{\frac{14}{3}}([0,T],L^{\frac{14}{5}})$.

On the other hand, from (39) it follows that

$$\|N(\psi) - N(\chi)\|_{L^{\frac{14}{11}}([0,T],L^{\frac{14}{9}})}$$

$$\leq C_2 \left(\|\psi\|_{L^{\frac{14}{3}}([0,T],L^{\frac{14}{5}})}^{\frac{8}{3}} + \|\chi\|_{L^{\frac{14}{3}}([0,T],L^{\frac{14}{5}})}^{\frac{8}{3}} \right) \|\psi - \chi\|_{L^{\frac{14}{3}}([0,T],L^{\frac{14}{5}})}$$

$$\leq C_2 2(2R)^{\frac{8}{3}} \|\psi - \chi\|_{L^{\frac{14}{3}}([0,T],L^{\frac{14}{5}})}, \qquad (41)$$

for every $\psi, \chi \in B_{2R}$. If we choose R such that

$$\varepsilon \leq \frac{1}{2C_0(4C_2)^{\frac{3}{8}}},$$

we finally obtain that the map $\Phi[\psi](t)$ is a strict contraction on the ball B_{2R}. Thus $\Phi[\psi]$ has a fixed point ψ, which is the unique solution of (1)–(3) in B_{2R}. So far, we have proved the statement of Theorem 1, as well as the part (i).

Notice, that the Strichartz estimate (21) implies a similar inequality, namely

$$\|\psi\|_{L^{\frac{14}{3}}([0,T],L^{\frac{14}{5}})} \leq C_0\|\psi_0\|_{L^2} + C_1\|\psi\|_{L^{\frac{14}{3}}([0,T],L^{\frac{14}{5}})}^{\frac{11}{3}} \qquad (42)$$

where we can take both the constants in (40) and (42) to be the same.

The second comment on the above proof is that from (40) and (42), the size of the ball B_{2R} depends directly on the size of the norm $\|U(t)\psi_0\|_{L^{\frac{14}{3}}([0,T],L^{\frac{14}{5}}(\mathbb{R}^3))}$ and it can be done small by taking either $\|\psi_0\|_{L^2(\mathbb{R}^3)}$ small or the interval $[0,T]$ small.

Let us denote by T^* the supremum of all $T > 0$ for which there exists a solution of (1)–(3) in $C^0([0,T],L^2(\mathbb{R}^3)) \cap L^{\frac{14}{3}}([0,T],L^{\frac{14}{5}}(\mathbb{R}^3))$. To prove (iii), observe that if ψ_0 is sufficiently small, then (40) holds regardless of the value of T. Thus we may accomplish the fixed point procedure in the ball $B_{2R}(\infty)$, providing $T^* = \infty$.

Further, we claim that if $T^* < \infty$, then $\|\psi\|_{L^q([0,T^*),L^r(\mathbb{R}^3))} = \infty$ for every $r > 14/5$. Indeed, on the contrary, let us assume that $T^* < \infty$ and $\|\psi\|_{L^{\frac{14}{3}}([0,T^*),L^{\frac{14}{5}})} < \infty$. For any $t \in [0,T^*)$ let $\tau \in [0, T^* - t)$. Using Duhamel's formula (10), we can write

$$\psi(t+\tau) = U(t+\tau)\psi_0 - i \int_0^{t+\tau} U(t+\tau-s)N(\psi)(s)ds$$

$$= U(\tau)U(t)\psi_0 - i \int_0^{t+\tau} U(t+\tau-s)N(\psi)(s)ds$$

$$= U(\tau)\psi(t) + iU(\tau)\int_0^t U(t-s)N(\psi)(s)ds - i\int_0^{t+\tau} U(t+\tau-s)N(\psi)(s)ds$$

$$= U(\tau)\psi(t) - i\int_t^{t+\tau} U(t+\tau-s)N(\psi)(s)ds. \quad (43)$$

From (43) and the estimate (39) in Lemma 8, we obtain

$$\|U(\cdot)\psi(t)\|_{L^{\frac{14}{3}}([0,T^*-t),L^{\frac{14}{5}})} \leq C\left(\|\psi(t)\|_{L^{\frac{14}{3}}([t,T^*),L^{\frac{14}{5}})} + \|\psi\|_{L^{\frac{14}{3}}([t,T^*),L^{\frac{14}{5}})}^{\frac{11}{3}}\right). \quad (44)$$

Observing now that $\|U(\cdot)\psi\|_{L^{\frac{14}{3}}([0,T],L^{\frac{14}{5}})} \to 0$ as $T \to 0$ and taking t close enough to T^*, it follows that $\|U(\cdot)\psi(t)\|_{L^{\frac{14}{3}}([0,T^*-t),L^{\frac{14}{5}})}$ can be made small enough and the assumptions in (iii) are fulfilled. Therefore, ψ can be extended after T^*, which contradicts the maximality. Let (q,r) be a Schrödinger-admissible pair with $r \geq \frac{14}{5}$. Then, from Hölder inequality, for $T < T^*$, we can write

$$\|\psi\|_{L^{\frac{14}{3}}([0,T],L^{\frac{14}{5}})} \leq \|\psi\|_{L^\infty([0,T],L^2)}^{1-\alpha}\|\psi\|_{L^q([0,T],L^r)}^{\alpha}, \quad \alpha \in (0,1). \quad (45)$$

Letting $T \to T^*$, we obtain that $\|\psi\|_{L^q([0,T],L^r)} = \infty$, which proves the statement (iv).

To prove (v), consider a sequence $\psi_0^k \in L^2(\mathbb{R}^3)$, such that $\psi_0^k \to \psi_0 \in L^2(\mathbb{R}^3)$ as $k \to \infty$. Thus, for k large enough, $\|\psi_0^k\|_{L^2(\mathbb{R}^3)} < \varepsilon$. We can use the Duhamel formula (10) to construct a sequence of solutions $\psi^k \in L^{\frac{14}{3}}([0,T],L^{\frac{14}{3}}(\mathbb{R}^3))$ to (5) with initial datum ψ_0^k. Applying the proof of (iii), we obtain that $\psi^k \to \psi$ in $C^0([0,T],L^2(\mathbb{R}^3)) \cap L^{14/3}([0,T],L^{14/5}(\mathbb{R}^3))$ as $k \to \infty$, and in fact in every $L^q([0,T],L^r(\mathbb{R}^3)))$ for (q,r) be an admissible pair. Thus, the proof of the Theorem is completed.

SMALL DATA SCATTERING THEORY IN $L^2(\mathbb{R}^3)$

In this section we shall prove Theorem 2. The arguments for proving the existence of the wave operator are standard and follows the exposition in [7, 9]. We shall prove only the $(+)$ case since the $(-)$ case can be proved similarly. Let $\psi_0 \in L^2(\mathbb{R}^3)$ with $\|\psi_0\|_{L^2} < \varepsilon$ and $\psi \in B_{2R}$, where the ball B_{2R} is defined in the previous section. Then, for $t > t_0$, using Duhamel's integral formula (10) we have

$$U(-t)\psi(t) = U(-t_0)\psi(t_0) - i\int_{t_0}^t U(-s)N(\psi)(s)ds. \quad (46)$$

Therefore, using the estimate (33), we have

$$\|U(-t)\psi(t) - U(-t_0)\psi(t_0)\|_{L^2} = \left\|\int_{t_0}^t U(-s)N(\psi)(s)ds\right\|_{L^2}$$

$$\leq C \|\psi\|_{L^{\frac{14}{3}}([t_0,t],L^{\frac{14}{5}})}^{\frac{11}{3}} \to 0, \tag{47}$$

as $t_0 \to \infty$. Since $U(-t_0)\psi(t_0) \in L^2$, then there exists a unique $\psi_0 \in B_\varepsilon$.

Now, assume that $\psi_+ \in L^2(\mathbb{R}^3)$, $\psi \in B_{2R}$ and consider the map

$$\Phi_+[\psi](t) = U(t)\psi_+ + i \int_t^\infty U(t-s)N(\psi)(s)ds, \quad t > T, \tag{48}$$

for some $T = T(\psi_+)$ large enough. Then, using the same arguments as in the proof of part (iii) of Theorem 1, we find that Φ_+ is a contraction on B_{2R} and has a unique fixed point if $\|\psi_+\|_{L^2} < \varepsilon$. Using the global well-posedness result established in Theorem 1 for small data, one can then extend this solution uniquely for any $t \in [0,\infty)$, and in particular ψ will take some value $\psi_0 = \psi(0) \in L^2$ at time $t = 0$.

This gives existence of the wave operators $\Omega_\pm$, defined by

$$\psi_0 = \Omega_\pm \phi_\pm = \psi_\pm \pm i \int_0^{\pm\infty} U(-s)N(\psi)(s)ds. \tag{49}$$

Finally, we observe that since the wave operators $\Omega_\pm$ are isometric in the space B_ε, it follows that the scattering operator $S : \phi_- \mapsto \phi_+$ is well defined as a map from B_ε onto itself and isometric in the L^2 norm, i.e., $\|S\psi\|_{L^2} = \|\psi\|_{L^2}$. This completes the proof of the Theorem 2.

ACKNOWLEDGMENTS

This work is partially supported by the Research and Development Sector of TUS, contract No 08094ni-10.

REFERENCES

1. J. Bourgain, *Int. Math. Res. Not.* **5**, 253–284 (1998).
2. T. Cazenave, *Semilinear Schrödinger equations*, Courant Lecture Notes **10**, American Mathematical Society, 2003.
3. J. Ginibre and G. Velo, *Math. Zeitschr.* **170**, 109–136 (1980).
4. J. Ginibre and G. Velo, *Comm. Math. Phys.* **151** 619–645, (1993).
5. H. Hayashi and T. Ozawa, *Comm. Math. Phys.* **110**, 467–478 (1987).
6. A. Ivanov and G. Venkov, *J. Evolution Equations* **8**, 217–229 (2008).
7. T. Kato, *Ann. Inst. H. Poincaré, Phys. Théor.* **46**, 113–129 (1987).
8. M. Keel and T. Tao, *Amer. J. Math.* **120**, 955–980 (1998).
9. H. Nawa and T. Ozawa, *Comm. Math. Phys.* **146**, 259–275 (1992).
10. E. Stein, *Harmonic Analysis,* Princeton Math. Ser. **43**, Princeton University Press, N.J., 1993.
11. R. Strichartz, *Duke Math. J.* **44**, 705–714 (1977).
12. T. Tao, *Proc. Centre Math. Appl. Austral. Nat. Univ.* **40**, 19–48, (2002).

On Stabilization of Nonautonomous Nonlinear Systems

A. Yu. Bogdanov

*Faculty of Mathematics and IT, Ulyanovsk State University, 42 Leo Tolstoy str.,
Ulyanovsk, 432970, Russia*

Abstract. The procedures to obtain the sufficient conditions of asymptotic stability for nonlinear nonstationary continuous-time systems are discussed. We consider different types of the following general controlled system:

$$\dot{x} = X(t,x,u) = F(t,x) + B(t,x)u, \quad x(t_0) = x_0. \qquad (*)$$

The basis of investigation is limiting equations, limiting Lyapunov functions, etc. The improved concept of observability of the pair of functional matrices is presented. By these results the problem of synthesis of asymptotically stable control nonlinear nonautonomous systems (with linear parts) involving the quadratic time-dependent Lyapunov functions is solved as well as stabilizing a given unstable system with nonlinear control law.

Keywords: Nonautonomous system, stabilization, Lyapunov function, limiting equations.
PACS: 02.30.Hq, 02.30.Yy

INTRODUCTION

Due to A.M. Lyapunov the stability and boundedness of solutions of wide class of dynamical systems can efficiently be studied with the aid of so called Lyapunov functions, namely, mappings which do not increase along solutions. There is no need to elaborate on the role that Lyapunov functions play in the analysis of asymptotic properties of solutions. We only mention here that for asymptotic stability and other attractivity properties the negative semidefiniteness of the derivatives of the Lyapunov functions is not enough. Negative definite derivatives in the appropriate domains would suffice. However, this much stronger property is not always possessed by the natural candidate for the Lyapunov function, e.g., the energy.

An effective tool to overcome the mentioned difficulty was developed by J.P. LaSalle and well known as the "invariance principle". By combining information on limiting level sets of the Lyapunov function with the topological dynamics of the solution flow it is often possible to localize the ω-limit set and thus establish attractivity of solutions. The technique is applicable to several important classes of dynamical systems and during the past 30 years well studied in the literature. We can only outline here the role of works of G. R. Sell [6], Z. Artstein [2, 3] and A. S. Andreev [1] where the reader can find the necessary references.

A different approach emerged mainly in control and system theory. By using observability-type arguments it is often possible to show that the Lyapunov function strictly decreases if enough time has passed, although locally it might only nonincrease.

CP1067, *Applications of Mathematics in Engineering and Economics '34—AMEE '08*, edited by M. D. Todorov
© 2008 American Institute of Physics 978-0-7354-0598-1/08/$23.00

In the presented work we intend to piece together the advantages of the two approaches, and show how they can complement each other. It is also worth to point out that, although the equations discussed here are controlled ordinary differential equations with special linear parts, the developed by author technique can be applied to the similar classes of functional differential equations and functional difference equations with vector control.

LINEAR NONAUTONOMOUS SYSTEMS

In this section we consider a class of linear continuous-time controlled systems of the following form

$$\dot{x} = X(t,x,u) = A(t)x + B(t)u \ , \quad x(t_0) = x_0 \ , \tag{1}$$

where $A(t)$ is a functional $n \times n$ matrix , $B(t)$ is a n-dimensional functional vector. Suppose that matrix $A(t)$ is bounded and continuous almost everywhere, i. e., $\|A(t)\| \leq a_1 = \mathrm{const}, t \geq t_0$, where $\| \cdot \|$ is a trace norm, namely $\|A\| = \sqrt{\mathrm{Tr}(A^T A)}$. All components of vector function $B(t)$ are also supposed bounded for every $t \geq t_0$. Let us consider the Lyapunov function

$$V(t,x) = x^T P(t)x \ , \tag{2}$$

where $P^T(t) = P(t) > 0$ is a functional $n \times n$ matrix, $P(t) \in C^1([t_0,+\infty);R^{n^2})$, such that $P(t) \geq p_0 I$, $\|P(t)\| \leq p_1$; $p_0, p_1 > 0$. So, according to well known terminology, $V(t,x)$ is positively definite and it admits infinitely small higher limit.

The upper estimate $-W(t,x)$ of derivative of Lyapunov function in the sense of system (1) we also consider as a quadratic form

$$W(t,x) = x^T Q(t)x \ , \tag{3}$$

where $Q^T(t) = Q(t) \geq 0$ is a functional $n \times n$ matrix, $\|Q(t)\| \leq q_0$, $q_0 > 0$.

Since $A(t)$, $Q(t)$ and vector $B(t)$ are bounded, we can define a limiting system with respect to the original system (1), and also limiting to $x^T Q(t)x$ function [3]:

$$\dot{x} = A^*(t)x + B^*(t)u, \quad A^*(t) = \frac{d}{dt} \lim_{t_n \to \infty} \int_0^t A(t_n + \tau)d\tau,$$

$$B^*(t) = \frac{d}{dt} \lim_{t_n \to \infty} \int_0^t B(t_n + \tau)d\tau,$$

$$\Omega(t,x) = x^T Q^*(t)x, \quad Q^*(t) = \frac{d}{dt} \lim_{t_n \to \infty} \int_0^t Q(t_n + \tau)d\tau.$$

By the assertions (1)-(3), the derivative of Lyapunov function in the sense of system (1) with control law $u \equiv 0$ can be written as [4]:

$$\beta(t,x,0) = x^T [\dot{P}(t) + P(t)A(t) + A^T(t)P(t) + Q(t)]x \leq 0 \tag{4}$$

353

if

$$\lambda(t,x) = x^T P(t)B(t) \equiv 0 . \qquad (5)$$

We reduce investigation of inequality (4) with the additional condition (5) to analysis of a single inequality by considering a scalar function $\lambda_1(t,x) = C^T(t)x$, where $C(t)$ is a n-th dimensional bounded functional vector to be chosen later. Then

$$\lambda(t,x)\,\lambda_1(t,x) = x^T(P(t)B(t)C^T(t) + C(t)B^T(t)P(t))x .$$

By addition this expression to function $\beta(t,x,0)$ we obtain, if $\lambda(t,x) = 0$,

$$\beta(t,x,0) = x^T[\dot{P}(t) + P(t)(A(t) + B(t)C^T(t)) + (A(t) + B(t)C^T(t))^T P(t) + Q(t)]x .$$

So, the initial problem is reduced to necessity of checking $\beta(t,x,0) \le 0$. From this one can easily establish the following stabilizing criteria by using the results of [1, 4].

Theorem 1 *The control law $u(t,x) \in D$ stabilizes system (1) (the solution $x = 0$) up to uniform asymptotic stability on the whole space, if*
1) there are n dimensional functional vector $C(t)$, positively definite for every $t \ge t_0$ functional matrix $P(t) = P^T(t)$ and semidefinite for every $t \ge t_0$ functional matrix $Q(t) = Q^T(t) \ge 0$ such that the matrices

$$\tilde{Q}(t) = -\{\dot{P}(t) + P(t)[A(t) + B(t)C^T(t)] + [A(t) + B(t)C^T(t)]^T P(t)\} \ge 0 \qquad (6)$$

and $\Delta Q(t) = \tilde{Q}(t) - Q(t) \ge 0$ are nonnegative for all $t \ge t_0$;
2) assume that

$$u(t,x)\mathrm{sgn}[x^T P(t)B(t)] \le C^T(t)x\,\mathrm{sgn}[x^T P(t)B(t)] + \frac{x^T \Delta Q(t)x}{2|x^T P(t)B(t)|} \qquad (7)$$

if $x^T P(t)B(t) \ne 0$;
3) for every limiting pair $(A^*(t)x + B^*(t)u^*(t,x), Q^*(t))$ the set of solutions of $\dot{x} = A^*(t)x + B^*(t)u^*(t,x)$, which are contained in the set $\{\Omega(t,x) = 0\}$, consists of zero solution $x = 0$.

It can be easily seen that inequality (7) besides the evident linear stabilizing control $u(t,x) = C^T(t)x$ allows and implies the possibility of stabilizing of linear system (1) by nonlinear control, which can be more effective than linear. To be more precise, the nonlinear stabilizing control can be chosen as follows

$$u(t,x) = C^T(t)x - f(t,x)\mathrm{sgn}\,x^T P(t)B(t) ,$$

where $f(t,x)$ is a nonnegative scalar function which provides $u(t,x) \in D$, i.e., it satisfies the precompactness conditions.

We can make the previous result (Theorem 1) more convenient in practical sense by introducing the following definition. But we have to restrict ourselves by considering linear controls.

Definition 1 *Let $U(t)$ and $V(t)$ be two $n \times n$ matrices, which have continuous derivatives up to $(n-1)$th order on the interval $(\tau - \delta, \tau + \delta)$, $\tau \in R$, $\delta > 0$. We shall say that the* *pair of matrices* $(U(t), V(t))$ *is observable if for the point* τ *the following equality holds* $\text{rank}[K_1\ K_2\ \ldots\ K_n]^T = n.$ *Here* $K_1 = U(\tau),\ K_i = K_{i-1}(\tau)V(\tau) + \frac{dK_{i-1}(\tau)}{d\tau},\ 2 \leq i \leq n.$

The matrix $K(t) = [K_1\ K_2\ \ldots\ K_n]^T$ *is called an observability matrix of the pair* $(U(t), V(t)).$

Suppose that for every limiting pair $(Q^*(t), A^*(t))$ there is a point $\tau \in R$ such that in its neighborhood $(\tau - \delta, \tau + \delta)$ the matrices $Q^*(t)$ and $A^*(t)$ have continuous derivatives up to $(n-1)$th order. Then the following result takes place.

Theorem 2 *The control* $u(t, x) = C^T(t)x$ *stabilizes the system (1) up to uniform asymptotic stability on the whole space, if*
  1) condition 1) of Theorem 1 is fulfilled;
  2) every limiting pair $(Q^*(t), A^*(t) + B^*(t)C^{*^T}(t))$ *is observable.*

Proof: Consider the set $E = \{\Omega(t, x) = 0\} = \{x^T Q^*(t)x = 0\} = \{Q^*(t)x = 0\}.$ Let M be a largest invariant subset of E with regard to uncontrolled system $\dot{x} = A^*(t)x.$ We are going to show that $M = \{0\}$ if every pair $(Q^*(t), A^*(t))$ is observable.

In fact, let a solution $x(t)$ of the system $\dot{x} = A^*(t)x$ belongs to $M.$ Then it implies $Q^*(t)x(t) \equiv 0.$ And as a clear consequence we derive the following chain of equalities

$$0 = \frac{d}{dt}(Q^*(t)x) = [\dot{Q}^*(t) + Q^*(t)A^*(t)]x,$$

$$0 = \frac{d^2}{dt^2}(Q^*(t)x) = [\ddot{Q}^*(t) + 2\dot{Q}^*(t)A^*(t) + Q^*(t)\dot{A}^*(t) + Q^*(t)A^{*2}(t)]x,$$

$$\ldots$$

$$0 = \frac{d^{(n-1)}}{dt^{(n-1)}}(Q^*(t)x) = \ldots$$

In matrix notation these equalities can be rewritten as $0 = Kx,$ where K is an observability matrix of the pair $(Q^*(t), A^*(t)).$

Since every pair $(Q^*(t), A^*(t))$ is observable then for a point $\tau \in R$ the rank of observability matrix K equals $n.$ Consequently, if $t = \tau$ then $x(\tau) = 0.$ Due to uniqueness of solution of linear system we obtain that $x(t) \equiv 0$ for every $t \in R.$

Taking under consideration that closed loop linear system has the following form

$$\dot{x} = (A(t) + B(t)C^T(t))x$$

on the basis of Theorem 1 we have the desired result of Theorem 2.

Definition 2 *The pair of matrices* $(U(t), V(t))$ *is said to be strictly observable if*
  1) there is nondiverging sequence of intervals $(t_n, t_n + s)$ *where* $t_n \to +\infty,$ $s = const > 0,$ $s < t_{n+1} - t_n \leq T = const > 0,$ *such that matrices* $U(t)$ *and* $V(t)$ *have continuous derivatives up to nth order for* $t \in (t_n, t_n + s);$
  2) there is a matrix $G_n(t)$ *of the type* $n \times n$ *which consists of arbitrarily chosen lines of the observability matrix* $K(t)$ *for the pair* $(U(t), V(t))$ *such that* $|\det [G_n(t)]| \geq \delta_0 = const > 0$ *for* $t \in [t_n, t_n + s].$

Let us note, that strictly observable property of $(U(t), V(t))$ implies observability of all limiting pairs $(U^*(t), V^*(t))$. Therefore, condition 2) of Theorem 2 can be replaced by the strictly observable property for the pair of matrices $(Q(t), A(t) + B(t)C^T(t))$.

We continue our investigation of stabilization problem for the system (*) on the whole space or in the given bounded region Γ using the same Lyapunov function (2).

Let $W(t, x)$ be as in (3), on the basis of [4] we can obtain the following condition for solving the stabilization problem:

$$\beta(t, x, 0) = x^T P(t) F(t, x) + F^T(t, x) P(t) x + x^T \dot{P}(t) x + x^T Q(t) x \leq 0 \qquad (8)$$

while

$$x^T P(t) B(t, x) = 0 . \qquad (9)$$

Suppose that equation (9) can be resolve with respect to one of it's state variables. Without loss of generality we can guess that the variable is x_1, i.e.,

$$x_1 = f(t, \tilde{x}, P(t)) , \qquad (10)$$

where $\tilde{x}^T = (x_2, x_3, ..., x_n)$ is a $(n-1)$-th dimensional vector. Then the sufficient condition for asymptotic stability of the controlled system (*) is inequality (8) for all $\tilde{x}$ after substitution in x the corresponding expression for x_1 from (10).

NONLINEAR NONAUTONOMOUS SYSTEMS WITH LINEAR PARTS

Consider a particular case of the system (*)

$$\dot{x} = A(t)x + D(t)\varphi(t, v) + B(t)u , \quad x(t_0) = x_0 , \qquad (11)$$

where $A(t)$ is a bounded functional $n \times n$ matrix , $D(t)$ and $B(t)$ are bounded n-dimensional vectors, $v = \tilde{C}^T(t)x$ ($\tilde{C}(t)$ is a functional n-dimensional vector), $\varphi(t, v)$ is a scalar nonlinearity which satisfies the following condition for every v and $t \geq t_0$

$$0 \leq \varphi(t, v)v \leq kv^2 , \quad k = \text{const} . \qquad (12)$$

Suppose that we are not sure of asymptotic stability of uncontrolled system and so we have to find a stabilizing control $u = u(t, x)$. For solving the mentioned problem we could check the inequality (8) under the condition (9). For system (11) the inequality (8) can be rewritten as

$$\beta(t, x, 0) = x^T [\dot{P}(t) + P(t)A(t) + A^T(t)P(t) + Q(t)]x + 2x^T P(t)D(t)\varphi(t, v) \leq 0 \quad (13)$$

while

$$x^T P(t) B(t) \equiv 0 . \qquad (14)$$

From the assertion (14) we find that

$$x_1 = G^T(t)\tilde{x} , \qquad (15)$$

where $G(t)$ is a functional $(n-1)$-th dimensional vector. Then

$$x = L(t)\tilde{x}, \tag{16}$$

where $L(t) = \begin{bmatrix} G^T(t) \\ I_{n-1} \end{bmatrix}$ is a functional $n \times (n-1)$ matrix.

After substitution (16) into the inequality (13) we see that the latter is fulfilled if

$$\tilde{\beta}(t,\tilde{x}) = \tilde{x}^T L^T(t)[\dot{P}(t) + P(t)A(t) + A^T(t)P(t) + Q(t)]L(t)\tilde{x} +$$

$$+ 2\tilde{x}^T L^T(t)P(t)D(t)\varphi(t,v) \leq 0 \tag{17}$$

for all $\tilde{x}$ and $v = \tilde{C}^T(t)L(t)\tilde{x}$.

V.M. Kuntsevich and M.M. Lychak in their work [5] suggested a method which allow to investigate the negative definiteness of Lyapunov function derivative $\dot{V}(t,x)$. This method reduces the initial problem to the task of finding the conditions when a family of ellipsoids does not intersect with a family of hyperplanes in R^n. The basic mathematical essence of this method was presented in [5] as a special geometrical lemma.

Lemma 1 *[5] Let in space* $R^n = \{x\}$ *for* $t \in [t_0, +\infty)$ *there is a family of ellipsoids*

$$x^T Q x + 2x^T H = \gamma \tag{18}$$

where $H = H(t,\sigma)$, $\gamma = \gamma(t,\sigma)$, $Q = Q(t,\sigma) > 0$ *for every* σ *and* $t \in [t_0, +\infty)$, *and also there is a family of hyperplanes*

$$x^T C = \sigma, \quad C = C(t,\sigma) \tag{19}$$

Then equations (18) and (19) have no solutions for every $t \in [t_0, +\infty)$ *and* $\sigma \in [\sigma_1, \sigma_2]$, $\sigma_2 > \sigma_1 \geq 0$ *if and only if*

$$\rho(t,\sigma) = \sqrt{(\gamma + H^T Q^{-1} H)C^T Q^{-1} C} - C^T Q^{-1} H < \sigma \tag{20}$$

for those $\sigma \in [\sigma_1, \sigma_2]$ *and* $t \in [t_0, +\infty)$ *when* $\gamma + H^T Q^{-1} H \geq 0$.

Geometrical lemma helps to check the inequality (17).

Theorem 3 *The control* $u(t,x) = C^T(t)x$ *stabilizes the system (11) up to uniform asymptotic stability on the whole space, if*

1) nonlinearity $\varphi(t,v)$ *belongs to a class of nonlinearities which is determined by the condition (12);*

2) there is n-dimensional functional vector $C(t)$, *positive definite for all* $t \geq t_0$ *functional matrix* $P(t) = P^T(t) > 0$ *and nonnegative for all* $t \geq t_0$ *functional matrix* $Q(t) = Q^T(t) \geq 0$ *such that the matrix*

$$\Delta Q(t) = -[\dot{P}(t) + P(t)(A(t) + B(t)C^T(t)) + (A(t) + B(t)C^T(t))^T P(t)] - Q(t) > 0$$

is positively definite for every $t \geq t_0$;

357

3) every limiting pair $(Q^*(t), A^*(t) + B^*(t)C^{*T}(t))$ is observable;
4) the following inequality holds for every $t \geq t_0$

$$k^{-1} \geq \sqrt{D^T(t)P(t)\tilde{P}(t)D(t)\tilde{C}^T(t)\tilde{P}(t)\tilde{C}(t) + \tilde{C}^T(t)\tilde{P}(t)D(t)} , \qquad (21)$$

where

$$\tilde{P}(t) = L^T(t)[L^T(t)\Delta Q(t)L(t)]^{-1}L(t) , \quad L^T(t) = [G(t) \mid I_{n-1}] .$$

In the case the condition (21) is valid for some $k = k_0$, but for this k the nonlinearity $\varphi(t, v)$ satisfies the condition (17) only for a certain interval of v which is defined as $|v| \leq \tilde{v} = $ const, then the system (11) is stabilized in the region $\tilde{\Gamma}$ according to the following estimate

$$\sup_{t \geq t_0}\{x^T P(t)x\} < \{\sup_{t \geq t_0}[\tilde{C}^T(t)P^{-1}(t)\tilde{C}(t)]\}^{-1}v^2 . \qquad (22)$$

Let us consider here more general type of nonlinear system of the form

$$\dot{x} = A(t)x + \tilde{D}(t)\Phi(t, v) + B(t)u , \quad x(t_0) = x_0 , \qquad (23)$$

where $A(t)$ and $B(t)$ are the same as in (11), $\tilde{D}(t)$ is a functional $n \times r$ matrix, $\Phi(t, v) = [\varphi_j(t, v_j)]$, $j = 1, 2, ..., r$ is r-dimensional vector function, $v_j = \tilde{C}_j^T(t)x$ ($\tilde{C}_j(t)$ are functional n-vectors), $\varphi_j(t, v_j)$ are scalar nonlinearities which satisfy the conditions

$$0 \leq \varphi_j(t, v_j)v_j \leq k_j v_j^2 , \quad k_j = \text{const.} \qquad (24)$$

Suppose again that we are not sure of asymptotic stability of uncontrolled system and so we have to find the stabilizing control $u = u(t, x)$. For the case of system (23) the inequality (8) can be rewritten as

$$\beta(t, x, 0) = x^T[\dot{P}(t) + P(t)A(t) + A^T(t)P(t) + Q(t)]x + 2x^T P(t)\tilde{D}(t)\Phi(t, v) \leq 0 \qquad (25)$$

under the condition (14).

After substitution (16) into (25) we see that the latter is fulfilled if

$$\tilde{\beta}(t, \tilde{x}) = \tilde{x}^T L^T(t)[\dot{P}(t) + P(t)A(t) + A^T(t)P(t) + Q(t)]L(t)\tilde{x} +$$

$$+ 2\tilde{x}^T L^T(t)P(t)\tilde{D}(t)\Phi(t, v) \leq 0 \qquad (26)$$

for all $\tilde{x} \in R^{n-1}$ and $v_j = \tilde{C}^T(t)L(t)\tilde{x}$.

We introduce some numbers $\alpha_j \geq 0$ such that $\sum_{j=1}^r \alpha_j = 1$. Then inequality (26) is equivalent the following one

$$\tilde{\beta}(t, \tilde{x}) = \sum_{j=1}^r \{\alpha_j \tilde{x}^T L^T(t)[\dot{P}(t) + P(t)A(t) + A^T(t)P(t) + Q(t)]L(t)\tilde{x} +$$

$$+ 2\tilde{x}^T L^T(t)P(t)\tilde{D}_j(t)\varphi_j(t, v_j)\} \leq 0 \qquad (27)$$

358

for all $\tilde{x} \in R^{n-1}$ and $v_j = \tilde{C}^T(t)L(t)\tilde{x}$, where $\tilde{D}_j(t)$ is jth column of the matrix $\tilde{D}(t)$.

By applying successively geometrical lemma to all components of (27) we get that inequality (27) holds if the following system of inequalities is valid

$$k_j \rho_j(t) \leq \alpha_j , \quad j \in [1, r] ,$$

where

$$\rho_j(t) = \sqrt{\tilde{D}_j^T(t)P(t)\tilde{P}(t)\tilde{D}_j(t)\tilde{C}_j^T(t)\tilde{P}(t)\tilde{C}_j(t)} + \tilde{C}_j^T(t)\tilde{P}(t)P(t)\tilde{D}_j(t) ,$$

$$\tilde{P}(t) = L^T(t)[L^T(t)\Delta Q(t)L(t)]^{-1}L(t) .$$

Due to arbitrary rule of choosing $\alpha_j \geq 0$ and taking under consideration that $\rho_j(t) \geq 0$ this system of inequalities can be replaced by one inequality:

$$\sum_{j=1}^{r} k_j \rho_j(t) \leq 1 .$$

Therefore we have proved the following result.

Theorem 4 *The control $u(t,x) = C^T(t)x$ stabilizes the system (23) up to uniform asymptotic stability on the whole space, if*
1) nonlinearities $\varphi_j(t, v_j)$ belong to the class of nonlinearities (24);
2) conditions 2) and 3) of Theorem 3 are fulfilled;
3) for every $t \geq t_0$ the following inequality holds

$$\sum_{j=1}^{r} k_j \rho_j(t) \leq 1 , \tag{28}$$

where

$$\rho_j(t) = \sqrt{\tilde{D}_j^T(t)P(t)\tilde{P}(t)\tilde{D}_j(t)\tilde{C}_j^T(t)\tilde{P}(t)\tilde{C}_j(t)} + \tilde{C}_j^T(t)\tilde{P}(t)P(t)\tilde{D}_j(t) .$$

In the case the condition (28) is valid for some $k_j = k_{0j}$, but for these k_j the nonlinearities $\varphi_j(t, v_j)$ satisfy the condition (24) only for the certain intervals v_j which are defined by

$$|v_j| \leq \tilde{v}_j = \mathrm{const} , \quad j = 1, ..., r ,$$

then system (23) is stabilized in the region $\tilde{\Gamma} \in R^n$ which is obtained by the following estimate

$$\sup_{t \geq t_0}\{x^T P(t)x\} \leq \inf_j \{ ([\sup_{t \geq t_0} \tilde{C}_j^T(t)P^{-1}(t)\tilde{C}_j(t))]^{-1} \tilde{v}_j^2 \} .$$

ACKNOWLEDGMENTS

This work was supported in part by Russian Foundation for Basic Research, projects No 08-01-97010, 08-08-97033.

REFERENCES

1. A. S. Andreev, *Appl. Math. and Mech.* **48**(2), 225–232 (1984).
2. Z. Artstein, *J. Differ. Equat.* **23**(2), 216–223 (1977).
3. Z. Artstein, *J. Differ. Equat.* **25**(2), 184–202 (1977).
4. A. Yu. Bogdanov, *Sci. Letters of USU* **8**(1), 31–38 (2000).
5. V. M. Kuntsevich and M. M. Lychak, *Synthesis of Automatic Control Systems via Lyapunov Functions*, Science, Moscow, 1977, pp. 41–43.
6. G. R. Sell, *Trans. Amer. Math. Soc.* **127**, 241–283 (1967).

Oscillation Criteria in First Order Neutral Delay Impulsive Differential Equations with Constant Coefficients

M.B. Dimitrova and V.I. Donev

Dept. of Mathematics, Technical University of Sofia, branch Sliven, 8800 Sliven, Bulgaria

Abstract. This paper is dealing with the oscillatory properties of first order neutral delay impulsive differential equations and corresponding to them inequalities with constant coefficients. The established sufficient conditions ensure the oscillation of every solution of this type of equations.

Keywords: Oscillation of solutions, neutral delay impulsive differential equations and inequalities, constant coefficients.
PACS: 02.30.Ks, 02.30.Hq

INTRODUCTION

In contrast to the theory of differential equations with deviating arguments (see, [7], [12]–[16]) and the theory of impulsive differential equations (see, [1] and [17]), the theory of impulsive differential equations with deviating arguments(IDEDA), due to theoretical and practical difficulties, is developing rather slowly. Let us pay attention, that IDEDA are adequate mathematical models for the simulation of processes that depend on their history and are subject to short-time disturbances. Such processes occur in the theory of optimal control, theoretical physics, population dynamics, biotechnology, industrial robotics, etc. We note here that [8] is the first work where IDEDA were considered. Among the numerous results, concerning the oscillation theory of IDEDA - with delayed or advanced arguments, we choose to refer to [2], [4], [9], [11] and [20]. Much less we know about the neutral impulsive differential equations (see, [5], [6] and [10]), i.e., equations in which the highest-order derivative of the unknown function appears in the equation with the argument t (the present state of the system), as well as with one or more retarded and/or advanced arguments (the past and/or the future state of the system). Note, that equations of this type appear in networks, containing lossless transmission lines. For example, such networks arise in high speed computers, where lossless transmission lines are used to interconnect switching circuits (see, [3] and [18]).

As it is known , the appearance of the neutral term (see [7]) and/or impulsive effects in a differential equation can cause or destroy the oscillation of its solutions. Moreover, the study of neutral differential equations in general, presents complications which are unfamiliar for non-neutral differential equations. As for a discussion on some more applications and some drastic differences in behavior of the solution of neutral differential equations see, for example, [14] and [19].

CP1067, *Applications of Mathematics in Engineering and Economics '34—AMEE '08*, edited by M. D. Todorov
© 2008 American Institute of Physics 978-0-7354-0598-1/08/$23.00

PRELIMINARIES

In this article we consider the first order neutral delay impulsive differential equation of the form

$$\frac{d}{dt}[y(t) - cy(t-h)] + py(t-\sigma) = 0, \ t \neq \tau_k, \ k \in N \tag{E_1}$$

$$\Delta[y(\tau_k) - cy(\tau_k - h)] + p_k y(\tau_k - \sigma) = 0, \ k \in N$$

as well as the corresponding to it inequalities

$$\frac{d}{dt}[y(t) - cy(t-h)] + py(t-\sigma) \leq 0, \ t \neq \tau_k, \ k \in N \tag{$N_{1,\leq}$}$$

$$\Delta[y(\tau_k) - cy(\tau_k - h)] + p_k y(\tau_k - \sigma) \leq 0, \ k \in N$$

and

$$\frac{d}{dt}[y(t) - cy(t-h)] + py(t-\sigma) \geq 0, \ t \neq \tau_k, \ k \in N \tag{$N_{1,\geq}$}$$

$$\Delta[y(\tau_k) - cy(\tau_k - h)] + p_k y(\tau_k - \sigma) \geq 0, \ k \in N.$$

Here, the deviations h and σ are positive constants and $\tau_k \in (0, +\infty)$, $k \in N$ are fixed moments of impulsive effect (the jump points). In order to manifest the jumps of a solution, we use the notation

$\Delta[y(\tau_k) - c_k y(\tau_k - h)] = \Delta y(\tau_k) - c_k \Delta y(\tau_k - h)$, where $\Delta y(\tau_k) = y(\tau_k + 0) - y(\tau_k - 0)$.

Denote by $P_\tau C(R, R)$ the set of all piecewise continuous on the intervals $(\tau_k, \tau_{k+1}]$, $k \in N$ functions $u: R \to R$ which at the jump points τ_k, $k \in N$ are continuous from the left, i.e., $u(\tau_k - 0) = \lim\limits_{t \to \tau_k - 0} u(t) = u(\tau_k)$, and may have discontinuities of first kind at the jump points τ_k, $k \in N$, which we characterize as down-jumps when $\Delta u(\tau_k) < 0$, $k \in N$ and as up-jumps when $\Delta u(\tau_k) > 0$, $k \in N$.

Introduce the following hypotheses :

H1. $0 < \tau_1 < \tau_2 < \ldots < \tau_k < \ldots$, $\lim\limits_{k \to +\infty} \tau_k = +\infty$, $\max\{\tau_{k+1} - \tau_k\} < +\infty$, $k \in N$;

H2. $h > \sigma$, $c > 1$, $p > 0$, $p_k \geq 0$, $k \in N$.

We will say that a function $y(t)$ is a *solution* of equation (E_1), if there exists a number $T_0 \in R$ such that $y \in P_\tau C([T_0, +\infty), R)$, the function $z(t) = y(t) - cy(t-h)$ is continuously differentiable for $t \geq T_0$, $t \neq \tau_k$, $k \in N$ and $y(t)$ satisfies equation (E_1) for all $t \geq T_0$.

Without other mention, we will assume throughout that every solution $y(t)$ of equation (E_1), that is under consideration here, is continuable to the right and is nontrivial. That is, $y(t)$ is defined on some ray of the form $[T_y, +\infty)$ and for each $T \geq T_y$, it is fulfilled $\sup\{|y(t)|: t \geq T\} > 0$. Such a solution is called a *regular solution* of (E_1).

We will say that a real valued function u defined on an interval of the form $[a, +\infty)$ has some property *eventually*, if there is a number $b \geq a$ such that u has this property on the interval $[b, +\infty)$.

A regular solution $y(t)$ of equation (E_1) is said to be *nonoscillatory*, if there exists a number $t_0 \geq 0$ such that $y(t)$ is of constant sign for every $t \geq t_0$. Otherwise, it is called *oscillatory*. Also, note that a *nonoscillatory* solution is called *eventually positive* (*eventually negative*), if the constant sign that determines its *nonoscillation* is positive (negative). Equation (E_1) is called oscillatory, if all its solutions are oscillatory.

Except that, in this article, when we write a functional expression, we will mean that it holds for all sufficiently large values of the argument.

And so, let consider $y(t)$ as a solution of equation (E_1) and construct

$$z(t) = y(t) - cy(t-h), \qquad \Delta z(\tau_k) = \Delta y(\tau_k) - c\Delta y(\tau_k - h), \ k \in N, \qquad (*)$$

$$w(t) = z(t) - cz(t-h), \qquad \Delta w(\tau_k) = \Delta z(\tau_k) - c\Delta z(\tau_k - h), \ k \in N. \qquad (**)$$

At the beginning, we introduce two lemmas (see [9], [10] and [16]), which investigate the asymptotic behavior of the auxiliary function $z(t)$, when $y(t)$ is a none-oscillatory solution of (E_1). First of them is formulated and proved for eventually positive solution $y(t)$ of the equation (E_1).

Lemma 1 *Let $y(t)$ be an eventually positive solution of (E_1) and the hypotheses $(H1) - (H2)$ are satisfied. Then*

(a) $z(t)$, defined by $(*)$, is an eventually decreasing function of t with down-jumps;

(b) $z(t)$, defined by $(*)$, is an eventually negative function, i.e., $z(t) < 0$ for enough large t and $\lim\limits_{t \to +\infty} z(t) = -\infty$.

Proof: (a) Let $y(t)$ be an eventually positive solution of the equation (E_1), i.e., there exists a number $\tilde{t} > 0$ such that $y(t)$ is defined for $t \geq \tilde{t}$ and $y(t) > 0, y(t-\sigma) > 0, y(t-h) > 0$ for $t \geq \tilde{t} + \max\{h, \sigma\} = t_0$. From (E_1) and $(*)$, it follows that

$$z'(t) = -py(t-\sigma) < 0, \ t \neq \tau_k, \ k \in N, \ t \geq t_0, \qquad (1)$$

$$\Delta z(\tau_k) = -p_k y(\tau_k - \sigma) < 0, \ k \in N, \ \tau_k \geq t_0.$$

Therefore, $z(t)$ is an eventually decreasing function $(z'(t) < 0)$ with "own-jumps" $(\Delta z(\tau_k) < 0)$ at the points of impulsive effect τ_k, for $t, \tau_k \geq t_0$. The proof of (a) is complete.

(b) It follows from (1) that $\lim\limits_{t \to +\infty} z(t)$ does exist, where $z(t)$ is an eventually strictly decreasing function with down-jumps. So, $\lim\limits_{t \to +\infty} z(t) = L$, where L could be positive constant, zero, negative constant, or $-\infty$.

Assume $\lim\limits_{t \to +\infty} z(t) = L > 0$. If integrate (E_1) from t_0 to t, we obtain

$$\int_{t_0}^{t} z'(r)dr + \int_{t_0}^{t} py(r-\sigma)dr = 0, \qquad \text{or}$$

$$z(t) - z(t_0) - \sum_{t_0 \leq \tau_k < t} \Delta z(\tau_k) + \int_{t_0}^{t} py(r-\sigma)dr = 0.$$

But $\Delta z(\tau_k) = -p_k y(\tau_k - \sigma)$, hence

$$z(t) = z(t_0) - \sum_{t_0 \leq \tau_k < t} p_k y(\tau_k - \sigma) - \int_{t_0}^{t} py(r-\sigma)dr \qquad (2)$$

363

Because $z(t) = y(t) - cy(t - h)$, we have in this case $L \leq z(t) < y(t)$, what determines $y(t)$ as a bounded function from below. Then (2) reduces to

$$z(t) \leq z(t_0) - L[\sum_{t_0 \leq \tau_k < t} p_k + \int_{t_0}^{t} p \, dr],$$
(3)

which implies $\lim_{t \to +\infty} z(t) = -\infty$ and contradicts our assumption.

Assume $\quad \lim_{t \to +\infty} z(t) = \lim_{t \to +\infty} [y(t) - cy(t - h)] = L = 0.$
(4)

It is obvious, that then $z(t) > 0$ eventually, i.e., there exists a number $t_1 \geq t_0$, such that we have $y(t) > cy(t - h) > y(t - h)$ for every $t \geq t_1$. Observe that the last inequality holds as well as for those moments of impulsive effect τ_k, for which $\tau_k > t_1$, $k \in N$. Therefore, our assumption implies that there will exist a strictly increasing sequence $\{y(t_n)\}_{n=1}^{\infty}$ (where $t_n = t_{n-1} + h$ and some t_n could be moments of impulsive effect τ_k), which is bounded by a positive number, i.e., $\lim_{n \to +\infty} y(t_n) = K$, $K > 0$, or which is unbounded, i.e., $\lim_{n \to +\infty} y(t_n) = +\infty$ and for which (4) has to be fulfilled. But,

$$\lim_{t_n \to +\infty} z(t_n) = \lim_{t_n \to +\infty} y(t_n) - c \lim_{t_n \to +\infty} y(t_n - h) = (1 - c) \lim_{t_n \to +\infty} y(t_n) < 0$$

and the contradiction with (4) is evident, because $c > 1$.

Assume $\lim_{t \to +\infty} z(t) = L < 0$. Then, because $z(t)$ is a decreasing function with down-jumps, for some $t_1 \geq t_0$ there will exist $\delta_v > 0$ such that $z(t) < -\delta_v$, for every $t \geq t_1$, $t \neq \tau_k$, $k \in N$, i.e.,

$$y(t) - cy(t - h) < -\delta_v, \; t \neq \tau_k, \; t \geq t_1.$$

Except that, because the sequence of eventually negative numbers $\{z(\tau_k)\}_{k=1}^{+\infty}$ is decreasing, for our $\delta_v > 0$, there will be such a term τ_v in the sequence of the impulsive moments $\{\tau_k\}$, whereafter $z(\tau_k) < -\delta_v$, for every $\tau_k \geq \tau_v$, when $k \geq v$, $k \in N$, $v \in N$. Hence,

$$y(\tau_k) - cy(\tau_k - h) < -\delta_v, \qquad \tau_k \geq \tau_v, \, k \geq v, \, k \in N, \, v \in N.$$

Denote $t_v = \max\{t_1, \tau_v\}$ and combine the last two inequalities as

$$y(t) < -\delta_v + cy(t - h), \qquad t \geq t_v.$$
(5)

It is obvious that the right side of (5) has to be positive , because $y(t)$ is positive. So, we obtain the inequality $0 < -\delta_v + cy(t - h)$, which clearly shows, that $y(t)$ is a bounded function from below. Hence, if integrate (E_1) from t_0 to t we can easily get (3), which will imply $\lim_{t \to +\infty} z(t) = -\infty$ and it will contradict our assumption again.

Thus, the above consideration approves $\lim_{t \to +\infty} z(t) = -\infty$. The proof of (b) and of the Lemma are completed.

The second lemma is only formulated for eventually negative solution $y(t)$ of the equation (E_1). The proof is carried out respectively to the proof of Lemma 1.

Lemma 2 *Let $y(t)$ be an eventually negative solution of (E_1) and the hypotheses $(H1)-(H2)$ are satisfied. Then* *(a)* $z(t)$, *defined by* $(*)$, *is an eventually increasing function of t with up-jumps;*

(b) $z(t)$, *defined by* $(*)$, *is an eventually positive function, i.e.,* $z(t) > 0$ *for enough large t and* $\lim_{t \to +\infty} z(t) = +\infty$.

Next lemma indicates, that the constructed from an eventually none-oscillatory solution $y(t)$ of equation (E_1), functions $z(t)$ and $w(t)$ are found to be auxiliary solutions of the same equation with useful characteristics.

Lemma 3 *Let* $y(t)$ *be an eventually none-oscillatory solution of the equation* (E_1) *and the hypotheses* $(H1) - (H2)$ *are satisfied. Then*
(a) the functions $z(t)$ *and* $w(t)$ *are also solutions of* (E_1);
(b) for eventually positive function $y(t)$, $w''(t)$ *is eventually positive function;*
(c) for eventually negative function $y(t)$, $w''(t)$ *is eventually negative function .*

Proof: (a) It is evident by direct substitution in (E_1) that $z(t)$, defined by $(*)$, is a solution of (E_1), as well as the same holds for $w(t)$, defined by $(**)$.

(b)-(c) From (a), it follows that $z(t)$, defined by $(*)$, is a solution of (E_1), i.e., we have $[z(t) - cz(t-h)]' = -[pz(t-\sigma)]$, $t \neq \tau_k$. From here, it is easy to see that

$$w(t)'' = [z(t) - cz(t-h)]'' = -p[z(t-\sigma)]', \ t \neq \tau_k. \tag{6}$$

From (6) and Lemma 1(a) one can easily derive (b). From (6) and Lemma 2(a) one can easily derive (c). The proof of the lemma is complete.

MAIN RESULTS

Utilizing the conclusions of the previous section, we establish some results, obtaining sufficient conditions under which the equation (E_1) is oscillatory.

Theorem 1 *Assume that the hypotheses* $(H1)$-$(H2)$ *are satisfied. Suppose also that:*
(i) $\liminf\limits_{t \to \infty} \left[\prod\limits_{t \leq \tau_k < t+h-\sigma} (1 + \frac{p_k}{c-1}) \right] \geq \frac{c-1}{ep(h-\sigma)}$.
Then the equation (E_1) *is oscillatory.*

Proof: Let assume, for the sake of contradiction, that equation (E_1) has a non-oscillatory solution. Since the negative of a solution of (E_1) is again a solution of (E_1), it suffices to prove the theorem considering an eventually positive solution.

So, let suppose that there exists a solution $y(t)$ of the equation (E_1) and a number $\tilde{t} > 0$, such that $y(t)$ is defined for $t \geq \tilde{t}$, $y(t) > 0$ for $t \geq \tilde{t}$ and $y(t-h) > 0$, $y(t-\sigma) > 0$ for $t \geq \tilde{t} + \max\{h, \sigma\} = t_0$.

Recalling $(*)$ and $(**)$, it follows by Lemma 3(a) and Lemma 1, that for every $t \geq t_0$ the function $z(t)$ is an eventually negative decreasing solution to (E_1), whereas by Lemma 3(a) and Lemma 2, $w(t)$ is an eventually positive increasing solution to (E_1). That is, $w(t)$ satisfies

$$\frac{d}{dt}[w(t) - cw(t-h)] + pw(t-\sigma) = 0, \ t \neq \tau_k, \tag{7}$$

$$\Delta[w(\tau_k) - cw(\tau_k - h)] + p_k w(\tau_k - \sigma) = 0, \ k \in N.$$

Note that, by Lemma 3(b), $w'(t)$ is an increasing function. Hence, from (7) we find that

$$w'(t-h) - cw'(t-h) + pw(t-\sigma) \le w'(t) - cw'(t-h) + pw(t-\sigma) = 0,$$

i.e., $\qquad w'(t-h) - cw'(t-h) + pw(t-\sigma) \le 0.$ \qquad (8)

Moreover, since $z(t)$ is a decreasing function, we have $z(\tau_k-\sigma) < z(\tau_k-\sigma-h)$. Hence, using the definitions of the functions $z(t)$ and $w(t)$, we can conclude that

$$\Delta w(\tau_k) = -p_k z(\tau_k-\sigma) > -p_k z(\tau_k-\sigma-h) = \Delta w(\tau_k-h), \ k \in N.$$

In view of the above observation, again from (7), it follows that for each $k \in N$

$$\Delta w(\tau_k-h) - c\Delta w(\tau_k-h) + p_k w(\tau_k-\sigma+h) \le \Delta w(\tau_k) - c\Delta w(\tau_k-h) + p_k w(\tau_k-\sigma) = 0,$$

i.e., $\qquad \Delta w(\tau_k-h) - c\Delta w(\tau_k-h) + p_k w(\tau_k-\sigma+h) \le 0.$ \qquad (9)

Now, from (8) and (9), it follows that $w(t)$ is an eventually positive function for which

$$(1-c)w'(t-h) + pw(t-\sigma+h) \le 0, \ t \ne \tau_k$$

$$(1-c)\Delta w(\tau_k-h) + p_k w(\tau_k-\sigma+h) \le 0, \ k \in N.$$

Divide the last two inequalities by $1-c<0$ and denote $s=h-\sigma$, $Q=\frac{p}{c-1}$, $q_k=\frac{p_k}{c-1}$. Then we see, that our assumption in the beginning leads us to the conclusion that $w(t)$ is an eventually positive increasing solution with "up-jumps" to the impulsive differential inequality with advanced argument of the form

$$w'(t) - Qw(t+s) \ge 0, \ t \ne \tau_k, \qquad (10)$$

$$\Delta w(\tau_k) - q_k w(\tau_k+s) \ge 0, \ k \in N,$$

for every $t \ge t_0$. Here, dividing the first inequality of (10) by $w(t)$ and using the increasing nature of $w(t)$ in the second one, we can rearrange (10), to obtain

$$\frac{w'(t)}{w(t)} > Q\frac{w(t+s)}{w(t)}, \qquad t \ne \tau_k, \ k \in N, \qquad (11)$$

$$\Delta w(\tau_k) > q_k w(\tau_k+s) > q_k w(\tau_k), \qquad k \in N.$$

Further, integrate (11) from t to $t+s$ and obtain

$$\int_t^{t+s} \frac{w(r)'}{w(r)} dr > \int_t^{t+s} Q\frac{w(t+s)}{w(t)} dr, \qquad \text{i.e.,}$$

$$\ln\frac{w(t+s)}{w(t)} + \sum_{t \le \tau_k < t+s} \ln\frac{w(\tau_k)}{w(\tau_k+0)} > Q\int_t^{t+s} Q\frac{w(t+s)}{w(t)} dr. \qquad (12)$$

From the second inequality of (11) we have $w(\tau_k+0) - w(\tau_k) = q_k w(\tau_k+s) > q_k w(\tau_k)$.

366

Hence, $w(\tau_k + 0) > (1 + q_k)w(\tau_k)$, i.e., $\frac{1}{1+q_k} > \frac{w(\tau_k)}{w(\tau_k+0)}$.

So, it is fulfilled $\ln\frac{1}{1+q_k} > \ln\frac{w(\tau_k)}{w(\tau_k+0)}$ and from (11) and (12) we get

$$\ln\left[\frac{w(t+s)}{w(t)}\prod_{t\le\tau_k<t+s}\frac{1}{1+q_k}\right] > Q\int_t^{t+s}\frac{w(t+s)}{w(t)}dr. \tag{13}$$

Bring to the mind, that $w(t)$ is an eventually strictly increasing function. Therefore, the function $\frac{w(t+s)}{w(t)}$ is bounded from below. That is why, we may denote $\liminf\limits_{t\to+\infty}\frac{w(t+s)}{w(t)} = w_0$, where $1 < w_0 < +\infty$. Thus, (13) implies

$$\ln\left[w_0\prod_{t\le\tau_k<t+s}\frac{1}{1+q_k}\right] > sQw_0,$$

from where, using the inequality $e^x > ex$, we obtain

$$\prod_{t\le\tau_k<t+s}(1+q_k) < \frac{1}{esQ}.$$

If resume $s = h - \sigma$, $Q = \frac{p}{c-1}$, $q_k = \frac{p_k}{c-1}$, we conclude from the last inequality

$$\prod_{t\le\tau_k<t+h-\sigma}\left(1+\frac{p_k}{c-1}\right) < \frac{c-1}{ep(h-\sigma)}, \tag{14}$$

which contradicts the condition (i) of the theorem. The proof is complete.

In the next theorem, using stronger condition for the coefficients p_k, we obtain more convenient result. For our needs in that theorem, we denote by $i[\tau_0,t)$ the number of fixed jump points τ_k that are situated in the interval $[\tau_0,t)$, $k \in N$, $t \in (\tau_0,+\infty)$. We clarify that

$$i[\tau_0,t) = \begin{cases} 0, & \text{for } t \in (\tau_0,\tau_1], \\ 1, & \text{for } t \in (\tau_1,\tau_2], \\ \cdots\cdots\cdots\cdots\cdots \\ k, & \text{for } t \in (\tau_k,\tau_{k+1}], \ k \in N. \\ \cdots\cdots\cdots\cdots\cdots \end{cases}$$

Theorem 2 *Assume that the hypotheses (H1)-(H2) are satisfied. Suppose also, that there exists a number $p_0 > 0$, such that $p_0 = \min\limits_{k\in N} p_k$, and*

(ii) $\left(1+\frac{p_0}{c-1}\right)^m \ge \frac{c-1}{ep(h-\sigma)}$, *where* $m = \min\limits_{t\in[t_0,+\infty)} i[t,t+h-\sigma)$.

Then the equation (E_1) is oscillatory.

Proof: Proceeding as in the proof of Theorem 1, we conclude (14). But, because $\min\limits_{k\in N} p_k = p_0 > 0$ and $m = \min\limits_{t\in[t_0,+\infty)} i[t,t+h-\sigma)$, it follows from (14), that

$$\left(1+\frac{p_0}{c-1}\right)^m \le \prod_{t\le\tau_k<t+h-\sigma}\left(1+\frac{p_0}{c-1}\right) \le \prod_{t\le\tau_k<t+h-\sigma}\left(1+\frac{p_k}{c-1}\right) < \frac{c-1}{ep(h-\sigma)},$$

which contradicts the conditions (ii) of the theorem. The proof is complete.

As a natural consequence, we have the following result:

Corollary 1 *Let the conditions of Theorem 1, or Theorem 2 are satisfied. Then*
(a) the inequality $(N_{1,\le})$ has no eventually positive solutions;
(b) the inequality $(N_{1,\ge})$ has no eventually negative solutions.

The proof of the corollary is similar to that of Theorem 1 and that is why, it is omitted.

Next theorem offers a different approach to obtain another sufficient condition for oscillation of the equation (E_1).

Theorem 3 *Assume that the hypotheses $(H1)$-$(H2)$ are satisfied. Suppose also that:*
(iii) $\liminf\limits_{t \to \infty}[p(h - \sigma) + \sum\limits_{t \le \tau_i < t+h-\sigma} p_i] \ge c.$
Then the equation (E_1) is oscillatory.

Proof: Let assume, for the sake of contradiction, that equation (E_1) has a non-oscillatory solution. Since the negative of a solution of (E_1) is again a solution of (E_1), it suffices to prove the theorem considering an eventually positive solution.

So, let suppose that there exists a solution $y(t)$ of the equation (E_1) and a number $\tilde{t} > 0$, such that $y(t)$ is defined for $t \ge \tilde{t}$, $y(t) > 0$ for $t \ge \tilde{t}$ and $y(t - h) > 0$, $y(t - \sigma) > 0$ for $t \ge t_0 = \tilde{t} + \max\{h, \sigma\}$.

Recalling (*), we conclude by Lemma 1(a), that $z(t)$ is a decreasing function for $t \ge t_0$ with "down-jumps" at the points of impulsive effect ($\Delta z(\tau_k) < 0$). Moreover, by Lemma 1(b), it follows that $z(t)$ is eventually negative.

Therefore, there exists some $t_1 \ge t_0$, such that $z(t) < 0$ for $t \ge t_1$ and $\Delta z(\tau_k) < 0$ for $\tau_k \ge t_1$. Obviously, then we have

$$y(t) - cy(t - h) = z(t) > -cy(t - h), \tag{15}$$

which implies $$z(t + h - \sigma) > -cy(t - \sigma).$$
Multiplying the both sides of the last inequality by $\frac{p}{-c} < 0$ we obtain

$$\frac{p}{-c}z(t + h - \sigma) < py(t - \sigma) = -z'(t).$$

Hence, $$z'(t) - \frac{p}{c}z(t + h - \sigma) < 0, \quad t \ne \tau_k, \ k \in N. \tag{16}$$

Again from (15) we have and $$z(\tau_k + h - \sigma) > -cy(\tau_k - \sigma).$$
Multiplying by $\frac{p_k}{-c} < 0$ the both sides of the last inequality, we obtain and

$$\frac{p_k}{-c}z(\tau_k + h - \sigma) < p_k y(\tau - \sigma) = -\Delta z(\tau_k),$$

i.e., $$\Delta z(\tau_k) - \frac{p_k}{c}z(\tau_k + h - \sigma) < 0, \ k \in N. \tag{17}$$

Denote $s = h - \sigma > 0$, $Q = \frac{p}{c}$, $q_k = \frac{p_k}{c}$.

We see from (16) and (17), that our assumption in the beginning leads us to the conclusion, that for every $t > t_1$, the decreasing negative function $z(t)$ satisfies the impulsive differential inequality with advanced argument of the form

$$z(t)' - Qz(t+s) < 0, \ t \neq \tau_k, \ k \in N \qquad (18)$$

$$\Delta z(\tau_k) - q_k z(\tau_k + s) < 0, \ k \in N.$$

Let now integrate (18) from t to $t+s$ and obtain

$$z(t+s) - z(t) - \sum_{t \leq \tau_k \leq t+s} \Delta z(\tau_k) - \int_t^{t+s} Qz(r+s)dr < 0.$$

From here, by omitting the second term and because $\Delta z(\tau_k) < q_k z(\tau_k + s)$ we find

$$-z(t+s) + \sum_{t \leq \tau_k \leq t+s} q_k z(\tau_k + s) + Q \int_t^{t+s} z(r+s)dr > 0.$$

Further, using the decreasing nature of $z(t)$ and the fact that

$$z(t+s) > z(\tau_k + s), \quad \text{when } t+s < \tau_k + s,$$

we obtain

$$-z(t+s) + z(t+s) \sum_{t \leq \tau_k \leq t+s} q_k + sQz(t+s) > 0,$$

i.e.,
$$z(t+s)(-1 + sQ + \sum_{t \leq \tau_k < t+s} q_k) > 0,$$

which, because $z(t)$ is supposed to be negative, leads us to the conclusion that

$$sQ + \sum_{t \leq \tau_k \leq t+s} q_k < 1.$$

Observe, that when $s = h - \sigma > 0$, $Q = \frac{p}{c}$, $q_k = \frac{p_k}{c}$, the last inequality looks like

$$[\frac{p}{c}(h-\sigma) + \sum_{t \leq \tau_i < t+h-\sigma} \frac{p_i}{c}] < 1 \qquad (19)$$

and contradicts the condition (iii) of the theorem. The proof is complete.

Theorem 4 *Assume that the hypotheses $(H1)$-$(H2)$ are satisfied. Suppose also, that there exists a number $p_0 > 0$, such that $p_0 = \min_{k \in N} p_k$, and*

$(iiii) \quad [p(h-\sigma) + mp_0] \geq c, \qquad \text{where} \qquad m = \min_{t \in [t_0, +\infty)} i[t, t+h-\sigma).$

Then the equation (E_1) is oscillatory.

Proof: Proceeding as in the proof of Theorem 3, we conclude (19), i.e.,

$$[p(h-\sigma) + \sum_{t \leq \tau_i < t+h-\sigma} p_i] < c. \qquad (1)$$

But, because $p_0 = \min_{k \in N} p_k$ and $m = \min_{t \in [t_0, +\infty)} i[t, t+h-\sigma)$, it follows from (1), that

$$[p(h-\sigma) + mp_0] \leq [p(h-\sigma) + \sum_{t \leq \tau_i < t+h-\sigma} p_0] \leq [p(h-\sigma) + \sum_{t \leq \tau_i < t+h-\sigma} p_i] < c,$$

which contradicts the condition (iiii) of the theorem. The proof is complete.

As immediate consequences of the above theorems, we formulate the following results.

Corollary 2 *Let the hypotheses* $(H1)$ - $(H2)$ *are satisfied and* $0 < \frac{p}{c}(h - \sigma) \leq \frac{1}{e}$. *Suppose also that either* $\displaystyle\liminf_{t \to +\infty} \sum_{t \leq \tau_k \leq t + (h - \sigma)} p_k \geq c$, *or* $mp_0 \geq c$, *where* $m = \displaystyle\min_{t \in [t_0, +\infty)} i[t, t + h - \sigma)$ *and* $p_0 = \displaystyle\min_{k \in N} p_k$. *Then the equation* (E_1) *is oscillatory.*

Corollary 3 *Let the conditions of Theorem 3, Theorem 4, or Corollary 2 are satisfied. Then*

(a) *the inequality* $(N_{1, \leq})$ *has no eventually positive solutions*
(b) *the inequality* $(N_{1, \geq})$ *has no eventually negative solutions*

Remark 1 *Note that, as it is well-known (see, for example [7], Corollary 3.1.6, or [13], Theorem 6.4.2), a sufficient condition for the oscillation of all solutions of the neutral delay differential equation*

$$\frac{d}{dt}[y(t) - cy(t - h)] + py(t - \sigma) = 0, \qquad (E_1^*)$$

without impulsive effects, is implied to be $\frac{p}{c}(h - \sigma) > \frac{1}{e}$. *Our result above, demonstrates the influence of the appearance of the impulsive effects, on the behavior of solutions of* (E_1). *Indeed, Corollary 2 shows the fact, that the neutral delay differential equation* (E_1), *is oscillatory even in the case, when* $\frac{p}{c}(h - \sigma) \leq \frac{1}{e}$.

EXAMPLES

We illustrate our results by two examples, where the coefficients p_k are set to be one and the same constant, for convenience.

Example 1 *Consider the neutral delay impulsive differential equation*

$$[y(t) - 2y(t - 2)]' + \tfrac{2}{10}y(t - 1) = 0, \, t \neq \tau_k,$$
$$\Delta[y(\tau_k) - 2y(\tau_k - 2)] + \tfrac{15}{10}y(\tau_k - 1) = 0, \, k \in N,$$

where $p = \tfrac{2}{10}$, $p_0 = p_k = \tfrac{15}{10}$, $c = 2$, $h = 2 > \sigma = 1$ *and* $\tau_{k+1} - \tau_k = 1$, $k \in N$, $e = \exp$.

Here, the hypotheses $(H1) - (H2)$ are satisfied, $m = \displaystyle\min_{t \in [t_0, +\infty)} i[t, t + h - \sigma) = 1$ and

$$(1 + \frac{p_0}{c - 1})^m = (1 + \frac{1.5}{2 - 1})^1 = 2.5 > \frac{10}{2e} = \frac{2 - 1}{e(2 - 1)\frac{2}{10}} = \frac{c - 1}{ep(h - \sigma)}.$$

It is easy to see that the conditions of Theorem 1, or of Theorem 2 are satisfied, whereas the conditions of Theorem 3, Theorem 4, or Corollary 3 are not. Indeed,
$$[p(h - \sigma) + mp_0] = [0.2(2 - 1) + 1 * 1.5] = 1.7 < 2 = c.$$

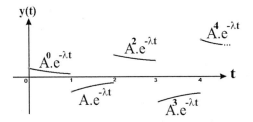

FIGURE 1. Oscillatory solution $y(t)$ of Example 1

Therefore, by Theorem 1, or Theorem 2, all solutions of the above equation are oscillatory. For example, one oscillatory solution of this equation is the "pulsatile exponent" (see, [5] for the definition of such "pulsatile" function).

$$y(t) = A^{i[t_0,t)}e^{-\lambda_* t}, \ t_0 \geq 0,$$

where $A = -1.6508013641$ is the "pulsatile" coefficient and $\lambda_* = +0.3534401819$ is the exponential argument (see Figure 1).

Example 2 *Consider the neutral delay impulsive differential equation*

$$[y(t) - (e^2 + 1)y(t - 2)]' + \tfrac{1}{e}y(t - 1) = 0, \ t \neq \tau_k,$$
$$\Delta[y(\tau_k) - (e^2 + 1)y(\tau_k - 2)] + 10y(\tau_k - 1) = 0, \ k \in N,$$

where $p = \frac{1}{e}$, $p_k = p_0 = 10$, $c = (e^2 + 1)$, $h = 2 > \sigma = 1$, $\tau_{k+1} - \tau_k = 1$, $k \in N$, $e = \exp.$

Here, the hypotheses $(H_1) - (H_2)$ are satisfied, $m = \min\limits_{t \in [t_0, +\infty)} i[t, t + h - \sigma) = 1$ and

$$\tfrac{p}{c}(h - \sigma) = \tfrac{1}{e(e^2 + 1)} < \tfrac{1}{e}, \ \liminf\limits_{t \to +\infty} \sum\limits_{t \leq \tau_k \leq t + (h - \sigma)} p_k = mp_0 = 10 > c = (e^2 + 1).$$

Note, that the conditions of Theorem 1, or Theorem 2 are not satisfied, whereas the conditions of Theorem 3, Theorem 4, or especially Corollary 2 are satisfied. Therefore, all solutions of the above equation are oscillatory. For example, one oscillatory solution of this equation is the "pulsatile exponent"

$$y(t) = A^{i[t_0,t)}e^{-\lambda_* t}, \ t_0 \geq 0$$

where $A = -1.7880967589$ is the "pulsatile" coefficient and $\lambda_* = +0.102618528$ is the exponential argument.

Observe, that the equation without impulses

$$[y(t) - (e^2 + 1)y(t - 2)]' + \frac{1}{e}y(t - 1) = 0,$$

admits the none-oscillatory solution $y(t) = e^t$. This demonstrates the influence of the appearance of the impulsive effects on the behavior of solutions of the differential equation, providing an occasion for their oscillations.

Remark 2 *The chart of the Example 2 is very similar to that of Example 1.*

REFERENCES

1. D. D. Bainov, V. Lakshmikantham, and P. S. Simeonov, *Theory of Impulsive Differential Equations* , World Scientific, Singapore, 1989.
2. D. D. Bainov, M. B. Dimitrova, and A. B. Dishliev, *Journ. of Appl. Anal.* **5**(2), 261–275 (1999).
3. R. K. Brayton and R. A. Willoughby, *J. Math.Anal. Appl.* **18**, 182–189 (1967).
4. M. B. Dimitrova, *Kyungpook Math.J.* **40**, 29–37 (2000).
5. M. B. Dimitrova and V. I. Donev, *Nonlinear Oscillations* **8**(3), 304–318 (2005).
6. V. I. Donev, "Existence and asymptotic behavior of the positive solutions of a class of neutral impulsive differential equations in their special cases," in *Proceedings of 28th Int. Math Conf. on Applications of Math.in Engin.& Econ., Sozopol, 2002*, Bulvest 2000, Sofia, 2003.
7. L. H. Erbe, Q. Kong, and B. G. Zhang, *Oscillation Theory for Functional Differential Equations*, Marcel Decker Inc., New York, 1995.
8. K. Gopalsamy and B. G. Zhang, *J. Math. Anal. Appl.* **139**(1), 110–122 (1989).
9. M. K. Grammatikopoulos, M. B. Dimitrova, and V. I. Donev, "Oscillations of first order delay impulsive differential equations," Technical Report, University of Ioannina, Greece, **16**, 2007, pp.171–182.
10. M. K. Grammatikopoulos, M. B. Dimitrova, and V. I. Donev, "Oscillations of first order neutral impulsive delay differential equations with constant coefficients," Technical Report, University of Ioannina, Greece, **16**, 2007, pp.183–193.
11. M. K. Grammatikopoulos, M. B. Dimitrova, and V. I. Donev, "Oscillations of first order impulsive differential equations with variable coefficients and advanced argument," in *Applications of Mathematics in Engineering and Economics'33 (AMEE'07)*, edited by M.D. Todorov, AIP CP**946**, New York, Melville, 2007, pp. 206–214.
12. M. K. Grammatikopoulos, E. A. Grove, and G. Ladas, *Applicable Analysis* **22**, 1–19 (1986).
13. I. Györi and G. Ladas, *Oscillation Theory of Delay Differential Equations with Applications*, Clarendon Press, Oxford, 1991.
14. J. K. Hale, *Theory of Functional Differential Equations*, Springer Verlag, New York, 1977.
15. G. S. Ladde, V. Lakshmikantham, and B. G. Zhang, *Oscillation Theory of Differential Equations with Deviating Arguments*, Pure and Applied Mathematics, **110**, Marcel Dekker, New York, 1987.
16. A. D. Myshkis, *Linear Differential Equations with Retarded Argument*, Nauka, Moscow, Nauka, 1972. (in Russian)
17. A. M. Samoilenko and N. A. Perestyuk, *Impulsive Differential Equations*, World Sci. Ser. A: Monograph and Treatises, **14**, 1995.
18. M. Slemrod and E. F. Infante, *J.Math.Anal.Appl.* **38**, 388–415 (1972).
19. W. Snow, "Existence, uniqueness and stability for nonlinear differential-difference equation in the neutral case," N.Y.U. Curant Inst. Math.Sci.Rep. IMM-NYU **328** (February) 1965.
20. A. Zhao and J. Yan, *J.Math.Anal.Appl.* **210**, 667–678 (1997).

Space-like Bézier Curves in the Three-dimensional Minkowski Space

G. H. Georgiev

Faculty of Mathematics and Informatics,
Konstantin Preslavski University of Shoumen, Bulgaria

Abstract. Bézier curves in the Euclidean spaces are widely used in CAGD. We study space-like curves in Minkowski 3-space which are presented in a Bernstein-Bézier form.

Keywords: Minkowski space, space-like curves, Bézier curves.
PACS: 02.40.Hw, 87.10.Ed, 87.57.Np

INTRODUCTION

The Bézier curves play an important role in geometric modeling and Computer Aided Geometric Design (CAGD). Their derivatives are usually expressed in terms of forward difference operator. Applications of Minkowski 3-space in CAGD are given in [1]. Polynomial curves in Bernstein-Bézier form also can be considered in Minkowski 3-space. We discuss conditions for such curves to be space-like. Formulas for curvature and torsion of space-like Bézier curves in Minkowski 3-space are obtained.

BÉZIER CURVES IN THE THREE-DIMENSIONAL EUCLIDEAN SPACE

The points in the Euclidean space $\mathbb{R}^3$ are presented by their Cartesian coordinates, for example $\mathbf{P} = (p_1, p_2, p_3)$, $p_i \in \mathbb{R}$. For any two vectors $x = (x_1, x_2, , x_3)$ and $y = (y_1, y_2, , y_3)$ from the associate vector space are defined the inner product and the vector cross product as follows: $\langle x, y \rangle = x_1 y_1 + x_2 y_2 + x_3 y_3$ and $x \times y = (x_2 y_3 - x_3 y_2, x_3 y_1 - x_1 y_3, x_1 y_2 - x_2 y_1)$.

Let us consider the Bézier curve $\mathscr{B}(t) : [0,1] \longrightarrow \mathbb{R}^3$ given by a parametric equation

$$\mathscr{B}(t) = \sum_{i=0}^{N} B_{N,i}(t) \mathbf{P}_i, \tag{1}$$

where $\mathbf{P}_0, \mathbf{P}_1, \ldots \mathbf{P}_N \in \mathbb{R}^3$ are N+1 control points in general position and

$$B_{N,i}(t) = \frac{N!}{i!(N-i)!} t^i (1-t)^{N-i}, \qquad t \in [0,1]$$

are the Bernstein polynomials. A detailed description of the Bézier curves is given in [1], [8] and [9].

CP1067, *Applications of Mathematics in Engineering and Economics '34—AMEE '08*, edited by M. D. Todorov
© 2008 American Institute of Physics 978-0-7354-0598-1/08/$23.00

Some Properties of the Bézier Curves

1. The shape of $\mathscr{B}(t)$ given by (1) depends only on the control points.
2. An affine invariance of $\mathscr{B}(t)$. This means that if

$$\Phi : \mathbb{R}^n \longrightarrow \mathbb{R}^n$$

is an affine transformation the image $\overline{\mathscr{B}}(t) = \Phi(\mathscr{B}(t))$ is the Bézier curve given by the equation

$$\overline{\mathscr{B}}(t) = \sum_{i=0}^{N} B_{N,i}(t)\,\Phi(\mathbf{P}_i), \qquad t \in [0,1].$$

The shapes of quadratic and cubic Bézier curves are studied in [3] and [4].

Derivatives of a Bézier Curve

Let Δ^k be the forward differencing operator defined as follows

$$\Delta^k \mathbf{P}_i = \Delta^{k-1}\mathbf{P}_{i+1} - \Delta^{k-1}\mathbf{P}_i \qquad k = 1,2\ldots \quad i = 0,1,\ldots,N-1$$

and

$$\Delta^0 \mathbf{P}_0 = \mathbf{P}_0, \ \Delta^0 \mathbf{P}_1 = \mathbf{P}_1,\ldots, \Delta^0 \mathbf{P}_N = \mathbf{P}_N.$$

Then the derivatives are given by

$$\mathscr{B}'(t) = \frac{d}{dt}\mathscr{B}(t) = N\sum_{i=0}^{N-1} B_{N-1,i}(t)\Delta^1 \mathbf{P}_i,$$

$$\mathscr{B}''(t) = \frac{d^2}{dt^2}\mathscr{B}(t) = N(N-1)\sum_{i=0}^{N-2} B_{N-2,i}(t)\Delta^2 \mathbf{P}_i, \tag{2}$$

$$\mathscr{B}'''(t) = \frac{d^2}{dt^2}\mathscr{B}(t) = N(N-1)(N-2)\sum_{i=0}^{N-3} B_{N-3,i}(t)\Delta^3 \mathbf{P}_i.$$

THREE-DIMENSIONAL MINKOWSKI SPACE

Let $\mathbb{R}_1^3 = \mathbb{R}_{1,2}^3$ be the Minkowski 3-space. Then, the points of $\mathbb{R}_1^3$ are presented by ordered triple of real numbers, i.e., $\mathbf{P} = (p_1, p_2, p_3) \in \mathbb{R}_1^3$, $p_i \in \mathbb{R}$. For two arbitrary vectors $x = (x_1, x_2, x_3)$ and $y = (y_1, y_2, y_3)$ of the associate vector space of $\mathbb{R}_1^3$, the inner product is the real number $\langle x, y \rangle_1 = -x_1 y_1 + x_2 y_2 + x_3 y_3$ and the vector cross product is defined by $(x \times y)_1 = (x_3 y_2 - x_2 y_3, x_3 y_1 - x_1 y_3, x_1 y_2 - x_2 y_1)$. In other words if vectors f_1, f_2, f_3 form an orthonormal basis of the associated vector space of $\mathbb{R}_1^3$, then

$$\langle f_i, f_i \rangle_1 = \begin{cases} -1, & i = 1 \\ 1, & i = 2,3 \end{cases} \quad \text{and} \quad \langle f_i, f_j \rangle_1 = 0 \quad \text{for} \quad i \neq j.$$

Moreover,

$$\begin{aligned}
(f_2 \times f_3)_1 &= -f_1 \\
(f_3 \times f_1)_1 &= f_2 \\
(f_1 \times f_2)_1 &= f_3.
\end{aligned}$$

The vector x is called

- space-like, if $\langle x, x \rangle_1 = -x_1^2 + x_2^2 + x_3^2 > 0$,
- time-like, if $\langle x, x \rangle_1 = -x_1^2 + x_2^2 + x_3^2 < 0$,
- light-like, if $\langle x, x \rangle_1 = -x_1^2 + x_2^2 + x_3^2 = 0$.

Note that for any three vectors x, y and z it is fulfilled $\langle (x \times y)_1, z \rangle_1 = \det(x, y, z)$. Hence, $\langle (x \times y)_1, x \rangle_1 = \langle (x \times y)_1, y \rangle_1 = 0$.

Regular Curves in Minkowski 3-space

Let $I \subset \mathbb{R}$ be an interval, and let

$$c(t) : I \longrightarrow \mathbb{R}_1^3 \tag{3}$$

be a smooth curve of class C^∞, i.e., all derivatives $c' = \frac{d}{dt}c(t)$, $c'' = \frac{d^2}{dt^2}c(t)$, $c''' = \frac{d^3}{dt^3}c(t), \ldots$ exist and $\langle c', c' \rangle_1 \neq 0$. Then, the smooth curve $c(t)$ is regular if

$$\langle c' \times c'', c' \times c'' \rangle_1 \neq 0$$

for any $t \in I$. The regular curve is called

- space-like, if $\langle c', c' \rangle_1 > 0$,
- time-like, if $\langle c', c' \rangle_1 < 0$,
- light-like, if $\langle c', c' \rangle_1 = 0$.

for any $t \in I$. Consider the regular space-like curve $c(t)$ given by (3) and denote

$$\varepsilon = \begin{cases} 1, & \text{for } \langle c' \times c'', c' \times c'' \rangle_1 > 0; \\ -1, & \text{for } \langle c' \times c'', c' \times c'' \rangle_1 < 0. \end{cases} \tag{4}$$

Then there exists an orthonormal moving frame

$$e_1(t) = \frac{c'(t)}{\sqrt{\langle c'(t), c'(t) \rangle_1}},$$

$$e_2(t) = \left(\frac{(c'(t) \times c''(t))_1}{\sqrt{\varepsilon \langle c'(t) \times c''(t), c'(t) \times c''(t) \rangle_1}} \times \frac{c'(t)}{\sqrt{\langle c'(t), c'(t) \rangle_1}} \right)_1,$$

$$e_3(t) = \frac{(c'(t) \times c''(t))_1}{\sqrt{\varepsilon \langle c'(t) \times c''(t), c'(t) \times c''(t) \rangle_1}}.$$

along the curve $c(t)$. It is clear that

- e_1, e_3 are space-like and e_2 is time-like, if $\varepsilon = 1$;
- e_1, e_2 are space-like and e_3 is time-like, if $\varepsilon = -1$.

Curvature and Torsion of Space-like Curves

Every regular space-like curve $c(t) : I \longrightarrow \mathbb{R}^3_1$ can be parameterized by an arc-length parameter $s(t) = \int_{t_0}^t \sqrt{\langle c'(u), c'(u) \rangle_1} \, du$, i.e., $c(s) : I_0 \longrightarrow \mathbb{R}^3_1$. Then,

$$\left\langle \frac{d}{ds}c(s), \frac{d}{ds}c(s) \right\rangle_1 = 1, \quad \left\langle \frac{dc(s)}{ds}, \frac{d^2c(s)}{ds^2} \right\rangle_1 = 0,$$

$$\left\langle \left(\frac{dc(s)}{ds} \times \frac{d^2c(s)}{ds^2} \right)_1, \left(\frac{dc(s)}{ds} \times \frac{d^2c(s)}{ds^2} \right)_1 \right\rangle_1 = - \left\langle \frac{d^2c(s)}{ds^2}, \frac{d^2c(s)}{ds^2} \right\rangle_1$$

and the orthonormal moving frame is

$$e_1(s) = \frac{d}{ds}c(s), \; e_2(s) = \frac{d^2c(s)}{ds^2} \cdot \frac{1}{\sqrt{-\varepsilon \left\langle \frac{d^2c(s)}{ds^2}, \frac{d^2c(s)}{ds^2} \right\rangle_1}}, \; e_3(s) = (e_1(s) \times e_2(s))_1,$$

where

$$\varepsilon = \begin{cases} 1, & \text{for } \left\langle \left(\frac{dc(s)}{ds} \times \frac{d^2c(s)}{ds^2} \right)_1, \left(\frac{dc(s)}{ds} \times \frac{d^2c(s)}{ds^2} \right)_1 \right\rangle_1 > 0; \\ -1, & \text{for } \left\langle \left(\frac{dc(s)}{ds} \times \frac{d^2c(s)}{ds^2} \right)_1, \left(\frac{dc(s)}{ds} \times \frac{d^2c(s)}{ds^2} \right)_1 \right\rangle_1 < 0. \end{cases}$$

According to [7], p. 36, the curvature of $c(s)$ is

$$\kappa_1(s) = \left\langle \frac{d}{ds}e_1(s), e_2(s) \right\rangle_1,$$

and the torsion of $c(s)$ is

$$\tau_1(s) = \left\langle \frac{d}{ds}e_2(s), e_3(s) \right\rangle_1.$$

Since $s'(t) = \frac{d}{dt}s(t) = \sqrt{\langle c'(t), c'(t) \rangle_1}$, for the derivatives of the unit vectors $e_1(t)$, $e_2(t)$, $e_3(t)$ we have

$$e_i'(t) = e_i'(s(t)) = \frac{d}{dt}e_i(s(t)) = s'(t) \cdot \frac{d}{ds}e_i(s) = \sqrt{\langle c'(t), c'(t) \rangle_1} \cdot \frac{d}{ds}e_i(s), \; i = 1,2,3.$$

376

Then, the formulas for curvature and torsion for a regular space-like curve $c(t) : I \longrightarrow \mathbb{R}_1^3$ which is parameterized by an arbitrary parameter t are

$$\kappa_1(t) = \frac{\sqrt{\varepsilon \langle c'(t) \times c''(t), c'(t) \times c''(t) \rangle_1}}{\left(\sqrt{\langle c'(t), c'(t) \rangle_1} \right)^3}$$

$$\tau_1(t) = \frac{\det(c'(t), c''(t), c'''(t))}{\langle c'(t) \times c''(t), c'(t) \times c''(t) \rangle_1},$$

(5)

where ε is determined by (4).

BÉZIER CURVES IN THE MINKOWSKI 3-SPACE

Many definitions and theorems for curves in the Euclidean space $\mathbb{R}^3$ have corresponding analogs in the Minkowski spaces $\mathbb{R}_1^3$ and $\mathbb{R}_1^4$ (see [2] and [7]). There is a natural way to determine Bézier curves in $\mathbb{R}_1^3$. Our considerations are restricted to the case of space-like curves.

Definition 1 *Let*

$$\mathbf{P}_0, \mathbf{P}_1, \dots \mathbf{P}_N$$

be different points in the Minkowski space $\mathbb{R}_1^3$ and the points $\mathbf{P}_{i-1}, \mathbf{P}_i, \mathbf{P}_{i+1}$ be non-collinear for $i = 1,2,\dots,N-1$. Then the curve $\mathscr{B}_M(t) : [0,1] \longrightarrow \mathbb{R}_1^3$ given by a parametric equation

$$\mathscr{B}_M(t) = \sum_{i=0}^{N} B_{N,i}(t)\, \mathbf{P}_i,$$

(6)

where

$$B_{N,i}(t) = \frac{N!}{i!(N-i)!}\, t^i(1-t)^{N-i}, \qquad i = 0,1,\dots,N, \qquad t \in [0,1]$$

are the Bernstein polynomials, is called a Bézier curve with a control polygon $\mathbf{P}_0, \mathbf{P}_1, \dots \mathbf{P}_N$.

From the properties of the Bernstein polynomials it follows that the polygons $\mathbf{P}_0, \mathbf{P}_1, \dots \mathbf{P}_N$ and $\mathbf{P}_N, \mathbf{P}_{N-1}, \dots \mathbf{P}_0$ determine one and the same Bézier curve. Moreover, the derivatives $\mathscr{B}_M'(t) = \dfrac{d}{dt}\mathscr{B}_M(t)$, $\mathscr{B}_M''(t) = \dfrac{d^2}{dt^2}\mathscr{B}_M(t)$, $\mathscr{B}_M'''(t) = \dfrac{d^3}{dt^3}\mathscr{B}_M(t)$ of $\mathscr{B}_M(t)$ can be expressed by the formulas (2) in which the vectors $\Delta^k \mathbf{P}_i$ are space-like, or time-like, or light-like.

Let us consider the coordinate functions $\pi_l : \mathbb{R}_1^3 \to \mathbb{R}$, $l = 1,2,3$, i.e., for any vector $v = (v_1, v_2, v_3) \in \mathbb{R}_1^3$ it is fulfilled $\pi_l(v) = v_l$.

Theorem 2 *Let $\mathscr{B}_M(t)$ be a Bézier curve given by (6). If all vectors*

$$\Delta^1 \mathbf{P}_0 = \mathbf{P}_1 - \mathbf{P}_0, \ \Delta^1 \mathbf{P}_1 = \mathbf{P}_2 - \mathbf{P}_1, \ \Delta^1 \mathbf{P}_2 = \mathbf{P}_3 - \mathbf{P}_2, \dots, \Delta^1 \mathbf{P}_{N-1} = \mathbf{P}_N - \mathbf{P}_{N-1}$$

377

are space-like, i.e.,

$$\left(\pi_1(\Delta^1\mathbf{P}_i)\right)^2 < \left(\pi_2(\Delta^1\mathbf{P}_i)\right)^2 + \left(\pi_3(\Delta^1\mathbf{P}_i)\right)^2 \quad \text{for} \quad i = 0, 1 \ldots, N-1, \tag{7}$$

then the curve $\mathscr{B}_M(t)$ is space-like.

Proof: The first derivative of $\mathscr{B}_M(t)$ is

$$\mathscr{B}'_M(t) = \frac{d}{dt}\mathscr{B}_M(t) = N \sum_{i=0}^{N-1} B_{N-1,i}(t)\Delta^1\mathbf{P}_i, \qquad t \in [0,1]. \tag{8}$$

Using the inequalities (7) and the equality

$$\left(\left(\pi_2(\Delta^1\mathbf{P}_i)\right)^2 + \left(\pi_3(\Delta^1\mathbf{P}_i)\right)^2\right)\left(\left(\pi_2(\Delta^1\mathbf{P}_j)\right)^2 + \left(\pi_3(\Delta^1\mathbf{P}_j)\right)^2\right)$$
$$= \left(\pi_2(\Delta^1\mathbf{P}_i)\pi_2(\Delta^1\mathbf{P}_j) + \pi_3(\Delta^1\mathbf{P}_i)\pi_3(\Delta^1\mathbf{P}_j)\right)^2$$
$$+ \left(\pi_2(\Delta^1\mathbf{P}_i)\pi_3(\Delta^1\mathbf{P}_j) - \pi_3(\Delta^1\mathbf{P}_i)\pi_2(\Delta^1\mathbf{P}_j)\right)^2$$

we obtain for $i \neq j$

$$\left|\pi_1(\Delta^1\mathbf{P}_i)\right| \cdot \left|\pi_1(\Delta^1\mathbf{P}_j)\right|$$
$$< \sqrt{\left(\left(\pi_2(\Delta^1\mathbf{P}_i)\right)^2 + \left(\pi_3(\Delta^1\mathbf{P}_i)\right)^2\right)\left(\left(\pi_2(\Delta^1\mathbf{P}_j)\right)^2 + \left(\pi_3(\Delta^1\mathbf{P}_j)\right)^2\right)}$$
$$< \left|\pi_2(\Delta^1\mathbf{P}_i) \cdot \pi_2(\Delta^1\mathbf{P}_j) + \pi_3(\Delta^1\mathbf{P}_i) \cdot \pi_3(\Delta^1\mathbf{P}_j)\right|$$

From here and (7) it follows that

$$\langle \mathscr{B}'_M(t), \mathscr{B}'_M(t) \rangle_1 > 0 \qquad \text{for any} \quad t \in [0,1].$$

$\square$

Theorem 3 *If $\mathscr{B}_M(t)$ is a Bézier curve in $\mathbb{R}^3_1$ with the property*

$$\pi_1(\Delta^1\mathbf{P}_i) = \pi_2(\Delta^1\mathbf{P}_i) \quad \left(or \quad \pi_1(\Delta^1\mathbf{P}_i) = \pi_3(\Delta^1\mathbf{P}_i)\right) \quad \text{for} \quad i = 0, 1 \ldots, N-1, \tag{9}$$

then $\mathscr{B}_M(t)$ is a space-like curve which is not regular.

Proof: For any $t \in [0,1]$ the condition (9) implies first the vector $\mathscr{B}'_M(t)$ is space-like and second the vector $(\mathscr{B}'_M(t) \times \mathscr{B}''_M(t))_1$ is light-like. $\square$

The last two theorems give sufficient conditions for the curve $\mathscr{B}_M(t)$ to be space-like. Now, we shall prove a necessary condition.

Theorem 4 *If the Bézier curve $\mathscr{B}_M(t)$ is space-like, then the vectors $\Delta^1\mathbf{P}_0 = \mathbf{P}_1 - \mathbf{P}_0$ and $\Delta^1\mathbf{P}_{N-1} = \mathbf{P}_N - \mathbf{P}_{N-1}$ are space-like.*

Proof: Using the expression (8) for the first derivative we obtain $\mathscr{B}'_M(0) = \Delta^1\mathbf{P}_0 = \mathbf{P}_1 - \mathbf{P}_0$ and $\mathscr{B}'_M(1) = \Delta^1\mathbf{P}_{N-1} = \mathbf{P}_N - \mathbf{P}_{N-1}$. Hence, both vectors are space-like. $\square$

Corollary 5 *A space-like Bézier curve $\mathscr{B}_M(t)$ with a control polygon $\mathbf{P}_0, \mathbf{P}_1, \ldots \mathbf{P}_N$, $(N > 3)$ is closed, if $\mathbf{P}_0 = \mathbf{P}_N$ and $\mathbf{P}_0$ is the midpoint of the segment $[\mathbf{P}_1, \mathbf{P}_{N-1}]$.*

378

The properties of closed space-like curves in $\mathbb{R}_1^3$ are studied in [5] and [6] .

Theorem 6 *Let $\mathscr{B}_M(t)$ be a space-like Bézier curve with a parametric representation (6). Define the functions*

$$
\begin{aligned}
Q_1(t) &= \langle \mathscr{B}'_M(t), \mathscr{B}'_M(t) \rangle_1 & t &\in [0,1] \\
Q_2(t) &= \langle (\mathscr{B}'_M(t) \times \mathscr{B}''_M(t))_1 , (\mathscr{B}'_M(t) \times \mathscr{B}''_M(t))_1 \rangle_1 & t &\in [0,1] \\
Q_3(t) &= \langle (\mathscr{B}'_M(t) \times \mathscr{B}''_M(t))_1 , \mathscr{B}'''_M(t) \rangle_1 & & \\
&= \det (\mathscr{B}'_M(t), \mathscr{B}''_M(t), \mathscr{B}'''_M(t)) & t &\in [0,1].
\end{aligned}
$$

If $\mathscr{B}_M(t)$ is regular for some $t_0 \in [0,1]$, then the curvature of $\mathscr{B}_M(t)$ at the point $\mathscr{B}_M(t_0)$ is

$$
\kappa_1(t) = \frac{\sqrt{|Q_2(t)|}}{\left(\sqrt{Q_1(t)}\right)^3} \tag{10}
$$

and the torsion of $\mathscr{B}_M(t)$ at the point $\mathscr{B}_M(t_0)$ is

$$
\tau_1(t) = \frac{Q_3(t)}{Q_2(t)} . \tag{11}
$$

Proof: Since $\mathscr{B}_M(t)$ is space-like, $Q_1(t) > 0$ for any t. Similarly, from regularity of $\mathscr{B}_M(t)$ it follows that either $Q_2(t_0) > 0$ or $Q_2(t_0) < 0$. Finally, using (5) we obtain formulas (10) and (11) . $\qquad \square$

Corollary 7 *Let $\mathscr{B}_M(t)$ be a space-like Bézier curve with a control polygon $\mathbf{P}_0, \mathbf{P}_1, \dots \mathbf{P}_N, (N \geq 3)$. Then, the curvature and the torsion of $\mathscr{B}_M(t)$ at the end point $\mathbf{P}_0$ are*

$$
\begin{aligned}
\kappa_1(0) &= \frac{N-1}{N} \cdot \frac{\sqrt{|\langle (\Delta^1\mathbf{P}_0 \times \Delta^1\mathbf{P}_1)_1 , (\Delta^1\mathbf{P}_0 \times \Delta^1\mathbf{P}_1)_1 \rangle_1 |}}{\left(\sqrt{\langle \Delta^1\mathbf{P}_0, \Delta^1\mathbf{P}_0 \rangle_1}\right)^3} , \\
\tau_1(0) &= \frac{N-2}{N} \cdot \frac{\det (\Delta^1\mathbf{P}_0, \Delta^1\mathbf{P}_1, \Delta^1\mathbf{P}_2)}{\langle (\Delta^1\mathbf{P}_0 \times \Delta^1\mathbf{P}_1)_1 , (\Delta^1\mathbf{P}_0 \times \Delta^1\mathbf{P}_1)_1 \rangle_1} .
\end{aligned} \tag{12}
$$

Proof: First, we observe that

$$
\begin{aligned}
\mathscr{B}_M{}'(0) &= N\Delta^1\mathbf{P}_0, \\
\mathscr{B}_M{}''(0) &= N(N-1)\Delta^2\mathbf{P}_0 \\
&= N(N-1)\left(\Delta^1\mathbf{P}_1 - \Delta^1\mathbf{P}_0\right), \\
\mathscr{B}_M{}'''(0) &= N(N-1)(N-2)\Delta^3\mathbf{P}_0 \\
&= N(N-1)(N-2)\left(\Delta^2\mathbf{P}_1 - \Delta^2\mathbf{P}_0\right) \\
&= N(N-1)(N-2)\left(\Delta^1\mathbf{P}_2 - 2\Delta^1\mathbf{P}_1 + \Delta^1\mathbf{P}_0\right).
\end{aligned}
$$

Then, applying Theorem 6 and doing some calculations, we get (12). $\qquad \square$

EXAMPLES

Example 1. If $\mathbf{P}_0 = (1, -1, 2)$, $\mathbf{P}_1 = (3, 1, 5)$, $\mathbf{P}_2 = (-2, -4, 3)$ are points in the Minkowski 3-space $\mathbb{R}_1^3$, then the quadratic Bézier curve

$$\mathscr{B}2_M(t) = (1-t)^2 \mathbf{P}_0 + 2(1-t)t\mathbf{P}_1 + t^2 \mathbf{P}_2, \quad t \in [0,1]$$

is space-like. According to Theorem 3 this curve is not regular.

Example 2. Let $\mathbf{P}_0 = (-3, -8, -9)$, $\mathbf{P}_1 = (-3, -3, -7)$, $\mathbf{P}_2 = (-2, 0, -5)$, $\mathbf{P}_3 = (2, 9, 4)$ be points in the Minkowski 3-space $\mathbb{R}_1^3$. Consider the cubic Bézier curve given by

$$\mathscr{B}3_M(t) = (1-t)^3 \mathbf{P}_0 + 3(1-t)^2 t\mathbf{P}_1 + 3(1-t)t^2 \mathbf{P}_2 + t^3 \mathbf{P}_3, \quad t \in [0,1].$$

Then, the vectors $\Delta^1 \mathbf{P}_0 = (0, 5, 2)$, $\Delta^1 \mathbf{P}_1 = (1, 3, 2)$ and $\Delta^1 \mathbf{P}_2 = (4, 9, 9)$ are space-like. From here and Theorem 2 it follows that $\langle \mathscr{B}3'_M(t), \mathscr{B}3'_M(t) \rangle_1 > 0$ for any $t \in [0, 1]$, i.e., $\mathscr{B}3_M(t)$ is a space-like curve. Moreover, $\langle \Delta^1 \mathbf{P}_0, \Delta^1 \mathbf{P}_0 \rangle_1 = 29$, $\left(\Delta^1 \mathbf{P}_0 \times \Delta^1 \mathbf{P}_1 \right)_1 = (-4, 2, -5)$ and $\det \left(\Delta^1 \mathbf{P}_0, \Delta^1 \mathbf{P}_1, \Delta^1 \mathbf{P}_2 \right) = -6$. Using formulas (12) we calculate the curvature and the torsion of $\mathscr{B}3_M(t)$ at the end point $\mathbf{P}_0$

$$\kappa_1(0) = \frac{2}{3} \cdot \frac{\sqrt{13}}{\left(\sqrt{29}\right)^3}, \qquad \tau_1(0) = -\frac{2}{13}.$$

ACKNOWLEDGMENTS

This work is partially supported by Konstantin Preslavski University of Shoumen under grant No RD-05-475/07.05.2008. The author is grateful to the referee for the useful suggestions.

REFERENCES

1. R. T. Farouki, *Pythagorean -Hodograph Curves: Algebra and Geometry Inseparable*, Springer-Verlag, Berlin, 2008.
2. J. B. Formiga and C. Romero, *American Journal of Physics* **74**, 1012–1016 (2006).
3. G. H. Georgiev, "Shapes of plane Bezier curves," in *Curve and Surface Design: Avignon 2006*, edited by P. Chenin, T. Lyche, and L. L.Schumaker, Nashboro Press, Brentwood, TN, 2007, pp. 143–152.
4. G. H. Georgiev, "On the shape of the cubic Bezier curve," in *Proc. of International Congress Pure and Applied Differential Geometry – Brussels 2007*, edited by F. Dillen and I. van de Woestyne, Shaker Verlag, Aachen, 2007, pp. 98–106.
5. S.Izumiya, M Kikuchi, and M Takahashi, *Journal of Knot Theory and its Ramifications* **15**, 869–881 (2006).
6. S. Izumiya and M. C. Romero Fuster, *Selecta Mathematica (NS)* **13**, 23–55 (2007).
7. W. Kühnel, *Differential geometry – Curves – Surfaces – Manifolds*, AMS, Providence, RI, 2006, 2nd edn.
8. D. Marsh , *Applied Geometry for Computer Graphics and CAD*, Spinger-Verlag, Berlin, 2005, 2nd edn.
9. H. Prautzsch, W. Boehm and M. Paluszny, *Bézier and B-spline Techniques*, Spinger-Verlag, Berlin, 2002.

Higher Order Convergence for Nonlinear Parabolic Integro-Differential Equations by Generalized Quasilinearization

T. G. Melton[*] and A. S. Vatsala[†]

[*]*Dept. of Mathematics and Physical Sciences, LSUA, LA, USA*
[†]*Dept. of Mathematics, University of Louisiana at Lafayette, LA, USA*

Abstract. In this paper the method of generalized quasilinearization is used in approach to the unique solution of the nonlinear problem from below and above by monotone convergent sequences of upper and lower solutions. We study a nonlinear parabolic integro-differential equation satisfying the following condition for the component functions $f(t,x,u) = f_1(t,x,u) + f_2(t,x,u)$, and $g(t,x,u)$ of the forcing function: (i) $\frac{\partial^{m-1} f_1(t,x,u)}{\partial u^{m-1}}$, $\frac{\partial^{m-1} f_2(t,x,u)}{\partial u^{m-1}}$ and $\frac{\partial^{m-1} g(t,x,u)}{\partial u^{m-1}}$ exist for $m > 2$ and they are one-sided Lipschitzian with respect to u; (ii) $\frac{\partial^{m-1} f_1(t,x,u)}{\partial u^{m-1}}$ and $\frac{\partial^{m-1} g(t,x,u)}{\partial u^{m-1}}$ are nondecreasing in u whenever $\frac{\partial^{m-1} f_2(t,x,u)}{\partial u^{m-1}}$ is nonincreasing in u for $m > 2$. Higher order convergence is shown. As an application of the developed results, a numerical example is constructed.

Keywords: Generalized quasilinearization, higher order convergence, integro-differential equation.
PACS: 01.10.Fv, 02.60.Nm

INTRODUCTION

The method of quasilinearization had been originally introduced fifty years ago in a theory of linear programming by Bellman and Kalaba [1, 2, 6] to solve individual or systems of non-linear differential equations. The modern developments of the method and its applications to different fields are given in monographs [7, 9]. In particular, for applications in quantum mechanics see [11, 12]. It has been extended recently and referred to as a generalized quasi-linearization method [10].

The neutron flux in the nuclear reactor model, the propagation of a voltage pulse through a nerve axon in the Hodgkin-Huxley model, etc. are governed by a Volterra type integro-differential equation [14]. The method of generalized quasilinearization has been applied to nonlinear integro-differential equations in [3, 5] in order to develop a quadratic order of convergence. However, rapid convergence has been obtained in [4, 8, 13].

The goal of this work is to extend the above results with certain conditions on the forcing function. Using an appropriate iterative scheme and lower and upper solutions, we obtain sequences which converge to the unique solution of the nonlinear integro-differential equations of Volterra's type rapidly. Finally, we provide a numerical example to illustrate the applicability of the obtained results.

CP1067, *Applications of Mathematics in Engineering and Economics '34—AMEE '08*, edited by M. D. Todorov
© 2008 American Institute of Physics 978-0-7354-0598-1/08/$23.00

PRELIMINARIES

We are concerned with the following nonlinear second order parabolic integro-differential equation.

$$\mathscr{L}u = f(t,x,u(t,x)) + \int_0^t g(t,x,s,u(s,x))ds \quad \text{in} \quad Q_T,$$

$$u(t,x) = \Phi(t,x), \quad x \in \partial\Omega, \qquad u(0,x) = u_0(x), \qquad x \in \Omega, \tag{1}$$

where Ω is a bounded domain in R^m with boundary $\partial\Omega \in C^{2+\overline{\alpha}}$ ($\overline{\alpha} \in (0,1)$) and closure $\overline{\Omega}$, $Q_T = (0,T) \times \Omega$, $\overline{Q}_T = [0,T] \times \overline{\Omega}$, $T > 0$. Let $\mathscr{L}$ be a second order differential operator defined by

$$\mathscr{L} = \frac{\partial}{\partial t} - L, \tag{2}$$

where

$$L = \sum_{i,j=1}^m a_{i,j}(t,x)\frac{\partial^2}{\partial x_i \partial x_j} + \sum_{i=1}^m b_i(t,x)\frac{\partial}{\partial x_i}. \tag{3}$$

We need to recall some preliminary definitions and theorems and to list the following assumptions (A_0).

(i) For each $i,j = 1,\ldots,m, a_{i,j}, b_j \in C^{\frac{\overline{\alpha}}{2},\overline{\alpha}}[\overline{Q}_T,R]$ and $\mathscr{L}$ is strictly uniformly parabolic in $\overline{Q}_T$;

(ii) $\partial\Omega$ belongs to the class $C^{2+\overline{\alpha}}$;

(iii) $f \in C^{\frac{\overline{\alpha}}{2},\overline{\alpha}}[[0,T] \times \overline{\Omega} \times R, R]$, $g \in C^{\frac{\overline{\alpha}}{2},\overline{\alpha}}[[0,T] \times \overline{\Omega} \times R^2, R]$ that is $f(t,x,u)$, $g(t,x,u)$ are Hölder continuous in t and (x,u) with exponents $\frac{\overline{\alpha}}{2}$ and $\overline{\alpha}$, respectively;

(iv) $\Phi \in C^{1+\frac{\overline{\alpha}}{2},2+\overline{\alpha}}[[0,T] \times \partial\Omega, R]$ and $u_0(x) \in C^{2+\overline{\alpha}}[\overline{\Omega},R]$;

(v) $u_0(x) = \Phi(0,x)$, $\Phi_t = Lu_0 + f(0,x,u_0)$ for $t = 0$ and $x \in \partial\Omega$.

Next we introduce the following definition.

Definition 1 *The functions α_0, $\beta_0 \in C^{1,2}[\overline{Q}_T,R]$ with $g(t,x,u)$ nondecreasing in u are said to be lower and upper solutions of (1), respectively, if*

$$\mathscr{L}\alpha_0 \leq f(t,x,\alpha_0(t,x)) + \int_0^t g(t,x,s,\alpha_0(s,x))ds \quad \text{in} \quad Q_T,$$

$$\alpha_0(t,x) \leq \Phi(t,x), \quad x \in \partial\Omega, \ \text{and} \ \alpha_0(0,x) \leq u_0(x), \qquad x \in \Omega;$$

$$\mathscr{L}\beta_0 \geq f(t,x,\beta_0(t,x)) + \int_0^t g(t,x,s,\beta_0(s,x))ds \quad \text{in} \quad Q_T,$$

$$\beta_0(t,x) \geq \Phi(t,x), \quad x \in \partial\Omega, \ \text{and} \ \beta_0(0,x) \geq u_0(x), \qquad x \in \Omega.$$

Let us state some theorems relative to the equation (1).

Theorem 1 *(see [5].) Assume that (A_0) holds. Then (1) has a unique smooth solution $u(t,x) \in C^{1+\frac{\overline{\alpha}}{2},2+\overline{\alpha}}[\overline{Q}_T,R]$.*

Theorem 2 *(see [5, 14].) Let $u(t,x) \in C^{\frac{1+\overline{\alpha}}{2},1+\overline{\alpha}}[\overline{Q}_T,R]$ be such that*

$$\mathscr{L}u + cu \geq 0 \quad \text{in} \quad Q_T,$$

$$u(t,x) \geq 0, \quad x \in \partial\Omega, \quad \text{and} \ u(0,x) \geq 0, \quad x \in \Omega,$$

where $c \equiv c(t,x)$ is a bounded function in Q_T. Then $u(t,x) \geq 0$ in $\overline{Q}_T$.

Theorem 3 *(see [15].) Assume that*

(i) $f_u(t,x,u)$ and $g_u(t,x,s,u)$ are bounded functions with $g(t,x,s,u)$ nondecreasing in u on $\overline{Q}_T$.

(ii) $\alpha(t,x)$ and $\beta(t,x)$ satisfy

$$\mathscr{L}\alpha \leq f(t,x,\alpha(t,x)) + \int_0^t g(t,x,s,\alpha(s,x))ds \quad in \quad Q_T,$$

$$\mathscr{L}\beta \geq f(t,x,\beta(t,x)) + \int_0^t g(t,x,s,\beta(s,x))ds \quad in \quad Q_T,$$

$$\alpha(t,x) \leq \beta(t,x), x \in \partial\Omega, \quad and \quad \alpha(0,x) \leq \beta(0,x), x \in \Omega.$$

Then $\alpha(t,x) \leq \beta(t,x)$ on $\overline{Q}_T$.

We also need the following comparison theorem which is a special case of Lemma 6.2 in [3].

Theorem 4 *Suppose that*

(i) $g(t,x,s,u)$ is monotone nondecreasing in u for each fixed point (t,x,s),

(ii) $\alpha(t,x)$ satisfies

$$\mathscr{L}\alpha \leq f(t,x,\alpha(t,x)) + \int_0^t g(t,x,s,\alpha(s,x))ds \quad in \quad Q_T,$$

$$\alpha(t,x) = 0, \quad x \in \partial\Omega, \quad and \quad \alpha_0(0,x) = u_0(x), \quad x \in \Omega,$$

(iii) $r(t)$ is the solution of the following ordinary integro-differential equation

$$r' = h_1(t,r) + \int_0^t h_2(t,s,r))ds, \quad r(0) = \max\{\max_{x\in\Omega} u_0(x), 0\},$$

$$h_1(t,r) \geq \max_{x\in\Omega} f(t,x,r), \quad and \quad h_2(t,r) \geq \max_{x\in\Omega} g(t,x,s,r).$$

Then $\alpha(t,x) \leq r(t)$ on $\overline{Q}_T$.

MAIN RESULTS

In this section we consider equation (1) with $f(t,x,u) = f_1(t,x,u) + f_2(t,x,u)$. We aim to prove a higher order of convergence m ($m > 2$) with iterates of $m-1$ order of nonlinearity by the generalized quasilinerisation method. We obtain two theorems depending on m being either an even or odd number. Let us assume at first that m is an odd number, say $m = 2k+1$.

Theorem 5 *Assume that all of (A_0) holds except (iii); further assume that*

(i) α_0, β_0 are lower and upper solutions of (1) with $\alpha_0(t,x) \leq \beta_0(t,x)$ on $\overline{Q}_T$.

(ii) $\dfrac{\partial^l f_1(t,x,u)}{\partial u^l}$, $\dfrac{\partial^l f_2(t,x,u)}{\partial u^l}$, and $\dfrac{\partial^l g(t,x,s,u)}{\partial u^l}$ exist and are bounded functions on $\overline{Q}_T$ for $l = 0,1,\ldots,2k$ such that $\dfrac{\partial^l f_1(t,x,u)}{\partial u^l}$, $\dfrac{\partial^l f_2(t,x,u)}{\partial u^l}$, and $\dfrac{\partial^l g(t,x,s,u)}{\partial u^l} \in C^{\frac{\alpha}{2},\alpha}[Q_T \times R, R]$.

383

(iii) Also $\dfrac{\partial^l g(t,x,s,u)}{\partial u^l}$ *are nondecreasing functions in u on* $\overline{Q}_T$ *for* $l = 0,1,\ldots,2k$ *such that*

$$g_u(\alpha_0) \geq g^{2k}(\beta_0)\frac{(\beta_0 - \alpha_0)^{2k-1}}{(2k-1)!} \quad and$$

$$0 \leq \frac{\partial^{2k} f_1(t,x,\eta_1)}{\partial u^{2k}} - \frac{\partial^{2k} f_1(t,x,\eta_2)}{\partial u^{2k}} \leq M_1(\eta_1 - \eta_2) \ on \ \overline{Q}_T,$$

$$0 \geq \frac{\partial^{2k} f_2(t,x,\zeta_1)}{\partial u^{2k}} - \frac{\partial^{2k} f_2(t,x,\zeta_2)}{\partial u^{2k}} \geq -M_2(\zeta_1 - \zeta_2) \ on \ \overline{Q}_T,$$

$$0 \leq \frac{\partial^{2k} g(t,x,\xi_1)}{\partial u^{2k}} - \frac{\partial^{2k} g(t,x,\xi_2)}{\partial u^{2k}} \leq M_3(\xi_1 - \xi_2) \ on \ \overline{Q}_T,$$

whenever
$$\alpha_0(t,x) \leq \eta_2(t,x) \leq \eta_1(t,x) \leq \beta_0(t,x), \qquad \alpha_0(t,x) \leq \zeta_2(t,x) \leq \zeta_1(t,x) \leq \beta_0(t,x),$$
$$\alpha_0(t,x) \leq \xi_2(t,x) \leq \xi_1(t,x) \leq \beta_0(t,x).$$

Then there exist monotone sequences $\{\alpha_n(t,x)\}$, $\{\beta_n(t,x)\}$, $n \geq 0$ *which converge uniformly and monotonically to the unique solution of (1) and the convergence is of order* $2k + 1$.

Proof: To prove the above theorem we consider the following simpler equations.

$$\mathcal{L}w = F_1(t,x,\alpha,\beta;w) + \int_0^t G_1(t,x,s,\alpha(s,x);w(s,x))ds = \sum_{i=0}^{2k} \frac{\partial^i f_1(t,x,\alpha)}{\partial u^i}\frac{(w - \alpha)^i}{i!}$$

$$+ \sum_{i=0}^{2k-1}\frac{\partial^i f_2(t,x,\alpha)}{\partial u^i}\frac{(w - \alpha)^i}{i!} + \frac{\partial^{2k} f_2(t,x,\beta)}{\partial u^{2k}}\frac{(w - \alpha)^{2k}}{(2k)!}$$

$$+ \int_0^t \sum_{i=0}^{2k}\frac{\partial^i g(t,x,s,\alpha(s,x))}{\partial u^i}\frac{(w(s,x) - \alpha(s,x))^i}{i!}ds \quad in \quad Q_T,$$

$$w(t,x) = \Phi(t,x), \quad x \in \partial\Omega, \ and \ w(0,x) = u_0(x), \quad x \in \Omega,$$

(4)

$$\mathcal{L}v = F_2(t,x,\alpha,\beta;v) + \int_0^t G_2(t,x,s,\beta(s,x);v(s,x))ds = \sum_{i=0}^{2k}\frac{\partial^i f_1(t,x,\beta)}{\partial u^i}\frac{(v - \beta)^i}{i!}$$

$$+ \sum_{i=0}^{2k-1}\frac{\partial^i f_2(t,x,\beta)}{\partial u^i}\frac{(v - \beta)^i}{i!} + \frac{\partial^{2k} f_2(t,x,\alpha)}{\partial u^{2k}}\frac{(v - \beta)^{2k}}{(2k)!}$$

$$+ \int_0^t \sum_{i=0}^{2k}\frac{\partial^i g(t,x,s,\beta(s,x))}{\partial u^i}\frac{(v(s,x) - \beta(s,x))^i}{i!}ds \quad in \quad Q_T,$$

(5)

$$v(t,x) = \Phi(t,x), \quad x \in \partial\Omega, \ and \ v(0,x) = u_0(x), \quad x \in \Omega,$$

where $\alpha(t,x) \leq v,w \leq \beta(t,x)$ and $\alpha(0,x) \leq u_0(x) \leq \beta(0,x)$.
We claim that (α_0,β_0) are lower and upper solutions of (4) and (5), respectively.
Letting $\alpha = \alpha_0$ and $\beta = \beta_0$ in (4) we get

$$\mathcal{L}\alpha_0 \leq f_1(t,x,\alpha_0) + f_2(t,x,\alpha_0) + \int_0^t g(t,x,s,\alpha_0(s,x))ds$$

$$= F_1(t,x,\alpha_0;\alpha_0) + \int_0^t G_1(t,x,s,\alpha_0(s,x);\alpha_0(s,x))ds,$$

(6)

$$\alpha_0(t,x) \leq \Phi(t,x), \quad x \in \partial\Omega, \ and \ \alpha_0(0,x) \leq u_0(x), \quad x \in \Omega,$$

$$\mathscr{L}\beta_0 \geq f_1(t,x,\beta_0) + f_2(t,x,\beta_0) + \int_0^t g(t,x,s,\beta_0(s,x))ds$$

$$\geq F_1(t,x,\alpha_0,\beta_0;\beta_0) + \int_0^t G_1(t,x,s,\alpha_0;\beta_0)ds, \tag{7}$$

$$\beta_0(t,x) \geq \Phi(t,x), \quad x \in \partial\Omega, \text{ and } \beta_0(0,x) \geq u_0(x), \quad x \in \Omega,$$

where $\alpha_0 \leq \xi_1, \xi_2, \xi_3 \leq \beta_0$. This proves that α_0 and β_0 are the lower and upper solutions of (4). One can show that $G_1(t,x,\alpha;w)$ and $F_1(t,x,\alpha;w)$ are Hölder continuous in t and (x,w) with exponents $\frac{\alpha}{2}$ and $\overline{\alpha}$, respectively. Hence there exists a unique solution α_1 of (4) by Theorem 1. Next we show that $\alpha_0 \leq \alpha_1 \leq \beta_0$. Set $\mu = \alpha_1 - \alpha_0$.

$$\mathscr{L}\mu = \mathscr{L}(\alpha_1 - \alpha_0) \geq F_{1u}(t,x,\alpha_0,\beta_0;\xi_1)\mu + \int_0^t G_{1u}(t,x,s,\alpha_0;\xi_2)\mu ds,$$

$$\mu(t,x) = 0, \quad x \in \partial\Omega, \text{ and } \mu(0,x) = 0, \quad x \in \Omega.$$

Thus $\mu \geq 0$ or $\alpha_0 \leq \alpha_1$ by Theorem 2. Similarly one can prove that $\alpha_1 \leq \beta_0$. Hence we have $\alpha_0 \leq \alpha_1 \leq \beta_0$. Using the same technique we can show that there exists a unique solution β_1 of (5) such that $\alpha_0 \leq \beta_1 \leq \beta_0$. In addition, we claim that $\beta_1 \geq \alpha_1$.

$$f_1(t,x,\alpha_1) + f_2(t,x,\alpha_1) + \int_0^t g(t,x,s,\alpha_1(s,x))ds$$

$$\geq F_1(t,x,\alpha_0,\beta_0;\alpha_1) + \int_0^t G_1(t,x,s,\alpha_0;\alpha_1) = \mathscr{L}\alpha_1, \tag{8}$$

$$\alpha_1(t,x) = \Phi(t,x), \quad x \in \partial\Omega, \text{ and } \alpha_1(0,x) = u_0(x), \quad x \in \Omega;$$

$$f_1(t,x,\beta_1) + f_2(t,x,\beta_1) + \int_0^t g(t,x,s,\beta_1(s,x))ds$$

$$\leq F_2(t,x,\alpha_0,\beta_0;\beta_1) + \int_0^t G_2(t,x,s,\beta_0;\beta_1) = \mathscr{L}\beta_1, \tag{9}$$

$$\beta_1(t,x) = \Phi(t,x), \quad x \in \partial\Omega, \text{ and } \beta_1(0,x) = u_0(x), \quad x \in \Omega.$$

Theorem 3 yield that $\beta_1 \geq \alpha_1$. Hence $\alpha_0 \leq \alpha_1 \leq \beta_1 \leq \beta_0$. The method of mathematical induction can be apply to prove that $\alpha_0 \leq \alpha_1 \leq ... \leq \alpha_n \leq \beta_n \leq ... \leq \beta_1 \leq \beta_0$ *for all* n. Let u be any solution of (1) such that $\alpha_0 \leq u \leq \beta_0$ with $\alpha_0(0) \leq u_0 \leq \beta_0(0)$ on $\overline{Q}_T$ and suppose that for some u, we have $\alpha_n \leq u \leq \beta_n$ on $\overline{Q}_T$. Then set $\Phi_1 = u - \alpha_{n+1}$ and $\Phi_2 = \beta_{n+1} - u$, respectively.

$$\mathscr{L}\Phi_1 = \mathscr{L}u - \mathscr{L}\alpha_{n+1} \geq f_{1u}(t,x,\eta_1)\Phi_1 + f_{2u}(t,x,\eta_2)\Phi_1 + \int_0^t [g_u(t,x,s,\eta_3)\Phi_1]ds$$

$$\Phi_1(t,x) = 0, \quad x \in \partial\Omega, \text{ and } \Phi_1(0,x) = 0, \quad x \in \Omega;$$

$$\mathscr{L}\Phi_2 = \mathscr{L}\beta_{n+1} - \mathscr{L}u \geq f_{1u}(t,x,\eta_4)\Phi_2 + f_{2u}(t,x,\eta_5)\Phi_2 + \int_0^t [g_u(t,x,s,\eta_6)\Phi_2]ds$$

$$\Phi_2(t,x) = 0, \quad x \in \partial\Omega, \text{ and } \Phi_2(0,x) = 0, \quad x \in \Omega,$$

where $\eta_1, \eta_2, \eta_4, \eta_5$ are between u and α_{n+1}, and η_3, η_6 are between u and β_{n+1}. Thus $\alpha_{n+1} \leq u \leq \beta_{n+1}$ by Theorem 2. Initially $\alpha_0 \leq u \leq \beta_0$. By applying the method of mathematical induction one can prove that $\alpha_n \leq u \leq \beta_n$ for all n, or $\alpha_0 \leq \alpha_1 \leq ... \leq \alpha_n \leq u \leq \beta_n \leq ... \leq \beta_1 \leq \beta_0$ for all n.

Since $\{\alpha_n(t,x)\}$ and $\{\beta_n(t,x)\}$ are in $C^{1+\frac{\alpha}{2},2+\overline{\alpha}}[\overline{Q}_T, R]$, it can be shown that these sequences converge to (ρ, r) using the same technique as in [14] or $\lim_{n\to\infty} \alpha_n(t,x) = \rho(t,x) \leq u \leq$

$r(t,x) = \lim_{n \to \infty} \beta_n(t,x)$. Using equations (4) and (5) and substitution $\Theta = r(t,x) - \rho(t,x)$ we have

$$\mathcal{L}\Theta \quad = \mathcal{L}r - \mathcal{L}\rho \leq L_1\Theta + L_2\Theta + \int_0^t L_3\Theta ds, \quad L_1, L_2, L_3 \geq 0,$$

$$\Theta(t,x) \quad = 0, \quad x \in \partial\Omega, \quad \text{and} \quad \Theta(0,x) = 0, \quad x \in \Omega.$$

By Theorem 2 we conclude that $r(t,x) \leq \rho(t,x)$. This proves that $r(t,x) = \rho(t,x) = u(t,x)$ is the unique solution of (1). Hence $\{\alpha_n(t,x)\}$ and $\{\beta_n(t,x)\}$ converge uniformly and monotonically to the unique solution of (1).

Next we consider the order of convergence of $\{\alpha_n(t,x)\}$ and $\{\beta_n(t,x)\}$ to the unique solution $u(t,x)$ of (1) by setting $p_n(t,x) = u(t,x) - \alpha_n(t,x) \geq 0$ and $q_n(t,x) = \beta_n(t,x) - u(t,x) \geq 0$. Using the definitions for α_n, β_n, the Taylor expansion with Lagrange remainder, and the mean value theorem, we get

$$\mathcal{L}p_{n+1} \quad = \mathcal{L}u - \mathcal{L}\alpha_{n+1} \leq K_{11}p_{n+1} + K_2 p_n^{2k+1} + K_{12}p_{n+1}$$

$$+ K_3 p_n^{2k}(q_n + p_n) + \int_0^t [K_4 p_{n+1} + K_5 p_n^{2k+1}]ds,$$

$$p_{n+1}(t,x) \quad = 0, \quad x \in \partial\Omega, \quad \text{and} \quad p_{n+1}(0,x) = 0, \quad x \in \Omega,$$

where $|f_{1u}| \leq K_{11}$, $|f_{2u}| \leq K_{12}$, $\dfrac{M_1}{(2k)!} = K_2$, $\dfrac{M_2}{(2k)!} = K_3$, $|g_u| \leq K_4$, and $\dfrac{M_3}{(2k)!} = K_5$. Assume now that $K_1 = K_{11} + K_{12}$ and let $r(t)$ be the solution of the following ordinary integro-differential equation.

$$r'(t) = K_1 r(t) + K_3(\max_\Omega p_n^{2k+1} + \max_\Omega p_n^{2k}q_n) + K_4 \int_0^t r(s)ds + (K_2 + K_5 T)\max_\Omega p_n^{2k+1}, r(0) = 0.$$

For the solution $r(t)$ of the above equation, we have

$$r(t) \leq \frac{2\exp(\sqrt{K_1^2 + 4K_4\,T})}{\sqrt{K_1^2 + 4K_4}}[(K_2 + K_5 T + K_3)\max_\Omega p_n^{2k+1} + K_3\max_\Omega p_n^{2k}q_n].$$

Since $p_{n+1}(t,x) \leq r(t)$ by Theorem 4, we obtain

$$\max_{\overline{Q}_T}|p_{n+1}| \leq \left[\frac{2\exp(\sqrt{K_1^2 + 4K_4\,T})}{\sqrt{K_1^2 + 4K_4}}\right][(K_2 + K_5 T)\max_{\overline{Q}_T}|p_n^{2k+1}|$$

$$+ K_3(\max_{\overline{Q}_T}|p_n^{2k+1}| + \max_{\overline{Q}_T}|p_n^{2k}q_n|)].$$

Similarly inequality holds for q_{n+1} or the order of convergence of the sequences $\{\alpha_n(t,x)\}$, $\{\beta_n(t,x)\}$ is $2k+1$. That completes the proof of Theorem 5.

Assume now that m is an even number, say $m = 2k$. Next we state our second main theorem.

Theorem 6 *Assume that all of* (A_0) *holds except* (iii); *further assume that*

(i) α_0, β_0 *are lower and upper solutions of (1) with* $\alpha_0(t,x) \leq \beta_0(t,x)$ *on* $\overline{Q}_T$.

(ii) $\dfrac{\partial^l f_1(t,x,u)}{\partial u^l}$, $\dfrac{\partial^l f_2(t,x,u)}{\partial u^l}$, *and* $\dfrac{\partial^l g(t,x,s,u)}{\partial u^l}$ *exist and are bounded functions on* $\overline{Q}_T$

for $l = 0,1,\ldots,2k-1$ *such that* $\dfrac{\partial^l f_1(t,x,u)}{\partial u^l}$, $\dfrac{\partial^l f_2(t,x,u)}{\partial u^l}$, *and* $\dfrac{\partial^l g(t,x,s,u)}{\partial u^l} \in$

$C^{\frac{\overline{\alpha}}{2},\overline{\alpha}}[Q_T \times R, R]$.

(iii) *Also* $\dfrac{\partial^l g(t,x,s,u)}{\partial u^l}$ *are nondecreasing functions in u on* $\overline{Q}_T$ *for* $l = 0,1,\ldots,2k-1$ *such*

that $g_u(\alpha_0) \geq [\dfrac{\partial^{2k-2} g(\beta_0)}{\partial u^{2k-2}} - \dfrac{\partial^{2k-2} g(\alpha_0)}{\partial u^{2k-2}}] \dfrac{(\beta_0 - \alpha_0)^{2k-3}}{(2k-3)!}$ *and*

$0 \leq \dfrac{\partial^{2k-1} f_1(t,x,\eta_1)}{\partial u^{2k-1}} - \dfrac{\partial^{2k-1} f_1(t,x,\eta_2)}{\partial u^{2k-1}} \leq M_1(\eta_1 - \eta_2)$ *on* $\overline{Q}_T$,

$0 \geq \dfrac{\partial^{2k-1} f_2(t,x,\zeta_1)}{\partial u^{2k-1}} - \dfrac{\partial^{2k-1} f_2(t,x,\zeta_2)}{\partial u^{2k-1}} \geq -M_2(\zeta_1 - \zeta_2)$ *on* $\overline{Q}_T$,

$0 \leq \dfrac{\partial^{2k-1} g(t,x,\xi_1)}{\partial u^{2k-1}} - \dfrac{\partial^{2k-1} g(t,x,\xi_2)}{\partial u^{2k-1}} \leq M_3(\xi_1 - \xi_2)$ *on* $\overline{Q}_T$,

whenever

$\alpha_0(t,x) \leq \eta_2(t,x) \leq \eta_1(t,x) \leq \beta_0(t,x), \qquad \alpha_0(t,x) \leq \zeta_2(t,x) \leq \zeta_1(t,x) \leq \beta_0(t,x),$
$\alpha_0(t,x) \leq \xi_2(t,x) \leq \xi_1(t,x) \leq \beta_0(t,x).$

Then there exist monotone sequences $\{\alpha_n(t,x)\}$, $\{\beta_n(t,x)\}$, $m \geq 0$ *which converge uniformly and monotonically to the unique solution of (1) and the convergence is of order 2k.*

Proof: In order to develop monotone sequences $\{\alpha_n(t,x)\}$ and $\{\beta_n(t,x)\}$, $n \geq 0$ which converge uniformly and monotonically to the unique solution of (1) when $m = 2k$ is an even number, we consider the following equations where $n = 1,2,\ldots$

$$\mathscr{L}\alpha_{n+1} = F_1(t,x,\alpha_n,\beta_n;\alpha_{n+1}) + \int_0^t G_1(t,x,s,\alpha_n(s,x);\alpha_{n+1}(s,x))ds$$

$$= \sum_{i=0}^{2k-1} \frac{\partial^i f_1(t,x,\alpha_n)}{\partial u^i} \frac{(\alpha_{n+1} - \alpha_n)^i}{i!}$$

$$+ \sum_{i=0}^{2k-2} \frac{\partial^i f_2(t,x,\alpha_n)}{\partial u^i} \frac{(\alpha_{n+1} - \alpha_n)^i}{i!} + \frac{\partial^{2k-1} f_2(t,x,\beta_n)}{\partial u^{2k-1}} \frac{(\alpha_{n+1} - \alpha_n)^{2k-1}}{(2k-1)!}$$

$$+ \int_0^t \sum_{i=0}^{2k-1} \frac{\partial^i g(t,x,s,\alpha_n(s,x))}{\partial u^i} \frac{(\alpha_{n+1}(s,x) - \alpha_n(s,x))^i}{i!} ds \text{ in } Q_T, \tag{10}$$

$$\alpha_{n+1}(t,x) = \Phi(t,x), \quad x \in \partial\Omega, \text{ and } \alpha_{n+1}(0,x) = u_0(x), \quad x \in \Omega,$$

$$\mathscr{L}\beta_{n+1} = F_2(t,x,\alpha_n,\beta_n;\beta_{n+1}) + \int_0^t G_2(t,x,s,\beta_n(s,x),\beta_n(s,x);\beta_{n+1}(s,x))ds$$

$$= \sum_{i=0}^{2k-2} \frac{\partial^i f_1(t,x,\beta_n)}{\partial u^i} \frac{(\beta_{n+1} - \beta_n)^i}{i!} + \frac{\partial^{2k-1} f_1(t,x,\alpha_n)}{\partial u^{2k-1}} \frac{(\beta_{n+1} - \beta_n)^{2k-1}}{(2k-1)!} \tag{11}$$

$$+ \sum_{i=0}^{2k-1} \frac{\partial^i f_2(t,x,\beta_n)}{\partial u^i} \frac{(\beta_{n+1} - \beta_n)^i}{i!}$$

$$+\int_0^t [\sum_{i=0}^{2k-2} \frac{\partial^i g(t,x,s,\beta_n(s,x))}{\partial u^i} \frac{(\beta_{n+1}(s,x)-\beta_n(s,x))^i}{i!}$$
$$+\frac{\partial^{2k-1} g(t,x,\alpha_n(s,x))}{\partial u^{2k-1}} \frac{(\beta_{n+1}(s,x)-\beta_n(s,x))^{2k-1}}{(2k-1)!}]ds \quad \text{in} \quad Q_T,$$
$$\beta_{n+1}(t,x) = \Phi(t,x), \quad x \in \partial\Omega, \quad \text{and} \quad \beta_{n+1}(0,x) = u_0(x), \quad x \in \Omega,$$

where $\alpha_n(t,x) \le \alpha_{n+1}(t,x), \beta_{n+1}(t,x) \le \beta_n(t,x)$ and $\alpha(0,x) \le u_0(x) \le \beta(0,x)$.
We omit the details of the proof, since it follows on the same lines as in Theorem 5.

NUMERICAL RESULT

Let us consider the following example.

$$u_t - u_{xx} = u^3 - 7u + \cos^2 t - ue^u + \int_0^t [0.5u^3(s,x) + 6.5u(s,x)]ds, \quad 0 \le x,t \le 1$$

$$u(t,0) = 0, \quad u(t,1) = 0.05, \quad 0 \le t \le 1 \tag{12}$$
$$u(0,x) = \cos(\pi x), \quad 0 \le x \le 1.$$

Choose $\alpha_0(t) \equiv 0, \beta_0(t) \equiv 1$, and denote

$$f_1(t,x,u) = u^3 - 7u + \cos^2 t, \quad f_2(t,x,u) = -ue^u, \quad \text{and} \quad g(t,x,u) = 0.5u^3 + 6.5u.$$

We can apply iterates of Theorem 5 to find the approximate solution of the equation (12) by using the finite-difference method and MATHEMATICA for each iterate. After only three iterates of α and β we can derive the approximate solution of (12) as shown in the following TABLE 1 for $x = 0.5$.

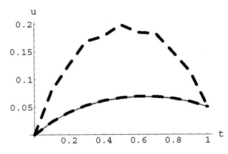

FIGURE 1. α, β - iterates for $x = 0.5$

In Figure 1 and Figure 2 one can see the α-iterates (with solid line), the β-iterates (with dashed line) for $x = 0.5$ and the approximate solution of (12), respectively.

CONCLUSION

In the above theorems we assumed that the $m - 1$-th derivative of the functions $f_1(t,x,u)$, $f_2(t,x,u)$ and $g(t,x,u)$ are monotone and one-sided Lipschitzian with respect to u. We have developed iterates of nonlinearity of order $m - 1$ which converge rapidly (order m) to the unique solution of nonlinear integro-differential equations of parabolic type.

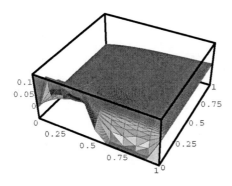

FIGURE 2. Approximate solution of (12)

TABLE 1. Table of Three α, β - Iterates and the Solution

t	$\alpha_1(t)$	$\alpha_2(t)$	$\alpha_3(t)$	u	$\beta_3(t)$	$\beta_2(t)$	$\beta_1(t)$
0.1	0.0227452	0.0232188	0.0232188	0.0232188	0.0232188	0.0232440	0.0789767
0.2	0.0403276	0.0403558	0.0403560	0.0403560	0.0403560	0.0405284	0.1256820
0.3	0.0515304	0.0528281	0.0528289	0.0528289	0.0528289	0.0529095	0.1683400
0.4	0.0612250	0.0612305	0.0612308	0.0612308	0.0612308	0.0615312	0.1776090
0.5	0.0648584	0.0665512	0.0665519	0.0665519	0.0665519	0.0666591	0.1982170
0.6	0.0687159	0.0687605	0.0687608	0.0687608	0.0687608	0.0690607	0.1843400
0.7	0.0669958	0.0684560	0.0684560	0.0684560	0.0684560	0.0685362	0.1825070
0.8	0.0651566	0.0652606	0.0652608	0.0652608	0.0652608	0.0654328	0.1487960
0.9	0.0586860	0.0591930	0.0591930	0.0591930	0.0591930	0.0593183	0.1135690

REFERENCES

1. R. Bellman, *Methods of Nonlinear Analysis*, Vol. 1, Academic Press, New York, 1970.
2. R. Bellman and R. Kalaba, *Quasilinearization and Nonlinear Boundary Value Problems*, Elsevier, New York, 1965.
3. S. G. Deo and C. Mcgloin Knoll, *Int. J. Nonl. Diff. Eqs: Theory Math. Appl.* **3**(1 and 2), 91–103 (1997).
4. A. Cabada and J. Nieto, *Applied Mathematics and Computations* **87**, 217–226 (1997).
5. J. Cannon and Y. P. Lin, *Differential and Integral Equations* **2**, 111–121 (1989).
6. R. Kalaba, *Journal of Mathematics and Mechanics* **8**, 519–574 (1959).
7. G. Ladde, V. Lakshmikantham and A. Vatsala, *Monotone Iterative Techniques for Nonlinear Differential Equations*, Pitman, Boston, 1985.
8. V. Lakshmikantham and S. Köksal, *Monotone Flows and Rapid Convergence for Nonlinear Partial Differential Equations*, Taylor & Francis Inc., London and New York, 2003.
9. V. Lakshmikantham and S. Leela, *Differential Integral Inequalities*, Vol II, Academic Press, New York, 1968.
10. V. Lakshmikantham and A. Vatsala, *Generalized Quasilinearization for Nonlinear Problems*, Kluwer Academic Publishers, Boston, 1998.
11. V. Mandelzweig, *Journal of Mathematical Physics* **40**(4), 6266–6291 (1999).
12. V. Mandelzweig and F. Tabakin, *Computer Physics Communications* **141**, 268–281 (2001).
13. R. Mohapatra K. Vajravelu, and Y. Yin, *Journal of Optimization Theory and Applications* **96**(3), 667–682 (1998).
14. C. Pao, *Nonlinear Parabolic and Elliptic Equations*, Plenum Publishers, Boston, 1992.
15. A. S. Vatsala and L. Wang, *Quarterly of Applied Mathematics* **LIX**(3), 459–470 (2001).

Galerkin's Method and Distributions of Zeros of Some Equations from the Mechanics I

Z. Petrova

Technical University of Sofia, 8 Kl.Ohridski Blvd., 1000 Sofia, Bulgaria

Abstract. This article presents sufficient conditions for the oscillation of a special type of solutions, obtained by Galerkin's method of the wide class of equations:

$$\sum_{i=0}^{n}\sum_{j=0}^{2} K_{ij} w_{x^i t^j}(x, t - \tau_{ij}) + c(x,t,w) = f(x,t), \quad x \in (0,l), \quad t > 0.$$

We suppose that $n \in \mathbf{N}$, $K_{ij} \in \mathbf{R}$ and $\tau_{ij} \geq 0$ are constants $\forall\, i = \overline{0,n}$ and $\forall\, j = \overline{0,2}$.

As we know, we made for the first time such a combination of oscillation results and Galerkin's method. More exactly, we treated an equation of a pipeline in the mentioned way and now we generalize this publication.

Keywords: Functional differential equations and inequalities, oscillation, pipeline.
PACS: 02.30.Jr,02.30.Ks,02.60.Lj,02.70.Dh, 03.40.Gc

INTRODUCTION

Oscillation theory of differential equations is a very important branch of mathematics. There are many publications in this field. We explained the essential role of the respective ordinary differential inequalities in almost all of our papers. Now we are concerned with three of them: [2], [3] and [5]. Here we generalize Petrova [2]. We point out that we made for the first time combination of oscillation results and Galerkin's method in [2]. In fact, there we proved sufficient conditions for the oscillation of these solutions of two equations of a pipeline, which were obtained by Galerkin's method in Vel'misov et al in [6] and [7] respectively. The present equation is

$$\sum_{i=0}^{n}\sum_{j=0}^{2} K_{ij} w_{x^i t^j}(x, t - \tau_{ij}) + c(x,t,w) = f(x,t), \quad x \in (0,l), \quad t > 0. \tag{1}$$

We suppose that $l > 0$, $n \in \mathbf{N}$, $K_{ij} \in \mathbf{R}$ and $\tau_{ij} \geq 0$ are constants $\forall\, i = \overline{0,n}$ and $\forall\, j = \overline{0,2}$ as well as that $\exists\, q > 0$:

$$\begin{aligned} c(x,t,w) &\geq qw, \quad \forall\, (x,t,w) \in (0,l) \times \mathbf{R}_+ \times \mathbf{R}_+, \\ c(x,t,w) &\leq qw, \quad \forall\, (x,t,w) \in (0,l) \times \mathbf{R}_+ \times \mathbf{R}_-. \end{aligned} \tag{2}$$

Here $\quad \tilde{M} = D w_{x^2}(x,t) + \xi w_{x^2 t}(x,t) \quad$ is the deflection moment

CP1067, *Applications of Mathematics in Engineering and Economics '34—AMEE '08*, edited by M. D. Todorov
© 2008 American Institute of Physics 978-0-7354-0598-1/08/$23.00

and $\tilde{Q} = \eta w_{xt^2}(x,t) - D w_{x^3}(x,t) + \xi w_{x^3 t}(x,t)$ is the cutting force.

We state the following boundary conditions at $x = 0$ or $x = l$:

$$w = w_x = 0 \quad \text{(rigid fixing)}; \qquad\qquad w = \tilde{M} = 0 \quad \text{(hinged fixing)}; \qquad (3)$$
$$w_x = \tilde{Q} = 0 \quad \text{(fixing to a free element)}; \qquad \tilde{M} = \tilde{Q} = 0 \quad \text{(free endpoint)}. \qquad (4)$$

Both parts of this work are inspired from the publications of Vel'misov et al, concerned to the Galerkin method. These authors treated the equation

$$L_1(w) \equiv M w_{t^2}(x,t) + D w_{x^4}(x,t) + \xi w_{x^4 t}(x,t-\tau_1) - \eta w_{x^2 t^2}(x,t) +$$
$$+ p w_{x^2}(x,t) + \gamma w_t(x,t-\tau_2) + \beta w(x,t-\tau_3) = 0, \quad x \in (0,l), \ t > 0 \qquad (5)$$

under the boundary conditions (3) at $x = 0$ or $x = l$ in [6]. The publication [7] was devoted to the equation

$$L_2(w) \equiv L_1(w) + \alpha w_{xt}(x,t) + \theta[w_t(x,t) + V w_x(x,t)] + \alpha_{30} w^3(x,t-\tau_3) +$$
$$+ \alpha_{21} w^2(x,t-\tau_3) w_t(x,t-\tau_2) + \alpha_{03} w_t^3(x,t-\tau_2) = 0, \quad x \in (0,l), \ t > 0 \qquad (6)$$

under all the boundary conditions (3) and (4).

According to [6] and [7] all the coefficients in (5) and (6) are constants such that

$$l > 0, \ M > 0, \ D > 0, \ \xi \geq 0, \ \eta \geq 0, \ p \geq 0, \ \gamma \geq 0, \ \beta > 0, \ \tau_i \geq 0 \qquad (7)$$

and $l > 0, \ M > 0, \ D > 0, \ \xi \geq 0, \ \eta \geq 0, \ \alpha \geq 0, \ p \geq 0, \alpha_{ij} \geq 0, \ \gamma \geq 0, \ \beta > 0, \ \tau_i \geq 0$

respectively. Hence, the equation (1) is a generalization of (5) because of (7). We shall consider a generalization of (6) in the second part.

In this publication we concentrate our attention on the four problems (1), (3) and (1), (4). More precisely, we establish sufficient conditions for the oscillation of their solution only, which are of the type

$$w(x,t) = w_1(x,t) = w_1(t) g_1(x), \qquad (8)$$

where $(\mathbf{H})$ $\begin{cases} \mathbf{H_1.} & w_1(t) \in C^2([0,\infty); \mathbf{R}); \\ \mathbf{H_2.} & g_1(x) \in C^n([0,l]; [0,\infty)), \quad g_1(x) \not\equiv 0; \\ \mathbf{H_3.} & g_1(x) \text{ satisfies the boundary conditions (3) and (4).} \end{cases}$

Our present oscillation results are based on the application of Petrova [5], where we continue the ideas of Yoshida [8] and [9]. The function $w_1(t)$ is a solution of respective particular case of ordinary differential equations and inequalities, whose oscillation behavior was considered in [5]. This function is not wellknown before to solve the problems (1), (3) and (1), (4), but the other function there $g_1(x)$ must be defined before.

Remark 1 *In general, the equation (1) is a non-linear one. Moreover, (1) is a linear one if and only if both assumptions in (2) hold simultaneously, i.e., if*

$$c(x,t,w) = q w \qquad (9)$$

and this important equation is

$$\sum_{i=0}^{n} \sum_{j=0}^{2} K_{ij} w_{x^i t^j}(x, t - \tau_{ij}) + qw = f(x,t), \quad x \in (0,l), \quad t > 0. \tag{10}$$

Here we continue the idea of Petrova [2], where the combination of oscillation arguments and Galerkin's method was realized for the first time. There we treated the boundary value problems for the equations (5) and (6) as they were stated in Vel'misov et al in [6] and [7] respectively.

Remark 2 *We point out that the present publication is not a direct corollary of [2] in the non-homogeneous case. It is essential that the equation (5) is linear.*

Although the type of the boundary conditions (3) and (4) appeared from the mathematical models in [6] and [7], there are many other authors, who considered particular cases of the equation (1), which do not coincide with (5). We give only one example:

$$\sum_{i=0}^{m} c_i w_{x^{2i}}(x, t - \tau_i) + \sum_{j=0}^{n} k_j w_{x^{2j}{}_t}(x,t) = f(x,t), \tag{11}$$

where $c_i \in \mathbf{R}$ and $k_j \in \mathbf{R}$ are constants $\forall\, i = \overline{0, m}$ and $\forall\, j = \overline{0, n}$ (see [1]), in order to make two interesting comments. The first one is about the choice of variables with delays and the second one is about the order of the partial derivatives, which consist delays.

Theoretically the delays could be with respect to both variables x and t. It is well-known that the delays with respect to t correspond to the fact, that the real processes are not simultaneous. We do not give any arguments about the absence of delays with respect to x in (1), but (5) and (11) make this absence intuitively clear. The question about the order of the partial derivatives with delays is also a very important one. Let us begin with some terminology.

Definition 1 *We say that a differential equation (ordinary or partial) is **functional**, if it consists at least one argument of the unknown function with a deviation.*

Definition 2 *We say that a differential equation (ordinary or partial) is **neutral**, if it consists at least one derivative of highest order of the unknown function with a deviation in its argument.*

Usually oscillation theory of neutral differential equations is much more complicated than oscillation theory of these functional differential equations, which are not neutral. We refer the reader to Yoshida [9] for an illustration of this fact. There Yoshida succeeded to explore a neutral hyperbolic equation via a non-neutral inequality. Since the publication [5] is devoted to functional ordinary differential equations, which are not neutral, then there are some restrictions of its applications. Especially here, we could made some additional comments to the above terminology. More concretely, we could reformulate the Definitions 1 and 2 in order to make clear the variable with delays. For example, there are three possible pairs of definitions for the equations (1), (5) and (11), i.e., if both delays are with respect to x and t or with respect to x only either with respect to t only. Obviously, the equations mentioned above satisfy the second possibility. Moreover, the assumption

$$\tau_{i2} = 0, \quad \forall\, i = \overline{0,n} \tag{12}$$

is a very natural one because of (5) and (11). Also, (12) will help us to apply [5].

PRELIMINARY RESULTS

We shall apply the results of Petrova [5]. So that, in the first subsection we repeat all the needed from there. Further, in the second one we obtain some formulas. More exactly, we apply Galerkin's method to find such ordinary differential equations and inequalities, which are the proper particular cases of these one from the first subparagraph.

Important Oscillation Results

First of all, we mention that always we are concerned with the classical solutions of the respective problems. In particular, it means that all the functions are continuous and sufficiently smooth both for the cases of one or more variables.

The present two definitions correspond to the problems (1), (3) and (1), (4). We suppose that $c \in \mathbf{R}$ is a constant.

Definition 3 *We say that the function* $\varphi(x,t) \in C((0,l) \times [c,\infty); R)$, $\varphi(\mathbf{x,t}) \not\equiv \mathbf{0}$, *is oscillating when* $t \to \infty$, *if there exist two sequences* $\{x_m\}_{m=1}^{\infty}$ *and* $\{t_m\}_{m=1}^{\infty}$ *such that*

$$x_m \in (0,l), \quad t_m > 0: \quad \lim_{m \to \infty} t_m = \infty \quad \text{and} \quad \varphi(x_m, t_m) = 0.$$

Definition 4 *We say that the function* $\varphi(x,t) \in C((0,l) \times [c,\infty); \mathbf{R})$ *is eventually positive (eventually negative), if there exists a constant* $\tilde{c} \geq c$ *such that*

$$\varphi(x,t) > 0 \quad (\varphi(x,t) < 0), \quad \forall\, x \in (0,l), \quad \forall\, t \in [\tilde{c},\infty).$$

We point out that every continuous function of two variables belongs to exactly one of these three types as well as that these definitions could become more complicated in the case of two variables since the arguments and the conclusions depend on the statement of the problems. Indeed, there is a difference in the case of two arguments if we deal with a characteristic initial value problem (see Petrova [4]) or with a boundary value problem, i.e., it is essential that both arguments could take infinitely large values or exactly one variable could be in a finite interval and the other one could take infinitely large values respectively. Here we are in the second situation.

In [5] we focused our attention on the following functional ordinary differential equations and inequalities:

$$Az''(t) + \sum_{i=1}^{n} \theta_i z'(t - \tau_i) + \sum_{k=1}^{\tilde{n}} \beta_k z(t - \sigma_k) + Bz(t) = F(t) \tag{13}$$

and

$$Az''(t) + \sum_{i=1}^{n} \theta_i z'(t - \tau_i) + \sum_{k=1}^{\tilde{n}} \beta_k z(t - \sigma_k) + Bz(t) \leq F(t), \tag{14}$$

where all the coefficients A, B, $\{\theta_i\}_{i=1}^{n}$, $\{\tau_i\}_{i=1}^{n}$, $\{\beta_k\}_{k=1}^{\tilde{n}}$ and $\{\sigma_k\}_{k=1}^{\tilde{n}}$ are constants. We suppose that

$$A > 0 \quad \text{and} \quad B > 0; \tag{15}$$

$$\theta_i \geq 0, \quad \tau_i \geq 0, \quad \forall i = \overline{1,n}, \, n \in \mathbf{N}, \quad \tau = \max_{1 \leq i \leq n} \tau_i \tag{16}$$

$$\text{and} \quad \beta_k \geq 0, \quad \sigma_k \geq 0, \quad \forall k = \overline{1,\tilde{n}}, \, \tilde{n} \in \mathbf{N}, \quad \sigma = \max_{1 \leq k \leq \tilde{n}} \sigma_k. \tag{17}$$

Remark 3 *According to [5] the most common conclusions for (13) and (14) were concerned with the property "monotonicity of the solution."*

Also, there we succeeded to omit this property for their subclasses:

$$Az''(t) + \theta z'(t) + \sum_{k=1}^{\tilde{n}} \beta_k z(t - \sigma_k) + Bz(t) = F(t) \tag{18}$$

and

$$Az''(t) + \theta z'(t) + \sum_{k=1}^{\tilde{n}} \beta_k z(t - \sigma_k) + Bz(t) \leq F(t) \tag{19}$$

under the same assumptions (15) and (17) for the constants A, B, $\{\beta_k\}_{k=1}^{\tilde{n}}$ and $\{\sigma_k\}_{k=1}^{\tilde{n}}$. Further, we replaced (16) in [5] with the proper condition for the constant $\theta \in \mathbf{R}$.

In several our papers we were interested in many important particular cases of the inequality (14) only (see [3] and the literature cited there), because these concrete inequalities were the main tools in the investigation of the oscillation behavior of certain partial differential equations. The publications of Yoshida and especially [8] and [9] were and still are essential in our scientific work. Let us explain the reason about it. More exactly, there figurate sufficient conditions for oscillation in the sense that there are written explicitly bounded sets, which consist at least one zero of the respective solutions. Let us concentrate our attention on both of them and remind these sets there.

Yoshida [8] considered the following problem of Goursat:

$$u_{xy}(x,y) + c(x,y,u) = f(x,y), \quad (x,y) \in Q_\rho, \tag{20}$$

$$\begin{aligned} u(x,0) &= p(x), \quad x \geq \rho, \\ u(0,y) &= q(y), \quad y \geq L^2 k^{-2} \rho, \end{aligned} \tag{21}$$

where $\rho \geq 0, k > 0, L > 0, 0 \leq t_1 \leq t_2$,

$$Q_\rho = \{x > 0, y > 0 : Lx + k^2 L^{-1} y > L\rho \} \tag{22}$$

and

$$Q(t_1, t_2] = \{x > 0, y > 0 : Lt_1 < Lx + k^2 L^{-1} y \leq Lt_2 \}. \tag{23}$$

There he stated proper assumptions:

A$_1$. $c(x,y,\xi) \in C(\overline{Q_\rho} \times \mathbf{R},\mathbf{R})$;

A$_2$. $\xi c(x,y,\xi) \geq 0, \quad \forall\, (x,y,\xi) \in \overline{Q_\rho} \times \mathbf{R}; \quad p(0) = q(0), \quad \text{if} \quad \rho = 0$;

A$_3$. $f(x,y) \in C(\overline{Q_\rho},\mathbf{R}), \; p(x), \; q(y) \in C^1(\mathbf{R},\mathbf{R})$;

A$_4$. $c(x,y,\xi) \geq k^2\xi, \quad \forall\, (x,y,\xi) \in \overline{Q_\rho} \times [0,\infty) \times \mathbf{R}^2$;

A$_5$. $c(x,y,\xi) \leq k^2\xi, \quad \forall\, (x,y,\xi) \in \overline{Q_\rho} \times (-\infty,0) \times \mathbf{R}^2$.

Lemma 1 ([8]) *If there is a number $s \geq \rho$ such that*

$$\int\limits_{s}^{s+\pi/L} F(t)\sin L(t-s)dt \leq 0, \tag{24}$$

then the ordinary differential inequality

$$z''(t) + L^2 z(t) \leq F(t) \tag{25}$$

has no positive solution in $(s, s+\pi/L]$.

We point out that Yoshida [8] applied his lemma and established that every solution of the problem (20), (21) has at least one zero in $Q(s_n, s_n + \pi/L]$, $n \in \mathbf{N}$ under proper additional assumptions, and one of them is that the function

$$\Phi(s) = \int\limits_{s}^{s+\pi/L} F(t)\sin L(t-s)dt \tag{26}$$

oscillates.

The papers of Yoshida and Lemma 1 inspired many our results both for hyperbolic equations (see Petrova [4] and the cited there joint results of Petrova & Mishev) and some equations from the fluid mechanics (see Petrova [2], [3], etc.). More exactly we include the derivative $z'(t)$ in [4], i.e., we treated the inequality

$$z''(t) + p_* z'(t) + q_* z(t) \leq F(t)$$

in order to explore the oscillation behavior of the equation:

$$u_{xy}(x,y) + au_x(x,y) + bu_y(x,y) + c(x,y,u,u_x,u_y) = f(x,y), \quad (x,y) \in Q_\rho,$$

where a and b are real constants under the same boundary value conditions (21). In Petrova [4] and [5] we explained in details the great role of this lemma of Yoshida as well as its importance for the distributions of the zeros of these ordinary differential equations, which have one and the same left hand side as the ordinary differential inequalities.

Corollary 1 ([5]) *Let the constants A, B, θ, $\{\beta_k\}_{k=1}^{\tilde{n}}$ and $\{\sigma_k\}_{k=1}^{\tilde{n}}$ satisfy (15), (17) and*

$$4AB > \theta^2 \quad \text{and then} \quad \tilde{L}_* = \frac{\sqrt{4AB - \theta^2}}{2A} > 0. \tag{27}$$

If there is a number $s \geq \rho$ such that

$$\int\limits_{s}^{s+\pi/\tilde{L}_*} F(t)e^{\frac{\theta t}{2A}} \sin \tilde{L}_*(t-s)dt \leq 0, \tag{28}$$

then the inequality (19) has no positive solution in $(s - \sigma, s + \pi/\tilde{L}_*]$.

Remark 4 *The above corollary is true if just $s \in \mathbf{R}$ satisfies (28) and the dependence of ρ could be omitted, but we prefer to write $s \geq \rho$ because usually the interval $[\rho, \infty)$ is the proper particular case of $[\tilde{c}, \infty)$ in Definition 4.*

The function

$$\tilde{\Phi}_*(s) = \int\limits_{s}^{s+\pi/\tilde{L}_*} F(t)e^{\frac{\theta t}{2A}} \sin \tilde{L}_*(t-s)dt \tag{29}$$

helped us to obtain important sufficient conditions for oscillation in [5].

Theorem 1 *[5] Let (15), (17) and (27) hold and let the function $\tilde{\Phi}_*(s)$ be oscillatory. Then every solution of the equation (18) oscillates.*

Remark 5 *The proof of Theorem 1 guarantees that if the function $\tilde{\Phi}_*(s)$ is oscillating:*

$$\exists \{s_n\}_{n=1}^{\infty}: \quad \lim_{n \to \infty} s_n = \infty \quad and \quad \tilde{\Phi}_*(s_n) = 0, \tag{30}$$

then the equation (18) has at least one zero in $(s_n - \sigma, s_n + \pi/\tilde{L}_*]$.

Remark 5 allows us to make mathematical parallels about the distributions of zeros of the equations (1) and (18) as well as about the influence of the delays for the type of the sets, which consist at least one zero of their solutions. In the paper [5], we did not treat the important particular cases of (18) and (19), where $\theta = 0$, i.e., we omit

$$Az''(t) + \sum_{k=1}^{\tilde{n}} \beta_k z(t - \sigma_k) + Bz(t) = F(t) \tag{31}$$

$$\text{and} \quad Az''(t) + \sum_{k=1}^{\tilde{n}} \beta_k z(t - \sigma_k) + Bz(t) \leq F(t). \tag{32}$$

We published the conclusions for (31) and (32) explicitly only in the Petrova PhD thesis.

Remark 6 *Here we mention that the hypothesis (15) guarantees that (27) is also fulfilled in the particular case, where $\theta = 0$ as well as that $\tilde{L}_* = \sqrt{B/A}$. So that, we could mention that according to Remark 5, the equation (13) has at least one zero in $(s_n - \sigma, s_n + \pi/\sqrt{B/A}]$ under proper additional assumptions.*

In fact, Remark 6 gives us possibility to make more mathematical parallels with the famous paper Yoshida [9], which is devoted to the following neutral hyperbolic equation:

$$u_{tt}(x,t) - [\Delta u(x,t) + \alpha \Delta u(x,t-\tau)] + c(x,t,u(x,t-\sigma)) = f(x,t), \qquad (33)$$

where $(x,t) \in \Omega \equiv G \times (0,\infty)$, G is a bounded domain in $\mathbf{R}^n$ and Δ is the Laplacian in $\mathbf{R}^n$. The author of [9] stated different boundary conditions on $\partial G \times (0,\infty)$. There he obtained sufficient conditions for oscillating of all the solutions of these boundary value problems in the sense that every such a solution has at least one zero in bounded domains of the type

$$G \times (s-T, s+\pi/L), \quad T = \max\{\sigma, \tau\}. \qquad (34)$$

In fact, in [9] the representation (34) comes from his previous publication [8]. More exactly, Yoshida applied Lemma 1 for the equation (33) since he succeeded to pass from (33) to an inequality, which is a particular case of (32), where

$$n = 2 \quad \text{and} \quad A = 1.$$

Further, Yoshida [9] established that this particular case leads to (25). Unfortunately, this nice result was not formulated in [9] separately from one hand and could be proved by Corollary 1 [5] from the other hand.

Traditionally Yoshida investigated the needed inequalities and did not continue his considerations for the respective ordinary differential equations. We followed this tradition almost everywhere and [5] is an exception. Also, many times we explained why we could omit the detailed conclusions about the other inequality, i.e., this one with just the opposite direction of (14), in particular.

Immediately from Remark 6 and (34) we could compare the bounded domains with at least one zero of both equations (31) and (33).

Remark 7 *Without any lost of generality we could have one and the same upper limits of summation in both sums in the left-hand sides of (13) and (14). Indeed, if $n < \tilde{n}$, then everything becomes correct if we denote $\theta_i = 0$, $i \in \{n+1, ..., \tilde{n}\}$. Similarly, if $n > \tilde{n}$, then we could suppose that $\beta_k = 0$, $k \in \{\tilde{n}+1, ..., n\}$. Also, without any lost of generality we could take one and the same lower limits of summation 0 instead of 1 there as well as one and the same index of summation i.*

Important Representations

Our maim tool in the last section will be Petrova [5]. There we shall treat the solutions of the four initial boundary problems under proper assumptions, which allow us to apply it. More exactly, we prefer to deal with the representation (8) and we shall explain the reason about it, but we take in attention the representation

$$w(x,t) = \sum_{k=1}^{v} w_k(t) g_k(x), \ v \in \mathbf{N}, \quad \text{Galerkin's method} \qquad (35)$$

in order to make some essential comments. We state the following conditions:

$$(\tilde{\mathbf{H}}) \begin{cases} \tilde{\mathbf{H}}_1. & w_k(t) \in C^2([0,\infty); \mathbf{R}), \ \forall\, k \in \mathbf{N}; \\ \tilde{\mathbf{H}}_2. & g_k(x) \in C^n([0,l]; \mathbf{R}), \ \forall\, k \in \mathbf{N}; \\ \tilde{\mathbf{H}}_3. & g_k(x) \ \text{satisfies the boundary conditions (3) and (4)}, \ \forall\, k \in \mathbf{N}; \\ \tilde{\mathbf{H}}_4. & \{g_k(x)\}_{k=1}^\infty \ \text{is a closed system over} \ \ [0,l]; \\ \tilde{\mathbf{H}}_5. & \{g_k(x)\}_{k=1}^v \ \text{is a linear independent system over} \ \ [0,l]; \\ & \forall\, v = \text{const} \in \mathbf{N}; \\ \tilde{\mathbf{H}}_6. & \{g_k(x)\}_{k=1}^\infty \ \text{is a full system over} \ \ C^n([0,l]; \mathbf{R}). \end{cases}$$

In fact, the applications of [5] will be concerned with concrete equations and inequalities, which are particular cases of (18) and (19). So that, we could go back to Remarks 1 and 2. We point out that we do not generalize automatically our previous paper [2] since there we treated the equations (5) and (6) via equations of the type (18), but we did not apply inequalities there at all and the reason about it is that both (5) and (6) are linear partial differential equations as well as that (18) is a linear ordinary differential equation. According to Remark 1, the linear subcase of (1) is (10), where (2) holds.

The immediate generalization of [2] means that we need of (18) only as well as that we reach conclusions for every function of the type

$$w_k(x,t) = w_k(t)g_k(x), \tag{36}$$

which is the addend in (35). In all the bellow we suppose that $0 \le t_1 \le t_2 \le \infty$.

Lemma 2 *Let* $(\tilde{\mathbf{H}})$ *hold for the function* $w(x,t)$ *of the type (35), which is a solution of the problems (10), (3) and (10), (4) in* $(0,l) \times (t_1,t_2)$. *Then the function* $w_k(t)$ *satisfies the equation*

$$\sum_{i=0}^n \sum_{j=0}^2 A_{ijk} z^{(j)}(t - \tau_{ij}) + qB_k z(t) = F_k(t) \tag{37}$$

in $(t_1,t_2], \ \forall\, k \in \{1,...,v\},$ *where*

$$A_{ijk} = K_{ij} \int_0^l g_k(x)g_k^{(i)}(x)dx, \ \ B_k = \int_0^l g_k^2(x)dx \ \text{ and } \ F_k(t) = \int_0^l f(x,t)g_k(x)dx. \tag{38}$$

Proof: We replace (35) in (10). Further, we multiply it by $g_k(x), \ \forall\, k = \overline{1,v}$. Finally, we integrate it over $(0,l)$ with respect to x and obtain the following formulas:

$$\sum_{i=0}^n \sum_{j=0}^2 K_{ij} w_k^{(j)}(t - \tau_{ij}) \int_0^l g_k(x)g_k^{(i)}(x)dx + qw_k(t) \int_0^l g_k^2(x)dx = \int_0^l f(x,t)g_k(x)dx. \tag{39}$$

Lemma 3 *Let (38) holds for* $k = 1$. *Further, let the function* $w(x,t)$ *of type (8) be a positive solution of one of the problems (10), (3) and (10), (4) in* $(0,l) \times (t_1,t_2)$, *where* *the functions $w_1(t)$ and $g_1(x)$ satisfy the conditions* (**H**). *Then the function $w_1(t)$ is a positive solution of the inequality*

$$\sum_{i=0}^{n}\sum_{j=0}^{2} A_{ij1} z^{(j)}(t-\tau_{ij}) + q B_1 z(t) \le F_1(t) \tag{40}$$

in (t_1, t_2).

Proof: We apply (2) and $\mathbf{H_2}$ to establish that

$$\sum_{i=0}^{n}\sum_{j=0}^{2} K_{ij} w_{x^i t^j}(x, t-\tau_{ij}) + q w(x,t) \le f(x,t), \quad x \in (0,l), \quad t > 0. \tag{41}$$

Then we multiply (41) by $g_1(x)$ and integrate over $(0,l)$ with respect to x.

MAIN RESULTS

In fact, we must comment if the equation (37) and the inequality (40) could be respective particular cases of one of the following pairs: (13) and (14) or (18) and (19). We apply Remark 7 in two directions. The first one is that we do not need to compare the lower and the upper limits of summation. The second one is that we avoid the present collision of the variable k. More exactly, we could have one and the same index of summation i and the symbol k will be applied only as an index in the representations of the function $w(x,t)$ in the sense of the previous subsection.

We point out that (37) and (40) could be of neutral type, if

$$\exists\, i_0 \in \{1,...,n\}: \quad K_{i_02} \ne 0 \quad \text{and} \quad \tau_{i_02} \ne 0, \tag{42}$$

but (13), (14), (18) and (19), which were the object of Petrova [5], are not of this type. So that, we could not apply [5], if (42) is fulfilled and then the oscillation behavior of the mentioned solutions of these four problems is still an open question. We avoid (42) in all the bellow and it means that we come back to the assumption (12) about the delays.

Also, according to Remark 3 as well as (8), (36) and (35), we omit (13) and (14), because we would obtain oscillation results, combined with the "monotonicity" of $w_k(t)$, which would not be very convenient for application since the function $w_k(t)$ is just a multiplier of $g_k(x)$. So that, finally we are interested in the present particular cases of (15), (17) and (27) for the present pair of equations and inequalities of the type (18) and (19). In other words, our main results are concerned with the equation

$$\sum_{i=0}^{n} \left(K_{i2} w_{x^i t^2}(x,t) + K_{i1} w_{x^i t}(x,t) + K_{i0} w_{x^i}(x, t-\tau_{i0}) \right) + c(x,t,w) = f(x,t),$$
$$x \in (0,l), \quad t > 0. \tag{43}$$

Theorem 2 *Let (2) holds and let the constants A_{ijk} and B_k as well as the functions $F_k(t)$ be defined by (38), $\forall\, k \in 0 \cup \mathbf{N}$, where the functions $\{g_k(x)\}_{k=1}^{\infty}$ satisfy* ($\tilde{\mathbf{H}}$). *Let we* *suppose that*

$$S_{jk} = \sum_{i=0}^{n} A_{ijk}, \quad \forall\, j \in \{0,1,2\}, \quad \forall\, k \in 0 \bigcup N \tag{44}$$

satisfy the following conditions : $\quad S_{2k} > 0;$ \hfill (45)

$$A_{i0k} \geq 0, \quad \forall\, k = \overline{0,n},\ n \in N, \quad \sigma = \max_{0 \leq k \leq n} \tau_{i0}; \tag{46}$$

$$S_{1k} \in R: \quad 4qB_k S_{2k} > S_{1k}^2 \quad and \quad \tilde{L}_k = \frac{\sqrt{4qB_k S_{2k} - S_{1k}^2}}{2S_{2k}} \tag{47}$$

and let the function $w(x,t)$ of type (35) be a solution of the problems (43), (3) and (43), (4). If we suppose that the function

$$\tilde{\Phi}_k(s) = \int_{s}^{s+\pi/\tilde{L}_k} F_k(t) e^{t_k} \sin \tilde{L}_k(t-s)\,dt, \quad t_k = \frac{t S_{2k}}{2S_{1k}}, \tag{48}$$

$k \in 0 \bigcup N$ *is oscillatory, then the function $w_k(x,t)$ oscillates.*

Proof: We begin with the fact that every such addend $w_k(x,t)$ could belong to one of the three types, described in Definitions 3 and 4. Hence, it is enough to establish that $w_k(x,t)$ is nor eventually positive, neither eventually negative one. Let us consider the first possibility. More exactly, we shall explain in details how the contrary assumption leads to a contradiction. So that, let us suppose that

$$\exists\, \tilde{w}_k(x,t) > 0, \quad \forall\, x \in (0,l), \quad \forall\, t \in [T_k,\infty), \quad T_k \geq 0 \tag{49}$$

and that $\tilde{w}_k(x,t)$ satisfies all the conditions in the formulation of Theorem 2. Then we have that (49) and Lemma 2 make us sure that $\tilde{w}_k(t)$ is an eventually positive solution of the equation

$$\left(\sum_{i=0}^{n} A_{i2k} \right) z''(t) + \left(\sum_{i=0}^{n} A_{i1k} \right) z'(t) + \sum_{i=0}^{n} A_{i0k} z(t - \tau_{i0}) + qB_k z(t) = F_k(t), \tag{50}$$

but (50) is a particular case of the equation (18), where

$$A = S_{2k} = \sum_{i=0}^{n} A_{i2k}, \quad B = qB_k;$$

$$\theta = S_{1k} = \sum_{i=0}^{n} A_{i1k}; \quad \beta_i = A_{i0k} \quad and \quad \sigma_i = \tau_{i0}.$$

Hence, we could apply Theorem 1 to the equation (50) and it leads to a contradiction. Similarly, the assumption about an eventually negative solution would guarantee a contradiction with the same theorem.

Remark 8 *We mention if $k = 1$, then that there is a parallel between the distributions of zeros of the equations (43) and (18) in the sense of Remark 5. More exactly, one of the* *assumptions of Theorem 2 is*

$$\exists \{\tilde{s}_m\}_{m=1}^{\infty} : \quad \lim_{m \to \infty} \tilde{s}_m = \infty \quad and \quad \tilde{\Phi}_k(\tilde{s}_m) = 0$$

and then every solution of the problems (43), (3) and (43), (4), which is of the type $w_1(x,t)$ oscillates if all the conditions of Theorem 2 hold. Moreover, the same function $w_1(x,t)$ has at least one zero in the sets $(0,l) \times (\tilde{s}_n - \sigma, \tilde{s}_n + \pi/\tilde{L}_1)$.

The last equation is

$$\sum_{i=0}^{n} \left(K_{i2} w_{x^i t^2}(x,t) + K_{i1} w_{x^i t}(x,t) + K_{i0} w_{x^i}(x,t - \tau_{i0}) \right) + qw = f(x,t),$$

$$x \in (0,l), \quad t > 0 \qquad (51)$$

because of Remarks 1 and 8.

Theorem 3 *Let the function $w_1(x,t)$ be a solution of the problems (51), (3) and (51), (4), which is of the type (8) and satisfies (**H**). Also, let the notations (38) and (44) — (48) be fulfilled for $k = 1$ and let the function $\tilde{\Phi}_1(s)$ be oscillatory. Then the function $w_1(x,t)$ oscillates.*

Proof: As before, we must obtain that $w_1(x,t)$ is nor eventually positive, neither eventually negative one. Also, it is enough to consider the first possibility, where the contrary assumption is the particular case of (49), where $k = 1$. Then we apply Lemma 3 to conclude that $\tilde{w}_1(t)$ is an eventually positive solution of the inequality

$$\left(\sum_{i=0}^{n} A_{i21} \right) z''(t) + \left(\sum_{i=0}^{n} A_{i11} \right) z'(t) + \sum_{i=0}^{n} A_{i01} z(t - \tau_{i0}) + qB_1 z(t) \le F_1(t). \qquad (52)$$

The fact that $\tilde{\Phi}_1(s)$ oscillates means that

$$\exists \{\tilde{s}_m\}_{m=1}^{\infty} : \quad \lim_{m \to \infty} \tilde{s}_m = \infty \quad and \quad \tilde{\Phi}_1(\tilde{s}_m) = 0.$$

Hence, we establish that

$$\tilde{\Phi}_1(\tilde{s}_m) \le 0 \quad and \quad \tilde{\Phi}_1(\tilde{s}_m) \ge 0. \qquad (53)$$

The first possibility in (53) guarantees that we could apply Corollary 1 to (52) in $(s_m - \sigma, s_m + \pi/\tilde{L}_1)$, $\forall m \in \mathbf{N}$ and this is the needed contradiction.

Again we succeeded to prove more in the sense that every function $w_1(x,t)$, which satisfies all the conditions of Theorem 3 has at least one zero in the sets $(0,l) \times (\tilde{s}_n - \sigma, \tilde{s}_n + \pi/\tilde{L}_1)$.

REFERENCES

1. A. E. El'sgol'ts, S. B. Norkin, *Introduction to the Theory and Application of Differential Equations with Deviating Arguments*, Academic Press, 1973.
2. Z. A. Petrova, "Application of Galerkin's method to the investigation of oscillation behaviour of two equations," in *Applications of Mathematics in Engineering & Economics*, edited by G. Venkov and M. Marinov, Proc. 28th International Conference, Sozopol, Bulvest 2000, Sofia, 2003, pp. 126–129.
3. Z. A. Petrova, *CFD Journal* **13**(1), 13–20 (2004).
4. Z. A. Petrova, *C. R. Bulg. Acad. Sci.* **58**(3), 251–256 (2005).
5. Z. A. Petrova, "Oscillations of some equations and inequalities of second order and applications in mechanics," in *Applications of Mathematics in Engineering & Economics*, edited by M. D. Todorov, Proc. 33rd Intenational Conference, AIP CP946, Melville, New York, 2007, pp. 224–234.
6. P. A. Vel'misov, L. V. Gârnefska, and S. D. Milusheva, "Investigation of the stability of the solutions of the equation of oscillations of an axis or a plate with a delay in time of the reactions and friction forces," in *Applications of Mathematics in Engineering & Economics*, edited by B. I. Cheshankov and M. D. Todorov, Proc. 24th Summer School, Sozopol, Heron Press, Sofia, (1999), pp. 83–88.
7. P. A. Vel'misov, L. V. Gârnefska, and S. D. Milusheva, "On the dynamical stability of nonlinear oscillation of axes and plates," in *Applications of Mathematics in Engineering & Economics*, edited by B. I. Cheshankov and M. D. Todorov, Proc. 24th Summer School, Sozopol, Heron Press, Sofia, 1999, pp. 89–95.
8. N. Yoshida, *Proc. Roy. Soc. Edinburgh A* **106**, 121–129 (1987).
9. N. Yoshida, *Differential and Integral Equations* **3**(1), 155–160 (1990).

On the Maximum Number of Periodical Solutions of a Class of Autonomous Systems

M. Raeva

Technical University of Sofia, 8, Kl. Ohridski Blvd, 1000 Sofia, Bulgaria

Abstract. The autonomous systems (1) are investigated on the condition (4), assuming the existence of inner resonances. Nonlinear parts are polynomials of a certain class. The present article is a continuation of the last one [16], which was published by American Institute of Physics, New York in 2007. We are interested both in the existence and the number of periodical trajectories of the system (1) with initial conditions (6).

Keywords: Autonomous systems, limit cycle, periodical trajectory, maximum number of periodical solutions.
PACS: 02.30.Hq, 02.30.Oz

INTRODUCTION

It is well known (see [1]) that the problem of finding the maximum number of limit cycles of polynomial autonomous systems of two equations is closely connected with Hilbert's 16th Problem. The autonomous systems with parameters have some interesting representations (see, for example [12, 14, 15]). We are interested in the range of the values of the parameters of the system where periodical trajectories exist or the range of values where these trajectories are missing. The existence of limit cycle is proved in [13]. The present article is a continuation of the last one [16], which was published by the American Institute of Physics, New York in 2007. In the present work we have investigated the maximum number of periodical trajectories of $G_{n,3}$ polynomial autonomous system of n ($n \in \mathbf{N}$) equations. We consider the system

$$\ddot{\psi} = C\psi + \varepsilon f(\varepsilon, \psi, \dot{\psi}), \tag{1}$$

where ε is a small positive parameter, $\psi = \mathrm{col}(\psi_1, \psi_2, ..., \psi_n)$, $C = ||c_{ij}||$ is a real square matrix of order n with eigenvalues

$$-k_1^2 < -k_2^2 < ... < -k_n^2. \tag{2}$$

The components of the vector function $f(\varepsilon, \psi, \dot{\psi})$ are

$$f_j(\varepsilon, \psi, \dot{\psi}) = \sum_{v=0}^{\infty} \varepsilon^v g_{jv}(\psi, \dot{\psi}), (j = 1, 2, ..., n) \tag{3}$$

where g_{jv} are analytical functions in some region $G \subset R^{2n}$, satisfying the conditions

$$g_{jv}(-\psi, \dot{\psi}) = -g_{jv}(\psi, \dot{\psi}). \tag{4}$$

CP1067, *Applications of Mathematics in Engineering and Economics '34—AMEE '08*, edited by M. D. Todorov

It is necessary to note, that the conditions (4) are satisfied for some dynamical systems in mechanics in articles [2], [7], [8], [3]. In addition, here it is worth to point out Hale's monograph [5] also. It is supposed that the infinite series (3) are convergent, when $(\psi, \dot{\psi}) \in G \subset R^{2n}$ in some vicinity of the point $\varepsilon = 0$. Let $\lambda^{(s)}(\lambda_{s1}, \lambda_{s2}, ..., \lambda_{sn})$ be a non-zero eigenvector of the matrix $C = ||c_{ij}||$ corresponding to the eigenvalue $(-1)k_s^2(s = 1, 2, ..., n)$. Besides, it is supposed that inner resonaces exist. That means

$$k_\eta = k(\eta)k_n(k(\eta) - \text{integer number} \geq 2, \eta = 1, ..., n-1, k(n) = 1) \qquad (5)$$

Such cases of inner resonance (5) when $n = 2$, are examined [4], [11].

In the present work we are interested both in the existence and the number of periodical solutions of the system (1), with initial conditions $(\varepsilon \neq 0)$

$$\psi^{(s)}(0, \beta_1, ..., \beta_{s1}, \beta_{s+1}, ..., \beta_n) = 0$$

$$\dot{\psi}^{(s)}(0, \beta_1, \beta_2, ..., \beta_{s-1}, \beta_{s+1}, ..., \beta_n) \qquad (6)$$

$$= \lambda^{(s)} + \sum_{i=1}^{s-1} p_{si}(N_i + \beta_i)\lambda^{(i)} + \sum_{i=s+1}^{n} p_{si}(N_i + \beta_i)\lambda^{(i)}.$$

Here p_{si} are real multipliers and the parameters $N_i(i = 1, 2, ..., s-1, s+1, ..., n)$ can be found out from the sufficient conditions for periodicity [16]

$$\varepsilon_{0s}\lambda_{sr} + \sum_{i=1}^{s-1} \varepsilon_{0i}p_{si}N_i\lambda_{ir} + \sum_{i=s+1}^{n} \varepsilon_{0i}p_{si}N_i\lambda_{ir})a_{0u}^{(1,s)}\left(\frac{\pi}{k_n}\right)$$

$$(7)$$

$$-\left[\varepsilon_{0s}\lambda_{su} + \sum_{i=1}^{s-1} \varepsilon_{0i}p_{si}N_i\lambda_{iu} + \sum_{i=s+1}^{n} \varepsilon_{0i}p_{si}N_i\lambda_{iu})a_{0r}^{(1,s)}\left(\frac{\pi}{k_n}\right)\right] = 0.$$

and $\det||l_{uj}|| \neq 0(n = 1, ..., n; j = 1, ..., s-1, s+1, ..., n)$ where l_{uj} is presented by the formula from [16].

The components of the vector $a_0^{(1,s)}(\frac{\pi}{k_n})$ in the conditions (7) can be found out by the formula [10]

$$a_{0,i}^{(1,s)}\left(\frac{\pi}{k_n}\right) = \sum_{q=1}^{n}\left[\frac{\varepsilon_q\lambda_{qi}}{k(q)k_n\Delta}\int_0^{\frac{\pi}{k_n}} G_q(u)\sin k(q)k_n u du\right]. \qquad (8)$$

Here $G_q(u)$ is a determinant obtained from

$$\Delta = \begin{vmatrix} \lambda_{11} & \lambda_{21} & \cdots & \lambda_{n1} \\ \lambda_{12} & \lambda_{22} & \cdots & \lambda_{n2} \\ \cdots\cdots\cdots\cdots\cdots \\ \lambda_{1n} & \lambda_{2n} & \cdots & \lambda_{nn} \end{vmatrix} \qquad (9)$$

404

after substitution of column q in (9) with

$$(g_{10}(u),...,q_{n0}(u))^T, \quad g_{i0} = g_{i0}(\psi_{10}^{(s)}(u),...,\psi_{n0}^{(s)}(u),\dot{\psi}_{10}^{(s)},...,\dot{\psi}_{n0}^{(s)}(u)) \quad (10)$$

where

$$\psi_{j0}^s(u) = \frac{\sin k(s)k_n u}{k(s)k_n}\lambda_{sj} + \sum_{i=1}^{s-1} p_{si}N_i \frac{\sin k(i)k_n u}{k(i)k_n}\lambda_{ij} + \sum_{i=s+1}^{n} p_{si}N_i \frac{\sin(i)k_n u}{k(i)k_n}\lambda_{ij}.$$

and $j = 1, 2, ..., n$).

MAIN RESULT

Now in order to solve the problem we study the class of polynomials $G_{n,3}$, which is obtained, when each of the functions g_{jv} is a polynomial of degree two of $2n$–variables and satisfy the conditions (4). First of all, it is proved that $g_{jv}(\psi_1, ..., \psi_n; \dot{\psi}_1...\dot{\psi}_n) \in G_{n,3}$ if and only if

$$g_{jv}(\psi_1, \psi_2, ..., \psi_n; \dot{\psi}_1, ..., \dot{\psi}_n) = \sum_{u=1}^{n}\sum_{v=u}^{n}\sum_{w=v}^{n} c_{jv}^{(uvw)}\psi_u\psi_v\psi_w + \sum_{u=1}^{n}\sum_{v=1}^{n} a_{jv}^{uv}\psi_u\psi_v$$

$$+ \sum_{u=1}^{n}\sum_{v=1}^{n}\sum_{w=v}^{n} d_{jv}^{(uvw)}\psi_u\dot{\psi}_v\dot{\psi}_w + \sum_{u=1}^{n} b_{jv}^{u}\psi_u. \quad (11)$$

In order to investigate the conditions for periodicity (7) for a class $G_{n,3}$ we use the formula (10) and obtain from (11)

$$g_{i0}(u) = \sum_{u=1}^{n}\sum_{v=u}^{n}\sum_{w=v}^{n} c_{i0}^{uvw}\left[\frac{\sin k(s)k_n u}{k(s)k_n}\lambda_{su} + \sum_{n=1}^{s-1} p_{si}N_i\frac{\sin k(i)k_n u}{k(i)k_n}\lambda_{iu}\right.$$

$$\left. + \sum_{m=s+1}^{n} p_{si}N_i\frac{\sin k(i)k_n u}{k(i)k_n}\lambda_{iu}\right]$$

$$\times \left[\frac{\sin k(s)k_n u}{k(s)k_n}\lambda_{sv} + \sum_{i=1}^{s-1} p_{si}N_i\frac{\sin k(i)k_n u}{k(i)k_n}\lambda_{iu} + \sum_{i=s+1}^{n} p_{si}N_i\frac{\sin k(i)k_n u}{k(i)k_n}\lambda_{iv}\right]$$

$$\times \left[\frac{\sin k(s)k_n u}{k(s)k_n u} + \sum_{i=1}^{s-1} p_{si}N_i\frac{\sin k(i)k_n u}{k(i)k_n}\lambda_{iw} + \sum_{i=s+1}^{n} p_{si}N_i\frac{\sin k(i)k_n u}{k(i)k_n}\lambda_{iw}\right] \quad (12)$$

$$+ \sum_{u=1}^{n}\sum_{v=1}^{n}\sum_{w=v}^{n} d_{i0}^{uvw}\left[\frac{\sin k(s)k_n u}{k(s)k_n}\lambda_{su} + \sum_{i=1}^{s-1} p_{si}N_i\frac{\sin k(i)k_n u}{k(i)k_n}\lambda_{iu} + \sum_{i=s+1}^{n} p_{si}N_i\frac{\sin k(i)k_n u}{k(i)k_n}\lambda_{iu}\right]$$

$$\times \left[\cos k(s)k_n u\lambda_{sv} + \sum_{i=1}^{s-1} p_{si}N_i\cos k(i)k_n u\lambda_{iv} + \sum_{i=s+1}^{n} p_{si}N_i\cos k(i)k_n u\lambda_{iv}\right]$$

$$\times \left[\cos k(s)k_n u\lambda_{sw} + \sum_{i=1}^{s-1} p_{si}N_i\cos k(i)k_n u\lambda_{iw} + \sum_{i=s+1}^{n} p_{si}N_i\cos k(i)k_n u\lambda_{iw}\right]$$

405

$$+ \sum_{u=1}^{n} \sum_{v=1}^{n} a_{i0}^{uv} \left[\frac{\sin k(s)k_n u}{k(s)k_n} \lambda su + \sum_{i=1}^{s-1} p_{si} N_i \frac{\sin k(i)k_n u}{k(i)k_n} \lambda_{iu} + \sum_{i=s+1}^{n} p_{si} N_i \frac{\sin k(i)k_n u}{k(i)k_n} \lambda_{iu} \right]$$

$$\times \left[\cos k(s)k_n u \lambda sv + \sum_{i=1}^{s-1} p_{si} N_i \cos k(i)k_n u k(i)k_n \lambda_{iv} + \sum_{i=s+1}^{n} p_{si} N_i \cos k(i)k_n u \lambda_{iv} \right]$$

$$+ \sum_{n=1}^{n} b_{i0}^{u} \left[\frac{\sin k(i)k_n u}{k(s)k_n} \lambda_{su} + \sum_{i=1}^{s-1} p_{si} N_i \frac{\sin k(i)k_n u}{k(i)k_n} \lambda_{iu} + \sum_{i=s+1}^{n} p_{si} N_i \frac{\sin k(i)k_n u}{k(i)k_n} \lambda_{iu} \right].$$

If we substitute formula from presented in the Appendix into (10) we can obtain for $G_q(u)$ an explicit relation.

Besides A_{ism}^q, B_{sis}^q and D_{is}^q have a similar structure of A_{sss}^q. In the expression for $G_q(u)$ (see the Appendix) the unknown quantities $N_1, N_2, ..., N_n$, participate in no more than degree three. Analogously to [16] in order to find out the coefficients of $N_1, N_2, ..., N_n$ in the conditions of existence of periodical trajectories (7) we will apply the following

Lemma 1 *The integral*

$$\int_0^{\frac{\pi}{k_n}} \sin k(i)k_n u \cos k(j)k_n u \sin k(m)k_n u \sin k(q)k_n u du \qquad (13)$$

is equal to $\frac{h\pi}{8k_n}$, *when* $h(h = 1,2,3,4)$ *of relations* $k(i) + k(j) = k(q) + k(m); k(i) + k(m) = k(j) + k(q); k(q) = k(i) + k(j) + k(m); k(i) = k(j) + k(q) + k(m)$ *are satisfied. The integral (13) is equal to* $-\pi h/8k_n$, *when* $h(h = 1,2,3)$ *of relations* $k(i) + k(q) = k(j) + k(m); k(j) = k(i) + k(q) + k(m); k(m) = k(i) + k(q) + k(j)$ *are satisfied. In the rest of the cases, the integral (13) is zero.*

Lemma 2 *The integral*

$$\int_0^{\frac{\pi h}{k_n}} \sin k(i)k_n u \sin k(j)k_n u \sin k(m)k_n u \sin k(q)k_n u du \qquad (14)$$

is equal to $\frac{h\pi}{8k_n}$, *when* $h(h = 1,2,3)$ *of relations* $k(i) + k(m) = k(j) + k(q); k(i) + k(j) = k(m) + k(q); k(i) + k(q) = k(m) + k(j)$. *The integral (14) is equal to* $-\pi h/8k_n$, *when* $h(h = 1,2,3,4)$ *of relations* $k(j) = k(i) + k(m) + k(q); k(q) = k(i) + k(j) + k(m); k(i) = k(q) + k(m) + k(j); k(m) = k(i) + k(q) + k(j)$ *are satisfied. In the rest of the cases, the integral (14) is zero.*

Similarly the other types of calculated integrals are coefficients of the unknown quantities $N_1, N_2,..., N_{s-1}, N_{s+1},...,N_n$ taking part in the sufficient conditions for periodicity (7). Actually (7) is an algebraic system of unknown quantities $N_1, N_2,..., N_{s-1},$

$N_{s+1},...,N_n$. In this case $G_{n,3}$ we have $n-1$ algebraic equations and each of them is of degree no more than four. So using [6] we get following

Theorem 1 *The number of the periodical trajectories of the autonomous systems (1) when inner resonances exist, at the initial conditions of the type (6),* $g_{jv}(\psi_1, \psi_2, ..., \psi_n, \dot{\psi}_1, ..., \dot{\psi}_n) \in G_{n,3}$ *is no more than* 4^{n-1}.

APPLICATION

In the particular case we study the mechanical system [7]

$$wa^2\ddot{\beta}_1 + wab\dot{\beta}_a^2 + \sin(\beta_1 - \beta_a) + wa\ddot{\beta}_a\cos(\beta_1 - \beta_a) + gwa\sin\beta_1 = 0 \tag{15}$$

$$(I + wb^2)\ddot{\beta}_a - wab\dot{\beta}_1^2(\beta_1 - \beta_a) + wab\ddot{\beta}_1\cos(\beta_1 - \beta_a) + gwb\sin\beta_a = 0$$

describing the motions of Centrofugal Compound Pendulum. Here is a compound pendulum of weight W and polar moment of inertia I about the center of gravity and g is an earth acceleration. It is attached at C by a frictionless pivot to a weightless rot of length a, the length of the pendulum between the pivot point C and its center of gravity is b, β_1 and β_a are angles of the vertical coordinates defining the position of the dynamical system. The system (16) with a substitution [11] gets the form

$$\ddot{\psi}_1 = -\frac{g}{a}\psi_1 - \frac{wb^2g}{aI}(\psi_1 - \psi_2) + \varepsilon \sum_{v=0}^{\infty} \varepsilon^v g_{1v}(\psi_1, \psi_2, \dot{\psi}_1, \dot{\psi}_2) \tag{16}$$

$$\ddot{\psi}_2 = \frac{wbg}{I}(\psi_1 - \psi_2) + \varepsilon \sum_{v=0}^{\infty} \varepsilon^v g_{2v}(\psi_1, \psi_2, \dot{\psi}_1, \dot{\psi}_2)$$

where $g_{j,0}(\psi_1, \psi_2, \dot{\psi}_1, \dot{\psi}_2)$, $(j = 1, 2)$ are expressed as

$$g_{1,0}(\psi_1, \psi_2, \dot{\psi}_1, \dot{\psi}_2) = -\frac{wb}{aI}\left\{\left(\frac{I}{w} + b^2\right)\left[\dot{\psi}_2^2(\psi_1 - \psi_2) - \frac{3}{2b}\psi_1^3\right]\right.$$
$$\left. + \frac{1}{2}gb\psi_2(\psi_1 - \psi_2)^2 + ab\left[\dot{\psi}_1^2(\psi_1 - \psi_2) + \frac{g}{6a}\psi_2^3\right]\right\}$$
$$+ \frac{wb}{aI^2}\left[\left(\frac{I}{w} + b^2\right)\frac{g}{b}\psi_1 - bg\psi_2\right]\left[wb^2(\psi_1 - \psi_2)^2\right]$$

$$g_{2,0}(\psi_1, \psi_2, \dot{\psi}_1, \dot{\psi}_2) = \frac{wb}{I}\left\{a\left[\dot{\psi}_1^2(-\psi_1 - \psi_2) + \frac{3}{2a}\psi_2^3\right]\right.$$
$$\left. - \frac{1}{2}g\psi_1(\psi_1 - \psi_2)^2 + b\left[\dot{\psi}_1^2(\psi_1 - \psi_2) - \frac{g}{6b}\psi_1^3\right]\right\}$$
$$- \frac{wb}{I^2}g(\psi_1 - \psi_2)wb(\psi_1 - \psi_2)^2.$$

In fact, in the case in question (16) we are interested in the existence of the periodical solutions and their maximum number. Actually when the indices in the sufficient conditions for periodicity for a class $G_{n,3}$ have fixed values and inner resonances exist, we get the following

Theorem 2 *Let the conditions* $s = 2$, $k_1 = kk_2$, $k \geq 2$, $k \neq 3$

$$k_2^2 = \frac{1}{2}\left[\frac{wg}{al}\left(\frac{I}{w} + b^2 + ab\right) - \sqrt{D}\right],$$

$$D = \left(\frac{g}{al}\right)^2\left[(I - wab)^2 + 2Iwb^2 + w^2b^3(2a + b)\right] > 0$$

hold, and N is the solution of the algebraic system

$$P_{11}N + P_{12}N^3 = 0 \qquad (17)$$

where the coefficients p_{11}, p_{12} *are calculated in [11], then the solutions* $\psi^{(2)}(t,\beta)$ *of the system (16) is periodical with initial conditions*

$$\psi^{(2)}(0,\beta,\varepsilon) = 0$$

$$\dot{\psi}^{(2)}(0,\beta,\varepsilon) = \lambda^{(2)} + p_{21}(N+\beta)\lambda^{(1)}$$

and with period $\frac{2}{k_2}(\pi + \delta)$.

CONCLUSION

In the above article the autonomous system has been investigated assuming the existence of internal resonances. The nonlinear parts are polynomials of a certain class. A condition of existence of periodical trajectories in nonlinear class $G_{n,3}$ is studied. The maximum number of periodical trajectories has been found out under concrete initial conditions. It is interesting to find the algebraic system (7) both for concrete values of the indices and the inner resonances existence. The algebraic structure of the sufficient condition for periodicity (17) enables us a computer simulation to be carried out at the concrete values of indices.

APPENDIX

$$G_q(u) = \frac{\sin^3 k(s)k_nu}{k^3(s)k_n^3}A_{sss}^q + \sum_{i=1}^{s-1}p_{si}N_i\frac{\sin k(i)k_nu\sin^2 k(s)k_nu}{k(i)k^2(s)k_n}A_{sss}^q$$

$$+ \sum_{i=s+1}^{n}p_{si}N_i\frac{\sin k(i)k_nu\sin^2 k(s)k_nu}{k(i)k^2(s)k_n^3}A_{sis}^q + \sum_{i=1}^{s-1}p_{si}N_i\frac{\sin k(i)k_nu\sin^2 k(s)k_nu}{k(i)k^2(s)k_n^3}A_{iss}^q$$

$$+ \sum_{i=1}^{s-1}\sum_{j=1}^{s-1}p_{si}p_{sj}N_iN_j\frac{\sin k(i)k_nu\sin k(j)k_nu\sin k(s)k_nu}{k(i)k(j)k(s)k_n^3}A_{ijs}^q$$

$$+ \sum_{i=1}^{s-1} \sum_{j=1}^{n-s} p_{si} p_{s,s+j} N_i N_{s+j} \frac{\sin k(i) k_n u \sin k(s+j) k_n u \sin k(s) k_n u}{k(s+j)k(i)k(s)k_n^3} A_{i,s+j,s}^q$$

$$+ \sum_{i=s+1}^{n} p_{si} N_i \frac{\sin k(i) k_n u \sin^2 k(s) k_n u}{k(i)k^2(s)k_n^3} A_{iss}^q$$

$$+ \sum_{i=s+1}^{n} \sum_{j=1}^{s-1} p_{si} p_{sj} N_i N_j \frac{\sin k(i) k_n u \sin k(j) k_n u \sin k(s) k_n u}{k(s)k(i)k(j)k_n^3} A_{ijs}^q$$

$$+ \sum_{i=s+1}^{n} \sum_{j=1}^{n-s} p_{s,s+j} p_{sj} N_i N_{s+j} \frac{\sin k(i) k_n u \sin k(s+j) k_n u \sin k(s) k_n u}{k(i)k(s+j)k(s)k_n^3} A_{i,s+j,s}^q$$

$$+ \sum_{m=1}^{s-1} p_{sm} N_m \frac{\sin k(u) k_n u \sin^2 k(s) k_n u}{k(m)k^2(s)k_n^3} A_{ssm}^q$$

$$+ \sum_{i=1}^{s-1} \sum_{m=1}^{s-1} p_{si} p_{sm} N_i N_m \frac{\sin k(i) k_n u \sin k(s) k_n u \sin k(m) k_n u}{k(i)k(s)k(m)k_n^3} A_{sim}^q$$

$$+ \sum_{i=1}^{s-1} \sum_{m=s+1}^{n} p_{si} p_{sm} N_i N_m \frac{\sin k(i) k_n u \sin k(s) k_n u \sin k(m) k_n u}{k(i)k(s)k(m)k_n^3} A_{sim}^q$$

$$+ \sum_{i=1}^{s-1} \sum_{m=1}^{s-1} p_{si} p_{sm} N_i N_m \frac{\sin k(i) k_n u \sin k(s) k_n u \sin k(m) k_n u}{k(i)k(s)k(m)k_n^3} A_{ism}^q$$

$$+ \sum_{i=1}^{s-1} \sum_{j=1}^{n-s} \sum_{m=1}^{s-1} p_{si} p_{sm} p_{sj} N_i N_m N_j \frac{\sin k(j) k_n u \sin k(i) k_n u \sin k(m) k_n u}{k(j)k(i)k(m)k_n^3} A_{ijm}^q$$

$$+ \sum_{i=1}^{s-1} \sum_{j=1}^{n-s} \sum_{m=1}^{s-1} p_{si} p_{s,s+j} p_{sm} N_i N_{s+j} N_m \frac{\sin k(i) k_n u \sin k(s+j) k_n u \sin k(m) k_n u}{k(i)k(s+j)k(m)k_n^3} A_{i,s+j,m}^q$$

$$+ \sum_{i=s+1}^{n} \sum_{m=1}^{s-1} p_{si} p_{sm} N_i N_m \frac{\sin k(i) k_n u \sin k(s) k_n u \sin k(m) k_n u}{k(i)k(s)k(m)k_n^3} A_{ism}^q$$

$$+ \sum_{i=1}^{s-1} \sum_{j=1}^{n-s} \sum_{m=1}^{s-1} p_{si} p_{s,s+j} p_{sm} N_i N_{s+j} N_m \frac{\sin k(i) k_n u \sin k(s+j) k_n u \sin k(m) k_n u}{k(s+j)k(m)k(i)k_n^3} A_{s+j,i,m}^q$$

$$+ \sum_{i=1}^{n-s} \sum_{j=1}^{n-s} \sum_{m=1}^{n-s} p_{s,s+i} p_{s,s+j} p_{s,s+m} N_{s+i} N_{s+j} N_m$$

$$\times \frac{\sin k(s+i) k_n u \sin k(s+j) k_n u \sin k(m) k_n u}{k(s+i)k(s+j)k(m)k_n^3} A_{s+i,s+j,m}^q$$

$$+ \sum_{m=s+1}^{n} p_{sm} N_m \frac{\sin^2 k(s) k_n u \sin k(m) k_n u}{k(m)k^2(s)k_n} A_{ssm}^q$$

409

$$+\sum_{i=1}^{s-1}\sum_{m=s+1}^{n}p_{si}N_ip_{sm}N_m\frac{\sin k(i)k_nu\sin k(s)k_nu\sin k(m)k_nu}{k(i)k(s)k(m)k_n^3}A_{sim}^q$$

$$+\sum_{i=s+1}^{n}\sum_{m=s+1}^{n}p_{si}p_{sm}N_iN_m\frac{\sin k(i)k_nu\sin k(s)k_nu\sin k(m)k_nu}{k(i)k(s)k(m)k_n^3}A_{ism}^q$$

$$+\sum_{i=1}^{s-1}\sum_{m=s+1}^{n}p_{si}p_{sm}N_iN_m\frac{\sin k(i)k_nu\sin k(s)k_nu\sin k(m)k_nu}{k(i)k(s)k(m)k_n^3}A_{ism}^q$$

$$+\sum_{i=1}^{s-1}\sum_{j=1}^{s-1}\sum_{m=s+1}^{n}p_{si}p_{s,s+j}p_{sm}N_iN_{s+j}N_m\frac{\sin k(i)k_nu\sin k(j)k_nu\sin k(m)k_nu}{k(i)k(j)k(m)k_n^3}A_{i,j,m}^q$$

$$+\sum_{i=1}^{s-1}\sum_{j=1}^{n-s}\sum_{m=s+1}^{n}p_{si}p_{s,s+j}p_{sm}N_iN_{s+j}N_m\frac{\sin k(i)k_nu\sin k(s+j)k_nu\sin k(m)k_nu}{k(i)k(s+j)k(m)k_n^3}A_{i,s+j,m}^q$$

$$+\frac{\sin k(s)k_nu\cos^2 k(s)k_nu}{k(s)k_n}B_{sss}^q+\sum_{i=s+1}^{n}p_{si}N_i\frac{\sin k(s)k_nu\cos k(i)k_nu\cos k(s)k_nu}{k(s)k_n}B_{sis}^q$$

$$+\sum_{i=1}^{s-1}p_{si}N_i\frac{\cos k(i)k_nu\sin k(s)k_nu\cos k(s)k_nu}{k(s)k_n}B_{sis}^q+\sum_{i=1}^{s-1}p_{si}N_i\frac{\sin k(i)k_nu\cos^2 k(s)k_nu}{k(i)k_n}B_{iss}^q$$

$$+\sum_{i=1}^{s-1}\sum_{j=1}^{s-1}p_{si}p_{sj}N_iN_j\frac{\sin k(i)k_nu\cos k(j)k_nu\cos k(s)k_nu}{k(i)k_n}B_{ijs}^q$$

$$+\sum_{i=1}^{s-1}\sum_{j=1}^{n-s}p_{si}p_{s,s+j}N_iN_{j+s}\frac{\sin k(i)k_nu\cos k(s+j)k_nu\cos k(s)k_nu}{k(i)k_n}B_{i,s+j,s}^q$$

$$+\sum_{i=s+1}^{n}p_{si}N_i\frac{\sin k(i)k_nu\cos^2 k(s)k_nu}{k(i)k_n}B_{iss}^q$$

$$+\sum_{i=1}^{s-1}\sum_{j=1}^{n-s}p_{si}p_{s,s+j}N_iN_{s+j}\frac{\cos k(i)k_nu\sin k(s+j)k_nu\cos k(s)k_nu}{k(s+j)k_n}B_{s+j,i,s}^q$$

$$+\sum_{i=1}^{n-s}\sum_{j=1}^{n-s}p_{s,s+i}p_{s,s+j}N_{s+i}N_{s+j}\frac{\sin k(s+i)k_nu\cos k(s+j)k_nu\cos k(s)k_nu}{k(s+i)k_n}B_{s+i,s+j,s}$$

$$+\sum_{i=1}^{s-1}p_{sm}N_m\frac{\sin k(s)k_nu\cos k(s)k_nu\cos k(m)k_nu}{k(s)k_n}B_{ssm}^q$$

$$+\sum_{i=1}^{s-1}\sum_{m=1}^{s-1}p_{si}p_{sm}N_iN_m\frac{\sin k(s)k_nu\cos k(i)k_nu\cos k(m)k_nu}{k(s)k_n}B_{sim}^q$$

$$+ \sum_{i=s+1}^{n} \sum_{m=1}^{s-1} p_{si} p_{sm} N_i N_m \frac{\sin k(i) k_n u \cos k(s) k_n u \cos k(m) k_n u}{k(i) k_n} B^q_{sim}$$

$$+ \sum_{i=1}^{s-1} \sum_{m=1}^{s-1} p_{si} p_{sm} N_i N_m \frac{\sin k(i) k_n u \cos k(s) k_n u \cos k(m) k_n u}{k(i) k_n} B^q_{ism}$$

$$+ \sum_{i=1}^{s-1} \sum_{j=1}^{s-1} \sum_{m=1}^{s-1} p_{si} p_{sj} p_{sm} N_i N_m N_j \frac{\sin k(i) k_n u \cos k(s) k_n u \cos k(j) k_n u}{k(i) k_n} B^q_{ijm}$$

$$+ \sum_{i=1}^{n-s} \sum_{j=1}^{n-s} \sum_{m=1}^{s-1} p_{s,s+j} p_{s,s+i} p_{sm} N_{s+i} N_m N_{s+j}$$

$$\times \frac{\sin k(s+i) k_n u \cos k(s+j) k_n u \cos k(m) k_n u}{k(s+i) k_n} B^q_{s+i,s+j,m}$$

$$+ \sum_{m=s+1}^{n} p_{sm} N_m \frac{\cos k(m) k_n u \sin k(s) k_n u \cos k(s) k_n u}{k(s) k_n} B^q_{ssm}$$

$$+ \sum_{i=1}^{s-1} \sum_{m=s+1}^{n} p_{si} p_{sm} N_i N_m \frac{\sin k(s) k_n u \cos k(i) k_n u \cos k(m) k_n u}{k(s) k_n} B^q_{sim}$$

$$+ \sum_{i=s+1}^{n} \sum_{m=s+1}^{n} p_{si} p_{sm} N_i N_m \frac{\sin k(s) k_n u \cos k(i) k_n u \cos k(s) k_n u}{k(i) k_n} B^q_{sim}$$

$$+ \sum_{i=1}^{s-1} \sum_{m=s+1}^{n} p_{si} p_{sm} N_i N_m \frac{\sin k(i) k_n u \cos k(s) k_n u \cos k(m) k_n u}{k(i) k_n} B^q_{ism}$$

$$+ \sum_{i=1}^{s-1} \sum_{j=1}^{s-1} \sum_{m=s+1}^{n} p_{si} p_{sj} p_{sm} N_i N_j N_m \frac{\sin k(i) k_n u \cos k(j) k_n u \cos k(m) k_n u}{k(i) k_n} B^q_{ijm}$$

$$+ \sum_{i=1}^{s-1} \sum_{j=1}^{n-s} \sum_{m=s+1}^{n} p_{si} p_{s,s+j} p_{sm} N_i N_{s+j} N_m \frac{\sin k(i) k_n u \cos k(s+j) k_n u \cos k(m) k_u n}{k(i) k_n} B^q_{i,s+j,m}$$

$$+ \sum_{i=s+1}^{n} \sum_{m=s+1}^{n} p_{si} p_{sm} N_i N_m \frac{\sin k(i) k_n u \cos k(s) k_n u \cos k(m) k_u n}{k(i) k_n} B^q_{ism}$$

$$+ \sum_{i=1}^{s-1} \sum_{j=1}^{n-s} \sum_{m=s+1}^{n} p_{si} p_{s,s+j} p_{sm} N_i N_{s+j} N_m \frac{\sin k(s+j) k_n u \cos k(m) k_n u \cos k(i) k_u n}{k(s+j) k_n} B^q_{s+j,i,m}$$

$$+ \sum_{i=1}^{s-1} \sum_{j=1}^{s-1} \sum_{m=s+1}^{n} p_{s,s+i} p_{s,s+j} p_{sm} N_{s+i} N_{s+j} N_m$$

$$\times \frac{\sin k(s+i) k_n u \cos k(s+i) k_n u \cos k(m) k_u n}{k(s+i) k_n} B^q_{s+i,s+j,m}$$

$$+ \frac{\sin k(s) k_n u \cos k(s) k_n u}{k(s) k_n} D^q_{ss} + \sum_{i=1}^{s-1} p_{si} N_i \frac{\cos k(i) k_n u \sin k(s) k_n u}{k(s) k_n} D^q_{si}$$

411

$$+ \sum_{i=s+1}^{n} p_{si} N_i \frac{\cos k(i)k_n u \sin k(s)k_n u}{k(s)k_n} D_{si}^q + \sum_{i=1}^{s-1} p_{si} N_i \frac{\sin k(i)k_n u \cos k(s)k_n u}{k(i)k_n} D_{is}^q$$

$$+ \sum_{i=1}^{s-1} \sum_{j=1}^{s-1} p_{si} p_{sj} N_i N_j \frac{\sin k(i)k_n u \cos k(j)k_n u}{k(i)} D_{ij}^q$$

$$+ \sum_{i=1}^{s-1} \sum_{j=1}^{n-s} p_{si} p_{s,s+j} N_i N_{s+j} \frac{\sin k(i)k_n u \cos k(s+j)k_n u}{k(i)k_n} D_{i,s+j}^q$$

$$+ \sum_{i=s+1}^{n} p_{si} N_i \frac{\sin k(i)k_n u \cos k(s)k_n u}{k(i)k_n} D_{is}^q$$

$$+ \sum_{i=1}^{s-1} \sum_{j=1}^{n-s} p_{si} p_{s,s+j} N_i N_{s+j} \frac{\cos k(i)k_n u \sin k(s+j)k_n u}{k(s+j)k_n} D_{s+j,i}^q$$

$$+ \sum_{i=1}^{n-s} \sum_{j=1}^{n-s} p_{s,s+i} p_{s,s+j} N_{s+i} N_{s+j} \frac{\sin k(i)k_n u \cos k(s+j)k_n u}{k(s+j)k_n} D_{s+i,s+j}^q$$

$$+ \frac{\sin k(s)k_n u}{k(s)k_n} E_s^q + \sum_{i=1}^{s-1} p_{si} N_i \frac{\sin k(i)k_n u}{k(i)k_n} E_i^q + \sum_{i=s+1}^{n} p_{si} N_i \frac{\sin k(i)k_n u}{k(i)k_n} E_i^q,$$

where

$$A_{sss}^q = \begin{vmatrix} \lambda_{11} & \lambda_{21} & \dots & \lambda_{q-1,1} & \sum_{u=1}^{n}\sum_{v=u}^{n}\sum_{w=v}^{n} c_{10}^{uvw} \lambda_{su}\lambda_{sv}\lambda_{sw} & \lambda_{q+1,1} \dots \lambda_{n,1} \\ \lambda_{12} & \lambda_{22} & \dots & \lambda_{q-1,2} & \sum_{u=1}^{n}\sum_{v=u}^{n}\sum_{w=v}^{n} c_{20}^{uvw} \lambda_{su}\lambda_{sv}\lambda_{sw} & \lambda_{q+1,2} \dots \lambda_{n,2} \\ \dots & & & & & \\ \lambda_{1n} & \lambda_{2n} & \lambda_{q-1,n} & \sum_{u=1}^{n}\sum_{v=u}^{n}\sum_{w=v}^{n} c_{n0}^{uvw} \lambda_{su}\lambda_{sv}\lambda_{sw} & \lambda_{q+1,n} \dots \lambda_{n,n} \end{vmatrix},$$

$$A_{sis}^q = \begin{vmatrix} \lambda_{11} & \lambda_{21} & \dots & \lambda_{q-1,1} & \sum_{u=1}^{n}\sum_{v=u}^{n}\sum_{w=v}^{n} c_{10}^{uvw} \lambda_{su}\lambda_{iv}\lambda_{sw} & \lambda_{q+1,1} \dots \lambda_{n,1} \\ \lambda_{12} & \lambda_{22} & \dots & \lambda_{q-1,2} & \sum_{u=1}^{n}\sum_{v=u}^{n}\sum_{w=v}^{n} c_{20}^{uvw} \lambda_{su}\lambda_{iv}\lambda_{sw} & \lambda_{q+1,2} \dots \lambda_{n,2} \\ \dots & & & & & \\ \lambda_{1n} & \lambda_{2n} & \lambda_{q-1,n} & \sum_{u=1}^{n}\sum_{v=u}^{n}\sum_{w=v}^{n} c_{n0}^{uvw} \lambda_{su}\lambda_{iv}\lambda_{sw} & \lambda_{q+1,n} \dots \lambda_{n,n} \end{vmatrix}.$$

REFERENCES

1. V. Amelkin, N. Lukashevich, and A. Sadovskii, *Nonlinear Oscillations in Second Order Systems*, BGU, Minsk, 1982. (in Russian)
2. G. Bradistilov, *God. VUZ* **2**(1), 2–11 (1956). (in Bulgarian)
3. G. Bradistilov and S. Manolov, "On an Algebraic Condition for Periodical Trajectories of Differential Equations with Nonlinear Parts of Polynomial Types,", *Theoretical and Applied Mechanics, BAS* **IV**(1), 9–18 (1973). (in Bulgarian)

4. W. Carnegie, *Centrufugal Compound Pendulum, Engineering Materials and Design*, **6** 423–231 (1963).
5. J. Hale, *Oscillations in Nonlinear System*, New York, 1963.
6. A. Kurosh, *Course of Higher Algebra*, Nauka, Moscow, 1965. (in Russian)
7. S. Manolov, *God. VUZ, Applied Mathematics* **10**(1), 19–27 (1974). (in Bulgarian)
8. S. Manolov, "Some Cases of Existence and Construction of Periodical Solutions for a Class of Nonlinear Systems with a Small Parameter,", *Izvestiya IM BAS* **11**, 197–212 (1970). (in Bulgarian)
9. M. Raeva, *God. VUZ, Applied Mathematics* **XII**(1), 109–223 (1976).
10. M. Raeva and S. Manolov, *God. VUZ, Applied Mathematics* **2**, 59–67 (1979).
11. M. Raeva, *God. VUZ, Applied Mathematics*, **21**(4) 37–44 (1985).
12. M. Raeva, "Perturbation of a Linear System with a Center Type Singular Point," in *Colloquia Mathematica Societatis Janos Bolyai, 53, Qualitative Theory of Differential Equations, Szeged, Hungary, 1988*, edited by B.Sz. Nagy and L. Hatvany, North Holland Publishing Co., Amsterdam-New York, 1990, pp.553–558.
13. M. Raeva, *Compt.Rend. de L'Academie Bulgare des Sciences* **51**(1-2), 16–21 (1998).
14. M. Raeva, *Mathematica Balkanica*, New Series **13**, 213–221 (1999).
15. M. Raeva, "Bifurcation Geberating Limit Cycle in a Class of Autonomous Systems," in *Applications of Mathematics in Engineering and Economics'28*, edited by M.Marinov and G.Venkov, Bulvest 2000, Sofia, 2003, pp.129–135.
16. M. Raeva, "On the Existence of a Maximum Number of Periodical Solutions of a Class of Autonomous Systems," in *Applications of Mathematics in Engineering and Economics'33*, edited by M.Todorov, AIP CP946, Melville, New York, 2007, pp.235–243.

Stability of the Solutions of One Class of Aerohydroelasticity Problems

P. A. Velmisov and A. V. Ankilov

Ulyanovsk State Technical University, 32 Severny Venets str., Ulyanovsk 432027, Russia.

Abstract. Dynamic stability of elastic elements (plates in this work) of thin-shelled constructions in interaction with fluid or gas flow is studied. Subsonic regime is considered. The definition of elastic body stability corresponds to Lyapunov's concept of dynamical system's stability. Aerodynamic load is determined by asymptotic aerohydromechanical equations [1]. The problems are studied both with consideration of time delay of the elements bases reactions and without it. The statements and investigation methods offered for dynamical damping elastic bodies, being in contact with subsonic fluid or gas flow, lead to the study of linked initial boundary problems for partial differential equation or system of partial differential equations. Being based on the construction of functionals, corresponding to this system, conditions of solutions stability are obtained for some aerohydroelastical problems, in particular, for dynamics of the plate elements; of elements of the plane channel through which fluid flows; of wing profile elements; of the pipeline. Similar problems without time delay of the plates bases reactions were earlier considered in works [2-5].

Keywords: Aerohydroelasticity; thin-shelled construction; dynamic stability; time delay; partial integro-differential equations and its system; retarded argument.
PACS: 46.40.Ff; 46.40.Jj

STABILITY OF ELASTIC PLATE WITH TIME DELAY OF BASE REACTION

Problem Statement

Dynamic stability of elastic plates in interaction with flow of fluid or gas is studied. It is expected that the plates are compressed by the longitudinal stress and are connected with elastic bases. Time delay of the plates bases reactions is taken into account.

Let us consider, as an example, the planar problem about dynamic stability of the elastic plate with non-circulatory flow of this plate by undisturbed subsonic gas stream with time delay of the base reaction.

Suppose on plane xOy, in which occur joint fluctuations of elastic plate and gas, segment $[0, l]$ on axis Ox corresponds to plate. In infinitely remote point a velocity of gas is V and has the direction, coinciding with the direction of axis Ox (Figure 1). It is marked $\bar{y} = y / \varepsilon$ on all the figures, where ε is the small parameter.

CP1067, *Applications of Mathematics in Engineering and Economics '34—AMEE '08*, edited by M. D. Todorov

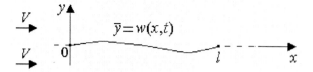

FIGURE 1. Elastic plate

Let perturbations of homogeneous gas stream directed along Ox-axis and deviation of the plate be small. Let us indicate: $\varphi(x,y,t)$ be the potential of gas velocity and $w(x,t)$ be the oscillation function of the plate.

Potential of velocity φ satisfies the Laplace equation

$$\Delta\varphi \equiv \varphi_{xx} + \varphi_{yy} = 0, \qquad (x,y)\in G = R^2 \setminus [0,l], \qquad (1)$$

Linearizing boundary condition

$$\varphi_y^{\pm}(x,0,t) = \lim_{y\to\pm 0} \varphi_y(x,y,t) = w_t(x,t) + Vw_x(x,t), \quad x\in[0,l], \qquad (2)$$

and condition of perturbations absence in infinitely removed point

$$|\nabla\varphi|_{\infty}^2 \equiv \left(\varphi_x^2 + \varphi_y^2 + \varphi_t^2\right)_{\infty} = 0. \qquad (3)$$

Linearizing the Lagrange-Cauchy integral, we present equations of small fluctuations of elastic plate in following way

$$L(w) = \rho\left(\varphi_t^+ - \varphi_t^-\right) + \rho V\left(\varphi_x^+ - \varphi_x^-\right), \ x\in(0,l), \qquad (4)$$

$$L(w) \equiv Dw''''(x,t) + M\ddot{w}(x,t) + Nw''(x,t) + \beta_0 w(x,t-\tau) + \beta_1 \dot{w}(x,t) + \beta_2 \dot{w}''''(x,t). \ (5)$$

Here dotted letters denote the time derivative, a prime is used for the derivative with respect to x or x_1, subindices x, y, t designate partial derivatives with respect to the corresponding variables; ρ is the density of the gas; V is velocity of undisturbed homogeneous flow; l is the length of plate; D and M are the moduli of rigidity and the specific mass of the plate; N is the force, compressing (spraining) the plate; β_1 is the rotational inertia coefficient, β_2 is damping coefficient of the plate material; β_0 are the stiffness coefficient of the base; τ – time of the delay of the plate base reaction.

Using methods of the theory of analytic functions [6], the solution of the problem (1)–(5) can be reduced to equation for unknown function of plate oscillations:

$$Dw''''(x,t) + M\ddot{w}(x,t) + Nw''(x,t) + \beta_0 w(x,t-\tau) + \beta_1 \dot{w}(x,t) + \beta_2 \dot{w}''''(x,t) =$$

$$= -\frac{\rho}{\pi} \int_0^l \left(\dot{w}(x_1,t) + V\dot{w}'(x_1,t) \right) K(x_1,x) dx_1 - \qquad (6)$$

$$- \frac{V\rho}{\pi} \int_0^l \left(\dot{w}(x_1,t) + V\dot{w}'(x_1,t) \right) \frac{\partial K(x_1,x)}{\partial x} dx_1, \quad x \in (0,l),$$

where $K(x_1,x) = 2ln \left| \dfrac{\sqrt{x(l-x_1)} + \sqrt{x_1(l-x)}}{\sqrt{x(l-x_1)} - \sqrt{x_1(l-x)}} \right|$, $x_1 \neq x$.

Stability Investigation

The dynamic stability of this integro-differential equation (6) is investigated supposing that at every end of plate one of the following boundary conditions is satisfied:

$$1) \; w(x,t) = w'(x,t) = 0; \quad 2) \; w(x,t) = w''(x,t) = 0, \quad x = 0, \; x = l, \qquad (7)$$

which corresponds to rigid or hinged fastenings. Write for this equation the functional of Lyapunov's type:

$$\Phi(t) = \int_0^l \left(M\dot{w}^2 + Dw''^2 - Nw'^2 + \beta_0 w^2 \right) dx + \beta_0 \int_0^l \left(\int_{t-\tau}^t dt_1 \int_{t_1}^t \dot{w}^2(x,s)ds \right) dx + I(t) + J(t),$$

$$\qquad (8)$$

$$I(t) = \frac{\rho}{\pi} \int_0^l dx \int_0^l \dot{w}(x,t)\dot{w}(x_1,t)K(x_1,x)dx_1, \quad J(t) = -\frac{\rho V^2}{\pi} \int_0^l dx \int_0^l w'(x,t)w'(x_1,t)K(x_1,x)dx_1.$$

Let us find differentiate Φ by t:

$$\dot{\Phi}(t) = \int_0^l \left(2M\dot{w}\ddot{w} + 2Dw''\dot{w}'' - 2Nw'\dot{w}' + 2\beta_0 w\dot{w} + \beta_0 \int_{t-\tau}^t \dot{w}^2(x,t)dt_1 - \beta_0 \int_{t-\tau}^t \dot{w}^2(x,s)ds \right) dx +$$

$$+ \dot{I} + \dot{J}.$$

Having in mind that $w(x,t-\tau) = w(x,t) - \int_{t-\tau}^t \dot{w}(x,s)ds$, for function $w(x,t)$, being

solution of equation (6), the following equality is obtained:

$$\dot{\Phi}(t) = \int_0^l \{-2\dot{w}\left(Dw'''' + Nw'' + \beta_0 w - \beta_0 \int_{t-\tau}^t \dot{w}(x,s)ds + \beta_1 \dot{w} + \beta_2 \dot{w}'''' + \right.$$

$$+ \frac{\rho}{\pi} \int_0^l \left(\dot{w}(x_1,t) + V\dot{w}'(x_1,t) \right) K(x_1,x)dx_1 + \frac{V\rho}{\pi} \int_0^l \left(\dot{w}(x_1,t) + V\dot{w}'(x_1,t) \right) \frac{\partial K(x_1,x)}{\partial x} dx_1 \right) + \qquad (9)$$

$$+ 2Dw''\dot{w}'' - 2Nw'\dot{w}' + 2\beta_0 w\dot{w} + \beta_0 \int_{t-\tau}^t w^2(x,t)dt_1 - \beta_0 \int_{t-\tau}^t \dot{w}^2(x,s)ds\}dx + \dot{I} + \dot{J}.$$

Taking into account the conditions (7) and integrating by parts, we get:

$$\int_0^l \dot{w}w''''dx = \int_0^l \dot{w}''w''dx, \quad \int_0^l \dot{w}w''dx = -\int_0^l \dot{w}'w'dx, \quad \int_0^l \dot{w}w''''dx = \int_0^l \dot{w}''^2 dx.$$

416

Using these equalities, we have:

$$\dot\Phi(t) = \int_0^l \{-2\beta_1\dot w^2 - 2\beta_2\dot w''^2 - \frac{2\rho}{\pi}\dot w\int_0^l(\dot w(x_1,t)+V\dot w'(x_1,t))K(x_1,x)dx_1 -$$

$$- \frac{2V\rho}{\pi}\dot w\int_0^l(\dot w(x_1,t)+Vw'(x_1,t))\frac{\partial K(x_1,x)}{\partial x}dx_1 + 2\beta_0\dot w\int_{t-\tau}^t \dot w(x,s)ds + \quad (10)$$

$$+ \beta_0\tau\dot w^2(x,t) - \beta_0\int_{t-\tau}^t \dot w^2(x,s)ds\}dx + \dot I + \dot J.$$

Considering that $K(x_1,x) = K(x,x_1) \geq 0$, changing order of integration, and using condition (7), we integrate by parts

$$\int_0^l dx\int_0^l \dot w(x,t)\dot w(x_1,t)\frac{\partial K(x_1,x)}{\partial x}dx_1 = \int_0^l dx_1\int_0^l \dot w(x,t)\dot w(x_1,t)\frac{\partial K(x_1,x)}{\partial x}dx =$$

$$= \int_0^l \dot w(x,t)\dot w(x_1,t)K(x_1,x).|_{x=0}^{x=l}\,dx_1 - \int_0^l dx_1\int_0^l \dot w'(x,t)\dot w(x_1,t)K(x_1,x)dx =$$

$$= -\int_0^l dx\int_0^l \dot w'(x_1,t)\dot w(x,t)K(x_1,x)dx_1,$$

where in the last equality the places of the intergration variables x and x_1 have been changed. Similarly we get

$$\int_0^l dx\int_0^l \dot w(x,t)w'(x_1,t)\frac{\partial K(x_1,x)}{\partial x}dx_1 = -\int_0^l dx\int_0^l \dot w'(x,t)w'(x_1,t)K(x_1,x)dx_1.$$

Using inequality $2ab \leq a^2+b^2$, we get $2\dot w(x,t)\dot w(x,s) \leq \dot w^2(x,t) + \dot w^2(x,s)$. Substituting these estimations in (10), we definitively establish

$$\dot\Phi(t) \leq -2\int_0^l\left(\beta_1\dot w^2 + \beta_2\dot w''^2 - \beta_0\tau\dot w^2\right)dx - \frac{2\rho}{\pi}\int_0^l\{\dot w(x,t)\int_0^l \dot w(x_1,t)K(x_1,x)dx_1\}dx +$$

$$+ \frac{2\rho V^2}{\pi}\int_0^l\{\dot w'(x,t)\int_0^l w'(x_1,t)K(x_1,x)dx_1\}dx + \dot I + \dot J. \quad (11)$$

Let us convert integral $\dot I(t)$, considering that $K(x_1,x) = K(x,x_1)$:

$$\dot I(t) = \frac{d}{dt}\frac{\rho}{\pi}\int_0^l dx\int_0^l \dot w(x,t)\dot w(x_1,t)K(x_1,x)dx_1 = \frac{\rho}{\pi}\int_0^l dx\int_0^l \dot w(x,t)\ddot w(x_1,t)K(x_1,x)dx_1 +$$

$$+ \frac{\rho}{\pi}\int_0^l dx\int_0^l \dot w(x_1,t)\ddot w(x,t)K(x_1,x)dx_1 = \frac{2\rho}{\pi}\int_0^l dx\int_0^l \dot w(x,t)\ddot w(x_1,t)K(x_1,x)dx_1.$$

Similarly

$$\dot J(t) = -\frac{2\rho V^2}{\pi}\int_0^l dx\int_0^l \dot w'(x,t)w'(x_1,t)K(x_1,x)dx_1.$$

Substituting, we get

417

$$\dot{\Phi}(t) \le -2\int_0^l \left\{ (\beta_1 - \beta_0\tau)\dot{w}^2 + \beta_2\dot{w}''^2 \right\}dx. \tag{12}$$

Let us consider the boundary-value problem for equation $\psi^{IV}(x) = \mu\psi(x)$, $x \in [0,l]$ with boundary conditions (7) [8, 9]. This problem is self-adjoint and completely defined. In fact, integrating by parts, it is not difficult to show, that

$$\int_0^l u(x)v^{IV}(x)dx = \int_0^l v(x)u^{IV}(x)dx, \quad \int_0^l u(x)u^{IV}(x)dx > 0, \quad \int_0^l u(x)u(x)dx > 0,$$

for any functions $u(x)$ and $v(x)$, satisfying considered boundary conditions and having on $[0,l]$ continuous derivatives up to the fourth order. We apply the Raley inequality for the function $\dot{w}(x,t)$:

$$\int_0^l \dot{w}(x,t)\dot{w}^{IV}(x,t)dx \ge \mu_1 \int_0^l \dot{w}(x,t)\dot{w}(x,t)dx,$$

where μ_1 is the least eigenvalue of the considered boundary problem. Integrating by parts, we present this inequality in the following way

$$\int_0^l \dot{w}''^2(x,t)dx \ge \mu_1 \int_0^l \dot{w}^2(x,t)dx. \tag{13}$$

Thereby, considering (13), the inequality (12) takes the form

$$\dot{\Phi}(t) \le -\frac{2}{\mu_1}\int_0^l (\beta_1 + \mu_1\beta_2 - \beta_0\tau)\dot{w}''^2 dx. \tag{14}$$

Let condition

$$\beta_0\tau - \beta_1 - \mu_1\beta_2 \le 0, \tag{15}$$

be satisfied. Then $\dot{\Phi}(t) \le 0$. Integrating from 0 till t, we get:

$$\Phi(t) \le \Phi(0). \tag{16}$$

According to (8)

$$\Phi(0) = \int_0^l \left\{ M\dot{w}^2(x,0) + Dw''^2(x,0) - Nw'^2(x,0) + \beta_0 w^2(x,0) \right\}dx +$$

$$+ \frac{\rho}{\pi}\int_0^l dx \int_0^l \dot{w}(x,0)\dot{w}(x_1,0)K(x_1,x)dx_1 - \frac{\rho V^2}{\pi}\int_0^l dx \int_0^l w'(x,0)w'(x_1,0)K(x_1,x)dx_1. \tag{17}$$

Based on the earlier proved theorem [5] it is possible to show

$$\int_0^l dx \int_0^l \dot{w}(x,t)\dot{w}(x_1,t)K(x,x_1)dx_1 \ge 0, \quad \int_0^l dx \int_0^l w'(x,t)w'(x_1,t)K(x,x_1)dx_1 \ge 0. \tag{18}$$

We estimate the iterated integral, using obvious inequality $2ab \leq a^2 + b^2$ and symmetry of kernel $K(x_1, x)$, as follows:

$$\int_0^l dx \int_0^l \dot{w}(x,0)\dot{w}(x_1,0)K(x,x_1)dx_1 \leq \int_0^l dx \int_0^l \dot{w}^2(x,0)K(x,x_1)dx_1. \tag{19}$$

Besides (18) and (19) we get the following

$$\Phi(t) \leq \int_0^l \left\{ M\dot{w}_0^2 + Dw_0''^2 - Nw_0'^2 + \beta_0 w_0^2 \right\} dx + \frac{\rho}{\pi} \int_0^l dx \int_0^l \dot{w}_0^2 K(x_1, x)dx_1,$$

where $\dot{w}_0 = \dot{w}(x,0)$, $w_0'' = w''(x,0)$, $w_0 = w(x,0)$, $w_0' = w'(x,0)$.

Let

$$K = \sup_{x \in (0,l)} K^1(x), \quad K^1(x) = \int_0^l K(x_1, x)dx_1, \tag{20}$$

then

$$\Phi(t) \leq \int_0^l \left\{ \left(M + \frac{\rho K}{\pi} \right) \dot{w}_0^2 + Dw_0''^2 - Nw_0'^2 + \beta_0 w_0^2 \right\} dx. \tag{21}$$

On the other hand,

$$\Phi(t) \geq \int_0^l \left[M\dot{w}^2(x,t) + Dw''^2(x,t) - Nw'^2(x,t) + \beta_0 w^2(x,t) \right] dx +$$

$$+ \frac{\rho}{\pi} \int_0^l dx \int_0^l \dot{w}(x,t)\dot{w}(x_1,t)K(x_1,x)dx_1 - \frac{\rho V^2}{\pi} \int_0^l dx \int_0^l w'(x,t)w'(x_1,t)K(x_1,x)dx_1. \tag{22}$$

We estimate the iterated integral, using obvious inequality $-2ab \geq -(a^2 + b^2)$ and symmetry of kernel $K(x_1, x)$, as follows:

$$-\int_0^l dx \int_0^l w'(x,t)w'(x_1,t)K(x,x_1)dx_1 \geq -\int_0^l dx \int_0^l w'^2(x,t)K(x,x_1)dx_1. \tag{23}$$

Estimates (18), (23) allow to write the inequality

$$\Phi(t) \geq \int_0^l \left[M\dot{w}^2(x,t) + Dw''^2(x,t) + \beta_0 w^2(x,t) \right] dx - \frac{\rho V^2 K}{\pi} \int_0^l w'^2(x,t)dx. \tag{24}$$

Thereby, besides (21) and (24) we get the following

$$\int_0^l \left[M\dot{w}^2(x,t) + Dw''^2(x,t) + \beta_0 w^2(x,t) \right] dx - \int_0^l \left(N + \frac{\rho V^2 K}{\pi} \right) w'^2(x,t)dx \leq$$

$$\leq \int_0^l \left\{ \left(M + \frac{\rho K}{\pi} \right) \dot{w}_0^2 + Dw_0''^2 - N_0 w_0'^2 + \beta_0 w_0^2 \right\} dx. \tag{25}$$

Consider the boundary-value problem for equation $\psi^{IV}(x) = -\lambda \psi''(x)$, $x \in [0, l]$ with boundary conditions (7) [6, 7]. This problem is self-adjoint and completely defined. In fact, integrating by parts, it is not difficult to establish

$$\int_0^l u(x) v^{IV}(x)dx = \int_0^l v(x) u^{IV}(x)dx, \quad \int_0^l u(x) v''(x)dx = \int_0^l v(x) u''(x)dx,$$

$$\int_0^l u(x) u^{IV}(x)dx > 0, \quad -\int_0^l u(x) u''(x)dx > 0,$$

for any functions $u(x)$ and $v(x)$, satisfying considered boundary conditions and having on $[0, l]$ continuous derivatives up to the fourth order. We apply the Raley inequality for the function $w(x,t)$:

$$\int_0^l w(x,t) w^{IV}(x,t)dx \geq -\lambda_1 \int_0^l w(x,t) w''(x,t)dx,$$

where λ_1 is the least eigenvalue of the considered boundary problem. Integrating by parts, we present this inequality in the form:

$$\int_0^l w''^2(x,t)dx \geq \lambda_1 \int_0^l w'^2(x,t)dx. \tag{26}$$

Hereinafter, using Bunyakowski inequality, we have

$$w^2(x,t) \leq l \int_0^l w'^2(x,t)dx. \tag{27}$$

Let

$$N < \lambda_1 D - \frac{\rho K V^2}{\pi}. \tag{28}$$

Using inequalities (26), (27), we estimate the left-hand side of (25) as follows

$$\int_0^l \left[M\dot{w}^2(x,t) + D w''^2(x,t) + \beta_0 w^2(x,t) \right]dx - \int_0^l \left(N + \frac{\rho V^2 K}{\pi} \right) w'^2(x,t)dx \geq$$

$$\geq \left(\lambda_1 D - N - \frac{\rho V^2 K}{\pi} \right) \frac{w^2(x,t)}{l}.$$

Taking into account this estimate and (26) also we get inequality

$$\left(\lambda_1 D - N - \frac{\rho V^2 K}{\pi} \right) \frac{w^2(x,t)}{l} \leq \int_0^l \left\{ \left(M + \frac{\rho K}{\pi} \right) \dot{w_0}^2 + \left(D - \lambda_1^{-1} N \right) w_0''^2 + \beta_0 w_0^2 \right\}dx,$$

from which follows

Theorem 1 *Let us assume, that function $w(x,t)$ satisfies the boundary conditions (7) and let inequalities (15) and (28) be obeyed. Then solution $w(x,t)$ of equation (6)* is stable with respect to perturbations of the initial values of quantities $\dot{w}(x,0), w(x,0), w''(x,0)$.

STABILITY OF ELASTIC ELEMENTS OF WING

Let us consider the planar problem of aerohydroelasticity about small fluctuations, appearing with non-circulatory flow around thin-shelled construction – the model of wing, which component parts are n elastic elements-insertions.

Suppose on plane xOy, in which occur joint oscillations of elastic insertions and gas, segment $[c, d]$ on axis Ox corresponds to wing, and segments $[a_{2k-1}, a_{2k}]$, $k = 1 \div n$, $-\infty < c \le a_{2k-1} < a_{2k} \le a_{2k+1} < a_{2k+2} \le d < +\infty$, $k = 1 \div n - 1$ to elastic insertions (Figure 2).

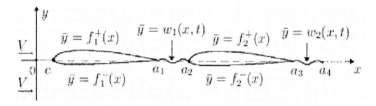

FIGURE 2. Wing profile

In infinitely remote point a velocity of gas is V and has the direction, coinciding with the direction of axis Ox. Let us indicate: $w_k(x,t)$ and $u_k(x,t)$ $(k = 1 \div n)$ are the functions of plate oscillations toward axis Oy and Ox correspondingly; $\varphi(x, y, t)$ is the potential of the gas velocity.

Potential of velocity $\varphi(x, y, t)$ satisfies the Laplace equation

$$\Delta\varphi \equiv \varphi_{xx} + \varphi_{yy} = 0, \quad (x, y) \in G = R^2 \setminus [c, d],\qquad(29)$$

with boundary conditions

$$\varphi_y^\pm(x,0,t) = \lim_{y \to \pm 0} \varphi_y(x,y,t) = V f_1^{\pm\,\prime}(x), \ x \in (c, a_1),\qquad(30)$$

$$\varphi_y^\pm(x,0,t) = w_{kt}(x,t) + V w_{kx}(x,t), \qquad x \in (a_{2k-1}, a_{2k}), k = 1 \div n,\qquad(31)$$

$$\varphi_y^\pm(x,0,t) = V f_{k+1}^{\pm\,\prime}(x), \ \ x \in (a_{2k}, a_{2k+1}), k = 1 \div (n-1),\qquad(32)$$

$$\varphi_y^\pm(x,0,t) = V f_{n+1}^{\pm\,\prime}(x), \ \ x \in (a_{2n}, d),\qquad(33)$$

where $f_k^\pm(x)$ $(k = 1 \div (n+1))$ are given functions determining the form of unstrained parts of a wing, and condition of undisturbed flow in infinitely removed point

$$|\nabla\varphi|_\infty^2 \equiv (\varphi_x^2 + \varphi_y^2 + \varphi_t^2)_\infty = 0.\qquad(34)$$

Let us present equations of small oscillations of elastic plates as

$$\begin{cases} -E_kF_k\left(u'_k+\dfrac{1}{2}w'^2_k\right)' +M_k\ddot{u}_k+g_k(t,x,u_k,w_k,\dot{u}_k,\dot{w}_k)=0, \\[2mm] -E_kF_k\left[w'_k\left(u'_k+\dfrac{1}{2}w'^2_k\right)\right]'+E_kJ_kw''''_k+M_k\ddot{w}_k+h_k(t,x,u_k,w_k,\dot{u}_k,\dot{w}_k)= \\[2mm] =\rho(\varphi^+_t(x,0,t)-\varphi^-_t(x,0,t))+\rho V(\varphi^+_x(x,0,t)-\varphi^-_x(x,0,t)), \\[2mm] x\in(a_{2k-1},a_{2k}),\ k=1\div n. \end{cases} \quad (35)$$

Here ρ is the density of the gas; E_k are the modules of elasticity of plates; F_k are the areas of cross-sections of plates; E_kJ_k are the rigidities of the plates; M_k are the specific masses of the plates; the functions $g_k(t,x,u_k,w_k,\dot{u}_k,\dot{w}_k)$, $h_k(t,x,u_k,w_k,\dot{u}_k,\dot{w}_k)$ represent nonlinear components of basic reactions or other nonlinear influences.

Using methods of the theory of analytic functions [6], the solution of the problem (29)–(34) can be reduced to a system of equations for unknown functions of plates oscillations, where right-hand sides of the second equation in system (35) will be the following:

$$\rho(\varphi^+_t(x,0,t)-\varphi^-_t(x,0,t))+\rho V(\varphi^+_x(x,0,t)-\varphi^-_x(x,0,t))=-\frac{\rho}{\pi}\sum_{k=1}^{n}\int_{a_{2k-1}}^{a_{2k}}(\ddot{w}_k(\tau,t)+V\dot{w}'_k(\tau,t))\times$$

$$\times K(\tau,x)d\tau-\frac{V\rho}{\pi}\sum_{k=1}^{n}\int_{a_{2k-1}}^{a_{2k}}(\dot{w}_k(\tau,t)+Vw'_k(\tau,t))\frac{\partial K(\tau,x)}{\partial x}d\tau+\frac{V^2\rho}{2\pi}\int_{c}^{a_1}\left(f^{+'}_1(\tau)+f^{-'}_1(\tau)\right)G(\tau,x)d\tau+$$

$$+\frac{V^2\rho}{2\pi}\sum_{k=1}^{n-1}\int_{a_{2k}}^{a_{2k+1}}\left(f^{+'}_{k+1}(\tau)+f^{-'}_{k+1}(\tau)\right)G(\tau,x)d\tau+\frac{V^2\rho}{2\pi}\int_{a_{2n}}^{d}\left(f^{+'}_{n+1}(\tau)+f^{-'}_{n+1}(\tau)\right)G(\tau,x)d\tau,$$

where $x\in(a_{2i-1},a_{2i})$, $\tau\neq x$,

$$K(\tau,x)=2ln\left|\frac{\sqrt{(x-c)(d-\tau)}+\sqrt{(\tau-c)(d-x)}}{\sqrt{(x-c)(d-\tau)}-\sqrt{(\tau-c)(d-x)}}\right|,\quad G(\tau,x)=\frac{\sqrt{(d-x)(x-c)}+\sqrt{(d-\tau)(\tau-c)}}{\sqrt{(d-x)(x-c)(x-\tau)}}.$$

In case of symmetric profiles the equation takes the following form

$$\rho(\varphi^+_t(x,0,t)-\varphi^-_t(x,0,t))+\rho V(\varphi^+_x(x,0,t)-\varphi^-_x(x,0,t))=$$

$$=-\frac{\rho}{\pi}\sum_{k=1}^{n}\int_{a_{2k-1}}^{a_{2k}}(\ddot{w}_k+V\dot{w}'_k)K(\tau,x)d\tau-\frac{V\rho}{\pi}\sum_{k=1}^{n}\int_{a_{2k-1}}^{a_{2k}}(\dot{w}_k+Vw'_k)\frac{\partial K(\tau,x)}{\partial x}d\tau. \quad (36)$$

The boundary conditions on the ends of plates under $x=a_{2k-1}$ or $x=a_{2k}$ can be:

I. The rigid sealing

$$w_k(x,t)=w'_k(x,t)=u_k(x,t)=0; \quad (37)$$

II. The hinged sealing

$$w_k(x,t)=w''_k(x,t)=u_k(x,t)=0; \quad (38)$$

III. The rigid fastening

$$w_k(x,t)=w'_k(x,t)=u'_k(x,t)=0; \quad (39)$$

IV. The hinged fastening

$$w_k(x,t) = w_k''(x,t) = u_k'(x,t) + \frac{1}{2}{w_k'}^2(x,t) = 0. \quad (40)$$

Let us introduce the functional

$$\Phi(t) = \sum_{k=1}^{n} \int_{a_{2k-1}}^{a_{2k}} \left\{ M_k(\dot{u}_k^2 + \dot{w}_k^2) + E_k J_k {w_k''}^2 + E_k F_k \left(u_k' + \frac{1}{2}{w_k'}^2 \right)^2 \right\} dx +$$

$$\quad (41)$$

$$+ \frac{\rho}{\pi} \sum_{i=1}^{n} \sum_{j=1}^{n} \int_{a_{2i-1}}^{a_{2i}} dx \int_{a_{2j-1}}^{a_{2j}} (\dot{w}_i(x,t)\dot{w}_j(\tau,t) - V^2 w_i'(x,t)w_j'(\tau,t)) K(\tau,x) d\tau.$$

By using the inequalities

$$\sum_{i=1}^{n} \sum_{j=1}^{n} \int_{a_{2i-1}}^{a_{2i}} dx \int_{a_{2j-1}}^{a_{2j}} \dot{w}_i(x,t)\dot{w}_j(\tau,t)K(x,\tau)d\tau \geq 0, \quad \sum_{i=1}^{n} \sum_{j=1}^{n} \int_{a_{2i-1}}^{a_{2i}} dx \int_{a_{2j-1}}^{a_{2j}} w_i'(x,t)w_j'(\tau,t)K(\tau,x)d\tau \geq 0,$$

$$\int_{a_{2i-1}}^{a_{2i}} {w_i''}^2(x,t)dx \geq \lambda_{1i} \int_{a_{2i-1}}^{a_{2i}} {w_i'}^2(x,t)dx, \qquad w_i^2(x,t) \leq (a_{2i-1} - a_{2i}) \int_{a_{2i-1}}^{a_{2i}} {w_i'}^2(x,t)dx, \; i = 1 \div n,$$

where λ_{1i} are the least eigenvalues of the boundary problem $\psi^{IV}(x) = -\lambda\psi''(x)$, $x \in [a_{2i-1}, a_{2i}]$ with boundary conditions, corresponding to mentioned types of fastening, the following theorem is proved based on the study of the functional (41)

Theorem 2 *Let us assume functions* $w_k(x,t)$, $u_k(x,t)$ $(k = 1 \div n)$ *satisfy one of the boundary conditions (37)–(40) and let inequalities be fulfilled*

$$\sum_{k=1}^{n} \int_{a_{2k-1}}^{a_{2k}} (\dot{u}_k \cdot g_k(t,x,u_k,w_k,\dot{u}_k,\dot{w}_k) + \dot{w}_k \cdot h_k(t,x,u_k,w_k,\dot{u}_k,\dot{w}_k)) dx \geq 0,$$

$$E_k J_k \lambda_{1k} > \frac{\rho V^2 K_k}{\pi}, \qquad K_k = \sup_{x \in (a_{2k-1}, a_{2k})} \sum_{i=1}^{n} \int_{a_{2i-1}}^{a_{2i}} K(\tau,x)d\tau, k = 1 \div n.$$

Then solutions $w_k(x,t)$ $(k = 1 \div n)$ *of equations systems (35), (36) are stable with respect to perturbation of the initial values of* $\dot{w}_k(x,0), w_k''(x,0), \dot{u}_k(x,0), u_k'(x,0)$.

STABILITY OF ELASTIC ELEMENTS OF A CHANNEL

We consider the planar problem about dynamic stability of elastic elements of the walls of an infinitely long channel along which an ideal incompressible fluid flows (Figure 3).

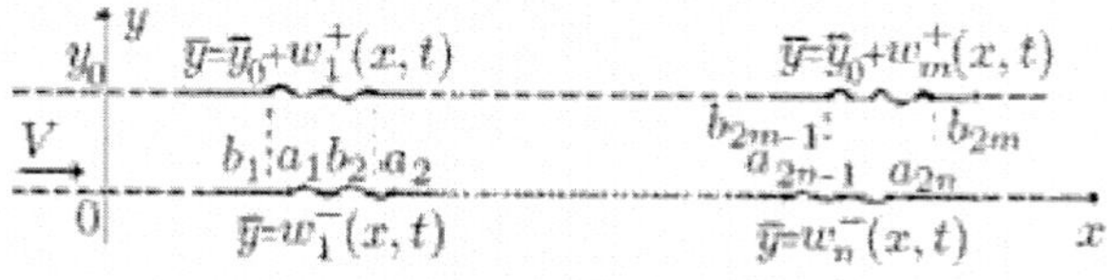

FIGURE 3. The channel

423

This problem is formulated in the following way

$$\Delta\varphi = 0; (x,y) \in R^2 : |x| < \infty, y \in [0, y_0],$$

$$\varphi_y(x, y_0, t) = w_{kt}^+(x,t) + Vw_{kx}^+(x,t), \quad x \in (b_{2k-1}, b_{2k}), k = 1 \div m,$$

$$\varphi_y(x, y_0, t) = 0, x \in R \setminus \left(\bigcup_{k=1}^{m} [b_{2k-1}, b_{2k}] \right),$$

$$\varphi_y(x, 0, t) = w_{kt}^-(x,t) + Vw_{kx}^-(x,t), \quad x \in (a_{2k-1}, a_{2k}), k = 1 \div n,$$

$$\varphi_y(x, 0, t) = 0, x \in R \setminus \left(\bigcup_{k=1}^{n} [a_{2k-1}, a_{2k}] \right),$$

$$\left(\varphi_x^2 + \varphi_y^2 \right)_{x=\pm\infty} = 0, (\varphi_t)_{x=-\infty} = 0, y \in (0, y_0),$$

$$\begin{cases} -E_k^+ F_k^+ \left(u_k^{+'} + \frac{1}{2} w_k^{+'2} \right)' + M_k^+ \ddot{u}_k^+ + g_k^+ (t, x, u_k^+, w_k^+, \dot{u}_k^+, \dot{w}_k^+) = 0, \\ \\ -E_k^+ F_k^+ \left[w_k^{+'} \left(u_k^{+'} + \frac{1}{2} w_k^{+'2} \right) \right]' + E_k^+ J_k^+ w_k^{+''''} + M_k^+ \ddot{w}_k^+ + \\ \\ + h_k^+ (t, x, u_k^+, w_k^+, \dot{u}_k^+, \dot{w}_k^+) = -\rho(\varphi_t(x, y_0, t) + V\varphi_x(x, y_0, t)), \\ x \in (b_{2k-1}, b_{2k}), k = 1 \div m. \end{cases} \tag{42}$$

$$\begin{cases} -E_k^- F_k^- \left(u_k^{-'} + \frac{1}{2} w_k^{-'2} \right)' + M_k^- \ddot{u}_k^- + g_k^- (t, x, u_k^-, w_k^-, \dot{u}_k^-, \dot{w}_k^-) = 0, \\ \\ -E_k^- F_k^- \left[w_k^{-'} \left(u_k^{-'} + \frac{1}{2} w_k^{-'2} \right) \right]' + E_k^- J_k^- w_k^{-''''} + M_k^- \ddot{w}_k^- + \\ \\ + h_k^- (t, x, u_k^-, w_k^-, \dot{u}_k^-, \dot{w}_k^-) = \rho(\varphi_t(x, 0, t) + V\varphi_x(x, 0, t)), \\ x \in (a_{2k-1}, a_{2k}), k = 1 \div n. \end{cases} \tag{43}$$

Here $w_k^\pm(x,t), u_k^\pm(x,t)$ are the oscillation functions of the plates on the bottom and top walls of the channel, respectively.

Using again the methods of analytic functions, the solution of the problem can be reduced to a system of equations for determination of $w_k^\pm(x,t), u_k^\pm(x,t)$, where right-hand sides of the second equations of system (42), (43) are

$$- \rho(\varphi_t(x,y_0,t) + V\varphi_x(x,y_0,t)) = -\frac{\rho}{\pi} \sum_{p=1}^{m} \int_{b_{2p-1}}^{b_{2p}} \overline{w}_{pt}^+(\tau,t) K^+(\tau,x) d\tau -$$

$$-\frac{V\rho}{\pi} \sum_{p=1}^{m} \int_{b_{2p-1}}^{b_{2p}} \overline{w}_p^+(\tau,t) K_x^+(\tau,x) d\tau -\frac{\rho}{\pi} \sum_{p=1}^{n} \int_{a_{2p-1}}^{a_{2p}} \overline{w}_{pt}^-(\tau,t) K^-(\tau,x) d\tau - \qquad (44)$$

$$-\frac{V\rho}{\pi} \sum_{p=1}^{n} \int_{a_{2p-1}}^{a_{2p}} \overline{w}_p^-(\tau,t) K_x^-(\tau,x) d\tau, \quad x \in (b_{2k-1}, b_{2k}), k = 1 \div m,$$

$$\rho(\varphi_t(x,0,t) + V\varphi_x(x,0,t)) = -\frac{\rho}{\pi} \sum_{p=1}^{m} \int_{b_{2p-1}}^{b_{2p}} \overline{w}_{p\,t}^+(\tau,t) K^-(\tau,x) d\tau -$$

$$-\frac{V\rho}{\pi} \sum_{p=1}^{m} \int_{b_{2p-1}}^{b_{2p}} \overline{w}_p^+(\tau,t) K_x^-(\tau,x) d\tau -\frac{\rho}{\pi} \sum_{p=1}^{n} \int_{a_{2p-1}}^{a_{2p}} \overline{w}_{p\,t}^-(\tau,t) K^+(\tau,x) d\tau - \qquad (45)$$

$$-\frac{V\rho}{\pi} \sum_{p=1}^{n} \int_{a_{2p-1}}^{a_{2p}} \overline{w}_p^-(\tau,t) K_x^+(\tau,x) d\tau, \quad x \in (a_{2k-1}, a_{2k}), k = 1 \div n,$$

where

$$\overline{w}_k^\pm(\tau,t) = w_{k\,t}^\pm(\tau,t) + V w_{k\,\tau}^\pm(\tau,t), \qquad K^\pm(\tau,x) = \ln \left| e^{-\frac{\pi a_1}{y_0}} + e^{-\frac{\pi b_1}{y_0}} \right| - \ln \left| e^{-\frac{\pi \tau}{y_0}} \mp e^{-\frac{\pi x}{y_0}} \right|.$$

Using a Lyapunov functional, the following theorem is proved

Theorem 3 *Let us assume functions* $w_k^\pm(x,t), u_k^\pm(x,t)$ *satisfy one of the boundary conditions (37)–(40) and let inequalities be executed*

$$\sum_{k=1}^{m} \int_{b_{2k-1}}^{b_{2k}} \left(\dot{u}_k^+ \cdot g_k^+(t,x,u_k^+,w_k^+,\dot{u}_k^+,\dot{w}_k^+) + \dot{w}_k^+ \cdot h_k^+(t,x,u_k^+,w_k^+,\dot{u}_k^+,\dot{w}_k^+) \right) dx +$$

$$+ \sum_{k=1}^{n} \int_{a_{2k-1}}^{a_{2k}} \left(\dot{u}_k^- \cdot g_k^-(t,x,u_k^-,w_k^-,\dot{u}_k^-,\dot{w}_k^-) + \dot{w}_k^- \cdot h_k^-(t,x,u_k^-,w_k^-,\dot{u}_k^-,\dot{w}_k^-) \right) dx \geq 0,$$

$$E_k^\pm J_k^\pm \lambda_{1k}^\pm > \frac{\rho V^2 K_{0k}^\pm}{\pi},$$

where

$$K_{0k}^+ = \sup_{x \in (b_{2k-1}, b_{2k})} K_1^+(x), \ K_{0k}^- = \sup_{x \in (a_{2k-1}, a_{2k})} K_1^-(x), \ K_1^\pm(x) = \sum_{k=1}^{m} \int_{b_{2k-1}}^{b_{2k}} K^\pm(\tau,x) d\tau + \sum_{k=1}^{n} \int_{a_{2k-1}}^{a_{2k}} K^\mp(\tau,x) d\tau.$$

Then solutions $w_k^+(x,t)(k = 1 \div m)$, $w_k^-(x,t)$ $(k = 1 \div n)$ *of the equations systems (42)–(45) are stable with respect to perturbations of the initial values of* $\ddot{w}_k^\pm(x,0), w_k^\pm{}''(x,0), \ddot{u}_k^\pm(x,0), u_k^\pm{}'(x,0)$.

425

ACKNOWLEDGMENTS

The work is supported financially by Grant RFBR-RA 07-01-91680.

REFERENCES

1. P. A. Velmisov, *Asymptotic Equations of Gas Dinamics*, Saratov University, Russia, 1986. (in Russian)
2. P. A. Velmisov and Yu. A. Reshetnikov, *Stability of Viscoelastic Plates at Aerohydrodynamic Affect*, Saratov University, Russia, 1994. (in Russian)
3. A. V. Ankilov and P. A. Velmisov, "Stability of Viscoelastic Elements of Thin-shelled Constructions Under Aerohydrodynamic Action", VINITI, Moscow, No 2522, 1998.
4. A. V. Ankilov and P. A. Velmisov, *Stability of Viscoelastic Elements of Channels Walls*, Ulyanovsk Technical University, Russia, 2000.
5. A. V. Ankilov, P. A. Velmisov, and N. A. Degtyareva, "Stability of elastic elements of wing profile" in *Applied Mathematics and mechanics*, edited by P. A. Velmisov et al, Ulyanovsk Technical University, Russia, Vol.7, 2007, pp. 9-18.
6. M. A. Lavrentev and B. V. Shabat, *Methods in Theory of Analytic Functions*, Nauka, Moscow, 1987. (in Russian)
7. F. D. Gahov, *Marginal Problems*, Nauka, Moscow, 1977. (in Russian)
8. L. Kollatz, *Problems on Eigenvalues*, Nauka, Moscow, 1968. (Russian translation)

Some Classes of the Solutions of Aerohydromechanic Equations

P.A. Velmisov[*], M.D. Todorov[†] and Yu.A. Kazakova[*]

[*]Ulyanovsk State Technical University, 32, Severny Venets str., Ulyanovsk 432027, Russia
[†]TechnicalUniversity of Sofia, 1000 Sofia, Bulgaria

Abstract. We suggest the method of constructing the solutions for systems of partial differential equations and consider its application to some aerohydromechanic equations. This method is parameterization-based and effective for finding exact particular solution of first-order systems of partial quasi-linear differential equations. The existence of the parametric solutions of polynomial type is proved, and the classifications for some aerohydromechanic equations are carried. Particularly, the solutions of asymptotic transonic equations describing flows with local supersonic zones are constructed.

Keywords: Aerohydromechanics, partial differential equations.
PACS: 02.30.Jr, 47.85.Gj

METHOD DESCRIPTION

Basic Ideas of Method

Let us consider a system of partial differential equations

$$F_k\left(x_1,...,x_m,u_1,...,u_n,u_{1x_1},...,u_{nx_1},...,u_{1x_m},...,u_{nx_m}\right)=0, \quad k=1\div n \tag{1}$$

where $u_k(x_1,...,x_m)$ are functions of m variables $x_1,x_2,...,x_m$. After transition to new variables $\xi_1,\xi_2,...,\xi_m$, which are functions of variables $x_1,x_2,...,x_m$, the solution of system (1) is found in the following kind

$$u_k=U_k(\xi_1,\xi_2,...,\xi_m), \quad k=1\div n; \quad x_l=X_l(\xi_1,\xi_2,...,\xi_m), \quad l=1\div m \tag{2}$$

The partial derivatives $\dfrac{\partial u_k}{\partial x_l}$ are found by the formulae

$$\frac{\partial u_k}{\partial x_l}=\sum_{j=1}^m\frac{\partial U_k}{\partial \xi_j}\frac{\partial \xi_j}{\partial x_l}, \quad \frac{\partial \xi_j}{\partial x_l}=\frac{\Delta_{jl}}{\Delta}, \quad \Delta=\begin{vmatrix}\dfrac{\partial X_1}{\partial \xi_1} & ...... & \dfrac{\partial X_1}{\partial \xi_m}\\ ... & ...... & ...\\ \dfrac{\partial X_m}{\partial \xi_1} & ...... & \dfrac{\partial X_m}{\partial \xi_m}\end{vmatrix}\neq 0 \tag{3}$$

CP1067, *Applications of Mathematics in Engineering and Economics '34—AMEE '08*, edited by M. D. Todorov
© 2008 American Institute of Physics 978-0-7354-0598-1/08/$23.00

In (3) Δ_{jl} is a determinant derived from the determinant Δ by replacement of the column with number j by a column with zero elements except the element with number l which is equal one. Then equation system (1) is rearranged in the form

$$F_k(\xi_1,...,\xi_m, X_1,...,X_m, X_{1\xi_1},...,X_{m\xi_m}, U_1,...,U_n, U_{1\xi_1},...,U_{n\xi_m}) = 0 \qquad (4)$$

In this system $U_k(k=1\div n), X_l(l=1\div m)$ are functions of variables $\xi_1, \xi_2,...,\xi_m$. The solution of given system of equations can be found in polynomial form

$$U_k = \sum_{i=0}^{\alpha_k} u_{ki}(\xi_1,...,\xi_{s-1},\xi_{s+1},...,\xi_m)\xi_s^i, \quad k=1\div n, \; s=1\div m$$

$$X_l = \sum_{j=0}^{\gamma_l} x_{lj}(\xi_1,...,\xi_{s-1},\xi_{s+1},...,\xi_m)\xi_s^j, \quad l=1\div m, \; s=1\div m \qquad (5)$$

where $\alpha_k, \gamma_l \in N$ (N is the set of natural numbers). Particularly, the problem is the determination of parameters $\alpha_k, \gamma_l \in N$, for which the system of differential equations for $u_{ki}(\xi_1,\xi_2,...,\xi_{m-1})$, $x_{lj}(\xi_1,\xi_2,...,\xi_{m-1})$ is determined or underdetermined.

Let us consider the particular case, when required functions depend on the coordinates x, y and the time t:

$$F_k(x,y,t,u_1,...,u_n,u_{1x},...u_{nx},u_{1y},...u_{ny},u_{1t},...u_{nt}) = 0, \; k=1\div n \qquad (6)$$

In this case the solution of system is found in the form

$$u_k = u_k(\xi,\eta,t), \; k=1\div n, \; x = x(\xi,\eta,t), \; y = y(\xi,\eta,t) \qquad (7)$$

Then formulae of conversion to new variables are

$$u_{kx} = \frac{u_{k\xi}y_\eta - u_{k\eta}y_\xi}{\Delta}, \; u_{ky} = \frac{u_{k\eta}x_\xi - u_{k\xi}x_\eta}{\Delta},$$

$$u_{kt} = u_{kt} + \frac{u_{k\xi}y_t x_\eta - u_{k\xi}y_\eta x_t}{\Delta} + \frac{u_{k\eta}y_\xi x_t - u_{k\eta}x_\xi y_t}{\Delta} \qquad (8)$$

where $\Delta = x_\xi y_\eta - x_\eta y_\xi$ $(\Delta \neq 0)$. The system of equations (6) is rearranged in the form

$$F_k(\xi,\eta,x,y,t,x_\xi,x_\eta,x_t,y_\xi,y_\eta,y_t,u_1,...,u_n,u_{1\xi},...,u_{n\xi},u_{1\eta},...,u_{n\eta},u_{1t},...,u_{nt}) = 0 \quad (9)$$

In this system $x, y, u_k(k=1\div n)$ are functions of variables ξ, η, t. The solution of system (9) can be found in polynomial form:

$$u_k = \sum_{i=0}^{\alpha_k} u_{ki}(\xi,t)\eta^i, \; x = \sum_{k=0}^{\gamma} x_k(\xi,t)\eta^k, \; y = \sum_{k=0}^{\omega} y_k(\xi,t)\eta^k \qquad (10)$$

where $\alpha_k, \gamma, \omega \in N$. For some types of equations (for example, for quasi-linear equations of first order, in which coefficients are the polynomials relative to dependent and independent variables) the relations for parameters $\alpha_k, \gamma, \omega \in N$ are obtained. These relations allow us to get all values of parameters $\alpha_k, \gamma, \omega \in N$, for which the differential equation system for $x_k(\xi,t), y_k(\xi,t), u_{ki}(\xi,t)$ is determined or underdetermined, that is $s \geq r$, $s = r + j$, where s is the number of unknown functions dependent on ξ, t; r is the number of equations; j is a degree of sub-definite. If r_k is the maximum degree of variable η (when we substitute (10) in the equation $F_k = 0 (k = 1 \div n)$), parameters s and r are

$$r = \sum_{k=1}^{n} r_k + n, \quad s = \gamma + \omega + \sum_{k=1}^{n} \alpha_k + n + 2 . \tag{11}$$

Software Implementation

The program has been developed, with the help of which can determine allowed values of parameters $\alpha_k, \gamma, \omega \in N$. Having substituted (10) in (9), we put in set all maximum degrees of variables η for each k th equation of system (9):

$$\left(J_{1,1}, J_{2,1}, ..., J_{i_1,1}, ..., J_{1,k}, J_{2,k}, ..., J_{i_k,k}, ..., J_{1,n}, J_{2,n}, ..., J_{i_n,n} \right),$$

where i_k the number of various degrees η in k-th equation, $k = 1 \div n$. Each of values $J_{1,1}, J_{2,1}, ..., J_{i_1,1}, ..., J_{1,k}, J_{2,k}, ..., J_{i_k,k}, ..., J_{1,n}, J_{2,n}, ..., J_{i_n,n}$ is a linear combination of parameters α_k, γ, ω. Then the following system is formed and solved:

$$\begin{cases} I_1 - J_{1,1} = x_{1,1} \\ .... \\ I_1 - J_{i_1,1} = x_{i_1,1} \\ .... \\ I_n - J_{1,n} = x_{1,n} \\ .... \\ I_n - J_{i_n,n} = x_{i_n,n} \\ \sum_{k=1}^{n} \alpha_k + \gamma + \omega + 2 + n = \sum_{k=1}^{n} I_k + j \end{cases} \tag{12}$$

where $I_k = \max(J_{1,k}, ..., J_{i_k,k})$ is the maximum degree of η in k-th equation; $x_{i_k,k} \in N$ are some natural numbers; j is a degree of sub-definite.

There exist $d = \prod_{k=1}^{n} i_k$ variants of expression choice I_k.

There are formed n matrixes $(S_1, S_2, ..., S_n)$ in this program. Each row of k-th matrix is coefficients α_k, γ, ω in degrees of variable η k th equation (for example, $S_k = \left(J_{1,k}, ..., J_{i_k,k} \right)^T$). The row number k th matrix is i_k. There is realized search cycle of maximum I_k degrees in this program, and there are choose parameters α_k, γ, ω for each set $\{ I_k \}_{k=1 \div n}$. From $\alpha_k, \gamma, \omega \in N$, $x_{i_k,k} \in N$ and the system (12), the parameters α_k, γ, ω are limited. At every step of search matrix (A) is formed, elements of which are coefficients of α_k, γ, ω in equations of system (12):

$$A[m_1, q] = S_1[l_1, q] - S_1[m_1, q], m_1 = 1 \div i_1$$

$$\cdots\cdots\cdots\cdots\cdots\cdots\cdots\cdots\cdots\cdots\cdots\cdots\cdots\cdots$$

$$A[m_n, q] = S_n[l_n, q] - S_n[m_n, q], m_n = 1 \div i_n \qquad , \qquad (13)$$

$$A\left[\sum_{k=1}^{n} i_k, q\right] = 1 - S_1[l_1, q] - S_2[l_2, q] - \ldots - S_n[l_n, q]$$

where $q = 1 \div (n+2)$; set $(l_1, l_2, \ldots, l_n)$ determine numbers of rows in matrixes $S_1, S_2, \ldots, S_n$, corresponding to maximum degrees $(I_1, I_2, \ldots, I_n)$. Values $x_{i_k,k}$ are put in vector b

$$b[i] = \sum_{k=1}^{n+2} A[i, k] \varphi_k, \varphi_k \in \{\alpha_k, \gamma, \omega\}, i = 1 \div \left(1 + \sum_{k=1}^{n} i_k\right)$$

The last element of vector b is

$$b\left[1 + \sum_{k=1}^{n} i_k\right] = b\left[1 + \sum_{k=1}^{n} i_k\right] + n + 2 .$$

Then, for $\gamma + \omega > 0$ and $b[i] \in N \ \forall i$ the given values of α_k, γ, ω are acceptable.

APPLICATION OF METHOD TO AEROHYDROMECHANIC EQUATIONS

Examples of Classification and Formation of Parametric Solutions of Some Aerohydromechanic Equation Systems

The system of aerohydromechanics for ideal gas in case of adiabatic process is:

$$\begin{cases} \rho(u_t + uu_x + vu_y + wu_z) = -p_x \\ \rho(v_t + uv_x + vv_y + wv_z) = -p_y \\ \rho(w_t + uw_x + vw_y + ww_z) = -p_z \\ \rho_t + (\rho u)_x + (\rho v)_y + (\rho w)_z = 0 \\ \left(\dfrac{p}{\rho^{\chi}}\right)_t + u\left(\dfrac{p}{\rho^{\chi}}\right)_x + v\left(\dfrac{p}{\rho^{\chi}}\right)_y + w\left(\dfrac{p}{\rho^{\chi}}\right)_z = 0 \end{cases} \qquad (14)$$

where $u(t,x,y,z)$, $v(t,x,y,z)$, $w(t,x,y,z)$ are projections of velocity vector, $\rho(t,x,y,z)$ is a density, $p(t,x,y,z)$ is a pressure, $\chi = \dfrac{c_p}{c_v} = const$, c_v, c_p are thermal coefficients at constant volume and constant pressure. The first three equations of system (14) are Euler equations of motion, the forth equation is continuity equation, the fifth equation is energy equation.

For plane flow $\dfrac{\partial}{\partial z} = 0, w = 0$. Let us suppose that the flow is isentropic: $\dfrac{p}{\rho^{\chi}} = c = const$, $p = c\rho^{\chi}$, $c = \dfrac{p_0}{\rho_0^{\chi}}$, where p_0, ρ_0 are some constant values of pressure and density. For these flows we get:

$$\begin{cases} u_t + uu_x + vu_y = -c\chi\rho^{\chi-2}\rho_x \\ v_t + uv_x + vv_y = -c\chi\rho^{\chi-2}\rho_y \\ \rho_t + \rho(u_x + v_y) + u\rho_x + v\rho_y = 0 \end{cases} \tag{15}$$

This is equation system of plane isentropic motion of gas for functions u, v, ρ.

Example 1. Let us consider the system:

$$\begin{cases} u_t + uu_x + vu_y = -\zeta w_x, \quad v_t + uv_x + vv_y = -\zeta w_y, \\ w(u_x + v_y) + \mu(w_t + uw_x + vw_y) = 0 \end{cases} \tag{16}$$

The system (16) is received from system (15), when in system (15) consider $w(t,x,y) = \rho^{\chi-1}, \zeta = \dfrac{c\chi}{\chi-1}, \mu = \dfrac{1}{\chi-1}$. Putting new variables ξ, η, t, the system (16) is

$$\begin{cases} u_t(x_\xi y_\eta - x_\eta y_\xi) + u_\xi(x_\eta y_t - y_\eta x_t) + u_\eta(v_\xi x_t - y_t x_\xi) + u(u_\xi y_\eta - u_\eta y_\xi) + \\ v(u_\eta x_\xi - u_\xi x_\eta) + \zeta(w_\xi y_\eta - w_\eta y_\xi) = 0 \\ v_t(x_\xi y_\eta - x_\eta y_\xi) + v_\xi(x_\eta y_t - y_\eta x_t) + v_\eta(v_\xi x_t - y_t x_\xi) + u(v_\xi y_\eta - v_\eta y_\xi) + \\ v(v_\eta x_\xi - v_\xi x_\eta) + \zeta(w_\eta x_\xi - w_\xi x_\eta) = 0 \\ w(u_\xi y_\eta - u_\eta y_\xi + v_\eta x_\xi - v_\xi x_\eta) + \mu(w_t(x_\xi y_\eta - x_\eta y_\xi) + w_\xi(x_\eta y_t - y_\eta x_t) + \\ w_\eta(v_\xi x_t - y_t x_\xi) + u(w_\xi y_\eta - w_\eta y_\xi) + v(w_\eta x_\xi - w_\xi x_\eta)) = 0 \end{cases} \tag{17}$$

The solution of system (17) is found in the form

$$u = \sum_{k=0}^{\alpha} u_k(\xi,t)\eta^k, \quad v = \sum_{k=0}^{\beta} v_k(\xi,t)\eta^k, \quad w = \sum_{k=0}^{\theta} w_k(\xi,t)\eta^k,$$

$$x = \sum_{k=0}^{\gamma} x_k(\xi,t)\eta^k, \quad y = \sum_{k=0}^{\omega} y_k(\xi,t)\eta^k \tag{18}$$

Here $\alpha, \beta, \theta, \gamma, \omega$ are natural numbers. Substituting expressions (18) in system (17), we obtain the following maximum degrees of variable η:

431

$$\begin{cases} J_1 = \alpha + \gamma + \omega - 1, & J_2 = 2\alpha + \omega - 1, & J_3 = \beta + \alpha + \gamma - 1, & J_4 = \theta + \omega - 1 \\ J_5 = \beta + \gamma + \omega - 1, & J_6 = \alpha + \beta + \omega - 1, & J_7 = 2\beta + \gamma - 1, & J_8 = \theta + \gamma - 1 \\ J_9 = \alpha + \omega + \theta - 1, & J_{10} = \beta + \gamma + \theta - 1, & J_{11} = \theta + \gamma + \omega - 1 \end{cases} \quad (19)$$

Parameters J_1, J_2, J_3, J_4 correspond to the first equation of system (17), J_5, J_6, J_7, J_8 correspond to the second equation of system (17), J_9, J_{10}, J_{11} correspond to the third equation of system (17). The number of coefficients in (18) is equal $r = \alpha + \beta + \theta + \gamma + \omega + 5$, and the number of equations in system (17) is determined by relation: $s = I_1 + I_2 + I_3 + 3$,
where $I_1 = \max(J_1, J_2, J_3, J_4)$, $I_2 = \max(J_5, J_6, J_7, J_8)$, $I_3 = \max(J_9, J_{10}, J_{11})$.
The maximum values I_1, I_2, I_3 can be selected 48 manners.

We use the program described above for determination of variables $\alpha, \beta, \theta, \gamma, \omega \in N$, for which system of equation is determine or underdetermined. The results of classification we write in the following summary table1:

Table 1. Table of acceptable values for system (16)

№	α	β	θ	γ	ω	j	No	α	β	θ	γ	ω	j
1	0	0	0	0	1	3	30	1	1	0	0	1	0
2	0	0	0	0	2	1	31	1	1	1	0	1	0
3	0	1	0	0	1	3	32	1	1	2	0	1	0
4	0	1	0	0	2	1	33	1	2	0	0	1	0
5	0	2	0	0	2	1	34	1	2	1	0	1	0
6	1	0	0	1	0	3	35	1	2	2	0	1	0
7	1	0	0	1	1	1	36	2	0	0	1	0	0
8	1	0	1	1	1	1	37	2	0	1	1	0	0
9	1	1	0	1	1	1	38	2	1	0	1	0	0
10	1	1	1	1	1	1	39	2	1	1	1	0	0
11	1	1	2	1	1	1	40	2	1	2	1	0	0
12	2	0	0	2	0	1	41	0	1	0	1	0	0
13	0	0	0	1	0	3	42	0	1	1	1	0	0
14	0	0	0	2	0	1	43	0	1	2	1	0	0
15	0	1	0	1	1	1	44	0	2	0	0	1	0
16	0	1	1	1	1	1	45	0	2	1	0	1	0
17	1	0	0	2	0	1	46	1	1	0	1	0	0
18	0	0	0	1	1	1	47	1	1	1	1	0	0
19	0	0	1	1	1	1	48	1	1	2	1	0	0
20	1	0	1	1	0	2	49	0	0	1	0	1	2
21	1	0	2	1	0	1	50	0	0	1	0	2	0
22	1	0	2	1	1	0	51	0	1	1	0	1	2
23	2	0	1	2	0	0	52	0	1	1	0	2	0
24	0	0	1	1	0	2	53	0	1	2	0	1	1
25	0	0	1	2	0	0	54	0	2	1	0	2	0
26	1	0	1	2	0	0	55	0	1	2	1	1	0

27	1	0	0	0	1	0	56	0	0	2	0	1	0
28	1	0	1	0	1	0	57	0	0	2	1	0	0
29	1	0	2	0	1	0							

Example 2. Let us consider the motion of ideal liquid describable by the following system of equations

$$\begin{cases} \rho(u_t + uu_x + vu_y) = -p_x \\ \rho(v_t + uv_x + vv_y) = -p_y \\ \rho_t + \rho_x u + \rho_y v + \rho(u_x + v_y) = 0 \\ \rho(p_t + up_x + vp_y) - \chi p(\rho_t + u\rho_x + v\rho_y) = 0 \end{cases} \qquad (20)$$

The system (20) is obtained from system (14) in case of when flows are considered plane $\left(\dfrac{\partial}{\partial z} = 0, w = 0 \right)$ and entropy is not constant.

Putting new variables ξ, η, t in system (20), we get the following system of equation

$$\begin{cases} \rho(u_t(x_\xi y_\eta - x_\eta y_\xi) + u_\xi(x_\eta y_t - x_t y_\eta) + u_\eta(x_t y_\xi - x_\xi y_t) + u(u_\xi y_\eta - u_\eta y_\xi) + \\ + v(x_\eta \mu_\eta - x_\mu \mu_\xi)) = p_\eta y_\xi - p_\xi y_\eta \\ \rho(v_t(x_\xi y_\eta - x_\eta y_\xi) + v_\xi(x_\eta y_t - x_t y_\eta) + v_\eta(x_t y_\xi - x_\xi y_t) + u(v_\xi y_\eta - v_\eta y_\xi) + \\ + v(x_\xi v_\eta - x_\eta v_\xi)) = p_\xi x_\eta - p_\eta x_\xi \\ \rho_t(x_\xi y_\eta - x_\eta y_\xi) + \rho_\xi(x_\eta y_t - x_t y_\eta) + \rho_\eta(x_t y_\xi - x_\xi y_t) + u(\rho_\xi y_\eta - \rho_\eta y_\xi) + \\ + v(x_\xi \rho_\eta - x_\eta \rho_\xi) + \rho(u_\xi y_\eta - u_\eta y_\xi + x_\xi v_\eta - x_\eta v_\xi) = 0, \\ \rho(p_t(x_\xi y_\eta - x_\eta y_\xi) + p_\xi(x_\eta y_t - x_t y_\eta) + p_\eta(x_t y_\xi - x_\xi y_t) + u(p_\xi y_\eta - p_\eta y_\xi) + \\ + v(x_\xi p_\eta - x_\eta p_\xi)) - \chi p(\rho_t(x_\xi y_\eta - x_\eta y_\xi) + \rho_\xi(x_\eta y_t - x_t y_\eta) + \\ + \rho_\eta(x_t y_\xi - x_\xi y_t) + u(\rho_\xi y_\eta - \rho_\eta y_\xi) + v(x_\xi \rho_\eta - x_\eta \rho_\xi)) = 0. \end{cases} \qquad (21)$$

The solution of system (21) is found in form

$$x(\xi,\eta,t) = \sum_{k=0}^{\gamma} x_k(\xi,t)\eta^k, \quad y(\xi,\eta,t) = \sum_{k=0}^{\omega} y_k(\xi,t)\eta^k, \quad u(\xi,\eta,t) = \sum_{k=0}^{\alpha} u_k(\xi,t)\eta^k,$$

$$v(\xi,\eta,t) = \sum_{k=0}^{\beta} v_k(\xi,t)\eta^k, \quad p(\xi,\eta,t) = \sum_{k=0}^{\theta} p_k(\xi,t)\eta^k, \quad \rho(\xi,\eta,t) = \sum_{k=0}^{\lambda} \rho_k(\xi,t)\eta^k. \qquad (22)$$

Having substituted (22) in system (21), we find the degrees of polynomials, incoming in each equation. The first equation contains polynomials of degrees:
$$J_1 = \gamma + \omega + \alpha + \lambda - 1, J_2 = \omega + 2\alpha + \lambda - 1, J_3 = \gamma + \alpha + \beta + \lambda - 1, J_4 = \omega + \theta - 1.$$
The second equation contains polynomials of degrees
$$J_5 = \gamma + \omega + \beta + \lambda - 1, J_6 = \omega + \alpha + \beta + \lambda - 1, J_7 = \gamma + 2\beta + \lambda - 1, J_8 = \gamma + \theta - 1.$$
The third equation contains polynomials of degrees
$$J_9 = \gamma + \omega + \lambda - 1, J_{10} = \omega + \alpha + \lambda - 1, J_{11} = \gamma + \beta + \lambda - 1.$$

433

The forth equation contains polynomials of degrees

$$J_{12} = \gamma + \omega + \theta + \lambda - 1, J_{13} = \omega + \alpha + \theta + \lambda - 1, J_{14} = \gamma + \beta + \theta + \lambda - 1$$

Let us use the maximum degrees of polynomials, incoming in equations:

$$I_1 = \max\{J_1, J_2, J_3, J_4\}, I_2 = \max\{J_5, J_6, J_7, J_8\}, I_3 = \max\{J_9, J_{10}, J_{11}\}, I_4 = \max(J_{12}, J_{13}, J_{14}).$$

The number of unknown functions is $s = \gamma + \omega + \alpha + \beta + \theta + \lambda + 6$, the number of equations is $r = I_1 + I_2 + I_3 + I_4 + 4$, a degree of sub-definite $j = s - r$.

Below is submitted the table of acceptable values of parameters $\gamma, \omega, \alpha, \beta, \theta, \lambda$, obtained in result of calculations with the help of the program described above (see Table2):

Table 2. Table of acceptable values for system (20)

No	γ	ω	α	β	θ	λ	j	No	γ	ω	α	β	θ	λ	j
1	0	1	0	0	0	0	3	19	1	0	1	0	0	0	3
2	0	1	0	0	0	1	0	20	1	0	1	0	0	1	0
3	0	1	0	0	1	0	2	21	1	0	1	0	1	0	2
4	0	1	0	0	1	1	0	22	1	0	1	0	1	1	0
5	0	1	0	0	2	0	0	23	1	0	1	0	2	0	1
6	0	1	0	1	0	0	3	24	1	1	0	0	0	0	0
7	0	1	0	1	0	1	0	25	1	1	0	0	1	0	0
8	0	1	0	1	1	0	2	26	1	1	0	1	0	0	0
9	0	1	0	1	1	1	0	27	1	1	0	1	1	0	0
10	0	1	0	1	2	0	1	28	1	1	1	0	0	0	0
11	0	2	0	0	0	0	0	29	1	1	1	0	1	0	0
12	0	2	0	1	0	0	0	30	1	1	1	1	0	0	0
13	0	2	0	2	0	0	0	31	1	1	1	1	1	0	0
14	1	0	0	0	0	0	3	32	1	1	1	1	2	0	0
15	1	0	0	0	0	1	0	33	2	0	0	0	0	0	0
16	1	0	0	0	1	0	2	34	2	0	1	0	0	0	0
17	1	0	0	0	1	1	0	35	2	0	2	0	0	0	0
18	1	0	0	0	2	0	0								

As an example we consider the particular case: $\lambda = \alpha = 1, \gamma = \beta = \omega = \theta = 0$, degree of sub-definite $j = 3$. Then the solution is

$$\begin{cases} x = x^0(\xi, t) + x^1(\xi, t)\eta, \quad y = y(\xi, t), \\ u = u^0(\xi, t) + u^1(\xi, t)\eta, \quad v = v(\xi, t), \\ p = p(\xi, t), \quad \rho = \rho(\xi, t) \end{cases} \Leftrightarrow \begin{cases} u = u_0(y, t) + u_1(y, t)x \\ v = v(y, t) \\ \rho = \rho(y, t) \\ p = p(y, t) \end{cases}$$

Substituting in (20), grouping summands at degrees of x and equating total coefficients to 0, we get system (consider, that $\rho \neq 0$):

434

$$\begin{cases} u_{1t} + u_1^2 + vu_{1y} = 0, u_{0t} + u_0 u_1 + vu_{0y} = 0, \\ \rho(v_t + vv_y) = -p_y, \rho_t + \rho_y v + \rho(u_1 + v_y) = 0, \\ \rho(p_t + vp_y) - \chi(\rho_t + v\rho_y) = 0. \end{cases} \tag{23}$$

Let us note that the solution of system (23) can be found in form:
$u_0 = u_0(\delta), u_1 = u_1(\delta), v = v(\delta), \rho = \rho(\delta), p = p(\delta)$,
where $\delta = y + ct$, c is an arbitrary constant. Then we get the system of five ordinary differential equations for five functions, dependent on δ.

Notation 1. Similarly we obtained full classifications of solutions in form (10) for the following systems.

1. Let us suppose that there is incompressible fluid ($\rho = const$), then in system (14) the energy equation is canceled. Then for plane flows we have system of equations:

$$u_t + uu_x + vu_y = -\frac{1}{\rho}p_x, \quad v_t + uv_x + vv_y = -\frac{1}{\rho}p_y, \quad u_x + v_y = 0 \tag{24}$$

2. Let us suppose that there is $\chi = 2$ in system (15), then we obtain:

$$\begin{cases} u_t + uu_x + vu_y = -2c\rho_x, \quad v_t + uv_x + vv_y = -2c\rho_y, \\ \rho_t + u\rho_x + \rho u_x + v\rho_y + \rho v_y = 0 \end{cases} \tag{25}$$

3. Let us suppose that there is $\chi = 1$ and $z = \ln \rho$ in system (15), then we get system of equations:

$$u_t + uu_x + vu_y = -cz_x, \quad v_t + uv_x + vv_y = -cz_y, \quad z_t + uz_x + vz_y + u_x + v_y = 0 \tag{26}$$

4. Stationary transonic flows of gas are described in the first approximation by asymptotic system of equations:

$$uu_x - v_y = 0, \quad u_y - v_x = 0 \tag{27}$$

Notation 2. Two-parameter method can be used for exact linearization of equation systems.

For example, in parametric form the system (27) is [8]

$$\begin{cases} u(u_\xi y_\eta - u_\eta y_\xi) - (v_\eta x_\xi - v_\xi x_\eta) = 0 \\ u_\eta x_\xi - u_\xi x_\eta - (v_\xi y_\eta - v_\eta y_\xi) = 0 \end{cases} \tag{28}$$

Substituting $u = \xi$, $v = \eta$ in system (28), we get:

$$\begin{cases} \xi y_\eta - x_\xi = 0 \\ -x_\eta + y_\xi = 0 \end{cases},$$

that is in a plane $(u, v) = (\xi, \eta)$ nonlinear system (27) becomes linear system for functions $x(u, v)$, $y(u, v)$. Let us note, that system (27) is a particular case of the quasi-linear system

$$\begin{cases} f_1(u,v)u_x + f_2(u,v)u_y + f_3(u,v)v_x + f_4(u,v)v_y = 0 \\ g_1(u,v)u_x + g_2(u,v)u_y + g_3(u,v)v_x + g_4(u,v)v_y = 0 \end{cases},$$

which becomes linear after transition to variables $\xi,\ \eta$ and at a choice $u = \xi, v = \eta$:

$$\begin{cases} f_1(\xi,\eta)y_\eta - f_2(\xi,\eta)x_\eta - f_3(\xi,\eta)y_\xi + f_4(\xi,\eta)x_\xi = 0 \\ g_1(\xi,\eta)y_\eta - g_2(\xi,\eta)x_\eta - g_3(\xi,\eta)y_\xi + g_4(\xi,\eta)x_\xi = 0 \end{cases}$$

Notation 3. Parametric method is useful for construction of solutions such as "simple wave".

As an example we consider the solution of system (16) in form of simple wave:

$$\begin{cases} u = u(\xi),\ v = v(\xi),\ w = w(\xi), \\ x = x_0(\xi) + x_1(\xi)y + x_2(\xi)t,\ y = \eta \end{cases}$$

Substituting in (17), then believe $u(\xi) = \xi$ and, considering that $w(\xi) \neq const$, we get:

$$\begin{cases} x_1 = -v'(\xi) \\ x_2(\xi) = \xi + v(\xi)v'(\xi) - \zeta w'(\xi) \\ w(\xi)\left(1 + (v'(\xi))^2\right) - \mu\zeta\left(w'(\xi)\right)^2 = 0 \end{cases}.$$

Two functions $w(\xi)$ and $v(\xi)$ are bound by one equation, therefore any of these functions is arbitrary. The function $x_0(\xi)$ is arbitrary too. Thus, the solution is obtained, and it depends on two arbitrary functions.

Let us find the solution in form of simple wave for system (27):

$$u = \xi,\ v = v(\xi),\ y = \eta,\ x = x_0(\xi) + x_1(\xi)\eta.$$

Substituting this solution in (28), we get

$$\begin{cases} \xi + v'(\xi)x_1(\xi) = 0 \\ -x_1(\xi) - v'(\xi) = 0 \end{cases} \Leftrightarrow \begin{cases} x_1(\xi) = -v'(\xi) \\ v'(\xi) = \pm\sqrt{\xi} \end{cases} \Rightarrow \begin{cases} v(\xi) = \pm\dfrac{2}{3}\xi^{3/2} + c \\ x_1(\xi) = \mp\xi^{1/2} \end{cases}. \qquad (29)$$

Then we get the following solution of (28):

$$u = \xi,\ v = \pm\frac{2}{3}\xi^{3/2} + c,\ y = \eta,\ x = x_0(\xi) \mp \xi^{1/2}y, \qquad (30)$$

where $x_0(\xi)$ is an arbitrary function.

Parametric Solutions of Equations of Transonic Gas Flows and Their Application

The unsteady transonic flows of ideal gas in the asymptotic approximation are described by system of equations

$$2u_\tau + uu_x - v_y - w_z = 0,\ u_y = v_x,\ u_z = w_x,\ v_z = w_y, \qquad (31)$$

Here u, v, w are projections of velocity vector to axes of rectangular coordinate system x, y, z; τ is the time. For velocity potential we have equation

$$2\varphi_{x\tau} + \varphi_x\varphi_{xx} - \varphi_{yy} - \varphi_{zz} = 0, \qquad (32)$$

Having differentiated the equation (32) by x, we get an equation for $u = \varphi_x$. Some solutions of equation system (31) and equation (32) have considered in [1-4]. The solution of equation (32), describing flow in Laval nozzles with local supersonic zones [5-7] (here and further we consider a nozzle with two transversely-spaced planes of symmetry), is

$$u = U(\xi,\tau) + a_1(\tau)y^2 + a_2(\tau)z^2, \quad x = m(\tau)\xi + n(\tau) + c_1(\tau)y^2 + c_2(\tau)z^2 \quad (33)$$

The equation for $U(\xi,\tau)$ are easily written. The primary intent is study of local supersonic zones (LSZ) with the course of time. Therefore we consider just plane and axisymmetric flows. Generalization of results, obtained below, on a three-dimensional case (33) is an easy make.

Equation system (31) admits the following form of solution:

$$u = \tau^{n-1}u_*(x_*, y_*, t) + 2\lambda'(\tau), \quad v = \tau^{\frac{3}{2}(n-1)}v_*(x_*, y_*, t) + \frac{4\lambda''(\tau)}{\omega + 1}y,$$

$$x_* = \frac{x - \lambda(\tau)}{\tau^n}, \quad y_* = y\tau^{-\frac{1}{2}(n+1)}, \quad t = \ln\tau \quad (34)$$

The solutions of u_*, v_* are found in form (33)

$$u_* = mU(\xi,t) + 2c(2c-1)y_*^2, \quad x_* = m\xi + cy_*^2$$

$$v_* = 2cm[2(2c-1)\xi - U(\xi,t)]y_* + \frac{8c(c-1)(2c-1)}{\omega+3}y_*^3 \quad (35)$$

Here m, c, n are arbitrary constants, $\lambda(\tau)$ is an arbitrary functions, $\omega = 0$ for plane flows and $\omega = 1$ for axisymmetric flows. For function $U(\xi,t)$ we get equation

$$2U_t + (U - 2n\xi)U_\xi + 2[n - 1 + (\omega + 1)c]U - 4c(2c-1)(\omega+1)\xi = 0 \quad (36)$$

At first let us consider self-similar solutions (that is $U_t = 0$). Then for U we have an ordinary differential equation, which contains two arbitrary parameters c and n. In this case the behavior of integral curves depends on values of quantities λ_1, λ_2

$$\lambda_{1,2} = q_{1,2} - 2n, \quad q_{1,2} = 1 - (\omega+1)c \mp \left[1 - (\omega+1)c(6 - 8c) + (\omega+1)^2c^2\right]^{1/2} \quad (37)$$

If λ_1, λ_2 are different and like-sign, in origin of coordinates of plane (U,ξ) we have critical point such as a knot (Figure 1a), if λ_1, λ_2 are unlike signs — a saddle (Figure 1b); if $\lambda_1 = \lambda_2 \neq 0$, we have degenerate node; in case of one of quantities λ_k (or both) is equal zero, the solution on plane (U,ξ) is shown by parallel lines. Notice that curves $U = U(\xi,t)$ give distribution of velocity (pressure) $u = u(x,t)$ on axis $y = 0$. The solutions, which is shown by straight lines through a critical point (dotted straight lines correspond to them), is

$$U = q_1\xi, \quad U = q_2\xi. \quad (38)$$

In plane case $(\omega = 0)$ $q_1 = 2(1 - 2c)$, $q_2 = 2c$.

In case of knot the curves concern in a critical point of a straight line $U = q_1\xi$, if $|\lambda_1| < |\lambda_2|$, and a straight line $U = q_2\xi$, if $|\lambda_1| > |\lambda_2|$. In case of when $\lambda_1 \neq \lambda_2$, the solution of the equation (36) is written in form

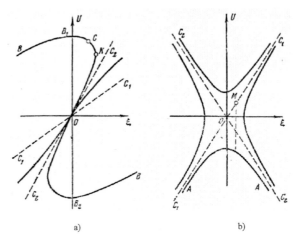

FIGURE 1. Integral curves

$$\left(U - q_1\xi\right)^{-\lambda_1}\left(U - q_2\xi\right)^{-\lambda_2} = A = const \tag{39}$$

or in parameter form

$$U = \frac{q_2}{q_2 - q_1}\eta + q_1 B\eta^{\chi}, \quad \xi = \frac{1}{q_2 - q_1}\eta + B\eta^{\chi}, \quad \chi = \frac{\lambda_1}{\lambda_2} \tag{40}$$

At $\lambda_1 = \lambda_2 = \lambda_0$ $\left(q_1 = q_2 = q\right)$ the solution is easily written also

$$U = \frac{q}{\lambda_0}\eta\ln\eta + \eta(1 + qB), \quad \xi = \frac{1}{\lambda_0}\eta\ln\eta + B\eta. \tag{41}$$

In formulae (39)-(41) A and B are arbitrary constants. The equation of sonic line for (35) in parametric form $y = y(\xi,\tau)$, $x = x(\xi,\tau)$ at $c \neq \frac{1}{2}, c \neq 0$ is

$$y^2 = \frac{m\tau^{n+1}}{2c(1-2c)}U(\xi,t) - \frac{\tau^2\lambda'(\tau)}{c(2c-1)}, \quad x = m\xi\tau^n + \frac{c}{\tau}y^2 + \lambda(\tau). \tag{42}$$

In case of $c = 0$ or $c = \frac{1}{2}$ a sonic line is $\xi = \xi_0(\tau)$. In what follows we consider a plane case $\omega = 0$ (in axisymmetric case the analysis is carried similarly) and suppose $\lambda'(\tau) = 0$. Then for (35) the equation of sonic line is

$$y^2 = \frac{m\tau^{n+1}}{2c(1-2c)}\left[\frac{c}{3c-1}\eta + 2(1-2c)B\eta^{\chi}\right] = \frac{m\tau^{n+1}}{2c(1-2c)}U(\eta),$$

$$x = m\tau^n\left[\frac{1-c}{2(3c-1)(1-2c)}\eta + 2B\eta^{\chi}\right], \chi = \frac{1-2c-n}{c-n}. \tag{43}$$

Let us consider $m = 1 > 0$ (at $m < 0$ reasoning is carried similarly). From the first formula (43) it is evident, that the sonic line can be constructed at $0 < c < \frac{1}{2}$, when $U > 0$, and at $c < 0$ or $c > \frac{1}{2}$, when $U < 0$. Analyzing the behavior of integral curves on Figure 1 and considering the first formula (35), we come to the conclusion, that curves AA and C_1OC_2, represented on Figure 1b at $c < 0$ or $c > \frac{1}{2}$ can described the flows with LSZ in Laval nozzle. The formulae (43) show us how LSZ change with the course of time.

Concerning, that in transonic approximation the equations of nuzzle walls (they are easily constructed in parametric form) is

$$y = y_0 + \varepsilon f(x, \tau), \frac{\partial f}{\partial x} = v(v_0, x, \tau), y_0 = const, \varepsilon \ll 1. \tag{44}$$

we see from (43), that at $n > -1$ LSZ, borrowing a part of nozzle throat in initial moment, with the course of time disappear, and flow becomes subsonic everywhere, that is the solutions at $n > -1$ describe flows with disappearing LSZ. As an example on Figure 2a the qualitative picture of this flow is shown us (at $n = 2$, $c = -3$, $B > 0$). At $n < -1$, on the contrary, the development of LSZ is observed (the example is on Figure 2b, $n = -\frac{3}{2}$, $c = 4$, $B < 0$). Similarly, from the second formula (43) it is evident, that at $n > 0$ LSZ expands, at $n < 0$ it is narrowed with growth of time.

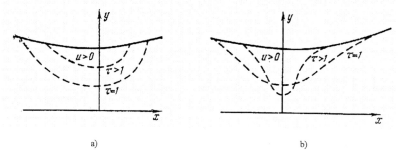

FIGURE 2. Local supersonic zones

Let us note, that in all formulae it is possible to replace τ on $(\tau + \tau_0)$, $0 \leq \tau < \infty, \tau_0 > 0$. At $\lambda'(\tau) = 0$, according to (43), flows with LSZ, closing on nozzle axis at $\tau \to \infty$ for $n < -1$ take place. If $\lambda'(\tau) \neq 0$, at same values of c $(c < 0, c > \frac{1}{2})$ LSZ is closed on nozzle axis at $\tau = \tau_1$, and then supersonic zone occupies the whole segment of an axis (or with growth of time there is a return process). Such LSZ are easy constructed under formulae (42), (43). Let us emphasize that solutions, which are given by curves, having closed parts $U > 0$ on axis $y = 0$, can describe flows with LSZ about profile (Figure 3). The solutions, having such parts, are given, for example, by curves BOB (Figure 1a), and also by curves AA, C_1OC_2 (Figure 1b), at $\lambda'(\tau) > 0$.

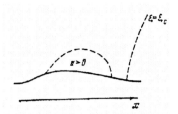

FIGURE 3. Flow about profile for $c = \frac{1}{4}$, $n = \frac{5}{8}, B > 0, y_0 = 3.35$

However, from the first formula (42) it results, that for such flows $0 < c < \frac{1}{2}$. But at such values c for the flows this type the solution is faulty (zones of ambiguity and zones of non-existence of solution). There is shown the qualitative picture of flow about profile at $c = \frac{1}{4}$, $n = \frac{5}{8}, B > 0, y_0 = 3.35$ on Figure 3. The integral curve BB_1C on Figure 1a corresponds to this solution ($\xi_C \le \xi_K, U'(\xi_K) = \infty$). Flow behind $\xi = \xi_C$ does not work to be constructed in a class of solutions (35). Apparently, this solution can be continued behind $\xi = \xi_C$, using more general, than (35), (39) solution, admissible by system (31), in the following form:

$$u_* = U_0(\xi) + U_2(\xi)y_*^2, \quad x_* = X_0(\xi) + X_2(\xi)y_*^2, \quad v_* = V_1(\xi)y_* + V_3(\xi)y_*^3. \quad (45)$$

Let us consider now the solution (35) in general case $U = U(\xi,t)$. For equation (36), using (40), the first two integrals are easily obtained. Then the general solution at $\lambda_1 \ne \lambda_2$ is written in form

$$F\left[(U - q_1\xi)^{-\lambda_1}(U - q_2\xi)^{\lambda_2}, (U - q_1\xi)\exp\left(-\tfrac{1}{2}\lambda_2 t\right)\right] = 0 \qquad (46)$$

Here F is an arbitrary function of two arguments. Having written the solution (46) in form solved with respect to the second argument, we get the solution in form $t = t(U, \xi)$. Let us note, that for function $t(U, \xi)$ the equation (36) is linear. In parametric form the solution (46) is written in form (40), where B is an arbitrary function of variable $\eta \exp\left(-\tfrac{1}{2}\lambda_2 t\right)$.

If $\lambda_1 = \lambda_2 = \lambda_0$ $(q_1 = q_2 = q)$, the general solution of equation (36) is

$$F\left[\ln(U - q\xi) - \frac{\lambda_0\xi}{U - q\xi}, (U - q\xi)\exp\left(-\tfrac{1}{2}\lambda_0 t\right)\right] = 0. \qquad (47)$$

In parametric form the solution is given by formulae (41), where $B(\eta \exp(-\lambda_0 t/2))$ is an arbitrary function.

REFERENCES

1. P.A. Velmisov, *Asymptotic Equations of Gas Dynamics*, Saratov University. Russia, 1986. (in Russian)
2. P.A. Velmisov, "Unsteady Motion of Gas in Laval Nozzles," in coll. *Aerodynamics. Saratov University, Russia*, No. 2, 1973. (in Russian)
3. O.S. Rizhov, *Study of Transonic Flows in Laval Nozzles*, Moscow, 1965. (in Russian)
4. O.S. Rizhov, *J. Comp. Math.& Math.Phys.* 7(4), 1967. (in Russian)

5. S. Tomotika and K. Tamada, *Quart. Appl. Math.* **7**(4), 1950.
6. S. Tomotika and Z. Hasimoto, *J. Math. Phys.* **29**(2), 1950.
7. T.C. Adamson, *J. Fluid Mech.* **52**(3), 437-449 (1972).
8. P.A. Velmisov and S.V. Falkovich, "Some Classes of Solutions of Transonic Equations and Equations of Short Waves," in coll. *Selected Problems of Applied Mechanics*, Moscow, 1974, pp. 215-223. (in Russian)

On The Solution of Stiff Ordinary Differential Equations

Y. Zhou[*,†] and S. Xiang[†]

[*]Dept. of Mathematics, Guangdong Ocean University, Zhanjiang, Guangdong 524088, China
[†]Dept. of Applied Mathematics and Software, Central South University
Changsha, Hunan 410083, China

Abstract. In this paper, we considered new method based on simplex integrals for stiff ordinary differential equations. In contrast to Kuntzmann-Butcher method, this method requires the solution of less simultaneous implicit equations. And numerical examples illustrate the efficiency of this technique.

Keywords: ODE, stiff, simplex integral, algorithm.
PACS: 02.30.Hq, 02.30.Mv

INTRODUCTION

Stiff ordinary differential equations(ODEs) are quite common, particularly in the study of vibrations, chemical reactions, and electrical circuits. We consider mth ODEs of the form

$$\begin{cases} \frac{dy_1}{dt} &= f_1(t, y_1, y_2, \ldots, y_m), \\ \frac{dy_2}{dt} &= f_2(t, y_1, y_2, \ldots, y_m), \\ \quad \vdots \\ \frac{dy_m}{dt} &= f_m(t, y_1, y_2, \ldots, y_m), \end{cases} \tag{1}$$

for $t > t_0$, with given initial conditions $y_1(t_0), y_2(t_0), \ldots, y_m(t_0)$.

If ODEs are stiff, many numerical analyst considered either v-stage singly diagonally implicit Runge-Kutta method (SDIRK [2, p.92]) usually of less than order v required only v-dimensional system or v-stage Kuntzmann-Butcher method [1, p.208] of order $2v$ but usually required the solution of mv-dimensional nonlinear system. In this paper, based on simplex integrals, we present a new v-stage discrete scheme of order $2v$ required only m-dimensional system.

POLYNOMIAL EXPANSION ABOUT SIMPLEX INTEGRALS

To discuss new algorithms for solving the above ODEs, we need introduce simplex integrals and their approximate polynomial expansions.

CP1067, Applications of Mathematics in Engineering and Economics '34—AMEE '08, edited by M. D. Todorov
© 2008 American Institute of Physics 978-0-7354-0598-1/08/$23.00

Simplex Integrals

For the sake of simplicity, we consider ODE

$$y'(t) = f(t, y(t)), \quad y(t_0) = y_0, \tag{2}$$

which solution $y(t) \in C^{v+1}[t_0, t_1]$. Obviously, we also know the initial value $y'(t_0), y''(t_0), \cdots, y^{(v)}(t_0)$. Similarly, we know $y'(t_1), y''(t_1), \cdots, y^{(v)}(t_1)$ if we obtain the solution $y(t_1)$. We define simplex integrals as

$$\phi_{-v}(t) = y^{(v)}(t),$$

$$\phi_{-k}(t) = \int_{t_0}^{t} \phi_{-(k+1)}(t)dt, \quad k = v-1, v-2, \ldots, 0.$$

It implies

$$\begin{cases} \phi_{-v}(t_1) &= y^{(v)}(t_1), \\ \phi_{-(v-1)}(t_1) &= y^{(v-1)}(t_1) - y^{(v-1)}(t_0), \\ \cdots \quad \cdots & \cdots\cdots\cdots \\ \phi_0(t_1) &= y(t_1) - \sum_{k=0}^{v-1} y^{(k)}(t_0) \frac{(t_1-t_0)^k}{k!}. \end{cases} \tag{3}$$

where $y^{(0)}(t_0)$ denotes $y(t_0)$.

Remark 1 *Integrating by parts repeatedly, it is well known that the multiple simplex integrals can also be written in the form of one-dimensional integrals*

$$\phi_{-k}(t) = \frac{1}{(-k+v-1)!} \int_{t_0}^{t} y^{(v)}(\xi)(t-\xi)^{-k+v-1}d\xi$$

with $k = v-1, v-2, \ldots, 0$.

Polynomial Expansion

It is well known Legendre polynomial

$$L_n(x) = \frac{1}{2^n n!} \frac{d^n}{dx^n}(x^2-1)^n.$$

Define shifted Legendre polynomial in $[t_0,t_1]$ as $\check{L}_n(t) = L_n(x)$ by substitution $x = \frac{2}{t_1-t_0}(t - \frac{t_0+t_1}{2})$ with $n = 0,1,\ldots$.

Theorem 1 *Suppose* $\phi_{-v}(t) \in C[t_0,t_1]$, *the polynomial* $Q_{v-1}(t) = \alpha_0 + \alpha_1 t + \alpha_2 t^2 + \cdots + \alpha_{v-1}t^{v-1}$, *which coefficients* $\alpha_0, \alpha_1, \cdots, \alpha_{v-1}$ *satisfy*

$$
\begin{cases}
\sum\limits_{k=0}^{v-1} \alpha_k \int\limits_{t_0}^{t_1} t^k dt &= \phi_{-(v-1)}(t_1) \\
\frac{1}{1!}\sum\limits_{k=0}^{v-1} \alpha_k \int\limits_{t_0}^{t_1} (t_1 - t)t^k dt &= \phi_{-(v-2)}(t_1) \\
\cdots\cdots\cdots \qquad \cdots \qquad \cdots \\
\frac{1}{(v-1)!}\sum\limits_{k=0}^{v-1} \alpha_k \int\limits_{t_0}^{t_1} (t_1 - t)^{v-1}t^k dt &= \phi_0(t_1)
\end{cases}
\tag{4}
$$

Then

$$
\phi_{-v}(t) - Q_{v-1}(t)
$$
$$
= \frac{\phi_{-v}(t_0) - Q_{v-1}(t_0) + (-1)^v(\phi_{-v}(t_1) - Q_{v-1}(t_1))}{2}\check{L}_v(t) + O(h^{2v+1})
$$

is true for all $t \in [t_0,t_1]$ *with* $h = t_1 - t_0$.

Proof: The formulas (3) and (4) imply that there exists a $2v - 1$ degree polynomial $P_{2v-1}(t)$ such that

$$
y^{(k)}(t_0) = P_{2v-1}^{(k)}(t_0), \quad y^{(k)}(t_1) = P_{2v-1}^{(k)}(t_1)
$$

with $k = 0,1,\ldots,v$, where $P_{2v-1}^{(v)}(t) = Q_{v-1}(t)$. In other words, $P_{2v-1}(t)$ is the Hermite interpolation polynomial of the exact solution $y(t)$. As for the interpolation error, the following estimate [3, p.349] holds

$$
y(t) - P_{2v-1}(t) = \frac{y^{(2v+1)}(\xi)}{(2v+1)!}(t - t_0)^{v+1}(t - t_1)^{v+1}
$$

with $\xi \in [t_0,t_1]$. The statement follows by derivatives.

Noting that $\check{L}_v(t_0) = \check{L}_v(t_1) = 1$, we have

Theorem 2 *Under the condition of Theorem 1,*

$$
\phi_{-v}(t_0) - Q_{v-1}(t_0) = (-1)^v [\phi_{-v}(t_1) - Q_{v-1}(t_1)] + O(h^{2v+1}).
$$

DISCRETE SCHEME

Using formulas (3) and (4), we can obtain the approximate solution of $y(t_1)$ related to ODE (2) according to Theorem 2. It follows

$$
y(t_1) - \frac{v}{2v}y'(t_1) + \frac{v(v-1)}{2v(2v-1)}\frac{y''(t_1)}{2!} - \cdots + (-1)^v \frac{v!v!}{(2v)!}\frac{y^{(v)}(t_1)}{v!}
$$
$$
= y(t_0) + \frac{v}{2v}y'(t_0) + \frac{v(v-1)}{2v(2v-1)}\frac{y''(t_0)}{2!} + \cdots + \frac{v!v!}{(2v)!}\frac{y^{(v)}(t_0)}{v!} + O(h^{2v+1}).
$$

444

It is very interesting that the left, right parts of above equation are similar to the denominator, numerator of (v,v)-Padé approximation to e^z.

Definition 1 *Suppose $y(t)$ is the solution of ODE (2) and satisfies $y(t) \in C^v[t_0,t_1]$, then we define v-stage simplex integral method as*

$$y(t_1) \;=\; y(t_0) + \frac{v}{2v}(y'(t_0) + y'(t_1))$$
$$+ \frac{v(v-1)}{2v(2v-1)}\,\frac{y''(t_0) - y''(t_1)}{2!} + \cdots + \frac{v!v!}{(2v)!}\,\frac{y^{(v)}(t_0) + (-1)^v y^{(v)}(t_1)}{v!}.$$

Noting that at every step we only need solve one simultaneous implicit equation which presents a striking contrast to vth implicit equations in v-stage Kuntzmann-Butcher method.

Remark 2 *One stage method is equivalent to trapezoidal rule*

$$y_1 = y_0 + \frac{h}{2}\left(f(x_0,y_0) + f(x_1,y_1)\right).$$

CONVERGENCE AND STABILITY

The convergence and stability theorem for simplex integral methods can now be established.

Theorem 3 *The v-stage method is of order $2v$.*

Proof: The result is an immediate consequence of Theorem 2.

Theorem 4 *The method is A-stable.*

Proof: Solving *Dahlquist test equation*

$$y'(t) = \lambda y(t), \quad y(0) = 1$$

after one step by v-stage simplex integral method, we obtain its *stability function*

$$R(z) = \frac{1 + \frac{v}{2v}z + \frac{v(v-1)}{2v(2v-1)}\frac{z^2}{2!} + \cdots + \frac{v!v!}{(2v)!}\frac{z^v}{v!}}{1 - \frac{v}{2v}z + \frac{v(v-1)}{2v(2v-1)}\frac{z^2}{2!} - \cdots + (-1)^v \frac{v!v!}{(2v)!}\frac{z^v}{v!}},$$

it is the (v,v)-Padé approximation to e^z. Obviously, $|R(iy)| \le 1$ is true for all real y and $R(z)$ is analytic for $\text{Re}\,z < 0$. Then according to the definition by Dahlquist [2, p.42], they imply the method A-stable.

Remark 3 *The v-stage simplex integral method performs more remarkable than v-stage Kuntzmann-Butcher method because same accuracy, less nonlinear equations for the former. Whereas it is not our claim that simplex integral methods are superior, since they need compute higher derivatives. Our aim is more modest, namely to argue that such an investigation might be very interesting [4, p.131].*

NUMERICAL EXPERIMENTS

Example 1 For a simple example let us consider a first order equation

$$y' = -\lambda(y - \cos t), \quad y(0) = 0$$

the stiffness of which was explained by Curtiss and hirschfelder [2, p.2] and its true solution is $-e^{-\lambda t}\lambda^2/(\lambda^2 + 1) + \lambda(\lambda \cos(t) + \sin(t))/(\lambda^2 + 1)$. More large is λ, more serious is stiffness. Figure 1 reflect numerical accuracy of 2-stage and 4-stage method compared with 5-stage singly diagonally implicit Runge-Kutta method (**SDIRK**, see [2, p.100]).

Remark 4 *The simplex integral method is explicit for linear ODEs.*

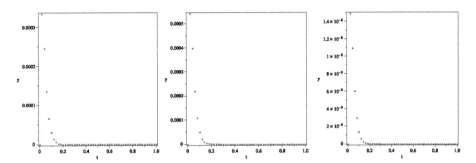

FIGURE 1. For Example 1. Let $\lambda = 50$, the left plot corresponds to the error of 5-stage SDIRK method with fixed step size $h = 1/50$ and $\text{Tol} = 10^{-25}$. And with same step size, the middle, right plots correspond to 2-stage, 4-stage simplex integral methods separately

Example 2 On the solution of *Van der Pol's equation*

$$y'_1 = y_2, \qquad y_1(0) = 2$$
$$y'_2 = \mu(1 - y_1^2)y_2 - y_1, \quad y_2(0) = 0.$$

The problem appears to be stiff if $\mu \gg 1$. Choosing $\mu = 50$, table 1 reflect numerical accuracy of new method with different stages and step size h, where referenced value computed by the Rosenbrock method [2, p.102] in Maple 11.

TABLE 1. The solution $y_1(t_1)$ of Van der Pol's equation after one step by new method

Stage	$h = 0.1$	$h = 0.01$	$h = 0.001$
2	1.9800000035	1.9999588222	1.9999986676
3	1.9999997754	1.9999986951	1.9999996254
4	1.9998979566	1.9999866755	1.9999986676
Maple 11	1.9998675483	1.9999875555	1.9999993572

446

ACKNOWLEDGMENTS

The authors are grateful to the helpful discussion on the 34th Conference "Applications of Mathematics in Engineering and Economics" held at Sozopol, Bulgaria. The authors are also grateful to the referees for their valuable comments and suggestions for improving this paper.

REFERENCES

1. E. Hairer, S. P. Norsett, and G. Wanner, *Solving Ordinary Differential Equations I. Nonstiff problems.* Springer-Verlag, Berlin, 1993, 2nd edn.
2. E. Hairer and G. Wanner, *Solving Ordinary Differential Equations II. Stiff and Differential-Algebraic problems.* Springer-Verlag, Berlin, 1993, 2nd edn.
3. A. Quarteroni, R. Sacco, and F. Saleri, *Numerical Mathematics.* Springer-Verlag, New York, 2000.
4. A. Iserles, H. Z. Munthe-Kaas, S. P. Norsett, and A. Zanna, *Acta Numerica*, 1–148 (2005).

OPERATIONS RESEARCH
AND STATISTICS

Multi-criterial Decision Making for Selection and Assignment of Sportsmen in Team-games

S. Baeva, L. Komarevska, C. Nedeva and L. Trenev

Technical University of Sofia, 1000 Sofia, Bulgaria

Abstract. The maximal common efficiency is the purpose in the team-games. This problem is formulated as a linear assignment problem. An example in baseball is solved and discussed.

Keywords: Multi-objective optimization, assignment problem.
PACS: 02.50.Le, 02.60.Pn

INTRODUCTION

The personal abilities and skills are of high importance for every single sport. In the individual sports the success depends on the person himself, but in the team games the problem of finale result is much more complicated matter: the success depends on the team efforts. The precise choice of a team, its selection and the place of every player on the pitch is very important. This is a very complex problem, connected with analysis of the sportsman's individual qualities.

The input data is a table of results for different tests made for every single player. The result is the selection of the team and the assignments in it.

DESCRIPTION OF THE PROBLEM

There is a finite number of alternatives – a finite number of players who can be included in the team and take corresponding position. The choice of the optimal alternative (the optimal team) is connected with the construction of the corresponding criteria. The optimal team is the one that possesses maximal common efficiency.

MATHEMATICAL MODEL

There are r players $x_1, x_2, ..., x_r$ and n different tests are made with them. These tests show some abilities and skills as: speed, reaction time, power, agility, *etc*. Every test is one criterion – $a_1, a_2, .., a_n$. The results are calculated in a way that presents the qualities of the players as numbers (for example, the number of obtained points is divided by the maximal possible). So we get the numbers $p_{ij}, i = 1, ..., r; j = 1, ..., n$ – measures of success of the ith player with respect to the jth criterion (test).

CP1067, *Applications of Mathematics in Engineering and Economics '34—AMEE '08*, edited by M. D. Todorov
© 2008 American Institute of Physics 978-0-7354-0598-1/08/$23.00

So we obtain the matrix P:

$$P = \begin{pmatrix} p_{11} & p_{12} & \cdots & p_{1n} \\ p_{21} & p_{22} & \cdots & p_{2n} \\ \cdots & \cdots & \cdots & \cdots \\ p_{r1} & p_{r2} & \cdots & p_{rn} \end{pmatrix}.$$

We have to choose a team in which the players take k number of positions ($k \leq r$). A weight vector is given for every position. It represents the importance of every criterion for corresponding position. For example, for the ith position a weight vector $w^j = (w_1^j, w_2^j, ..., w_n^j)$, $w_i^j \geq 0$, $j = 1, 2, ..., n$, $\sum_{j=1}^n w_j^i = 1$, $i = 1, ..., k$. The choice of the weight vectors is very important and is taken by the person-expert in the corresponding sport.

So we have the weight matrix W:

$$W = \begin{pmatrix} w_1^1 & w_1^2 & \cdots & w_1^k \\ w_2^1 & w_2^2 & \cdots & w_2^k \\ \cdots & \cdots & \cdots & \cdots \\ w_n^1 & w_n^2 & \cdots & w_n^k \end{pmatrix}.$$

As in the weight method [1] for the multiple objective optimization new generalized criterion is obtained as a linear combination of the measures of success of the players for the different tests.

Let Q_i^j be the generalized criterion that represents the efficiency of the jth player for ith position, $i = 1, 2, ..., k$, $j = 1, 2, ..., r$. These numbers are obtained as

$$Q = P \cdot W = \begin{pmatrix} p_{11} & p_{12} & \cdots & p_{1n} \\ p_{21} & p_{22} & \cdots & p_{2n} \\ \cdots & \cdots & \cdots & \cdots \\ p_{r1} & p_{r2} & \cdots & p_{rn} \end{pmatrix} \cdot \begin{pmatrix} w_1^1 & w_1^2 & \cdots & w_1^k \\ w_2^1 & w_2^2 & \cdots & w_2^k \\ \cdots & \cdots & \cdots & \cdots \\ w_n^1 & w_n^2 & \cdots & w_n^k \end{pmatrix} = \begin{pmatrix} Q_1^1 & Q_1^2 & \cdots & Q_1^k \\ Q_2^1 & Q_2^2 & \cdots & Q_2^k \\ \cdots & \cdots & \cdots & \cdots \\ Q_r^1 & Q_r^2 & \cdots & Q_r^k \end{pmatrix}.$$

The mathematical model is:

$$\max_Y Z(Y) = \sum_{i=1}^r \sum_{j=1}^k y_{ij} Q_i^j$$

subject to

$$\sum_{i=1}^r y_{ij} = 1, \; j = 1, ..., k, \qquad \sum_{j=1}^k y_{ij} \leq 1, i = 1, ..., r, \qquad y_{ij} = \{0, 1\}.$$

This problem is a type of a linear assignment problem [2]. Here $k \leq r$ and the problem is an open one. In order to get a classical assignment problem we take into consideration that $Q_i^j \geq 0$ and we add $r - k$ columns of 0 in the matrix Q. These columns will correspond to the players who will not be included in the team. The objective function Z represents the global efficiency of the team and every of the unknown parameters y_{ij} is equal to 1, when ith player takes the position j and 0 otherwise.

SOLUTION

The model includes 3 stages: calculation of data of the tests for the sportsmen; description of a general criterion for global efficiency of the team; solution of the optimization problem. This problem is solved by the Hungarian algorithm or by another method.

EXAMPLE

Choosing an Optimal Baseball Team Using Exact Data

For the choice of an optimal baseball team, we will use data gathered in the examination of 20 children in the 9-11 age group. They have all been subjected to 5 tests:

1. *"Upper limbs strength" test (batting from a stand)*
   The maximal number of points for 10 attempts is 40.
2. *"Agility" test (catching a soft ball with bare hands)*
   The maximal number of points for 5 attempts is 15.
3. *"Agility and reaction time" (a combined test)*
   The maximal number of points for 10 attempts is 10.
4. *"Reaction time and lower limbs strength"*
   The time taken for the test is measured.
5. *"Control" test (throwing at a target)*
   Each accurate throw from 10 gives 1 point.

The results from the tests are noted in Table 1.

TABLE 1.

player	test 1	test 2	test 3	test 4	test 5
player 1	28	13	21	7.9	8
player 2	28	13	24	7.89	10
player 3	28	15	20	7.9	10
player 4	32	12	25	7.65	10
player 5	34	13	28	7.75	8
player 6	26	15	25	8.12	8
player 7	26	12	25	7.84	10
player 8	34	15	20	7.6	8
player 9	27	15	22	7.95	10
player 10	29	15	27	7.7	10
player 11	23	8	17	8.2	6
player 12	18	11	19	8.42	6
player 13	18	10	19	8.8	6
player 14	18	11	18	8.25	4
player 15	24	10	19	8.91	6
player 16	19	12	17	8.3	6
player 17	19	9	19	8.42	4
player 18	16	9	15	8.42	6
player 19	22	9	15	8.19	6
player 20	18	11	18	8.95	6

Each of these tests shows the personal qualities of the players. We process the data, in order to get a numerical measurement for the success rate of each player in each test (numbers), which will be expressing the level of possession of the qualities, associated with the test. The results from test 1, 2, 3 and 5 are divided by the maximum amount of points. For test number 4, we insert a maximum of 200 points, granted if the player has finished in 7 seconds. For each hundredth above 7 seconds, a point is taken. If the time is equal or more than 8 seconds, the player receives 0 points. The processed data is noted in Table 2.

TABLE 2.

player	test 1	test 2	test 3	test 4	test 5
player 1	0.7	0.867	0.7	0.55	0.8
player 2	0.7	0.867	0.8	0.555	1
player 3	0.7	1	0.677	0.55	1
player 4	0.8	0.8	0.833	0.675	1
player 5	0.85	0.867	0.933	0.625	0.8
player 6	0.65	1	0.833	0.44	0.8
player 7	0.65	0.8	0.833	0.58	1
player 8	0.85	1	0.667	0.7	0.8
player 9	0.675	1	0.733	0.525	1
player 10	0.725	1	0.9	0.635	1
player 11	0.575	0.6	0.567	0.4	0.6
player 12	0.45	0.733	0.633	0.29	0.6
player 13	0.45	0.677	0.633	0.1	0.6
player 14	0.45	0.733	0.6	0.675	0.4
player 15	0.6	0.667	0.633	0.4	0.6
player 16	0.475	0.8	0.567	0.35	0.6
player 17	0.475	0.6	0.633	0.29	0.4
player 18	0.4	0.6	0.5	0.285	0.6
player 19	0.55	0.6	0.5	0.405	0.6
player 20	0.45	0.733	0.6	0.1	0.6

In baseball defense of the entire team is on the field. That is why, in the selection of an optimal team (see Figure 1), we would be interested in positioning of the players over all nine positions [3]:

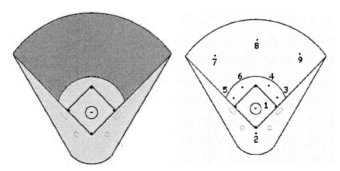

FIGURE 1.

454

- Position 1 — Pitcher
- Position 2 — Catcher
- Position 3 — First baseman
- Position 4 — Second baseman
- Position 5 — Third baseman
- Position 6 — Shortstop
- Position 7 — Left outfielder
- Position 8 — Central outfielder
- Position 9 — Right outfielder

Unlike most collective sports, in baseball players play both in defense and attack. That is why we will insert a 10th position – batter – which we will use in summing of the general effectiveness of the players. The expert has given Table 3.

TABLE 3.

	test 1	test 2	test 3	test 4	test 5
w1	0	0	0.1	0	0.9
w2	0	0.7	0.2	0.1	0
w3	0	0.3	0.3	0.3	0.1
w4	0	0.3	0.3	0.3	0.1
w5	0	0.3	0.3	0.3	0.1
w6	0	0.25	0.25	0.25	0.25
w7	0	0.25	0.25	0.25	0.25
w8	0	0.25	0.25	0.25	0.25
w9	0	0.25	0.25	0.25	0.25
w10	1	0	0	0	0

Efficiencies of each player on each position are calculated and are shown in Table 4. A possible but trivial solution of the initial problem can be obtained as follows: The general efficiency:

$$F_i = \sum_{j=1}^{10} y_j Q_i^j, \quad i = 1,...,20 \qquad u = (0.1\ 0.1\ 0.1\ 0.1\ 0.1\ 0.1\ 0.1\ 0.1\ 0.1\ 0.1),$$

is calculated for every player, then 9 (12) players whose numbers are biggest are selected and they form the team.

Their positions are determined by solving the linear assignment problem and its solution is:

$K=9$:

Player 10 on position 4,
Player 4 on position 6,
Player 5 on position 9,
Player 9 on position 8,
Player 8 on position 5,

Player 3 on position 2,
Player 2 on position 1,
Player 7 on position 3,
Player 6 on position 7.
The general efficiency is *7.4578*.

TABLE 4.

	position 1	position 2	position 3	position 4	position 5	position 6	position 7	position 8	position 9	batter
player 1	0.79	0.8017	0.715	0.715	0.715	0.7292	0.7292	0.7292	0.7292	0.7
player 2	0.98	0.8222	0.7665	0.7665	0.7665	0.8054	0.8054	0.8054	0.8054	0.7
player 3	0.9667	0.8883	0.765	0.765	0.765	0.8042	0.8042	0.8042	0.8042	0.7
player 4	0.9833	0.7942	0.7925	0.7925	0.7925	0.8271	0.8271	0.8271	0.8271	0.8
player 5	0.8133	0.8558	0.8075	0.8075	0.8075	0.8062	0.8062	0.8062	0.8062	0.85
player 6	0.8033	0.9107	0.762	0.762	0.762	0.7683	0.7683	0.7683	0.7683	0.65
player 7	0.9833	0.7847	0.764	0.764	0.764	0.8033	0.8033	0.8033	0.8033	0.65
player 8	0.9833	0.9033	0.79	0.79	0.79	0.7917	0.7917	0.7917	0.7917	0.85
player 9	0.9833	0.8992	0.7775	0.7775	0.7775	0.8146	0.8146	0.8146	0.8146	0.675
player 10	0.9833	0.9435	0.8605	0.8605	0.8605	0.8838	0.8838	0.8838	0.8838	0.725
player 11	0.9833	0.5733	0.53	0.53	0.53	0.5417	0.5417	0.5417	0.5417	0.575
player 12	0.9833	0.669	0.557	0.557	0.557	0.5442	0.5442	0.5442	0.5442	0.45
player 13	0.9833	0.6033	0.48	0.48	0.48	0.5	0.5	0.5	0.5	0.45
player 14	0.9833	0.6708	0.5525	0.5525	0.5525	0.5271	0.5271	0.5271	0.5271	0.45
player 15	0.9833	0.5978	0.4635	0.4635	0.4635	0.4862	0.4862	0.4862	0.4862	0.6
player 16	0.9833	0.7083	0.575	0.575	0.575	0.5792	0.5792	0.5792	0.5792	0.475
player 17	0.9833	0.5757	0.497	0.497	0.497	0.4808	0.4808	0.4808	0.4808	0.475
player 18	0.9833	0.5485	0.4755	0.4755	0.4755	0.4962	0.4962	0.4962	0.4962	0.4
player 19	0.9833	0.5605	0.5115	0.5115	0.5115	0.5263	0.5263	0.5263	0.5263	0.55
player 20	0.9833	0.6358	0.4675	0.4675	0.4675	0.4896	0.4896	0.4896	0.4896	0.45

When we mean 3 duplicate players the solution is:
K=12, with 3 duplicate players:

Player 10 on position 3,
Player 4 on position 4,
Player 5 on position 5,
Player 9 on position 8,
Player 3 on position 9,
Player 2 on position 7,

Player 7 on position 6,
Player 6 on position 2,
Player 12 on position 1,
Players 1, 8, 16 are duplicating ones.
The general efficiency is upper: *8.432*.

The exact solution is obtained when all 20 players are included in solving the assignment problem described above and it is:

Player 2 on position 7,
Player 3 on position 6,
Player 4 on position 3,
Player 5 on position 5,
Player 6 on position 2,
Player 7 on position 8,

Player 9 on position 9,
Player 10 on position 4,
Player 19 on position 1.
All other players are duplicating ones.
The general efficiency is again *8.432*.

The numerical solutions were obtained by LINGO version 10.

CONCLUSION

Players with lower general effectiveness may replace players with higher ones which leads to the improvement of the defensive team. With $k = 10$, contrary to the expectations, player 6 does not take up a replacement position, but takes the players 4, who has better general effectiveness. The reason for this is the usage of higher coefficient for the position of batter. In this way a good defending player may receive a lower general effectiveness and may not be placed within the first 9 players for the optimal defending team. But since the effectiveness for batters are not used when solving assignment problem, such a player has higher effectiveness on the remaining position and he may take the place of someone else. Thus improving the team's performance in defense, but reducing it in attack - batting the decision - taking person should judge whether he wants to defense in favor of attack.

REFERENCES

1. I.M. Stancu-Minasian, *Stochastic Programming with Multiple Objective Functions*, D.Reidel Publishing Company, 1984.
2. R.E.Burkard, E.Cela, *Linear Assignment Problems and Extensions, Handbook of Combinatorial Optimization*, Vol. 4, Kluwer Publishers, 1999.
3. J. McFarland, *Coaching Pitchers*, Leisure Campaign, Illinois, 1994.

Statistical Characteristics of Single Sort of Grape Bulgarian Wines

D. Boyadzhiev

Dept. of Applied Mathematics and Modeling, University of Plovdiv, Bulgaria

Abstract. The aim of this paper is to evaluate the differences in the values of the 8 basic physicochemical indices of single sort of grape Bulgarian wines (white and red ones), obligatory for the standardization of ready production in the winery. Statistically significant differences in the values of various sorts and vintages are established and possibilities for identifying the sort and the vintage on the base of these indices by applying discriminant analysis are discussed.

Keywords: ANOVA, post hoc tests, discriminant analysis.
PACS: 87.16.dt

INTRODUCTION

The purpose of this work is to study the influence of the sort and vintage factors on the basic physicochemical indices of the final product of Bulgarian winery. Eight of the values of these indices are given in the standardization certificate of ready production which obligatory accompanies the separate lots. It follows that their values can be used without doing additional measurements. If such statistically proved differences exist then we study the possibility to model the vintage and sort of single sort of grape Bulgarian wines when using the values of these indices. A model for the geographical origin of Bulgarian wines with the same indices is done in [2] and [3]. The same or similar technology for determining the origin of wines from three European countries, Australia and South Africa with a chosen subset of over 100 analytical parameters is proposed in [4] and [5]. By the help of data from atomic spectroscopy and the same technology in [6] it is proved the origin of white Spanish wines from a certain area. Analogous model for studying the authenticity of the origin of potatoes is given in [7] and a similar model for investigating the authenticity of raw material when manufacturing vegetal oils is described in [8].

MATERIALS AND METHODS

The database includes values of eight physicochemical indices for 475 and 444 lots of white and red wines respectively distributed by sort and vintage as shown on Table 1. These wines have been presented at five (1996-2000) consecutive annual wine-tasting competitions "Vinaria" in Bulgaria. The analysis of wine samples has been performed in the respective manufacturing research and control laboratories. Considering the fact that excellent wines are subject to compulsory additional control

CP1067, *Applications of Mathematics in Engineering and Economics '34—AMEE '08*, edited by M. D. Todorov
© 2008 American Institute of Physics 978-0-7354-0598-1/08/$23.00

and if any differences are found they are disqualified we can assume the data as reliable.

TABLE 1. Sorts, Vintages and Quantities of White and Red Wines

White Wines	Vint. 95	Vint. 96	Vint. 97	Vint. 98	Vint. 99	Total
Aligote	6	7	10	5	5	33
Dimiat	6	9	6	5	11	37
Misket	9	6	9	10	4	38
Muskat	12	10	9	11	14	56
Traminer	10	15	20	17	15	77
Sauvignon blanc	5	8	21	13	12	59
Chardonnay	16	25	43	41	50	175
Total	64	80	118	102	111	475
Red Wines	Vint. 95	Vint. 96	Vint. 97	Vint. 98	Vint. 99	Total
Gamza	5	3	5	2	4	19
Cabernet Sauvignon	26	29	50	55	70	230
Mavrud	2	2	4	3	6	17
Merlot	15	22	45	44	52	178
Total	48	56	104	104	132	444

In Table 2 all indices' names, dimensions and codes are given. All measurements are performed due to the requirements of the Bulgarian State Standard (BSS).

TABLE 2. Standard physicochemical indices of wine lots

No	Indices	Dimension	Code
1	Alcohol - BSS 6071-88	vol. %	ALCOHOL
2	Sugar – BSS 6410-85	g/dm^3	SUGAR
3	Common Acidity – BSS 6409-86	g/dm^3	ACID
4	Volatile Acids – BSS 6280-86	g/dm^3	VOLATIL
5	Total Extract – BSS 6070-85	g/dm^3	EXT_TOT
6	Sugar Free Extract – BSS 6070-85	g/dm^3	S_FR_E
7	Free SO_2 – BSS 2847-85	mg/dm^3	FREE_SO2
8	Total SO_2 – BSS 2847-85	mg/dm^3	SO2TOTAL

For the statistical processing of ANOVA and Discriminant analysis procedures in the software package "Statistica 6.0" [9] with standard level of confidence $\alpha=0.05$ have been used.

RESULTS AND DISCUSSION

WHITE WINES. Statistical analysis results of single sort of grape white wines are shown below, namely in Tables 3 and 4 respectively analysis results by factor sort and vintage are listed. These tables (as well as Tables 6 and 7) are typical ANOVA tables where in the first three columns sum of the squares (SS Eff), the degrees of freedom (Df Eff) and the variance (MS Eff) between the different groups are given and in the following three columns sum of the squares (SS Err), the degrees of freedom (Df Err) and the variance (MS Err) within the different groups are shown. In the next column (F Calc) the calculated value of the index (as a ratio of the two variances) for the Fisher test is given and in the last one the probability of first order error is calculated. The differences in the degrees of freedom of the last two indices in Table 2, namely

FREE_SO2 and **SO2TOTAL** in the fifth column (Df Err) are due to missing values of these indices for some wine lots.

TABLE 3. Analysis results by factor SORT
The bold typed **p**-value means significant differences between the factor levels

White wines	SS Eff	Df Eff	MS Eff	SS Err	Df Err	MS Err	F Calc	p
ALCOHOL	35.83	6	5.972	145.5	468	0.311	19.209	**0.000**
SUGAR	83.05	6	13.842	1388.9	468	2.968	4.664	**0.000**
ACID	34.95	6	5.824	210.3	468	0.449	12.961	**0.000**
VOLATILE	0.06	6	0.010	3.6	468	0.008	1.251	0.279
EXTRACT	188.58	6	31.430	3036.9	468	6.489	4.8435	**0.000**
S_FR_E	224.21	6	37.369	1565.9	468	3.346	11.168	**0.000**
FREE_SO2	113.05	6	18.841	16255.9	406	40.039	0.471	0.830
SO2TOTAL	14370.12	6	2395.021	242595.6	406	597.526	4.008	**0.001**

TABLE 4. Analysis results by factor VINTAGE
The bold typed **p**-value means significant differences between the factor levels

White wines	SS Eff	Df Eff	MS Eff	SS Err	Df Err	MS Err	F Calc	p
ALCOHOL	16.680	4	4.170	164.7	470	0.350	11.901	**0.000**
SUGAR	34.234	4	8.559	1437.7	470	3.059	2.798	**0.026**
ACID	46.999	4	11.750	198.3	470	0.422	27.854	**0.000**
VOLATILE	0.124	4	0.031	3.5	470	0.007	4.126	**0.003**
EXTRACT	132.794	4	33.198	3092.7	470	6.580	5.045	**0.001**
S_FR_E	187.226	4	46.806	1602.9	470	3.410	13.725	**0.000**
FREE_SO2	325.832	4	81.458	16043.1	408	39.321	2.072	0.084
SO2TOTAL	5296.177	4	1324.044	251669.6	408	616.837	2.147	0.074

From the tables above it is seen that there are six statistically significant different indices by factor "**SORT**" (Table 3) and the same number of indices by factor "**VINTAGE**" (Table 4), where five of them are common for both factors. It is not needed to study the joint influence of both factors since they are significant for at least seven (except **FREE_SO2**) of the indices (see the tables above). This fact is confirmed by the performed analysis where the results from it are omitted.

Detailed investigation of the influence of the factors on the respective indices can be done by the use of post hoc comparison test (for example, "Unequal N HSD test" modification of the Tukey honest significant difference test for unequal numbers in the groups [9] within package „Statistica" can be applied). Similar conclusions can be made when the mean values and their confidence intervals for different levels of the respective factor are compared. Statistically significant differences in the index' values can be expected for these levels of the factor for which the confidence intervals are not overlapped.

Indices' tendency alteration will be explained with the **Common Acidity** (Figure 1) and the **Sugar Free Extract** (Figure 2) of different sorts and vintages. On the left hand side of Figures 1 and 2 the mean values of the discussed indices by vintages for all the white wines are shown. On the right hand side of Figures 1 and 2 differences in the sorts regardless the tendencies characterizing the vintage can be seen.

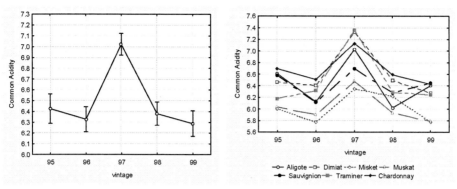

FIGURE 1. Plot of Common Acidity by Vintages: Means and Conf. Intervals 95.00% (left) and Means by Sorts (right)

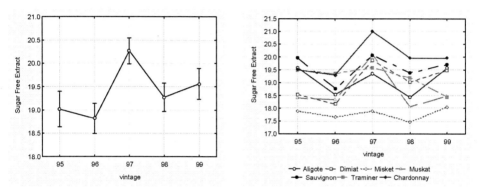

FIGURE 2. Plot of Sugar Free Extract by Vintages: Means and Conf. Intervals 95.00% (left) and Means by Sorts (right)

From Figures 1 and 2 (left panels) it is seen that for indices **Common Acidity** and **Sugar Free Extract** Vintage 97 is significantly different from the other vintages which on the other hand are indistinguishable amongst themselves (except vintages 96 and 99 by **Sugar Free Extract**). Considering the right hand sides of the same figures the following conclusions can be made: during the years *Misket* and *Muskat* sorts have mainly low values of **Common Acidity** which fact makes them distinguishable by this index from the other sorts but not between themselves (Figure 1). Moreover *Misket* and *Chardonnay* sorts differ from the other white wines by traditionally low (for *Misket*) and high (for *Chardonnay*) values of **Sugar Free Extract** for different vintages. Based on these conclusions we can state that due to comparatively large number (5 from 8 in total) of indices for which both factors are statistically significant it is hardly possible to develop a general model for the recognition of any vintage without considering the respective sort and vice versa. On the contrary it is possible to model separate vintages independently of the respective sort and separate sorts independently of the respective vintage.

RED WINES. Statistical analysis results of single sort of grape red wines are given in Tables 5 and 6 namely results by factor sort and vintage are shown. The tables below are organised analogously to the tables for the white wines and thus comments about their structure are omitted.

TABLE 5. Analysis results by factor SORT
The bold typed **p**-value means significant differences between the factor levels

Red wines	SS Eff	Df Eff	MS Eff	SS Err	Df Err	MS Err	F Calc	p
ALCOHOL	5.360	3	1.787	94.6	440	0.2150	8.311	**0.000**
SUGAR	14.897	3	4.966	411.0	440	0.9342	5.316	**0.001**
ACID	5.871	3	1.957	131.3	440	0.2984	6.558	**0.000**
VOLATILE	0.016	3	0.005	3.7	440	0.0083	0.630	0.596
EXTRACT	70.431	3	23.477	1951.1	440	4.4343	5.294	**0.001**
S_FR_E	28.007	3	9.336	1580.8	440	3.5927	2.599	0.052
FREE_SO2	4.823	3	1.608	10680.1	334	31.9763	0.050	0.985
SO2TOTAL	2478.192	3	826.064	201196.0	334	602.3833	1.371	0.251

TABLE 6. Analysis results by factor VINTAGE
The bold typed **p**-value means significant differences between the factor levels

Red wines	SS Eff	Df Eff	MS Eff	SS Err	Df Err	MS Err	F Calc	p
ALCOHOL	2.687	4	0.672	97.3	439	0.222	3.032	**0.017**
SUGAR	10.014	4	2.504	415.9	439	0.947	2.643	**0.033**
ACID	10.722	4	2.681	126.5	439	0.288	9.306	**0.000**
VOLATILE	0.289	4	0.072	3.4	439	0.008	9.356	**0.000**
EXTRACT	148.406	4	37.101	1873.1	439	4.267	8.696	**0.000**
S_FR_E	89.268	4	22.317	1519.5	439	3.461	6.447	**0.000**
FREE_SO2	312.599	4	78.150	10372.3	333	31.148	2.509	**0.043**
SO2TOTAL	7836.339	4	1959.085	195837.9	333	588.102	3.332	**0.011**

Analysing the results listed in Tables 5 and 6 we see that there are four statistically significant different indices by factor "**sort**" and eight by factor "**vintage**". Again the combination of both factors (without showing the exact values) influence significantly over seven of the indices except **FREE_SO2** whose values of **p**-level are highest for either of the factors. Here, similarly to the white wines a detailed investigation for the influence of both factors (separately and together) is not done but the described methodology is applicable for the red wines too. We can again conclude that it is hardly possible to develop general models for the recognition of any vintage without considering the respective sort and vice versa but it is possible to model separate vintages independently of the respective sort and separate sorts independently of the respective vintage. With other words effective mathematical models for description of single sort of grape red wines can be developed as well.

The graphs of the different indices by vintages are not given since for the white and for the red wines separate sorts follow the year dependences. Indices' tendency alteration by sort and vintage is depicted with the parameters **Common Acidity** and **Volatile Acids** (Figure 3), and **Total Extract** and **Sugar Free Extract** (Figure 4).

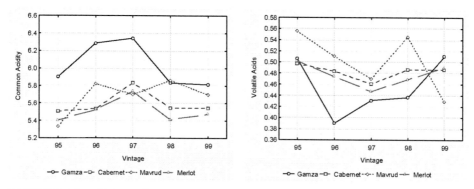

FIGURE 3. Plot of Means of Common Acidity (left) and Volatile Acids (right) by Vintages and by Sorts

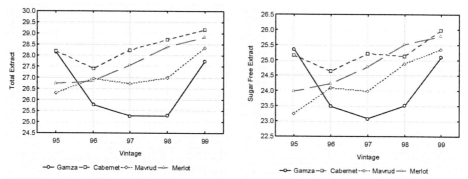

FIGURE 4. Plot of Means of Total Extract (left) and Sugar Free Extract (right) by Vintages and by Sorts

From the figures above one can immediately see the similarity between *Gamza* and *Mavrud* sorts and between *Cabernet* and *Merlot* respectively. Moreover the indices' values of the last two sorts of wines are comparatively close. This means that when developing red wine models we can expect efficient models describing the first two sorts but it is obligatory additional indices to be considered in order to better distinct the *Cabernet* and *Merlot* sorts.

Discriminant Analysis for Sorts and Vintages of Wines

In the following two examples we shall tray to show the possibilities of modelling Bulgarian wines by sort and vintage by the help of discriminant analysis. In both cases we shall use a classical linear model. At the beginning we develop a model by the help of **learning set** of wines and we test it with another group of wines namely the so called **training set** of wines which group has not been used for its development. As it was already mentioned it is impossible to create "good" models when using only data for the eight standard indices. That is why in the following examples as a learning set we shall use only high quality wines from a certain geographical region in Bulgaria called **authentic** for which it is sure that possible deviations from sort and vintage are minimal. As training set we shall utilize ordinary wines from the same region in Bulgaria as the authentic ones if they are typical for it (as it is done in the example for the white wines), or authentic wines from other regions (see the example for the red wines).

For the first example an attempt for modelling the *Aligote, Chardonnay* and *Sauvignon* sorts of white wines is done where as learning sets authentic wines from **Danube** region are used and as test sets – commercial wines of the same sorts and region. Classification results are given in Table 7.

TABLE 7. Posteriori Classification by Factor SORT for the White Wines

Sorts	Learning set classification				Training set classification			
	% correct	AL	CH	SO	% correct	AL	CH	SO
Aligote (AL)	100.00	2	0	0	57.14	4	2	1
Chardonnay (CH)	90.48	0	19	2	60.61	6	20	7
Sauvignon (SO)	90.91	0	1	10	66.67	0	4	8
Total	**91.18**	**2**	**20**	**12**	**61.54**	**10**	**26**	**16**

For the second example an attempt for modelling the 95, 98 and 99 vintages of authentic red wines from the two most widely spread sorts *Cabernet* and *Merlot* is done. As learning sets wines from **Thrace** region are used and as test sets – wines of the same quality and sorts from **Danube** region are utilized. The same linear model is applied. Classification results are given in Table 8.

TABLE 8. Posteriori Classification by Factor VINTAGE for the Red Wines

Sorts	Learning set classification				Training set classification			
	% correct	95	98	99	% correct	95	98	99
Vintage 1995	58.33	7	3	2	30.43	7	8	8
Vintage 1998	77.78	2	7	0	70.00	0	7	3
Vintage 1999	88.89	1	0	8	87.50	0	1	7
Total	**73.33**	**10**	**10**	**10**	**51.22**	**10**	**16**	**18**

After obtaining the respective discriminant model it is possible to visualise the results when performing the so called canonical analysis. Our aim is to project the initial data on a sub-space in order to achieve maximum separation among the given classes. The magnitude of this sub-space is with 1 less than the number of groups in the model (or of the number of variables if it is smaller than the number of the groups). The projection coordinates have no particular sense but by there help we can estimate the distance between the starting values [1].

In this case we plot the projections of separate wine lots as points of 8-dimensional space on 2-dimensional space. The discriminant regions of the learning set for the above two examples after performing canonical analysis are given on the Figure 5. The straight lines show the discriminant borders of separate wine groups.

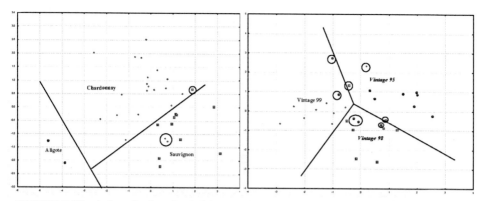

FIGURE 5. Discriminant Regions for Training Sets by **Sorts** of White Wines (left) and by **Vintages** of Red Wines (right). The surrounded lots are misclassified

Squared Mahalanobis distances between centroids of different groups and p-values for the significance of these distances for the white and red wines respectively from the examples above are given in Table 9.

TABLE 9. Squared Mahalanobis Distances and p-values between different groups
The bold typed **p**-value means significant differences between the groups

Examples	White wines - sorts			Red wines - vintages		
Groups	*AL-CH*	*AL-SO*	*CH-SO*	*95-98*	*95-99*	*98-99*
Distance	22.87	34.96	4.65	2.57	5.30	4.72
p-value	**0.0098**	**0.0012**	**0.0014**	0.0584	**0.0029**	**0.0103**

From this table it is seen that the white wines model is very efficient, the probability of error when separating in groups the learning set is less than 1%. Worse are the results for the red wines. The error probability between the groups from vintage 95 and vintage 98 is nearly 6% and the one for the groups from vintage 98 and vintage 99 is bigger than 1%.

CONCLUSIONS

The application of discriminant analysis allows comparatively good (75-90%) differentiation of certain sorts and vintages of authentic wines and a satisfactory one (up to 50%) of commercial wines or wines from other regions when only obligatory standardization indices accompanying the lots are used. Including additional, objective, independent of the technological process indices will increase the reliability of the models, as it is done for the distinction of geographical regions in [2], [4] and [5].

ACKNOWLEDGEMENTS

The presented work was partly supported by the Project PU-NPD, IS-M-4/2008.

REFERENCES

1. B. G. M. Vandeginste and S. C. Rutan (Eds), *Handbook of Chemometrics and Qualimetrics: Part B*, Data Handling in science and Technology, Vol. 20B, Elsevier, 1997.
2. S. Kemilev, P. Georgieva, D. Boyadzhiev, and S. Dimov, "Caractéristiques et origine de vins bulgares de région determine," *Annals. UFT - Plovdiv*, Vol. LII, 2005, pp. 238-243.
3. D. Boyadzhiev, S. Kemilev, and P. Georgieva, "Methodology for Evaluating the Origin of Bulgarian Wines II. Modelling the Origin of Young Wines," *Annals. USB branch Plovdiv*, Vol. B, 2005, pp. 118-125. (in Bulgarian)
4. U. Roemisch, D. Vandev, and K. Zur, *Austrian Journal of Statistics* **35**(1), 45-55 (2006).
5. U. Roemisch, H. Jaeger, and D. Vandev, *Pliska Stud. Math. Bulgar.* **18**, 327-339 (2007).
6. M. Latorre, C. Garcia-Jares, B. Medina, and C. Herrero, *J. Argic. Food Chem.* **42**, 1451-1455 (1994).
7. P. M. Padin, R. M. Pena, S. Garcia, R. Iglesias, S. Barro, and C. Herrero, *Analyst* **126**, 97-103 (2001).
8. F. Carsten, F. Reniero, and C. Guillou, *Magnetic Resonance in Chemistry* **38**, 436-443 (2000).
9. Statistics: Methods and Applications: *A Comprehensive Reference for Science, Industry and Data Mining (Electronic Version)*, StatSoft Inc., Tulsa, 2006, *Electronic Statistics Textbook*, StatSoft. WEB: http://www.statsoft.com/textbook/stathome.html

Realization of Ridge Regression in MATLAB

S. Dimitrov, S. Kovacheva and K. Prodanova

Technical University of Sofia, 1000 Sofia, Bulgaria

Abstract. The least square estimator (LSE) of the coefficients in the classical linear regression models is unbiased. In the case of multicollinearity of the vectors of design matrix, LSE has very big variance, i.e., the estimator is unstable. A more stable estimator (but biased) can be constructed using ridge-estimator (RE). In this paper the basic methods of obtaining of Ridge-estimators and numerical procedures of its realization in MATLAB are considered. An application to Pharmacokinetics problem is considered.

Keywords: Unbiased estimators, ridge regression.
PACS: 02.50.Ey

INTRODUCTION

Let us consider the linear regression model

$$y = X\alpha + \varepsilon \tag{1}$$

where the response has size $n \times 1$, the design matrix has size $n \times m$, the error ε has size $n \times 1$ and α is vector of unknown parameters with size $m \times 1$. The following assumptions are satisfied [1]:

- There are no restrictions for the vector α;
- $E(\varepsilon_k) = 0, \quad k = 1,...,n; \quad E(\varepsilon) = 0$;
- $\text{cov}(\varepsilon) = \sigma^2 I_n$;
- $\text{rank}(X) = m$.

It is known that the estimator of α constructed using the least square method (LSM) is

$$\alpha = (X'X)^{-1}Xy'. \tag{2}$$

The estimator $\tilde{a}$ is unbiased, i.e., $E\hat{a} = \alpha$. Let us denote the ith vector-column of the matrix by x_i.

Definition 1 *The vectors* $x_1,...,x_m$ *are* **multicollinear** *if there exist real numbers* $v_1,...,v_m$ *with* $\sum_{i=1}^{m} v_i^2 \neq 0$, *which satisfy* $v_1 x_1 + ... + v_m x_m \approx 0$, *i.e., the vectors are "almost linear dependent".*

The multicollinearity grows with the growth of the exactness of the final equality. The LSM estimator has large variance in the case of multicollinearity and at the growth of multicollinearity it becomes unstable. In the class of biased estimators there exist more stable estimators. One of them are so-called ridge-estimators, introduced for first time

CP1067, *Applications of Mathematics in Engineering and Economics '34—AMEE '08*, edited by M. D. Todorov
© 2008 American Institute of Physics 978-0-7354-0598-1/08/$23.00

from A. Hoerl [2]. The ridge-estimator of the vector α from the linear regression model is

$$\alpha(K) = (X'X + K)^{-1}Xy'$$ (3)

where K is a nonnegative-definite matrix with size $n \times m$. The addition to the matrix $X'X$ of nonnegative-definite matrix makes it better definite, and the estimator – more stable.

ABOUT RIDGE-ESTIMATORS

Normally, the matrix K is diagonal and its diagonal elements are proportional of the diagonal elements of the matrix $X'X$, i.e.,

$$K_{ii} = k(X'X)_{ii}, \qquad K_{ij} = 0 \quad \text{for} \quad i \neq j$$ (4)

where $k \geq 0$. The most simple expression for the matrix K one can obtain by $K = kI_m$, where k is a nonnegative constant and I_m is an identity matrix.

One can prove that case (4) reduces to the ridge-estimator (3).

Let us denote

$$D = \begin{pmatrix} (X'X)_{11} & & 0 \\ & \ldots & \\ 0 & & (X'X)_{mm} \end{pmatrix}$$

$$Z = XD^{-\frac{1}{2}}, \qquad \beta = D^{\frac{1}{2}}\alpha.$$ (5)

Then the model (1) reduces to the following model:

$$y = Z\beta + \varepsilon$$ (6)

The matrix $Z'Z = D^{-\frac{1}{2}}XX'D^{-\frac{1}{2}}$ is the correlation matrix of the independent variables $x_1, \ldots, x_m$. A ridge-estimator (3) with matrix $K = kI_m$ for model (6) is

$$b(k) = (Z'Z + kI)^{-1}Z'y = (D^{-\frac{1}{2}}XX'D^{-\frac{1}{2}} + kI)^{-1}D^{-\frac{1}{2}}X'y = D^{-\frac{1}{2}}(XX' + K)X'y$$

where $K_{ii} = K(X'X)_{ii}$. Therefore

$$a(k) = D^{-\frac{1}{2}}b(k)$$ (7)

is ridge-estimator (3) of initial model (1) with choice of by the rule (4). Thus instead of (1) we can consider the reduced regression model (6) at whom the ridge-estimator has simpler structure

$$b(k) = (Z'Z + kI)^{-1}Z'y.$$ (8)

Here are some properties of this estimator:

1) $b(0)$ is LSM estimator;
2) $b(k)$ is linear transformation of the LSM estimator:

$$b(k) = Bb, \qquad B = [I + k(Z'Z)^{-1}]^{-1},$$

where b is LSM estimator of (6). Since $Eb = \beta$, and $B \neq I_m$, then the ridge-estimator is biased;

3) In the class of the estimators with fixed length the ridge-estimator (8) minimizes the error sum of squares (ESS).

4) In the class of the estimators with given ESS the ridge-estimator has minimal length.

5) The length of the ridge-estimator (8) is a decreasing function of k.

In the case of multicollinearity the mean of the error sum of squares of LSM estimator is large. Let us verify whether that is true for the estimator (8). By definition

$$L(\beta) = E[b(k) - \beta)'(b(k) - \beta)].$$

Let P is an orthogonal matrix, whose columns are formed by eigenvectors of the matrix ZZ'. Then $P'Z'ZP = \Lambda$, $\Lambda_{ii} = \lambda_i > 0$. We denote

$$\gamma = P'\beta, \qquad W = ZP, \qquad y = W\gamma + \varepsilon. \tag{9}$$

After computations based on linear algebra one obtains:

$$L(k) = k^2 \sum_{i=1}^{m} \frac{\gamma_i^2}{(\lambda_i + k)^2} + \sigma^2 \sum_{i=1}^{m} \frac{\lambda_i}{(\lambda_i + k)^2} = \sum_{i=1}^{m} \frac{\gamma_i^2 + \sigma^2 \lambda_i}{(\lambda_i + k)^2}.$$

It is clear that $L(0) = \sigma^2 \sum_{i=1}^{m} \frac{1}{\lambda_i} = \sigma^2 \mathrm{tr}(Z'Z)^{-1}$ is the mean of the error sum of squares of LSM estimator. For approximate computation of k^* the Newton-Raphson method is used. Approximately $L(k)$ in the neighborhood of the point $k = 0$:

$$L(k) \approx L(0) + kL'(0) + \frac{1}{2}k^2 L''(0).$$

The value of k at which $L(k)$ reaches the minimum we obtain from the equation

$$L'(k) \approx L'(0) + kL''(0) = 0.$$

It follows

$$k^* = -\frac{L'(0)}{L''(0)} = \sigma^2 \frac{\sum\limits_{i=1}^{m} \frac{1}{\lambda_i^2}}{\sum\limits_{i=1}^{m} (3\sigma^2 + \gamma_i^2 \lambda_i)/\lambda_i^3}. \tag{10}$$

PROCEDURE FOR OBTAINING OF OPTIMAL VALUE OF k

For finding k^* the parameters $\sigma^2, \gamma_1, ..., \gamma_2$ must be known. So, the immediately application of (10) is impossible. Let us apply the following iterative procedure:

1) From the initial equation by LSM we obtain the estimator a. With the help of (5) and (9) we reduce it in the estimator g of the parameter γ.

2) We shall find the value of k^* by formula (10), after replacing σ^2 and γ with their estimators s^2 and g_i, where

$$g = (W'W)^{-1}W'Y, \qquad s^2 = \frac{(Y - X\hat{a})(Y - X\hat{a})}{n - m}.$$

3) Using k^* we find ridge-estimator $b(k^*)$ by formula (8) and return to second step. The computations continue while the results of two successive iterations coincide. The ridge-estimator with matrix K from the type $K = kI_m$ is not very different from the least-squares estimator.

Let us see that it is not true in the more general case when

$$K = \begin{pmatrix} k_1 & & 0 \\ & \ddots & \\ 0 & & k_m \end{pmatrix}, \qquad k_i \geq 0.$$

The ridge-estimator for the orthogonalized model (9) is

$$g(K) = (W'W + K)^{-1}W'y = (\Lambda + K)^{-1}Wy.$$

Analaogous to the estimator (8) one can find the mean of the error sum of squares

$$L(k_1, \ldots, k_m) = \sum_{i=1}^{m} \frac{\sigma^2 \lambda_i + \gamma_i^2 k_i^2}{(\lambda_i + k)^2}.$$

From the necessary condition for minimum of $L(k_1, \ldots, k_m)$ it follows

$$k_i = \sigma^2 / \gamma_i^2 \qquad i = 1, \ldots, m. \tag{11}$$

Based on of (11) Hoerl and Kennard [3] proposed the following iterative procedure for estimation: Use LSM we create the estimator s^2, than find the value of k_i by formula (11), afterwards the ridge-estimator, the next value of k_i, etc.

The conditions for convergence of this process and the analytic limit estimator were found from Hemmerle [4]: Let us denote:

$$e_i = \frac{s^2}{\lambda_i g_i^2}, \qquad i = 1, \ldots, m,$$

where $g = (g_1, \ldots, g_m)'$ is least-squares estimator of orthogonalized model (9).

Theorem 1 (Hemmerle) *If $e_i \leq 1/4$ then the sequence $k_i^1, k_i^2, \ldots$ has a limit, which is equal to*

$$k_i^* = \frac{1 - 2e_i - \sqrt{1 - 4e_i}}{2e_i}, \qquad i = 1, \ldots, m. \tag{12}$$

If $e_i > 1/4$, then $k_i^r \to \infty$, $r \to \infty$ $(k_i^* = \infty)$.

On the basis of this theorem and taking into account (10) one can find analytically the limit estimator:

$$g_i^* = \begin{cases} g_i(1 + k_i^*), & e_i \leq 1/4 \\ 0, & e_i > 1/4 \end{cases}. \tag{13}$$

APPLICATION TO PHARMACOKINETICS

Let us consider one concrete example. We are interested in the result of the interaction of two drugs M_1 and M_2. The outcome of the interaction is the metabolite M_3.

The reaction between M_1 and M_2 occurs in the presence of two inhibitors (catalysts) T and C. Let the quantity of M_2 is the same in all experiments and the quantity of M_1 will change. We carry out 15 experiments as the quantity of M_1 varies in each of them. Each experiment occurs at some quantities of T and C. Let us introduce the notations:

y_k – outcome of the interaction (y_k – the quantity of M_3 in the kth experiment);
x_{k1} – the quantity of M_1 in the kth experiment;
x_{k2} – the quantity of T in the kth experiment;
x_{k3} – the quantity of C in the kth experiment.

We suppose that the model is linear, i.e., $y_k = \alpha_1 x_{k1} + \alpha_2 x_{k2} + \alpha_3 x_{k3} + \alpha_4 x_{k4} + \varepsilon_k$, $k = 1, ..., 15$ where each $x_{k4} = 1$ for any k.

The experimental data are presented in the vector Y and the matrix :

$$
Y = \begin{bmatrix} 140.28 \\ 142.02 \\ 149.90 \\ 147.12 \\ 163.62 \\ 173.40 \\ 178.86 \\ 186.26 \\ 183.53 \\ 198.76 \\ 205.30 \\ 206.77 \\ 198.76 \\ 216.48 \\ 221.45 \end{bmatrix}, \quad
X = \begin{bmatrix}
252.36 & 96.67 & 8.37 & 1 \\
262.54 & 100.07 & 9.07 & 1 \\
285.70 & 96.78 & 9.35 & 1 \\
277.52 & 101.30 & 9.67 & 1 \\
307.95 & 100.35 & 9.45 & 1 \\
322.44 & 104.8 & 10.12 & 1 \\
334.88 & 106.17 & 10.35 & 1 \\
350.11 & 109.2 & 11.03 & 1 \\
346.10 & 104.48 & 10.38 & 1 \\
374.91 & 106.88 & 12.15 & 1 \\
378.49 & 113.14 & 12.98 & 1 \\
397.48 & 112.38 & 11.34 & 1 \\
378.39 & 109.07 & 10.95 & 1 \\
393.44 & 114.45 & 12.89 & 1 \\
403.84 & 115.23 & 13.71 & 1
\end{bmatrix}.
$$

Using LSM and package STAISTICA 6 we obtain the estimator:

$$\hat{a} = (X'X)^{-1}X'y = [0.398, 0.4534, 2.942, -33.4237]'.$$

The regression equation is

$$\hat{Y} = 0.398x_1 + 0.4534x_2 + 2.9412x_3 - 33.4237x_4.$$

On the Figure 1 the graphics of Y and $\hat{Y}$ are presented.

For the estimators g and s^2 of the parameters γ and σ^2 we calculate the following:

$$g = [-274.8539, 112.9103, -360.8819, 353.8421]', \qquad s^2 = 6.9946.$$

The module "Ridge-regression" in the package STATISTICA 6 required to insert the optimal value k^*. Because there is not possibility to make iteration procedure of

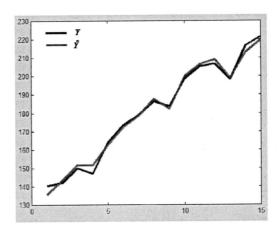

FIGURE 1. The graphs of Y and $\hat{Y}$

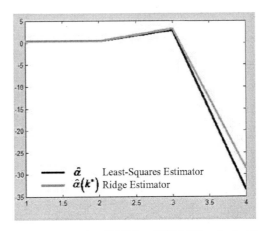

FIGURE 2. The graphs of LSE $\hat{\alpha}$ and RE $\hat{\alpha}(k^*)$ at the first iteration

finding k^* we realized this procedure in MATLAB. The application of formula (9) to the experimental data reduces to convergence process in the third iteration. The value of k^* in the first iteration is equal to $k^* = 0.000081$. Using this value instead k in formula (7) we find the estimator

$$\hat{\beta} = [523.1457, 158.4482, 136.344, -110.5912]'.$$

The mean of the error sum of squares (MESS) of the vector $\hat{\beta}$ is

$$L(k^* = 0.000081) = 23527 < 41942 = L(0).$$

Now using (6) we calculate the ridge-estimator

$$\hat{a}(0.000081) = (0.3958, 0.3851, 3.2315, -28.5545)'.$$

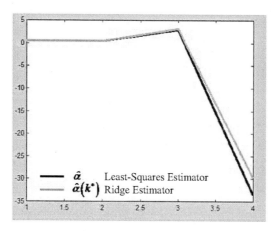

FIGURE 3. The graphs of LSE $\hat{\alpha}$ and RE $\hat{\alpha}(k^*)$ after the third iteration

TABLE 1. Limit RE g^*

Estimator	α_1	α_2	α_3	α_4
$g(0)$	-274.8532	112.9103	-360.8819	353.8421
e	0.0491	2.7432	0.0038	0.000014023
k^*	0.0546	∞	0.0039	0.000014023
g^*	-260.6231	0	-359.4799	353.8372
$\hat{\alpha}(k^*)$	0.4271	0.0002	3.7376	-3.7899

On the Figure 2 the least-squares estimator and ridge estimator are presented.

The value of k^* in the third iteration is equal to $k^* = 0.000056$. Using this value we find the estimator

$$\hat{\beta} = (524.4069, 165.1208, 132.9806, -115.1734)'.$$

The mean of the error sum of squares (MESS) of the vector $\hat{\beta}$ is

$$L(k^* = 0.000056) = 27051 < 41942 = L(0).$$

So, the ridge-estimator decrease MESS. Now using (6) we calculate the ridge-estimator

$$\hat{a}(0.000056) = (0.3967, 0.4013, 3.1518, -29.7376)'.$$

On the Figure 3 the least-squares estimator and ridge estimator after the third iteration are presented.

The computation shows that the ridge-estimator with matrix K from the type $K = kI_m$ is not very different from the least-squares estimator.

We realized in MATLAB also the numerical procedure to find the limit ridge-estimator for the regression example using Hemmerle's theorem. The results are presented in the Table 1.

473

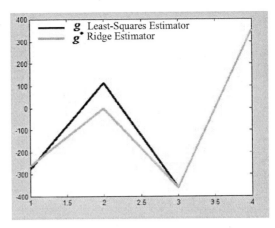

FIGURE 4. The graphs of LSE g and RE g^*

On the Figure 4 the least-squares estimator and ridge estimator using Hemmerle's theorem are presented.

One can see that the ridge-estimator is very different than the least-squares estimator.

ACKNOWLEDGMENTS

This research is supported by the Technical University of Sofia under Grant No. 8071NI-10/2008.

REFERENCES

1. E.Z. Demidenko, *Linear and Nonlinear Regressions*, Finance&Statistics, Moscow, 1981.
2. A.E. Hoerl, *Chemical Engineering Progress* **58**(1), 1962.
3. A.E. Hoerl and R.W. Kennard, *Technometrics* **12**(1), (1970).
4. W.J. Hemmerle, *Technometrics* **17**(3), (1975).

On the Application of the Multidimensional Statistical Techniques for Exploring Copper Bromide Vapor Laser

S. G. Gocheva-Ilieva* and I. P. Iliev†

*Faculty of Mathematics and Informatics, "Paisii Hilendarski" University of Plovdiv,
24 Tsar Assen Str, 4000 Plovdiv, Bulgaria
†Dept. of Physics, Technical University of Plovdiv, 25 Tzanko Djusstabanov Str,
4000 Plovdiv, Bulgaria,

Abstract. A big amount of experimental data for copper bromide vapor lasers with wavelengths 510.6 nm and 578.2 nm, obtained in Georgi Nadjakov Institute of Solid State Physics, Bulgarian Academy of Sciences are examined. Multidimensional statistical methods, such as factor analysis, and cluster analysis were used for classifying the laser parameters. It is established a good accordance between the results from these techniques. Some solutions for designing and planning the experiment in order to enhance the outgoing laser power generation are discussed.

Keywords: Copper bromide laser, multivariate factor analysis, cluster analysis.
PACS: 02.50.Sk, 42.60.Lh, 42.55.Lt

INTRODUCTION

The copper bromide vapor lasers (CBVL) are the more perspective among the metal vapor lasers. They are the most powerful sources in the visible region. This type of laser finds large applications in medicine, atmospheric and submarine location, in the investigation of the pollution of the atmosphere, in technological treatment of different materials etc. [1]. In order to enhance the laser working and output characteristics and study the physical processes different analytical and numerical models as well as computer simulations have been developed [2,3]. All these methods can be classified as structured methods. In recent papers [4-6] we have explored a different approach – a phenomelogical independent study, based on the experimental data. To this end a big amount of data for copper bromide laser with wave lengths 510.6 nm and 578.2 nm, obtained in Metal Vapour Lasers Department Georgi Nadjakov Institute of Solid State Physics, Bulgarian Academy of Sciences - Sofia in the recent decades were examined (see [4-6] and quoted there literature). Different statistical methods, such as multidimensional factor analysis, principal components analysis, multiple regression and others were applied for studying the influence of ten basic input laser parameters on the output laser power and laser efficiency. It was established that the most considerable contribution to increase the output laser power have six of these parameters: the inside diameter of the laser tube, the diameter of the internal rings, the length of the active area and the input electrical power, the input electrical power per unit length and the hydrogen gas pressure.

CP1067, *Applications of Mathematics in Engineering and Economics '34—AMEE '08*, edited by M. D. Todorov
© 2008 American Institute of Physics 978-0-7354-0598-1/08/$23.00

In this paper we extend the upper results, including the hierarchical cluster analysis in order to classify the lasing parameters. The results from the multidimensional factor analysis and cluster analysis are compared together and a good accordance between them is established. The physical interpretation is presented.

The investigation was performed by means of the statistical package SPSS.

DESCRIPTION OF DATA AND SOME PRELIMINARY RESULTS

We consider eleven basic physical parameters which characterize the CuBr laser. The initial variable parameters are as follows: D - the inside diameter of the laser tube, dr – the inside diameter of the internal rings in the tube, L – the length of the active area (electrode separation), Pin – the input electrical power, $P_L = 0.75Pin/L$ – the input electrical power per unit length with 25% of losses, Prf – the pulse repetition frequency, Pne - the neon gas pressure, P_{H2} – the hydrogen gas pressure, C – the equivalent capacity of the capacitor bank, Tr – the temperature of the CuBr reservoirs. The main dependant variable we observe in this study is the output laser power $Pout$.

It total, the data from more than 300 experiments from literature have been observed (see [4-6] and quoted there references). Many carefully randomized samples were examined in order to cover the strong requirements of the used statistical techniques. All the results in this paper are based on a 25% sample of all experiments for the above mentioned eleven variables.

An Overview of the Basic Results from the Factor Analysis

In factor analysis the aim is to predetermine the unknown quantity of macro-variables (factors), which group the independent input variables according to their degree of correlation with one another. The resulting factors are usually uncorrelated, which allows them to be used later for analysis, for example, in order to construct regression models and to predict system behavior.

In papers [4,6] our main goal was to apply the factor analysis to examine the initial variables with respect to the output power $Pout$. To this end the correlation matrix of all variables, represented by the selected sample of data was obtained (see Table 1). It must be specified, that all variables are standardized. In Table 1 we can observe that only the first six variables correlate with $Pout$, but in the same time, they correlate to each other. Statistically, it is also important that the corresponding levels of significance are sufficiently small, which is satisfied only for the first six variables. We use the significance level $\alpha = 0.05$. At these restrictions, we neglected the rest of the parameters and consider the parameters D, dr, L, Pin, P_L and P_{H2} as more important. For our data in particular the Kaiser-Meyer-Olkin (KMO) test is equal to 0,665 and Bartlett's test of sphericity has a value of 0,00. This is usually considered enough for the adequacy of multidimensional factor analysis.

Further we have chosen to use three factors and have obtained the corresponding factor model which explains 92.99 % of all data being considered. As a result of the extraction of the factors by the method of principal components and factor rotation by the standard VARIMAX method (see [7]) we got the classification of independent variables, expressed by three factors according to their correlation level. In the first factor the following variables are grouped together: D, dr, L, Pin, P_L and P_{H2}. The second factor includes P_L, and the third - P_{H2} (see Table 2).

TABLE 1. Correlation matrix showing the intercorrelations among all initial variables. The upper half part of the table represents the cross-correlation coefficients and the lower half part shows their corresponding levels of significance

	Variable	D	dr	L	Pin	PL	PH2	Pfr	Pne	C	Tr	Pout
Correlation	D	1.00	.836	.749	.694	-.558	.294	-.089	-.186	.215	.113	.723
	dr	.836	1.00	.916	.862	-.533	.411	-.169	-.211	.190	.221	.915
	L	.749	.916	1.00	.855	-.697	.539	-.197	-.127	.086	.124	.914
	Pin	.694	.862	.855	1.00	-.296	.372	-.170	-.095	.179	.116	.954
	PL	-.558	-.533	-.697	-.296	1.00	-.493	.132	.310	-.065	-.004	-.426
	PH2	.294	.411	.539	.372	-.493	1.00	-.236	-.062	-.188	-.304	.469
	Prf	-.089	-.169	-.197	-.170	.132	-.236	1.00	.262	-.059	.087	-.209
	Pne	-.186	-.211	-.127	-.095	.310	-.062	.262	1.00	-.307	.041	-.124
	C	.215	.190	.086	.179	-.065	-.188	-.059	-.307	1.00	.292	.129
	Tr	.113	.221	.124	.116	-.004	-.304	.087	.041	.292	1.00	.100
	Pout	.723	.915	.914	.954	-.426	.469	-.209	-.124	.129	.100	1.00
Sig. (1-tai led)	D		.000	.000	.000	.000	.005	.225	.056	.033	.169	.000
	dr	.000		.000	.000	.000	.000	.075	.036	.053	.029	.000
	L	.000	.000		.000	.000	.000	.046	.140	.234	.147	.000
	Pin	.000	.000	.000		.005	.001	.074	.209	.063	.163	.000
	PL	.000	.000	.000	.005		.000	.131	.004	.291	.486	.000
	PH2	.005	.000	.000	.001	.000		.022	.299	.054	.004	.000
	Prf	.225	.075	.046	.074	.131	.022		.016	.318	.242	.045
	Pne	.056	.036	.140	.209	.004	.299	.016		.006	.370	.158
	C	.033	.053	.234	.063	.291	.054	.318	.006		.008	.148
	Tr	.169	.029	.147	.163	.486	.004	.242	.370	.008		.211
	Pout	.000	.000	.000	.000	.000	.000	.045	.158	.148	.211	

TABLE 2. Rotated component matrix for Varimax rotation method with Kaiser normalization. Extraction method: principal component analysis

Variable	Component 1	Component 2	Component 3
Pin	.942		
dr	.905		
L	.789		
D	.744		
P_L		-.913	
P_{H2}			.943

Analyzing the Results of Factor Analysis

The first factor (component) has a mixed dimension. Here geometric size of the laser tube and the electrical power applied are predominant. We could consider these values crucial for the behavior of the laser source. What is more it is clarified that in the process of the experiment they should be considered as a group. For example, the increase of laser power output we can get by simultaneously increasing the geometric size of the tube and the supplied electrical power.

The second factor is described by means of the quantity P_L. In reality this variable has no separate physical significance but it expresses a particular tendency towards an increase of variables from the first factor. From Table 2 it is obvious that P_L has a negative significance (-0.913). This means that Pin and L are inversely proportional, i.e. the increase of the length of the active zone should be greater than the increase of the electrical power supplied. The last requirement has its physical justification and is experimentally confirmed in [8]. When electrical power is greater, electrodes overheat and the loss of power in the electrode zone increases. The increase of the distance between electrodes leads to the reduction of temperatures and respectively to a longer service life for the laser tube. The relative participation of electrical power in the active laser volume rises, which leads to an increase of laser generation and efficiency.

The third factor is P_{H2} - hydrogen pressure. The effect of adding hydrogen in order to increase laser power output and efficiency of a CuBr laser has been thoroughly explored both experimentally and theoretically [9] and that is why we will not be discussing it.

The resulting classification of independent variables made by means of factor analysis is partial because the remaining four variables *Prf, Pne, C* and *Tr* are not scrutinized due to their weak correlation. In order to account for their role we will provide a cluster analysis.

CLUSTER ANALYSIS

Unlike factor analysis the procedures of cluster analysis are based on the classification of objects in view of their homogeneousness and proximity. The formation of the groups (clusters) following the given criteria is carried out by bringing together similar objects and clusters themselves remain heterogeneous. The quantitative measure for proximity is considered in regard to previously chosen metric, usually Euclidean distance.

A great number of methods for clustering have been developed [10]. When the number of objects being examined is small, as it is in our case, some of the most suitable are hierarchical agglomeration methods. Results are in the form of tables and dendrograms (tree diagrams) which express the hierarchical structure of the similarity matrix and the rules for the formation of clusters. There is a large number of different strategies for the grouping together of objects into clusters and later on for the grouping of the clusters themselves. In case of clusterization in the form of a "chain", between-groups linkage and the nearest neighbor methods are used. Choosing a metric, a suitable number of clusters and the methods for cluster formation is an important and sometimes difficult step in the carrying out of cluster analysis [10].

Cluster Analysis Results

Initially we will conduct a partial cluster analysis for the first six variables (*D, dr, L, Pin, P_L* and *P_{H2}*), participating in the previous consideration. Our task is to compare the results with the calculations from factor analysis.

The first stage is to construct a matrix containing the results from the comparison of the objects (Table 3). In our case the squared Euclidean distance is used as the indicator for similarity (or difference). It has to be noted that in Table 3 are given only the comparative results for stage one when each object is considered a cluster. Using the between group linkage method, independent variables are grouped into three clusters as indicated in Table 4. The first cluster includes variables D, dr, L and Pin, the second - P_L, and the third $-P_{H2}$. There is complete correspondence with the results from factor analysis (see Table 3).

TABLE 3. Proximity matrix of the first six variables

Variable	D	dr	L	Pin	P_L	P_{H2}
D	0	23.9	36.7	44.7	227.5	103.0
dr	23.9	0	12.2	20.1	223.8	86.0
L	36.7	12.2	0	21.2	247.8	67.3
Pin	44.7	20.1	21.2	0	189.3	91.7
P_L	227.5	223.8	247.8	189.3	0	217.9
P_{H2}	103.0	86.0	67.3	91.7	217.9	0

TABLE 4. Cluster membership in 3 clusters

Variable	3 clusters
D	1
dr	1
L	1
Pin	1
P_L	2
P_{H2}	3

Dendrogram

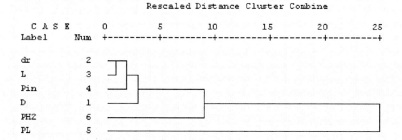

FIGURE 1. Dendrogram of variables from Tables 3 and 4

Table 3 shows that the minimum significance of the coefficient characterizing the homogeneousness of formed clusters (in our case the squared Euclidean distance) is equal to 12.2 and links variables dr and L. Respectively this is the first similarity link observed in Figure 1. The second biggest coefficient is 20.1 and it links dr and Pin.

For this reason in the next stage the variable *Pin* is grouped with the already formed first cluster and so on. Following this procedure we obtain the full structure of Figure 1. The squared distance is indicated along the horizontal axis in a normed scale from 0 to 25, where 25 corresponds to the maximum value in Table 3 - 247.8.

The next stage is classifying all ten variables. Table 5 shows their similarity matrix. Basis Table 6 shows the simultaneous classification of two, three, four and five clusters. In every column opposite each independent variable is given its corresponding cluster number. The optimal number of clusters has to be established. This problem can be solved by means of the dendrogram in Figure 2. It is the result of the same method which was used in Figure 1.

TABLE 5. Proximity matrix of ten input variables

Variable	D	dr	L	Pin	P_L	P_{H2}	Prf	Pne	C	Tr
D	0	19.5	33.0	39.4	202.4	95.9	145.0	158.4	101.5	115.9
dr	19.5	0	9.5	16.9	201.1	79.7	155.3	161.0	105.8	101.6
L	33.0	9.5	0	19.4	223.3	58.9	159.0	149.5	120.2	115.0
Pin	39.4	16.9	19.4	0	169.3	83.1	155.1	145.0	107.6	116.3
P_L	202.4	201.1	223.3	169.3	0	196.0	113.4	88.3	141.2	132.6
P_{H2}	95.9	79.7	58.8	83.1	196.0	0	164.8	140.7	158.2	174.3
Prf	144.7	155.3	159.0	155.1	113.4	164.9	0	97.4	139.8	120.5
Pne	158.4	161.0	149.5	145.0	88.3	140.7	97.4	0	172.6	126.5
C	101.5	105.8	120.2	107.6	141.2	158.2	139.8	172.6	0	93.4
Tr	115.9	101.6	115.0	116.3	132.6	174.3	120.5	126.5	93.4	0

A careful review of the sequence of the clusterization procedure reveals that all ten independent variables form three clusters. The first cluster includes *D*, *dr*, *L*, *Pin* and P_{H2}. The second includes variables P_L, *Pne* and *Prf*, and the third - *C* and *Tr*. This grouping corresponds to the column of three clusters in Table 6. In the end we get three clusters for the classification of all ten independent variables.

The next stage is to determine the position of the dependent variable (*Pout*) among the independent variables. The proximity with them is visible in Figure 3. As expected *Pout* is nearer to variables *D*, *dr*, *L*, *Pin* and P_{H2} forming together with them the first cluster. The latter serves to confirm the influence of these variables on *Pout*.

TABLE 6. Cluster membership in 2 to 5 clusters of all input variables

Variable	5 clusters	4 clusters	3 clusters	2 clusters
D	1	1	1	1
dr	1	1	1	1
L	1	1	1	1
Pin	1	1	1	1
P_L	2	2	2	2
P_{H2}	1	1	1	1
Prf	3	3	2	2
Pne	2	2	2	2
C	4	4	3	1
Tr	5	4	3	1

Dendrogram

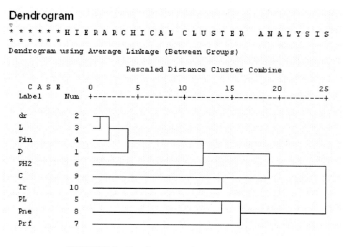

FIGURE 2. Dendrogram of ten input variables

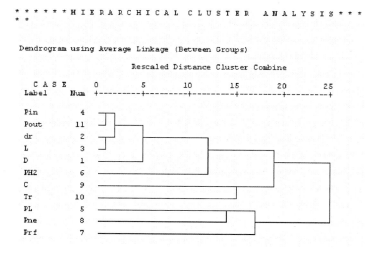

FIGURE 3. The dendrogram of all input variables and the laser power *Pout*

Discussion of the Cluster Analysis Results

The results (Table 6 and Figure 2) are evidence that five of the physical quantities examined display a high level of homogeneousness. What is more Table 6 indicates that they cannot be separated even when the number of clusters is increased. This conclusion is reached also by comparing with Figure 2. Their relative position remains unchanged. This leads to the conclusion that these quantities are the most important when it comes to determining the behavior of the laser source.

The resulting classification can be used when planning a filtering experiment in which out of the totality of independent quantities the main group of quantities has to be separated, so as to be used later on for a more detailed examination [11]. Also under the conditions of an extremal experiment with the goal of optimizing the object being studied, it is necessary to begin varying essential variables in accordance with their homogeneousness, i.e., dr, L, Pin, D and P_{H2}.

CONCLUSION

This work uses classification analysis of variables in the field of metal vapor lasers. Ten independent variables are examined, nine of them being actual physical quantities and one dependent variable – laser power $Pout$. Based on the factor analysis conducted for a sample of a general totality of all available experiments, following the correlation principle, two groups of variables are differentiated - significant and insignificant. The significant ones are classified into three groups (factors). After that they are classified using statistical techniques for cluster analysis following the principle of homogeneity. The result is a hierarchical linking of variables. Three clusters and the order of classification are established.

Problems, connected with the application of these results for the planning of a filtering and extremal experiment, are solved. The obtained results could be used as a basis of multiple regression analysis and prognosis of further experiment in order to enhance the laser power generation.

ACKNOWLEDGMENTS

This work is supported by the National Scientific Fund of Bulgarian Ministry of Education and Science under grant VU-MI-205/2006 and Plovdiv University "Paisii Hilendarski" project IS-M-4/2008.

REFERENCES

1. N. V. Sabotinov, "Copper Bromide Lasers," in *Pulsed Metal Vapour Lasers,* edited by C. E. Little and N. V. Sabotinov, NATO ASI Series, Disarmament Technologies-5, Kluwer Academic Publishers, Dordrecht, 1996, pp. 113-124.
2. D. V. Shiyanov, G. S. Evtushenko, V. B. Sukhanov, and V. F. Fedorov, *Quantum Electronics* **37**, 49-52 (2007).
3. R. G. Carman, D. J. W. Brown, and A. Piper, *IEEE J. of Quanttum Electronics* **30**(8), 1876-1895 (1994).
4. I. P. Iliev and S. G. Gocheva-Ilieva, "Statistical techniques for examining copper bromide laser parameters," in *Int. Conf. of Numerical Analysis and Applied Mathematics, ICNAAM-2007,* edited by T. E. Simos et al, AIP CP936, Melville, New York, 2007, pp. 267-270.
5. I. P. Iliev, S. G. Gocheva-Ilieva, D. N. Astadjov, N. P. Denev, and N. V. Sabotinov, *Optics and Laser Technology* **40**(4), 641-646 (2008).
6. I. P. Iliev, S. G. Gocheva-Ilieva, D. N. Astadjov, N. P. Denev, and N. V. Sabotinov, *Quantum Electronics* **38**(5), 436-440 (2008).
7. J-O. Kim and Ch. W. Mueller, *Factor Analysis: Statistical Methods and Practical Issues,* Sage Publication Inc., Iowa, 1986, 11th Printing.
8. D. N. Astadjov, N. K. Vuchkov, and N. V. Sabotinov, *IEEE J. of Quantum Electronic* **24**(9), 1926-1935 (1988).
9. D. N. Astadjov, K. D. Dimitrov, D. R. Jones, V. K. Kirkov, C. E. Little, N. V. Sabotinov, and N. K. Vuchkov, *IEEE J. of Quantum Electronics* **33**(5), 705-709 (1997).
10. M. S. Aldenderfer and R. K. Blanshfield, *Cluster Analysis,* Sage Publication Inc., Iowa, 1984.
11. D. C. Montgomery, *Design and Analysis of Experiments,* Wiley, New York, 2004.

Bayesian Updating of Demand and Backorder Distributions in a Newsvendor Inventory Problem

Ü. Gürler *, E. Berk[†] and Ü. Akbay**

*Bilkent University, Dept. of Industrial Engineering, 06800 Bilkent Ankara, Turkey
[†]Bilkent University, Faculty of Business Administration, 06800 Bilkent Ankara, Turkey
**Columbia University, Business School, 2960 Broadway New York, NY 10027-6900, USA

Abstract. We consider Bayesian updating of demand and backorder distributions in a partial back-order Newsvendor model. In inventory problems usually the demand distribution is assumed to be known and when stock-outs occur it is commonly assumed that the excess demand is either lost or fully backordered. In this paper we consider a partial backorder setting, where the unsatisfied demand is backordered with a certain probability. Both demand and backorder probabilities are assumed to be random variables and Bayesian estimation methods are used to update the distributions of these variables as data accumulates. We develop expressions for the exact posteriors where the prior distributions are chosen from natural conjugate families. In particular, we assume that the demand within a period is Poisson and the bacorder probability has a Beta distribution.

Keywords: Inventory, newsvendor, partial backordering, Bayesian updating.
PACS: 02.50.Cw, 02.50.Le, 02.50.Tt

INTRODUCTION

Inventory control is one of the most challenging activities in retail management. The challenge mainly stems from the uncertainties involved in the process regarding the demand and other parameters such as lead time, backordering and substitution behavior of customers, etc. The essential component in any inventory problem is the demand for the products; accurate estimation of future demand is crucial. Especially, with short product cycles and fiercer competition on market shares, demand estimation becomes one of the popular research areas in operations management. Hence a vast literature is devoted to inventory models in operations research and management science studies. Most of these analytical models assume that the stochastic demand follows a known probability distribution. For some products where demand elasticity is low and for which data for long past behavior is available, this assumption may be realistic. However, for a number of other products with short product cycles, such assumptions may be highly risky since they may not reflect the true and dynamic market situations.

In this paper, we consider an inventory problem where the demand during a period is a random variable with Poisson distribution, with the parameter of this distribution being also a random variable. It is also assumed that the unsatisfied demands are partially backordered; that is, they are backordered with a certain probability which is also assumed to be a random variable. Initially, we assume some probability distributions for the variables involved; but, as data accumulates, we dynamically update these distribu-

CP1067, *Applications of Mathematics in Engineering and Economics '34—AMEE '08*, edited by M. D. Todorov
© 2008 American Institute of Physics 978-0-7354-0598-1/08/$23.00

tions using a Bayesian estimation approach.

Although there are some inventory models with uncertain demand distributions under Bayesian or frequentest updating, they usually assume either full backordering, or full lost sales. For instance, Lariviere and Porteus [5], consider a periodic review inventory problem under unknown demand distribution and Bayesian updating of the demand parameter. However, they assume a lost sales environment, where an unsatisfied demand is totally lost. Although this assumption makes the analysis more tractable, it is not very realistic in practice especially with expensive items, since usually the customers are offered to backorder the item when it is not immediately available in the store. Some customers may indeed accept to backorder while the others may not. We therefore consider partial backordering with a certain probability. Partial backordering modeling in inventory literature has started with the works [6] and [1], and was followed by several authors including [2], [4], [8] and recently [7].

In this paper, we consider an inventory problem where the demand during a period is a random variable with Poisson distribution, where the parameter of this distribution is also a random variable. It is also assumed that unsatisfied demands are partially backordered, that is they are backordered with a certain probability which is also assumed to be a random variable. Initially we assume some probability distributions for the variables involved but as data accumulates we dynamically update these distributions using a Bayesian estimation approach.

Introduction of Bayesian updating of demand parameter(s) involves estimation of the parameters after some related data is observed. This results in an expansion of the state space; as the number of periods N increases, so does the dimension of the state space. Hence, the the complexity of the problem increases. Azoury [3] made a significant contribution by achieving a "state space reduction" for the dynamic inventory problem with Bayesian updating. Using the theory of sufficient statistics, she showed that if the demand distribution that is dependent on the unknown parameter and the prior density function belong to conjugate families, then the dimensionality of the optimization problem reduces to that of a single period. As discussed below, we also show that if the demand information is fully recorded even in the case of lost sales, then the joint posterior distribution of the demand parameter and the substitution probability is decomposed to the product of the marginal distributions, implying that independence is retained after Bayesian updating. Furthermore the posterior distributions stay in the conjugate families with modified parameters. This result implies a practical approach for the estimation of demand and backordering probabilities dynamically as data accumulates in the course of the business horizon.

THE MODEL

In this study, we consider an inventory problem, where the system is inspected periodically and ordering decisions are made accordingly. The replenishment lead time is assumed negligible. It is assumed that unit demands are faced at the beginning of a period and when the inventory level reaches zero, an arriving customer is offered a backorder. The total demand in a period is assumed to be Poisson distributed with a mean rate to be defined below. In case of a stockout, a customer backorders with probability P; conversely, the sale is lost with probability $1 - P$. Since backordering is not a customer's first choice, a unit cost of π_b is incurred every time a backorder is placed. The probability that a customer will accept a backorder, P, is a random with a prior probability density function.

The beginning inventory level in a period is denoted by Q and the total random demand observed in a period is denoted by D, where d corresponds to the realized demand. Demand in a period has a Poisson distribution with mean Λ, where Λ also changes randomly with Gamma distribution. The shape parameter of the gamma distribution is denoted by α and the scale parameter by β.

When demand exceeds the maximum stock level, the number of backordered demands is a random variable denoted by B. It is assumed that the backorder behavior of customers are independent of each other. When the backorder probability is fixed, then, the number of backorders in a period has a Binomial distribution with parameters $(D - Q)$, which is the difference between the total demand in one period and the corresponding beginning inventory level, and P, the backorder parameter. Then, $B = b$ indicates that the realized number of backorders in that period is b. Also, the lost-sales is denoted by L, the realized value of which is denoted by l. Since a demand that is unsatisfied immediately is either backordered, or lost, it follows that $L = d - Q - b$ if $d > Q$.

For the model described above, distributions of the underlying random variables are as follows. The conditional demand distribution when Λ is given $f_D(d|\lambda)$ is Poisson(λ), with the probability density function given as

$$
f_D(d|\lambda) = \begin{cases} \frac{e^{-\lambda}\lambda^d}{d!} & \text{if} \quad d \geq 0 \\ 0 & \text{otherwise.} \end{cases}
\tag{1}
$$

The demand rate parameter Λ is also assumed to be random due to several effects, which is assumed to have a gamma distribution denoted as Gamma(α, β) and with probability density function

$$
f_\Lambda(\lambda) = \begin{cases} \frac{\alpha e^{-\alpha\lambda}(\alpha\lambda)^{\beta-1}}{\Gamma(\beta)} & \text{if} \quad \lambda \geq 0 \\ 0 & \text{otherwise.} \end{cases}
\tag{2}
$$

The conditional distribution of the backorders when the difference between the total demand and the beginning inventory level $(d - Q)$ and P are given is Binomial$(d - Q, p)$ with probability mass function:

$$
f_B(b|d - Q, p) = \begin{cases} \binom{d-Q}{b}p^b(1-p)^{d-Q-b} & \text{if} \quad b \leq d - Q \\ 0 & \text{otherwise.} \end{cases}
\tag{3}
$$

Finally, the backorder probability p is distributed with Beta(γ, θ) given as

$$
f_P(p) = \begin{cases} \frac{\Gamma(\gamma+\theta)}{\Gamma(\gamma)\Gamma(\theta)}p^{\gamma-1}(1-p)^{\theta-1} & \text{if} \quad 0 < p < 1 \\ 0 & \text{otherwise.} \end{cases}
\tag{4}
$$

BAYESIAN UPDATING

As stated above we want to update the distributions of the random variables for demand and substitution probability, as data becomes available from the sales of the items across the periods. We say that a shortage or stock-out occurs if the total demand D is greater than the beginning inventory level Q. When shortages occur, a random number of back-orders will take place according to the choice of the customers. Hence we will have two types of additional information available at the end of the period: The information that the demand was larger than the stock quantity and that a number of ecxess demand customers accepted to backorder and the others have not. This information is used to obtain the posterior distributions of both the demand and the substitution probabilities conditioned on the observed data. If shortages do not occur however, we observe the exact demand and there are no backorders. Hence only demand distribution is updated in this case since no additional information is available for substitution behavior. Initially the demand parameter Λ and the backorder parameter P are independent of each other. However, since their posterior distributions are updated conditional on the same set of realized data, they will no longer be independent in the following periods. Therefore, their joint distribution needs to be considered. This bivariate distribution will be a mixture distribution with both continuous and discrete parts corresponding to the substitution probability and demand respectively. With some abuse of terminology we will refer to this joint probability function as joint density. In the beginning of the horizon, since it is assumed that Λ and the backorder parameter P are independent random variables, their joint density function $\pi_1(\lambda, p)$ is simply obtained as the product of their marginal densities $f_\Lambda(\lambda)$ and $f_P(p)$ as:

$$\pi_1(\lambda, p) = f_\Lambda(\lambda) f_P(p).$$

Using the Gamma density for Λ and the Beta density for P as given in the previous section, and multiplying the two densities we obtain:

$$\pi_1(\lambda, p) = \frac{\alpha e^{-\alpha\lambda}(\alpha\lambda)^{\beta-1}}{\Gamma(\beta)} \frac{\Gamma(\gamma+\theta)}{\Gamma(\gamma)\Gamma(\theta)} p^{\gamma-1}(1-p)^{\theta-1}. \tag{5}$$

As backordered demand depends on the total demand, we have to pursue a different procedure to find the joint probability mass function of the total demand and the backordered demand

$$f_{D,B}(d, |\lambda, p) = f_D(d|\lambda) f_B(b|D-Q, p).$$

Using the Poisson distribution for the conditional demand distribution and Binomial distribution for the conditional backorder distributions and multiplying the two densities, we obtain:

$$f_{D,B}(d, b|\lambda, p) = \begin{cases} \frac{e^{-\lambda}\lambda^d}{d!}\binom{d-Q}{b} p^b (1-p)^{d-Q-b} & \text{if} \quad d \geq Q_i \text{ and } b \leq d-Q \\ 0 & \text{otherwise.} \end{cases} \tag{6}$$

Proposition 1 *If the joint prior of the demand parameter Λ and the backorder parameter P is given by (5), and the joint conditional probability mass function of demand D and backorder B is given by (6), then the joint predictive distribution of demand and backorder is given by:*

$$f_{D,B}(d,b) = \binom{d-Q}{b} \frac{\alpha^{\beta}\Gamma(\gamma+\theta)}{d!\,\Gamma(\gamma)\Gamma(\theta)\Gamma(\beta)}$$

$$\times \frac{\Gamma(\gamma+b)\Gamma(d-Q-b+\theta)}{\Gamma(d-Q+\gamma+\theta)} \frac{1}{(\alpha+1)^{\beta+d}}\Gamma(\beta+d). \quad (7)$$

Proof: The joint distribution of demand, backorder, demand parameter and backorder parameter can be obtained by multiplication of $\pi_1(\lambda,p)$ (5) $f_{D,B}(d,b|\lambda,p)$ and (6).

$$f_{D,B,\Lambda,P}(d,b,\lambda,p) = \pi_1(\lambda,p)f_{D,B}(d,b|\lambda,p)$$

$$= \frac{\alpha e^{-\alpha\lambda}(\alpha\lambda)^{\beta-1}}{\Gamma(\beta)} \frac{\Gamma(\gamma+\theta)}{\Gamma(\gamma)\Gamma(\theta)}p^{\gamma-1}(1-p)^{\theta-1}\frac{e^{-\lambda}\lambda^{d}}{d!}\binom{d-Q}{b}p^{b}(1-p)^{d-Q-b}$$

$$= \binom{d-Q}{b}\frac{\alpha^{\beta}\Gamma(\gamma+\theta)}{d!\,\Gamma(\gamma)\Gamma(\theta)\Gamma(\beta)}e^{-(\alpha+1)\lambda}\lambda^{\beta+d-1}p^{\gamma+b-1}(1-p)^{d-Q-b+\theta-1}.$$

Taking the integral of the above expression with respect to λ and p, we obtain the marginal distribution of demand and backorder.

$$f_{D,B}(d,b) = \int_{\lambda=0}^{\infty}\int_{p=0}^{1}\binom{d-Q}{b}\frac{\alpha^{\beta}\Gamma(\gamma+\theta)}{d!\,\Gamma(\gamma)\Gamma(\theta)\Gamma(\beta)}$$

$$\times e^{-(\alpha+1)\lambda}\lambda^{\beta+d-1}p^{\gamma+b-1}(1-p)^{d-Q-b+\theta-1}dpd\lambda$$

$$= \int_{\lambda=0}^{\infty}\binom{d-Q}{b}\frac{\alpha^{\beta}\Gamma(\gamma+\theta)}{d!\,\Gamma(\gamma)\Gamma(\theta)\Gamma(\beta)}e^{-(\alpha+1)\lambda}\lambda^{\beta+d-1}\int_{p=0}^{1}p^{\gamma+b-1}(1-p)^{d-Q-b+\theta-1}dpd\lambda$$

$$= \binom{d-Q}{b}\frac{\alpha^{\beta}\Gamma(\gamma+\theta)}{d!\,\Gamma(\gamma)\Gamma(\theta)\Gamma(\beta)}\frac{\Gamma(\gamma+b)\Gamma(d-Q-b+\theta)}{\Gamma(d-Q+\gamma+\theta)}\int_{\lambda=0}^{\infty}e^{-(\alpha+1)\lambda}\lambda^{\beta+d-1}d\lambda$$

$$= \binom{d-Q}{b}\frac{\alpha^{\beta}\Gamma(\gamma+\theta)}{d!\,\Gamma(\gamma)\Gamma(\theta)\Gamma(\beta)}\frac{\Gamma(\gamma+b)\Gamma(d-Q-b+\theta)}{\Gamma(d-Q+\gamma+\theta)}\frac{1}{(\alpha+1)^{\beta+d}}\Gamma(\beta+d).$$

Theorem 1 *If the joint prior of the demand parameter Λ and the backorder parameter P is given by (5), and the joint conditional probability mass function of demand D and backorder B is given by (6), then the joint posterior distribution of Λ and P is given by:*

$$\pi_2(\lambda,p) = \begin{cases} \dfrac{\alpha_2 e^{-\alpha_2\lambda}(\alpha_2\lambda)^{\beta_2-1}}{\Gamma(\beta_2)}\dfrac{\Gamma(\gamma_2+\theta_2)}{\Gamma(\gamma_2)\Gamma(\theta_2)}p^{\gamma_2-1}(1-p)^{\theta_2-1} & \text{if} \quad 0<p<1 \text{ and } \lambda\geq 0 \\ \\ 0 & \text{otherwise,} \end{cases}$$

$$(8)$$

where $\alpha_2 = \alpha + 1$, $\beta_2 = \beta + d$, $\gamma_2 = \gamma + b$, $\theta_2 = \theta + d - Q - b = \theta + l$.

Proof: Using the result of Proposition, and pursuing the typical Bayesian updating procedure, the joint posterior of Λ and P can be derived by dividing the joint distribution of demand, backorder, Λ and P; $f_{D,B,\Lambda,P}(d,b,\lambda,p)$, by the joint distribution of demand and backorder; $f_{D,B}(d,b)$

$$
\begin{aligned}
\pi_2(\lambda,p) &\equiv \pi_2(\lambda,p|d,b)\\[4pt]
&= \frac{f_{D,B,\Lambda,P}(d,b,\lambda,p)}{f_{D,B}(d,b)}\\[6pt]
&= \frac{\binom{d-Q}{b}\frac{\alpha^\beta \Gamma(\gamma+\theta)}{d!\Gamma(\gamma)\Gamma(\theta)\Gamma(\beta)}e^{-(\alpha+1)\lambda}\lambda^{\beta+d-1}p^{\gamma+b-1}(1-p)^{d-Q-b+\theta-1}}{\binom{d-Q}{b}\frac{\alpha^\beta \Gamma(\gamma+\theta)}{d!\Gamma(\gamma)\Gamma(\theta)\Gamma(\beta)}\frac{\Gamma(\gamma+b)\Gamma(d-Q-b+\theta)}{\Gamma(d-Q+\gamma+\theta)}\frac{1}{(\alpha+1)^{\beta+d}}\Gamma(\beta+d)}\\[6pt]
&= \frac{\Gamma(d-Q+\gamma+\theta)}{\Gamma(\gamma+b)\Gamma(d-Q-b+\theta)}p^{\gamma+b-1}(1-p)^{d-Q-b+\theta-1}\frac{(\alpha+1)e^{-(\alpha+1)\lambda}\lambda^{\beta+d-1}}{\Gamma(\beta+d)}.
\end{aligned}
$$

Rearranging the terms, we finally obtain:

$$
\pi_2(\lambda,p) = \frac{(\alpha+1)e^{-(\alpha+1)\lambda}\lambda^{\beta+d-1}}{\Gamma(\beta+d)}\frac{\Gamma(d-Q+\gamma+\theta)}{\Gamma(\gamma+b)\Gamma(d-Q-b+\theta)}(1-p)^{d-Q-b+\theta-1}.
$$

For the No-Shortage case, since there will be no backorders and lost-sales realized, the backorder parameter P will not be updated, and its distribution parameters; γ and θ will be the same as the prior distribution. The results of Theorem conforms with this requirement in the sense that in a no-shortage case b and l are going to be 0, and thus the backorder parameter is going to remain not updated. Hence, the parameters of the posterior distribution for the no-shortage case are: $\alpha_2 = \alpha + 1$, $\beta_2 = \beta + d$, $\gamma_2 = \gamma$ and $\theta_2 = \theta$.

The results of the theorem above indicate that for the full information case, where demand and substitutions are fully recorded, the posterior distributions of the demand rate and substitution probabilities remain in the same families of gamma and beta distributions respectively. However the parameters of these distributions are updated to take into account the observed data. Namely, we see that as is well known from the conjugate properties of gamma and Poisson distributions, the scale parameter of gamma distribution is modified by increasing it by one, as $\alpha_2 = \alpha + 1$, and the shape parameter is modified by adding the observed number of demands d as $\beta_2 = \beta + d$. For the parameters of the beta distribution, the parameter related to backordering probability is modified by adding the number of backorders as $\gamma_2 = \gamma + b$ and the parameter related to the complement of the substitution probability is modified by adding the observed number of lost sales as $\theta_2 = \theta + d - Q_1 - b = \theta + l$. From the analysis of this one period behavior we can show by induction that for the n period problem in general, where the number of demands, backorders and lost sales for the ith period are denoted by d_i, b_i and

l_i respectively, similar results will hold and we can get the joint posterior densities of Λ and and P as given below, for which the proof is skipped.

Corollary 1 *The joint posterior of Λ and P for the nth period can be derived as:*

$$\pi_n(\lambda,p) = \begin{cases} \dfrac{\alpha_n e^{-\alpha_n \lambda}(\alpha_n \lambda)^{\beta_n-1}}{\Gamma(\beta_n)} \dfrac{\Gamma(\gamma_n+\theta_n)}{\Gamma(\gamma_n)\Gamma(\theta_n)} p^{\gamma_n-1}(1-p)^{\theta_n-1} & \text{if} \quad 0<p<1 \ and \ \lambda \geq 0 \\[2mm] 0 & otherwise. \end{cases}$$

where $\alpha_n = \alpha+n-1$, $\beta_n = \beta + \sum\limits_{i=1}^{n-1} d_i$, $\gamma_n = \gamma + \sum\limits_{i=1}^{n-1} b_i$, $\theta_n = \theta + \sum\limits_{i=1}^{n-1} l_i$.

CONCLUSION

In this paper we have focused on a periodic review inventory problem with stochastic demand and partial backorder. When the units in stock depletes, the customers are offered backordering and they accept this with a stochastic probability. In most of the existing inventory models, although randomness is allowed for quantities such as demand, lead time or backordrer probabilities, it is commonly assumed that the probability distributions of these quantities are completely satisfied. This assumption may not be realistic for some products for which there is not sufficient historical data. Such cases are encountered more in modern retail industry since life cycles of products become shorter with increasing technological developments and the number of product that enter the market due to competition and the attempts to increase the market share. We therefore introduce a model where the distributions of the stochastic quantities are updated in the course of the business horizon as related information accumulate from the observed data. In particular, based on the observed demand and backorder quantities, the distributions of the demand and substitution probability parameters are modified using a Bayesian approach. To this end, at the end of each period, it is proposed to update these distributions according to the observed data on the number of demands and backordered quantities. The demand during a period is assumed to have a Poisson distribution and the parameter of the Poisson demand is assumed to have gamma distribution which has the property of being the conjugate prior. Similarly, the number of backorders is assumed to have a binomial distribution where the backorder probability has a beta distribution, a conjugate prior for Binomial. The results obtained for the posterior joint distribution of the demand parameter and backorder probability indicated that these quantities remain to be independent variables after their distributions are modified according to the observed data. This result provides an easily implementable method for dynamically updating the demand and backorder probability distributions in a business horizon of n periods. A challenging and important extension of these results would be to consider the case where demand or backorder informations are incomplete or censored which is encountered when only the purchase data not the exact demand data is recorded. Such incomplete information creates complications in the Bayesian methods since the posterior distributions do not retain the conjugacy properties. Furthermore the bivariate nature of the problem introduces additional complication to the problem and such extensions would call more sophisticated methods both for the statistical analysis and computational aspects in practice.

REFERENCES

1. P. L. Abad, *Management Science* **42** , 1093–1104 (1996).
2. P. L. Abad, *European Journal of Operations Research* **144** , 677–685 (2003).
3. K. S. Azoury, *Management Science* **31**, 1150–1160 (1985).
4. C. W. Chu, B. E. Patuwo, A. Mehrez et al, *Computers and Operations Research* **28** , 935–953 (2001).
5. M. A. Lariviere and E. L. Porteus, *Management Science* **45**, 346–363 (1999).
6. G. Rabinowitz, A. Mehrez, C-W. Chu, and B. E. Patuwo, *Computers and Operations Research* **22**, 689–700 (1995).
7. A. Thangam and R. Uthayakumar, *European Journal of Operations Research* **187**, 228–242 (2008).
8. A. Z. Zeng, *Production Planning&Control* **12**, 660–668 (2001).

Ontological Model of Business Process Management Systems

G. Manoilov[*] and B. Deliiska[†]

[*]Product and Service Manager in Imbility Ltd., Sofia, Bulgaria
[†]Dept. of Computer Systems and Informatics, University of Forestry, Sofia, Bulgaria

Abstract. The activities which constitute business process management (BPM) can be grouped into five categories: design, modeling, execution, monitoring and optimization. Dedicated software packets for business process management system (BPMS) are available on the market. But the efficiency of its exploitation depends on used ontological model in the development time and run time of the system. In the article an ontological model of BPMS in area of software industry is investigated. The model building is preceded by conceptualization of the domain and taxonomy of BPMS development. On the base of the taxonomy an simple online thesaurus is created.

Keywords: Business process, ontology, conceptualization, taxonomy, BPMS.
PACS: 89.65.Gh, 07.05.Mh, 07.05.Tp, 07.05.Kf

INTRODUCTION

Business process (BP) is a set of one or more linked procedures or activities which collectively realize a business objective or policy goal, normally within the context of an organizational structure defining functional roles and relationships [12].

There are three types of BP [en.wikipedia.org/wiki/Business process]:

- management processes, the processes that govern the operation of a system;
- operational processes, that constitute the core business and create the primary value stream. Typical operational processes are: purchasing, manufacturing, marketing, and sales;
- supporting processes, which support the core operational processes.

BP should be described in some specification language, which includes the activities that need to be performed, the participants who could or should perform them, and the interdependencies that exist between these activities [9].

Business process management (BPM) is a structured approach that models an enterprise's human and machine tasks and the interactions between them as processes [14]. It includes:

- organizing the business around processes and focusing on customer satisfaction;
- clarifying and documenting processes;
- monitoring process performance and compliance;

CP1067, *Applications of Mathematics in Engineering and Economics '34—AMEE '08*, edited by M. D. Todorov
© 2008 American Institute of Physics 978-0-7354-0598-1/08/$23.00

- continuously identifying opportunities for improvement and deploying them.

Organization and management of the BP can use a software system (Business Process Management System, BPMS) that is capable of ensuring that the process description is realized in practice. This system must: allow the manual activities to be assigned appropriately, provide access to the software tools required to complete the tasks and ensure that dependencies between the tasks are satisfied. Moreover, the software system should transparently support multiple instances of a given process and a given task.

An important notion of the BPMS is a workflow. In [13] a workflow is defined as the automation of a business process, in whole or part, during which documents, information or tasks are passed from one participant to another for action, according to a set of procedural rules defined as workflow. So the workflow can be any business process, which consists of two or more tasks (sub-processes or actions) performed in serial or concurrently by two or more people.

BMPS, ideally, is used to define (design and modeling), execute (implement), monitor and optimize human to human, human to system, and system to system workflows which makes evolution of BPs more smooth and close to the regulatory, market, competitive and conformance challenges faced by businesses.

BPMSs are being increasingly used both in traditional and in Internet-based enterprises to support administrative and production processes, execute commerce transactions, and monitor business operations. In fact, BPMSs typically allow companies to reduce costs and to improve the speed and quality of BP. One of the main features of BPMS tools is the availability of a BP modeling facility, which enables BP designers to describe the many aspects involved in a BP execution, such as tasks, execution flows, data flows, resources, constraints, and exceptions. In addition, BPMSs also provide support for BP modification and versioning.

A BPMS may comprise a variety of independent packages or a comprehensive business process management suite, which includes tools for modeling and analysis, application integration, business rules support, business intelligence (BI), activity monitoring and optimization [14]. Advanced BPMSs provide a development tool for creating forms-based applications, which are often the start of many BPs.

Today's BPMS development suffers from lack of business semantics. Semantics can be add by domain ontology development. Ontologies, which are a vital part of the Semantic Web, can be used as a semantic backbone for BPM. An ontology is explicit formal specification of the concepts and terms in given area of knowledge and the relationships among them. It serves two distinct purposes. Firstly, it makes knowledge explicit and allows for knowledge sharing among domain experts and IT people engaged in software design and development. Secondly, since it includes machine-readable definitions of concepts, it serves as a requirements specification from which a number of software artifacts can be generated.

In the world of information systems an ontology is a software (or formal language) artifact designed with a specific set of uses and computational environments in mind, and often ordered up by a specific client or customer or application program in a specific context [11]. Basic elements of ontology are concepts (classes), slots (relations), facets (restrictions of slots) and instances (individuals). An concept can be examined as unary predicate generalizing set of instances. By creating instances of concepts, a knowledge base can be easily constituted. Hence, the knowledge base can be used for rapid prototyping without having to write a single line of code. Prototyping can be performed by non-IT experts, such as business analysts.

For example, [10] claims that building a Business Process Management Ontology (BPMO) and providing adequate tools to business analysts will be possible to minimize the "understanding gap" and speed up software development significantly. In [8] an hierarchical semantic model of BPM based on ontologies about BPs, organizations and law norms is given.

Here an ontological model of BPMS is developed. On the base of the model an application ontology of BPMS in area of software industry is investigated.

This paper is organized as follows. A method of BPMS ontology building in Section 2 is given. In Section 3 BPMS ontology in software industry is described. Section 4 discusses the conclusions and future work.

BPMS ONTOLOGY BUILDING

There are no commonly agreed methodologies about the development of shared and consensual ontologies [2]. Generally, all known methodologies comprise following consecutive stages [4]:
- gathering of terminology in a controlled vocabulary;
- defining external relations with another ontologies;
- taxonomy creating;
- building thesaurus and
- converting the thesaurus into formal ontology.

Creating of Controlled Vocabulary

Controlled vocabulary [1] is a list of terms in given area that have been explicitly enumerated. All terms in a controlled vocabulary must have an unambiguous, non-redundant definitions. Moreover the definition of a each new concept is on the base of other previously defined concepts in the same or other related vocabulary. It can be phrase (sentence) according to prescribed standard or formal expression in terms of description logics (DL) used in knowledge representation systems.

For example, the definition of BP can be regarded as DL declaration:

$$BP \equiv Business \cap Process$$

A controlled vocabulary of BPMS including about 100 most important (preferred) terms (concepts or classes of the following ontology model) in the domain is created.

External Relations

General knowledge representation is impossible without investigation of relationships with another ontologies. In any case, common ontology is on the top of each ontology hierarchy. A common (core or general) ontology is minimal ontology including universal classes available for all the rest ontologies as *State, Process, Action, Event, Part*, etc. Moreover, in the concrete case, BPMS ontology imports classes (or namespaces) from:

- BP ontology;
- domain-specific ontologies about the knowledge areas related to concrete

BPMS applications and

- organization ontology including classes about structure, roles (actors), rules, objectives, resources etc. in an organization implementing BP (Figure 1).

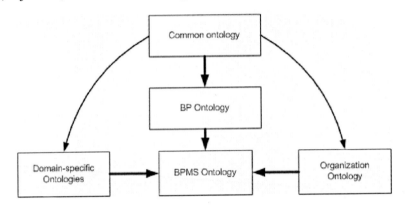

FIGURE 1. External relations of BMPSO

BPMS Taxonomy Building

A taxonomy designates hierarchical relations between preferred terms (classes or concepts) in controlled vocabulary. The hierarchical relations are of type *is-subclass-of/has-subclass, is-superclass-of/has-superclass, is-a/has-a, is-part-of/has-part, is-instance-of/has-instance*.

The several upper levels of the BMPS taxonomy of on Figure 2 is given. The direct subclasses of the top class BPMS are: *BP workflow, Interface, BP definition tool, Execution engine, Member* and *Business object*.

The class *BP dashboard* denotes monitoring feature and the class *BP Execution engine* – the transactional control mechanism in BPMS for executing a business process from start to finish. *BP definition tool* includes subclasses *BP metrics, BP modeler* (a software or person) etc. *BP workflow* consists of *BPs* described in BP ontology which subclasses as *Business action* (sub-process of BP), *Business document* (order, invoice), *Business product* (software, information, article), *Business event, Resource* etc. Further, one of subclasses of *Business action* is *BP regulation* comprising *BP rule* (declarative statement described by IF...THEN expression), etc.

An excerpt of the upper levels of BP taxonomy on Figure 3 is shown.

A class *Member* denotes one or more individuals managing or participating in BPMS. Each member has some *Role* and *Skills*. The *Role* defines prototypical job functions in an organization [6]. Within the organization, there is usually a hierarchy of roles. Subclasses of *Role* are *Purchasing officer*, *Customer*, *Manager*, *Sales clerk*, etc.

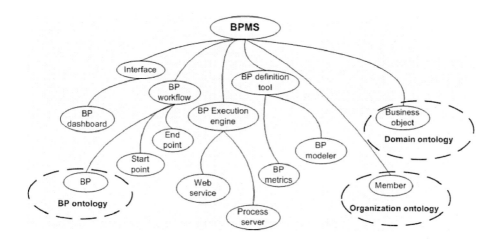

FIGURE 2. Excerpt of BMPS taxonomy

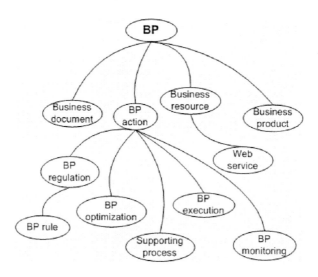

FIGURE 3. Excerpt of BP taxonomy

An organization ontology has the following upper classes (Figure 4):

By analogy, the class *Business object* is the top class of domain ontology in concrete business area (for example, software industry).

FIGURE 4. Excerpt of Organization taxonomy

BPMS Thesaurus Building

A thesaurus is taxonomy that also includes non-hierarchical relations with associated and related terms. Introducing hierarchical and non-hierarchical relations between terms in the controlled vocabulary convert it into thesaurus. The non-hierarchical relations are two types – equivalent and associative. Equivalent relations are between preferred and related (non-preferred) terms and express mainly synonymy (antonymy), near synonymy and lexical variants. The associative relations are specific for given domain and are defined by experts. For example:

$$Role \xrightarrow{\ is\ synonym\ of\ } Position \xrightarrow{\ is\ synonym\ of\ } Appointment$$

BPMS Ontology

In the final stage thesaurus is transformed to OWL (Web Ontology Language) or RDF (Resource Description Framework) ontology code using the method and algorithms developed in [5]. The ontology can be instantiated thus creating knowledge base.

BPMS APPLICATION ONTOLOGY IN SOFTWARE INDUSTRY

Using the domain BMPS ontology in area of software industry (SI) leads to application BMPS ontology building. Thus, software industry BPs may be grouped into three process areas [7] – development or lifecycle processes (analogous to core business processes), supporting business processes and economical indices calculating processes.

Primary lifecycle processes in Table 1 are illustrated. The supporting processes include: defect repairs, minor enhancements, versioning dictated by evolution of underlying technologies, etc.

Additional business processes calculating indices of economical effectiveness and estimating costs of each primary business process typically comprise: project planning; estimating (or forecasting) effort, duration, delivered quality; tracking and reporting status; defining standards; measuring and monitoring process performance; training; controlling versions and releases of software work products.

All software processes can be appropriately characterized by some combination of the following metrics:

- production time (in office hours, person-days or calendar time);
- relisbility (failure and defect rates);
- size (e.g., function points, lines of code, storage volume, etc.);
- capability (efficiency) of the software;
- data quality;
- result accuracy;
- file formats, etc.

The *Software development dashboard* (or scorecard) corresponding to *BP dashboard* in BPMS ontology (see Figure 3) presents a set of measures in table view. An instance of this concept on Table 1 is given.

The upper levels of the taxonomy of software industry BPMS application ontology on Figure 5 are given.

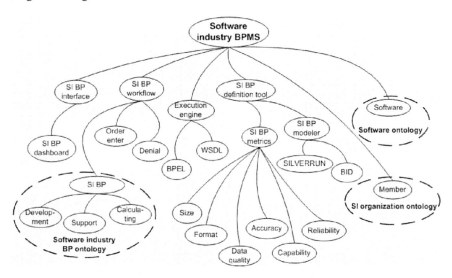

FIGURE 5. Excerpt of Software industry BMPS taxonomy

For example, in the above taxonomy the languages BPEL (Business Process Execution Language) and WSDL (Web Services Description Language) are instances of the class *Execution engine*. Analogically, SILVERRUN and BID are instances of software packets belonging to the class *SI BP modeler*, etc.

TABLE 1. Software development BPs dashboard model

BP	Cycle Time (duration/size)	Defect Rate (number/percent)	Effort (price)	Participants (name,position)
Requirement defining	...	...	...	...
Designing solution	...	...	...	...
software coding	...	...	...	...
software validation	...	...	...	...
Software documentation	...	...	...	...
Software implementation	...	...	...	...
Software deployment	...	...	...	...

The following set of specific associative relations are proposed (Table 2):

TABLE 2. A part of associative relations

Direct relation	Inverse relation
isDataFor	*hasData*
isSourceOf	*hasSource*
isLanguageFor	*hasLanguage*
isLocatedIn	*hasLocation*
isMethodOf	*hasMethod*
isModeOf	*hasMode*
isImplementedOn	*hasImplementation*
isReferencedBy	*hasReference*
isPropertyOf	*hasProperty*
isVersionOf	*hasVersion*
isFormatOf	*hasFormat*

CONCLUSION AND FUTURE WORK

The proposed domain and application BMPS ontologies are one of multiple works investigating the semantics of BP. The future work will direct efforts to expand the size of this ontologies with specific functions and axioms. Moreover, another application BPMS ontologies can be developed.

REFERENCES

1. ANSI/NISO Z39.19-2005 "Guidelines for the Construction, Format, and Management of Monolingual Controlled Vocabularies", NISO Press, Bethesda, Maryland, USA, ISBN: 1-880124-65-3, 2005
2. V. R. Benjamin and A. Gómez Pérez, "Knowledge-System Technology: Ontologies and Problem-Solving Methods", 1999, online at
http://hcs.science.uva.nl/usr/richard/pdf/kais.pdf

3. P. Dalmaris, E. Tsui, B. Hall, and B. Smith, "A Framework for the improvement of knowledgeintensive business processes", 2005, online at
http://www.futureshock.com.au/docs/KBPI-BPMJ.pdf
4. B. Deliiska, "Domain Ontology Design," in *Proc. of Computer Science'2005*, Kassandra, Greece, 30.09–02.10.2005.
5. B. Deliiska, "Ontology-based e-Learning in Area of Computer Networks and Ecoinformatics," research work qualifying for an academic rank, University of Forestry, Sofia, 2008, pp.100-102. (in Bulgarian)
6. M.S. Fox, M. Barbuceanu, M. Gruninger, and J. Lin, "An Organization Ontology for Enterprise Modelling," 2007, online at
http://web.unicam.it/matinf/Dispense/ruffino/Dispense/Esempio%20ont ologia%20 organizzativa.pdf7
7. G.A. Gack, "Business Process Management for Software Development" in *SixSigma Software/IT* - July 6, 2005-Vol. 3, #14-ISSN: 1530-7603.
8. H.-J. Happel and L. Stojanovic, "Ontoprocess – a prototype for semantic business process verification using SWRL rules," 2006, online at http://www.eswc2006.org/demo-papers/FD40-Happel.pdf
9. J. Grundspenkis and D. Pozdnyakov, "An Overview of the Agent Based Systems for the Business Process Management," in *International Conference on Computer Systems and Technologies - CompSysTech'2006*, pp.221-228.
10. D.E. Jenz, *Defining a Private Business Process in a Knowledge Base*, Jenz & Partner GmbH, 2003, online at http://www.bpiresearch.com/WP_BPMOntology.pdf
11. M. K. Smith, Ch. Welty, and D. L. McGuinness (Eds), *OWL Web Ontology Language Guide*, World Wide Web Consortium Recommendation, 2004, online at http://www.w3.org/TR/owl-guide/
12. "Workflow Glossary of Definitions" at
http://ais.its.psu.edu/workflow/media/glossary.doc
13. "WfMC: Workflow Management Coalition Terminology & Glossary" at
http://www.wfmc.org/standards/docs/TC-1011_term_glossary_v3.pdf
14. "ZDNet Dictionary" at http://dictionary.zdnet.com/definition/BPM.html

SOFTWARE INNOVATIONS
AND ALGEBRAIC METHODS

Some Applications of Matrix - and Computer Algebra in Relation Theory

G. Bijev

Technical University of Sofia, 1000 Sofia, Bulgaria

Abstract. Boolean matrices, i.e., matrices over a Boolean algebra, do not satisfy many of the fundamental properties of matrices over the real or complex field. Algebraic aspects of the Boolean pseudoinverse matrix concept are investigated. Maps operating on Boolean vector spaces by analogy with those on linear spaces are defined. Some properties of them are obtained. Most of these properties remain the same in the case of linear spaces and they are essential for solving linear systems. In particular orthogonal projections in linear spaces receive their algebraic analogue in Boolean vector spaces. The maps defined are useful for solving equations in the Boolean algebra, if no pseudoinverse matrix to a given matrix exists. We also share our experience in using computer algebra systems for teaching discrete mathematics and linear algebra. Some examples for computations with binary relations and Boolean matrices using computer algebra systems are given.

Keywords: pseudoinverse matrix, computer algebra system, Maple, Boolean matrix, orthogonal projection, linear equation, binary relation, Boolean lattice, Boolean semilattice.
PACS: 02.10.Yn

INTRODUCTION

Boolean matrices are matrices over a Boolean algebra. They do not satisfy many of the fundamental properties of matrices over the real or complex field. For example, a Boolean matrix need not always possess a pseudoinverse and row and column rank need not to be equal [1]. Some singular matrices over a field with operation matrix multiplication can be regarded as semigroup elements which are invertible as belonging to a subgroup in a semigroup [2]. In some algebraic sense every Boolean matrix can be considered as a homomorphism operator, whose restriction to a row (column) subspace is a bijection, i.e., it is invertible as a map. The same property has every matrix over a field.

For investigations of algebraic structures computer methods can be useful. The development of software supporting mathematics requires enlarged boundaries of the engineer's education. One of the successful applications of CAS in this area is [5]. Maple can be used in teaching Physics [6], Calculus [5], [7], etc. One of the aims of this paper is to share our experience in using Maple in linear algebra for students at the Faculty of Applied Mathematics and Informatics and the Faculty of German Engineering Education and Industrial Management, Technical University of Sofia.

CP1067, *Applications of Mathematics in Engineering and Economics '34—AMEE '08*, edited by M. D. Todorov
© 2008 American Institute of Physics 978-0-7354-0598-1/08/$23.00

APPLICATIONS IN BOOLEAN MATRIX – AND RELATION THEORY

We consider some algebraic properties of Boolean matrices and investigate several maps, connected with them. We give some examples for computations with binary relations using Maple.

Pseudoinverse Matrix

There are many different kinds of generalized inverses of Boolean matrices. Three of them are discussed in [1], Chapter 3. The Moore – Penrose inverses are called also pseudoinverses. The pseudoinverse A^+ of a matrix A over the real or complex field is the unique matrix satisfying the following criteria: $A A^+A = A$, $A^+A A^+ = A^+$, $(A^+A)*$ $= A^+A$, $(AA^+)* = AA^+$. Here $M*$ denotes the conjugate transpose of a matrix M or the transpose M^T for matrices whose elements are real numbers. This concept was independently described by E. H. Moore [3] in 1920 and Roger Penrose [4] in 1955. If there exists a pseudoinverse matrix A^+ to a Boolean matrix A then $A=A^T$ holds [1], p.135. Furthermore A needs to be of a very special kind.

In this paper we use an algebraic approach to the pseudoinverses generalization problem, which is similar to those applied in [2]. We denote both Boolean binary operations with (+) and (.) and the complement of an element x is denoted by x'. We use also (+) and (.) for the operations component-wise addition and multiplication of Boolean vectors respectively.

Let V_m (V_n) denotes the set of all m-tuples (n-tuples) together with the operation (+), called the Boolean vector space of dimension m (n). Let A be a Boolean matrix with m rows and n columns generating subspaces U_n and U_m respectively.

Let f be the map defined as follows:

$$f: V_n \to V_m, \quad f(x) = Ax'. \tag{1}$$

Let Φ be the restriction of f to U_n. Then Φ is a bijection from U_n onto U_m, see [1], the proof of Theorem 1.2.3.

It is easily to see that Φ is an anti-isomorphism of lattices from U_n onto U_m.

Let x_1 and x_2 be elements of U_n, for which $x_1 \le x_2$ holds. Then, since $x_1' \ge x_2'$ and $A(x_1)' \ge A(x_2)'$ hold, we obtain $\Phi(x_1) \ge \Phi(x_2)$. Conversely, let $\Phi(x_1) \ge \Phi(x_2)$. Then $A(x_1)' \ge A(x_2)'$ implies $A(x_1)'+A(x_2)' =A(x_1)'$, $A(x_1)' =A(x_1'+x_2')= A(x_1 . x_2)'$, $\Phi(x_1)$ $= \Phi(x_1 . x_2)$. Since Φ is a bijection, we obtain $x_1 = x_1 . x_2$, i.e., $x_1 \le x_2$. Furthermore, from

$$f(x_1 . x_2)= A(x_1 . x_2)'=A(x_1'+x_2')= A(x_1)'+A(x_2)'= f(x_1)+ f(x_2),$$

where x_1 and x_2 belong to V_n, we obtain the following result.

Lemma 1 *The map $f: V_n \to V_m$, $f(x)= Ax'$ is a semilattice homomorphism from the meet-semilattice (V_n , .) to the join-semilattice (V_m , +).*

Kernel Equivalence (π) of Map f

The equivalence relation π, defined as $x_1 \pi x_2$ if and only if $f(x_2) = f(x_1)$, is known as kernel equivalence. Let $x_1 \pi x_2$. Then lemma 1 implies $f(x_1 . x_2) = f(x_1) + f(x_2) = f(x_1) + f(x_1) = f(x_1)$, i.e., $(x_1 . x_2) \pi x_1$. So we have the following statement.

Lemma 2 *Every (kernel) equivalence class of π is a subsemilattice of (V_n, .).*

Orthogonal Projections in Linear Spaces and Their Algebraic Analogue in Boolean Vector Spaces

By analogy with the matrices over the real or complex field we can define a map, which is quite similar to orthogonal projections in the linear algebra.

Let us consider the equation

$$A x' = b. \tag{2}$$

Since $f(V_n) = U_m$, we have the statement: (2) has a solution if and only if the vector b is in U_m. If (2) has a solution, then there exists an exactly one vector x_0 in U_n such that

$$f(x_0) = \Phi(x_0) = A (x_0)' = b. \tag{3}$$

Let (2) be an equation in a linear algebra, i.e., V_n and V_m denote linear spaces with U_n and U_m as their linear subspaces respectively and A denotes a matrix over a field. In this case we have exactly the same result: the solution, denoted as a vector $(x_0)'$ and belonging to U_n, is unique and it is given by the formula $(x_0)' = A^+ b$, where A^+ is the pseudoinverse matrix. In addition to that, the linear map (called orthogonal projection) $p: V_n \rightarrow V_n$, defined by the formula

$$p(x) = A^+ A \, x, \tag{4}$$

maps the whole vector space V_n onto the subspace U_n.

If V_n denotes a Boolean vector space, then we can define the map $p : V_n \rightarrow V_n$ in the following way:

$$p(x) = \Phi^{-1}(A x'), \quad p : V_n \rightarrow V_n . \tag{5}$$

Since $\Phi^{-1}(f x) = x$ holds for all x in U_n, the map p in (5) is evidently an idempotent as well as p is in (4) (case of linear space V_n).

The kernel equivalence (π) of p is the same as those of f. Since Φ is a bijection, there exists in every equivalence class of π exactly one vector belonging to the subspace U_n.

So we obtain the following result.

Theorem 1 *Let V_m and V_n denote the (Boolean) lattices $(V_m, + , .)$ and $(V_n, + , .)$ respectively. Let A be a Boolean matrix with m rows and n columns generating subsemilattice (Boolean vector subspaces) $(U_n, +)$ and $(U_m, +)$ respectively. Then the map $f: V_n \rightarrow V_m$, $f(x) = Ax'$ with the kernel equivalence π has the following properties:*

    (i)   *f is a semilattice homomorphism from (V_n, .) to (V_m, +). The equivalence classes of π form subsemilattices of (V_n, .).*

    (ii)   *The restriction Φ of f to U_n is a bijection.*

(iii)   There is a one-to-one correspondence between the set of all equivalence classes of π and U_n. Every class contains exactly one vector belonging to U_n, i.e., this vector can be considered as a representative of the class.

(iv)   The map $p: V_n \to V_n$, defined as $p(x) = \Phi^{-1}(A\,x')$, is an idempotent with the same kernel equivalence π as those of f and $p(x)$ is for every element x of V_n the representative of the equivalence class containing x.

The statements (ii), (iii) and (iv) in Theorem 1 remain true, if V_n and V_m denote linear spaces over the real or complex field, U_n and U_m denote the corresponding linear subspaces and the maps f and p are defined as follows: $f(x) = Ax$ and $p(x) = \Phi^{-1}(Ax)$.

An Algebraic Analogue of Pseudoinverse in Boolean Vector Spaces

Let $Iso(U_n, U_n)$ be the set of all square Boolean matrices with n rows and columns generating one and the same subspace U_n. For example to $Iso(U_n, U_n)$ belongs any symmetric matrix. Let A be a square matrix in $Iso(U_n, U_n)$. Let (o) denote the symbol for map composition, i.e., $(f \circ g)(x)$ means $f(g(x))$.

We define a map g (pseudoinverse to f) in the following way:

$$g : V_n \to V_n, \quad g(x) = \Phi^{-1}(\Phi^{-1}(Ax')). \tag{6}$$

It follows from (5) and (6) that $g(x) = \Phi^{-1}(p(x))$ and

$$f(p(x)) = f(\Phi^{-1}(f(x))) = f(x), \tag{7}$$

$$p(f(x)) = \Phi^{-1}(f(f(x))) = f(x), \tag{8}$$

$$g(p(x)) = \Phi^{-1}(\Phi^{-1}(f(p(x)))) = \Phi^{-1}(p(x)) = g(x), \tag{9}$$

$$g(f(x)) = \Phi^{-1}(p(f(x))) = \Phi^{-1}(f(p(x))) = p(x), \tag{10}$$

$$f(g(x)) = f(\Phi^{-1}(p(x))) = p(x). \tag{11}$$

$$f(g(f(x))) = f(p(x)) = f(x), \tag{12}$$

$$g(f(g(x))) = g(p(x)) = g(x). \tag{13}$$

Since $g(x)$ is in U_n, we have evidently

$$p(g(x)) = g(x). \tag{14}$$

From (5) - (14) we obtain the following result.

Theorem 2 Let A be a matrix in $Iso(U_n, U_n)$ and let f, p and g be maps, defined by (1) for $m=n$, (5) and (6) respectively. Then it holds:

(i) $p \circ p = p, \quad f \circ p = p \circ f = f, \quad g \circ p = p \circ g = g;$

(ii) $f \circ g = p, \quad g \circ f = p;$

(iii) $f \circ g \circ f = f, \quad g \circ f \circ g = g.$

Let us consider the linear equation $Ax=b$ and let A^+ denote the pseudoinverse matrix to A. The vector $x_0 = A^+ b$ belongs to the linear subspace U_n. If there exists a solution, then x_0 is the only vector in U_n satisfying the equation and it is the orthogonal projection of all other vectors, satisfying the equation. The vector (Ax_0-b) is the only vector, which norm is the least possible in case of no solutions. By analogy with the case of a Boolean vector space V_n we define maps f, p, and g in much the same manner as those in Theorem 2. Let us interpret U_n, V_n, A etc. as the corresponding concepts in the linear algebra. Let $f(x)=Ax$, $g(x)=A^+x$ and $p(x)=AA^+x$, for all x in V_n. The maps f, p, and g satisfy all conditions in Theorems 1 and 2 except the first one in Theorem 1. So we obtain $x_0=g(b)$.

The Boolean vector space V_n is finite as a set. If there exists no pseudoinverse Boolean matrix A^+ with the property $g(x)=A^+x$ for all x in V_n, the solution

$$x_0=g(b)=\Phi^{-1}(\Phi^{-1}(Ab')) \tag{15}$$

could be calculated by an appropriate combinatorial algorithm.

Computations in Discrete Structures Using Maple

Some results in semigroup theory using computer methods are published in [12,13,14]. Information concepts concerning redundancy in groups are published in [11], which main result is obtained by computer methods. Some of our experience in using Maple for educational purposes is shared in [10]. A majority of the students are interested mainly in applications in science or engineering. A large proportion of them has serious difficulties with the assimilation of subjects involving formalism [8, 9]. Using Maple they can solve some nontrivial problems involving symbolic calculations.

Computations with Binary Relations Using Maple

We give some examples for computations with binary relations. Tables in Maple can represent successfully binary relations. Let us convert the Boolean matrix A with different row and column rank, which is given in [1], p.15:

$$A = \begin{bmatrix} 1 & 0 & 1 & 0 \\ 0 & 1 & 1 & 0 \\ 1 & 1 & 0 & 0 \\ 1 & 0 & 0 & 0 \end{bmatrix}$$

```
>BB:=bmat_rel(A);  BB := table([1 = { 1, 3 }, 2 = { 2, 3 }, 3 = { 1, 2 }, 4 = { 1 }])

>R_sp_A:=row_space(A,4,4): print(R_sp_A);
      table([1 = { 1, 3 }, 2 = { 2, 3 }, 3 = { 1, 2 }, 5 = { }, 4 = { 1 }, 6 = { 1, 2, 3 }])

> R_sp_A_transposed:=row_space(_ (A,4,4),4,4):
print(R_sp_A_transposed);

      table([1 = { 1, 3, 4 }, 2 = { 2, 3 }, 3 = { 1, 2 }, 5 = { 1, 2, 3, 4 }, 4 = { }, 6 = { 1, 2, 3 }])
```

Converting relations into matrices: `>R_sp_A_matr:=rel_bmat(R_sp_A,6,4):`
`R_sp_A_transposed_matr:=rel_bmat(R_sp_A_transposed,6,4):`
`print(R_sp_A_matr),print(R_sp_A_transposed_matr);`

$$\begin{bmatrix} 1 & 0 & 1 & 0 \\ 0 & 1 & 1 & 0 \\ 1 & 1 & 0 & 0 \\ 1 & 0 & 0 & 0 \\ 0 & 0 & 0 & 0 \\ 1 & 1 & 1 & 0 \end{bmatrix} \begin{bmatrix} 1 & 0 & 1 & 1 \\ 0 & 1 & 1 & 0 \\ 1 & 1 & 0 & 0 \\ 0 & 0 & 0 & 0 \\ 1 & 1 & 1 & 1 \\ 1 & 1 & 1 & 0 \end{bmatrix}$$

CONCLUSIONS

An algebraic approach to the pseudoinverse generalization problem is used, which is similar to those applied in [2]. By analogy with the linear spaces three maps f, g, and p operating on Boolean vector spaces are defined. In the linear algebra the maps, corresponding to f and g, are connected with a matrix and its pseudouinverse respectively and the map, corresponding to p, is the orthogonal projection onto the vector space, generated by the rows of the matrix. Important properties of these maps, which are essential for solving linear systems, remain the same in the case of Boolean vector spaces, see Theorems 1 and 2. The maps could be useful for solving equations in the Boolean algebra, if no pseudoinverse matrix to a given matrix exists.

We also share our experience in using computer algebra systems for teaching discrete mathematics and linear algebra. Some examples for computations with binary relations and Boolean matrices using computer algebra systems are given.

REFERENCES

1. K.H. Kim, *Boolean Matrix Theory And Applications*, Marsel Dekker, Inc., New York, Basel, 1982.
2. G. Bijev, "Some Applications of Semigroups and Computer Algebra in Matrix Theory," in *Applications of Mathematics In Engineering and Economics'33'*, edited by M.D.Todorov, AIP CP946, Melville, New York, 2007, pp. 263-270.
3. E. H. Moore, "On the reciprocal of the general algebraic matrix," Bulletin of the American Mathematical Society 26, 394-395 (1920).
4. R. Penrose, *Proc. of the Cambridge Philosophical Society* **51**, 406-413 (1955).
5. T. Westermann, *Mathematik fuer Ingenieure mit Maple*, Vol.1, Vol.2, Springer verlag, Berlin, Hedelberg, New York, 1996.
6. M. Komma, *Moderne Physik mit Maple*, Int. Thomson Publishing, Bonn, 1996.
7. R.J. Lopez, *Maple via Calculus*, Birdhouse, Boston, 1994.
8. A. Croft, "Innovations in Engineering Mathematics Education in The UK," in *Applications of Mathematics in Engineering and Economics'27*, edited by D.T.Ivanchev and M.D.Todorov, Conference Proceedings, Heron Press, Sofia, 2002, pp. 680-685.
9. Y. Velinov, "Design of Multimedia Lectures for Mathematical Subjects in Computer Science", in *Applications of Mathematics in Engineering and Economics'27*, edited by D.T. Ivanchev and M.D. Todorov, Conference Proceedings, Heron Press, Sofia, 2002, pp. 701-709.
10. G.T. Bijev, "Discrete Mathematics with Maple" in *Applications of Mathematics in Engineering and Economics'31*, edited by M.S. Marinov and M.D. Todorov, Conference Proceedings, Softtrade Ltd., Sofia, 2006, pp. 61-65.
11. N.K. Kasabov and G.T. Bijev, *Cybernetics* **3**, 135-136 (1980). (in Russian)

12. G.T. Bijev and K.J. Todorov, *Semigroup Forum* **31**, 118-122 (1985).
13. G.T. Bijev and K.J. Todorov, *Semigroup Forum* **43**, 253-257 (1991).
14. H. Jürgensen, *Semigroup Forum* **15**, 1-20 (1977).

Hybrid Inverter Analysis Using Mathematical Software

N. Hinov, D. Vakovsky and G. Kraev

Technical University of Sofia; 8 Kliment Ohridski Blvd., 1000 Sofia, Bulgaria

Abstract. This paper examines the transient processes at turning on of the hybrid (series-parallel) current source inverter, using the program MatLab and Internet technologies. The obtained results allow the transient process to be assessed and its design to be done to obtain favourable turn on transient process.

Keywords: Current source inverters, MatLab, research, induction heating, CAD.
PACS: 07.05.Tp, 01.50.hv

PREFACE

The contemporary mechanical engineering is unimaginable without using electrical technologies, based on the induction heating principle. The current inverters are widely distributed power supplies for the realisation of various electrical technologies [1]. Many analyses are well-known in the transient mode [3, 4, 6], as well as the steady mode of operation [3, 4, 5].

Theoretical analysis of resonant inverters may be performed using different methods according to the equivalent schematics which are in effect between two consequent switches of the power elements [3, 4, 7, 8]. Unlike the analytic methods for analyzing the current inverters, the mathematical methods are using computers and permit the accomplishment of the analysis, the design and the verification of the reliability of the results using arbitrary schematics, along with giving account of the secondary elements, the Control System, etc. In that way during the study we make minimum assumptions and therefore we assure maximum accuracy and reliability of the obtained results.

The mathematical modelling begins with preparing an assignment, in which we should include the schematics, the scheme elements, the initial conditions of the currents and voltages of the reactive elements. The actual methods of mathematical modelling permit doing analysis of the entire conduct of line-frequency converter, which consists of phase-controlled rectifier, current inverter and load. And todays IT gives the perfect chance of data exchange and visualization of different types of data used during the modelling.

The present work proposes the creation of mathematical model describing the operation of a hybrid current inverter, operating in transient or steady modes. The system of differential equations describing those operations is solved using specialized

CP1067, *Applications of Mathematics in Engineering and Economics '34—AMEE '08*, edited by M. D. Todorov
© 2008 American Institute of Physics 978-0-7354-0598-1/08/$23.00

mathematical software – MatLab [10], without the help of the visualization package Simulink.

MODEL OF HYBRID CURRENT INVERTER

The schematic of a hybrid current inverter is shown on Figure 1. Semiconductor elements are represented by one-way conductivity switches S1-S4.

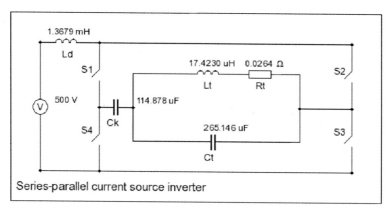

Series-parallel current source inverter

FIGURE 1. Full-bridge current inverter with hybrid compensated load

Brief Mathematical Description of the Model

The operation of the hybrid-resonant inverter is described by the following two systems of differential equations:

$$\dot{x}_i = \sum_{j=1}^{4} a_{ij}^1 x_j + b_i. \qquad (1)$$

Equation (1) with initial conditions $x_i(0) = x_i^1$, for the odd half-periods of circuit operation $i = 1,2,3$. The coefficients in this system are:

$$\{a_{ij}^1\} = \begin{pmatrix} 0 & 0 & a_{13} & a_{14} \\ 0 & 0 & a_{23} & 0 \\ a_{31} & a_{32} & 0 & 0 \\ a_{41} & 0 & 0 & a_{44} \end{pmatrix} \text{ and } \{b_i\} = \begin{pmatrix} 0 & 0 & b_3 & 0 \end{pmatrix}^T,$$

and

$$\dot{x}_i = \sum_{j=1}^{4} a_{ij}^2 x_j + b_i. \qquad (2)$$

Equation (1) with initial conditions $x_i(0) = x_i^2$, for the even half-periods of inverter bridge operation $i = 2, 4, 6$, where

$$\{a_{ij}^2\} = \begin{pmatrix} 0 & 0 & -a_{13} & a_{14} \\ 0 & 0 & -a_{23} & 0 \\ -a_{31} & -a_{32} & 0 & 0 \\ a_{41} & 0 & 0 & a_{44} \end{pmatrix} \text{ and } \{b_i\} = \begin{pmatrix} 0 & 0 & b_3 & 0 \end{pmatrix}^T.$$

We calculate the coefficients

$a_{13} = 1/C$, $a_{14} = -1/C$, $a_{23} = 1/C_k$, $a_{31} = 1/L$, $a_{32} = -1/L_d$, $a_{41} = 1/L$ and $a_{44} = -R/L$, $b_3 = U/L_d$.

The model works with arbitrary number of steps. On each step we consecutively calculate system of equations One or system of equations Two, while at the odd steps we solve the system One and at the even steps – we solve system Two. The peculiarity here is that at every step the initial conditions are changing following the plan: at the first step they are zero, and at every next step – they are the result of solving of the previous system of equations; at the moment $t = T/2$, i.e., in the end of the previous half-period.

Method Used for Solving the Aforementioned Systems of Equations

Lots of authors [4, 6] tend to solve the aforementioned systems of equations using the method of Runge-Kutta, but considering the potential of modern information technology we believe that the matrix exponent method to be more suitable. The usage of the method of the matrix exponent assures an insignificant inaccuracy [9]. This can be proved by comparing the results obtained by the proposed model and those obtained by the computer simulation program PSPICE. While with the method of Runge-Kutta we are obliged to evaluate the inaccuracy, because it could be significant.

We represent the system as a matrix, i.e., (3), with initial condition $x(0) = x^0$

$$\dot{x} = Ax + b. \tag{3}$$

The a.m. matrix $A = \{a_{ij}\}_{i,j=1}^4$ is the matrix of coefficients, while x, b and x^0 are column vectors of the functions to find, the free members and the initial conditions, i.e.,

$x = (x_1(t), x_2(t), x_3(t), x_4(t))^T$, $b = (0, 0, b_3, 0)^T$ and $x^0 = (x_1^0, x_2^0, x_3^0, x_4^0)^T$.

Without describing the matrix exponent method in details, we will remind 2 important properties of that exponent: $\dfrac{d}{dt} e^{At} = A e^{At}$, $e^0 = E$.

Using the a.m. properties we can show that the formula (4), give us the general solution of the system of equations

$$x = e^{At} x^0 + A^{-1}(e^{At} - E)b. \tag{4}$$

512

We calculate

$$\dot{x} = Ae^{At}x^o + A^{-1}Ae^{At}b = Ae^{At}x^o + e^{At}b,$$

$$Ax + b = Ae^{At}x^o + AA^{-1}(e^{At} - E)b + b = Ae^{At}x^o + e^{At}b.$$

Because the right parts of the above expression are equal, therefore formula (4) gives the solution of the system of equations (3).

The precision of the initial condition can be checked directly. Really

$$x(t = 0) = Ex^o + A^{-1}(E - E)b = x^o.$$

Using the a.m. description of the mathematical model we have made software, which simulates the starting process of a hybrid current inverter.

The module for the mathematical calculation is based on the software product MatLab 5.3. The results of the simulation are presented as diagrams describing the changes over time of the load voltage, the input inductivity current, the load voltage and the load current.

SPECIALIZED WEB-BASED PROGRAM FOR THE STUDY OF HYBRID CURRENT INVERTER

A special web-based program has been developed which allows input of initial data and output of the corresponding results as diagrams or tables. The user interface has been developed as a series of web pages which have been published in a web server and are accessible for every Internet or intranet user.

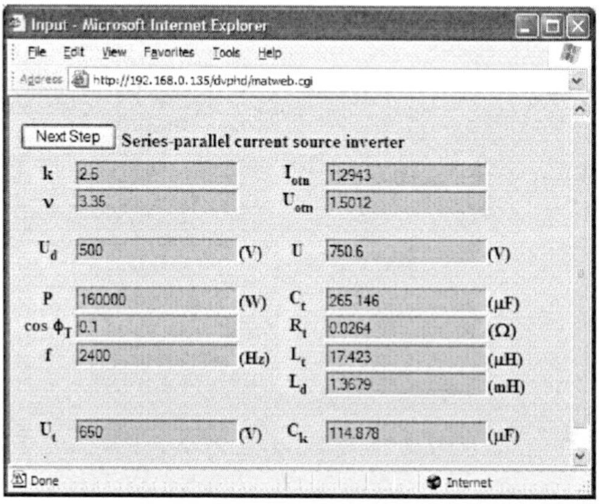

FIGURE 2. The Control Panel of the program

Figure 2 shows the program control panel which is used for the input for initial data. Figure 3 shows the result of the study in a diagram. The obtained results from the solution of the system of equations permit us to determine all remaining currents and voltages in the power circuit of the inverter and their visualization. The figure shows the input current waveform with specific values of the input quantities.

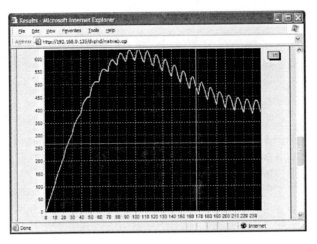

FIGURE 3. Graphic result

COMPARISON OF RESULTS AND ACCURACY ASSESSMENT

In order to check the accuracy of the received results they were compared with results received from simulation of a hybrid current-source inverter using the computer simulator "PSPICE" and also with results received from analysis of the steady mode of operation. The initial data used for the design of the parallel current-source inverter is: output active load power $P=160$kW; load power factor $\cos\varphi_T=0.1$; effective value of load voltage $U_T =650$V; output frequency $f=2400$Hz.

Table 1 is used to compare the results estimated using the method for design of a hybrid current-source inverter with the results computed by the simulator and the MatLab application.

TABLE 1. Results

Results	Estimated	PSPICE	MATLAB
R_T, Ω	0.0264	0.0264	0.0264
L_T, μH	17.4234	17.4234	17.4234
C_T, μF	265.146	265.146	265.146
C_K, μF	114.8787	114.8787	114.8787
L_d, mH	1.3669	1.3669	1.3669
U_T, V	650	652	651
U_{Dm}, V	1060.66	1056.6	1058.63
t_q, μs	48.874	48.4	48.637
f, Hz	2400	2400	2400
I_d, A	320	324.15	322.075

Table 1 shows that the results received from the program created by the authors are similar to the results received from the computer simulator and the steady mode analysis (the difference is less than 2%).

CONCLUSION

The proposed program allows the evaluation of the transition process occurring on start-up of the parallel resonant inverter and enables designing an inverter with the most favourable start-up transition process which is essential with this type of converting devices. The usage of specialized mathematical software and the up-to-date IT enables the integration of the proposed program as a module in a web based allocated system for research of autonomous inverters. This system may include education, choosing of power circuit schematic, estimation of circuit elements' values [2, 11], examination of transition and steady modes of operation using the computer simulator PSPICE and the mathematical software MATLAB, design enhancement based on certain criteria, evaluation of design accuracy and preparation of project documentation.

REFERENCES

1. E. Bercovich, G. Ivenskii, Y. Yoffe, A. Matchak, and V. Morgun, *High-frequency Thyristor Converters for Electrical Networks*, Energoatomizdat, Leningrad, 1983. (in Russian)

2. D. Vakovsky, "CAD of autonomous resonant inverters operating in forced switching mode," in *Report collection of the Second National applied-science Conference with International participation Electronic Technique "ELECTRONIKA '02, 17–18 October 2002*, Sofia, 2002, pp.213-218. (in Bulgarian)

3. V. I. Senko and T. S. Todorov, *Power Electronic Devices*, Gabrovo, 1975. (in Bulgarian)

4. T. S. Todorov, "Thyristor autonomous inverters in the frequency converters in the electrical technology installations ," Ph.D. Dissertation, Gabrovo, 1987. (in Bulgarian)

5. N. Gradinarov, "Analysis and development of autonomous resonant inverters with electrical application," Ph.D. Dissertation, Sofia, 2002. (in Bulgarian)

6. E. Popov, "Examination of the transient process in the single-phased full-bridge thyristor inverters," in *Report collection of the Second National applied-science Conference with International participation Electronic Technique ET'93, Sozopol , Bulgaria*, Vol. 2, 1993, pp.9-17. (in Bulgarian)

7. F. Labrique F., G. Seguier, and R. Bausiere. *Les convertisseurs de l'electronique de puissance - tome 4 "La conversion continu-alternatif*, TEC&DOC – Lavoisier, 1995. (in French)

8. N. Mohan, T.M. Undeland, and W.P. Robbins. *Power Electronics – Converters, Applications and Design*, John Wiley&Sons Inc, New York, 1995, 2nd edn.

9. L. Zade and Tch. Dezoer, *Linear Schematics Theory*, Nauka, Moscow, 1970. (in Russian)

10. Y. Ketkov, A. Ketkov, and M. Shultz, *MATLAB 6.x*, Saint Petersburg, 2004. (in Russian)

11. D. Vakovsky. "Module for automated synthesis of resonant inverters with complicated output circuit," in *Report collection of the National Conference with International participation Electronic Technique "ELECTRONIKA '08*, Sofia, 2008, pp.318-323. (in Bulgarian)

Solving Max-min Relational Equations. Software and Applications

Z. Zahariev

Faculty of Applied Mathematics and Informatics, Technical University of Sofia
1000 Sofia, Bulgaria

Abstract. Method, algorithm and software for solving max-min relational equations are presented. They are based on an existing algorithm, but with some significant improvements, developed by the author. Applications with software implementation are also provided.

Keywords: Fuzzy equations, max-min, algorithm, software, applications.
PACS: 01.30.Cc, 01.50.hv, 02.10.-v, 07.05.Mh

INTRODUCTION

This paper is focused on solving fuzzy max-min linear systems of equations (FLSE) with some applications. More specific, here is presented a new software, intended to deal with this problems. This software is based on algorithm presented in [5], but introduces a lot of improvements, which highly rise its productivity. Since there is also available software, based strictly on the original algorithm [6], in the article is provided comparison between this two different programs.

The studied applications are actually problems which are solved by use at first the base algorithm and software, and then using presented here modifications and new software.

PRELIMINARIES

In this section all needed preliminaries will be defined. The main problem, studied in this paper, will be promoted here, as well. All notions and symbols will be used according to definitions given in this section.

Basic Definitions

Definition 1 A matrix $A = (a_{ij})_{m \times n}$, where $a_{ij} \in [0,1]$ for each $i = \overline{1,m}$, $j = \overline{1,n}$, is called *membership* or *fuzzy matrix*. Fuzzy matrix with $n = 1$ is called *fuzzy vector*.

In the further statement 'matrix' is used instead of 'membership matrix' or 'fuzzy matrix', 'vector' is used instead of 'fuzzy vector'.

Definition 2 Matrices $A = (a_{ij})_{m \times p}$ and $B = (b_{jk})_{p \times n}$, where p is the number of the columns in the matrix A and the number of the rows in the matrix B, are called *conformable*.

CP1067, *Applications of Mathematics in Engineering and Economics '34—AMEE '08*, edited by M. D. Todorov
© 2008 American Institute of Physics 978-0-7354-0598-1/08/$23.00

Definition 3 Inequality between vectors $X = (x_j)_{n \times 1}$ and $Y = (y_j)_{n \times 1}$, marked as $X \leq Y$ or $X \geq Y$ is defined as follows:

$$X \leq Y \text{ if } x_j \leq y_j \text{ for each } j = \overline{1, n}. \tag{1}$$

or

$$X \geq Y \text{ if } x_j \geq y_j \text{ for each } j = \overline{1, n}. \tag{2}$$

Operations

Definition 4 Operation $\vee$ is defined as:

$$a \vee b = \max(a, b). \tag{3}$$

Definition 5 Operation $\wedge$ is defined as:

$$a \wedge b = min(a, b). \tag{4}$$

Definition 6 Operation α is defined as:

$$a \alpha b = \left\{ \begin{array}{l} 1, \text{ if } a \leq b \\ b, \text{ if } a > b \end{array} \right. \tag{5}$$

Products

Definition 7 $Max - min$ product, marked as $\bullet$ ($1 \leq i \leq m$, $1 \leq j \leq n$):

$$A \bullet B = X = (x_{ij})_{m \times n} \quad \text{where} \quad x_{ij} = \bigvee_{k=1}^{n} (a_{ik} \wedge b_{kj}) \tag{6}$$

Definition 8 $Min - \alpha$ product, marked as α ($1 \leq i \leq m$, $1 \leq j \leq n$):

$$A \alpha B = X = (x_{ij})_{m \times n} \quad \text{where} \quad x_{ij} = \bigwedge_{k=1}^{n} (a_{ik} \alpha b_{kj}) \tag{7}$$

Max-min Fuzzy Linear System of Equations

Max-min FLSE is a system of the following form:

$$\begin{vmatrix} (a_{11} \wedge x_1) \vee (a_{12} \wedge x_2) \vee \ldots \vee (a_{1m} \wedge x_m) = b_1 \\ (a_{21} \wedge x_1) \vee (a_{22} \wedge x_2) \vee \ldots \vee (a_{2m} \wedge x_m) = b_2 \\ \ldots \\ (a_{n1} \wedge x_1) \vee (a_{n2} \wedge x_2) \vee \ldots \vee (a_{nm} \wedge x_m) = b_n \end{vmatrix} \tag{8}$$

where $a_{ij}, b_i, x_j \in [0, 1]$, $i = \overline{1, n}$, $j = \overline{1, m}$, here x_i marks the unknown values in the system.

Same system can be represented by its matrix form, written as follows:

$$A \bullet X = B \qquad (9)$$

where $A = (a_{ij})_{m \times n}$ is a matrix with the coefficients of the system (8), $B = (b_i)_{m \times 1}$ is the right hand side of (8) and $X = (x_j)_{1 \times n}$ is a matrix with all unknown values of (8)

Types of Solutions for the a FLSE

Definition 9 $X^0 = (x_j^0)_{n \times 1}$ with $x_j^0 \in [0,1]$ and $j = \overline{1,n}$ is called point solution of the system (9) if $A \bullet X^0 = B$ holds.

Definition 10 The set of all point solutions of (9) is called complete solution set and is denoted by $\mathbb{X}^0$.

Definition 11 A solution $X_{low}^0 \in \mathbb{X}^0$ is called a lower solution of (9) if there is not such a $X^0 \in \mathbb{X}^0$ for which $X^0 \leq X_{low}^0$.

Definition 12 A solution $X_u^0 \in \mathbb{X}^0$ is called an upper solution of (9) if there is not such a $X^0 \in \mathbb{X}^0$ for which $X_u^0 \leq X^0$.

Definition 13 When there is unique upper solution, it is called greatest (or maximum) solution.

It is well known fact that every $max - min$ FLSE has unique upper solution (i.e. greatest solution) and many lower solutions [5].

Definition 14 $(X_1, ..., X_n)$ with $X_j \subset [0,1]$ for each j, $1 \leq j \leq n$, is called an interval solution of the system (9) if any $X^0 = (x_j^0)_{n \times 1}$ with $x_j^0 \in X_j$ for each j, $1 \leq j \leq n$, implies $X^0 = (x_j^0)_{n \times 1} \in \mathbb{X}^0$.

Definition 15 Any interval solution of (9) whose components (interval bounds) are determined by a lower solution from the left and by the greatest solution from the right, is called maximal interval solution of (9).

Definition 16 If for a given system (9), $\mathbb{X}^0 \neq \emptyset$ then the system is called consistent, otherwise it is called inconsistent.

SOLVING FLSE

In spite of that in the last years the interest in solving FLSE increases rapidly and many results were published in this area, there are still only few relatively simple and strain forward algorithms and less than few implemented programs [5], [6], [2], [1], [4], [7]. Software implementation of such algorithms can be found in [6]. Most of the algorithms are still too complicated and sometimes confusing, thus it is hard to write software based on them.

In this work the algorithm presented in [5] is used, because it is strain forward, and a prove for this is the available software implementation of it in [6]. However by this moment a lot of improvements has been done to the original algorithm in order to decrease its complexity and increase its simplicity. Although the original software was used at the beginning of the work discussed here, a complete new program is used for it at this moment.

In this section just brief description of the original algorithm, is given. Many of the notions will not be introduced as well as most of the details. The point of the section is just to present the algorithm and its steps for producing the complete solution set of a FLSE and to show where the improvements are made. More descriptive information of the algorithm can be found in [5], [6], [7]. The original as well as the new software can be obtained for free and are distributed over GPL license [13].

Original Algorithm

Step 1 Rearrangement of the equations

Step 2 Obtain associated system

Step 3 Obtain *IND* vector and the greatest solution of the system. If they do not exist then the system is incompatible - Go to End. If they exist the system is compatible - Go to Step 4.

Step 4 Obtain help matrix

Step 5 Obtain dominance matrix form the help matrix.

Step 6 Use the algebraic-logical method described in [5], [6], [7] to find the minimal solutions of the system from its dominance matrix.

Step 7 Obtain all maximal interval solutions from the greatest solution and from the set of all minimal solution.

Improvements of the Original Algorithm

Step 1 Instead of rearranging the whole system (matrix A with the coefficients and vector B with the right hand side of the system) it is with less computational complexity to arrange only vector B, save the indexes of arrangement in another vector and use it for any further indexing in the algorithm.

Step 2 Although it is really easier for humans to solve a FLSE using its associated system, from programmers point of view there is no difference. That is why this step is totally omitted in the new program.

Step 3 This step is kept as is.

Step 4, Step 5 and Step 6 In general these three steps are kept too, but they are performed simultaneously and in the end, the dominance matrix is produced by obtaining the help matrix and use the algebraic-logical method used in Step 6 on every row of the help matrix, row by row. In this manner the program makes less cycles with bigger complexity but in the end it is much faster, especially for bigger systems.

Step 7 This step is kept as is.

TABLE 1. Execution time comparison

Size	Base algorithm	New algorithm
5×5	0.0195s	0.0016s
9×9	15.9913s	0.0064s
18×14	4.4776s	0.0116s
32×13	1143.9373s	0.1055s

Many other "little" improvements were made to the original program on order to get it faster. Some of them are preallocating of most of the used variables, reuse of variables, etc.

In addition, the way in which the algebraic-logical method from Step 6 is used, is partially changed in the new program, which constitutes the biggest part of the improvement of the execution time of the program.

Programs Comparison

The result of all these little improvements is one great improvement of the whole program. Table 1 contains comparison results of running the two programs on same machine and same examples. This results are obtained on a computer with AMD Sempron 2800+ processor working on 1.6 GHz with 512Mb RAM, using MATLAB7.

APPLICATIONS

All described here applications have implemented programs also available under GPL license. More specific these are:

- Solving fuzzy linear system of inequalities (FLSI) of $\geq$ form.
- Check for linear dependence and linear independence in fuzzy algebra.
- Computing behavior matrix of finite fuzzy machines (FFM), and minimization of FFM.
- Fuzzy optimization.

Fuzzy Linear Systems of Inequalities (FLSI) [13]

This problem is actually easy solvable once we had technique to solve system of equations. The only difference between FLSE and FLSI is the maximum solution. While the maximum solution of FLSE is a vector which have to be computed by one or other technique, the maximum solution of FLSI is always vector from ones. Keeping this in mind, we can use vector from ones for the greatest solution and obtain all lower solutions and all maximal interval solutions of the system using described algorithm and software.

Check for Linear combination, Linear Dependence or Linear Independence over Max-min Fuzzy Algebra [7], [10]

Definition 17 An expression of the form

$$(\lambda_1 \wedge v_1) \vee (\lambda_2 \wedge v_2) \vee \ldots \vee (\lambda_n \wedge v_n) \tag{10}$$

where v_i are fuzzy vectors and $\lambda_1, \ldots, \lambda_n \in [0,1]$ for $i = \overline{1,n}$ is called $max - min$ linear combination of the fuzzy vectors v_i with coefficients λ_i for $i = \overline{1,n}$.

Definition 18 The fuzzy vector B is called $max - min$ linear combination of the fuzzy vectors $A = a_1, a_2, \ldots, a_n$ if there exist X such that $A \bullet X = B$.

With direct usage of the above definition and presented here software, it is easy to determine if some vector is a linear combination of set of vectors.

Definition 19 The set $A = a_1, \ldots, a_n$ of fuzzy vectors is called $max - min$ linearly dependent if one of the vectors from A can be expressed as a $max - min$ linear combination of the others.

Below is a simple algorithm for checking for linear dependence or independence, which uses described algorithm for solving FLSE:

Step 1 Enter $A = a_1, \ldots, a_n$.

Step 2 Solve n-times a system of the form $A \bullet X = B$. Every time omit (initialize with zeros) a vector from the matrix A and set it to the right-hand side. If each system is inconsistent, then the set $A = a_1, \ldots, a_n$ is max-min linearly independent, otherwise it is max-min linearly dependent.

Computing Behavior Matrix and Minimization of Finite Fuzzy Machines (FFM) [7] [11], [12], [8]

Definition 20 A finite fuzzy machine (FFM) is a quadruple $\mathscr{A} = (X, Q, Y, \mathscr{M})$, where:
- X, Q, Y are nonempty finite sets of input letters, states and output letters, respectively.
- $\mathscr{M}$ is the set of transition-output matrices of $\mathscr{A}$, that determines its stepwise behavior. Each matrix $M(x|y) = (m_{qq'}(x|y)) \in \mathscr{M}$ is a square matrix of order $|Q|$ and $x \in X$, $y \in Y$, $q, q' \in Q$, $m_{qq'}(x|y) \in [0,1]$.

Definition 21 Let $\mathscr{A} = (X, Q, Y, \mathscr{M})$ be max-min FFM.
For any $(u|v) \in (X|Y)^*$ the extended input-output behavior of $\mathscr{A}$ is determined by the square matrix $M(u|v)$ of order $|Q|$:

$$M(u \,|v) = \begin{cases} M(x_1|y_1) \bullet \ldots \bullet M(x_k|y_k), \\ \quad \text{if } (u|v) = (x_1 \ldots x_k|y_1 \ldots y_k), k \geq 1 \\ U, \quad \text{if } (u|v) = (e|e) \end{cases} \tag{11}$$

where $U = (\delta_{ij})$ is the square matrix of order $|Q|$ with elements δ_{ij} determined as follows:

$$\delta_{ij} = \begin{cases} 1, & \text{if } i = j \\ 0, & \text{if } i \neq j \end{cases} . \tag{12}$$

Definition 22 Complete input-output behavior matrix $(T_{\mathscr{A}})$ for a FFM $(\mathscr{A})$ is determined as follows:

$$T(u|v)_{|Q| \times 1} = \left(t_q(u|v)\right) = \begin{cases} M(u|v) * E, & \text{if } (u|v) \neq (e|e); \\ E, & \text{if } (u|v) = (e|e), \end{cases} \tag{13}$$

where E is the $|Q| \times 1$ column-matrix with all elements equal to 1.

Definition 23 Behavior matrix $(B_{\mathscr{A}})$ for a FFM $(\mathscr{A})$ is obtained from $T_{\mathscr{A}}$ by removing all columns in $T_{\mathscr{A}}4$ which are linear combination of the columns before them.

From Definition 23 it is visible, that all linear dependent columns of the matrix $T_{\mathscr{A}}$ have to be determined and removed. This is possible by using the algorithm (and software) for checking for linear dependence. By using this algorithm, next follows an algorithm for obtaining $B_{\mathscr{A}}$:

Step 1 Enter the set of matrices $\mathscr{M}$.
Step 2 Find $k \in \mathbb{N}$, such that $T(k) \cong T(k+1)$.
Step 3 Obtain $B(k) = B_{\mathscr{A}}$ excluding all linear combinations from $T(k)$.
Step 4 End.

Definition 24 A FFM $\mathscr{A}$ is in minimal form if there are not any rows in $B_{\mathscr{A}}$ which are linear combination of the other rows.

Step 1 Enter the set of matrices $\mathscr{M}$ and obtain $B_{\mathscr{A}}$.
Step 2 Exclude all rows which are linear combination of the other rows from $B_{\mathscr{A}}$.
Step 3 End.

Fuzzy Optimization [3] [9]

Fuzzy optimization problem is optimization problem with linear objective function:

$$Z = \sum_{j=1}^{n} c_j x_j \tag{14}$$

whit traditional addition and multiplication, where $c_j \in \mathbb{R}$, $0 \leq x_j \leq 1$, $1 \leq j \leq n$, which is subject to a FLSE $(A \bullet X = B)$ of FLSI $(A \bullet X \geq B)$ as constraint.

In general the algorithm for optimization needs to split the negative and non negative coefficients of the objective function into two sub functions:

$$Z' = \sum_{j=1}^{n} c_j^+ x_j \tag{15}$$

$$Z'' = \sum_{j=1}^{n} c_j^- x_j \qquad (16)$$

Then depending on the type of optimization (minimization or maximization) one of the function reach its extremum from the greatest solution of the system of constraint and the other reach its extremum among the minimal solutions. This meant that to solve this optimization problem there is a need of finding the complete solution set for the system of constraint for the optimization problem. Described in this paper algorithm can be used here as well as in the other showed examples.

REFERENCES

1. B. De Baets, "Analytical solution methods for fuzzy relational equations", in the series *Fundamentals of Fuzzy Sets, The Handbooks of Fuzzy Sets Series*, Vol. 1, edited by D. Dubois and H. Prade, Kluwer, Academic Publishers, 2000, pp. 291–340.
2. L. Cheng and P. Wang, *Soft Computing* **6**, 428–435 (2002).
3. S.-G. Fang and G. Li, *Fuzzy Sets and Systems* **103**, 107–113 (1999).
4. M. Miyakoshi and M. Shimbo, *Fuzzy Sets and Systems* **19**, 37–46 (1986).
5. K. Peeva, *Italian Journal of Pure and Applied Mathematics* **19**, 9–20 (2006).
6. K. Peeva and Y. Kyosev "Fuzzy Relational Calculus-Theory, Applications and Software (with CD-ROM)," in the series *Advances in Fuzzy Systems – Applications and Theory*, Vol. 22, World Scientific Publishing Company, 2004, Software downloadable from http://www.mathworks.com/
7. K. Peeva, "Fuzzy linear systems – Theory and applications in Artificial Intelligence areas", DSc Thesis, Sofia, 2002. (in Bulgarian)
8. K. Peeva and Z. Zahariev, "Computing Behavior of Finite Fuzzy Machines – Algorithm and its Application to Reduction and Minimization", *Information Sciences*, Manuscript number: INS-D-07-727R1.
9. K. Peeva, Z. Zahariev, and I. Atanasov, "Software for optimization of linear objective function with fuzzy relational constraint," in *Intern. Conf. on Intelligent Systems, September 2008, Varna, Bulgaria.* (accepted)
10. K. Peeva and Z. Zahariev, "Software for testing linear dependence in Fuzzy Algebra", in *Intern. Sci. Conf. Computer Science, September 2005, Chalkidiki, Greece*, 2005, pp. 294–299.
11. E. S. Santos, *J. Math. Anal. Appl* **24**, 246–259 (1968).
12. E. S. Santos, *J. Math. Anal. Appl* **40**, 60–78 (1972).
13. http://www.gnu.org/copyleft/gpl.html

WORKSHOP ON GRID AND SCIENTIFIC ENGINEERING APPLICATION (GRID AND SEA)

Software Agents in *ADAJ*: Load Balancing in a Distributed Environment

M. Drozdowicz*, M. Ganzha*, W. Kuranowski*, M. Paprzycki*, I. Alshabani†, R. Olejnik†, M. Taifour†, M. Senobari** and I. Lirkov ‡

*Systems Research Institute, Polish Academy of Sciences, Warsaw, Poland
†Computer Science Laboratory of Lille (UMR CNRS 8022), University of Sciences and Technologies of Lille, Lille, France
**Dept. of Computer Science, Tarbiat Modares University, Tehran, Iran
‡Institute for Parallel Processing, Bulgarian Academy of Science, Sofia, Bulgaria

Abstract. *Adaptive Distributed Applications in Java* (*ADAJ*) is a platform developed for execution of distributed applications in Java. The objectives of this platform is to facilitate application design and to efficiently use the power of distributed computing. The *ADAJ* offers both a programming and an execution environment. In the latter it implements object observation and load balancing mechanisms. The observation mechanism allows estimating of the JVM load for each node running the *ADAJ* client. The load balancing mechanism dynamically adapts the workload across the system according to this information. Here we discuss how the original design based on JavaParty is going to be superseded by utilization of software agents.

Keywords: Distributed applications, adaptive load balancing, software agents.
PACS: 07.05.Bx, 07.05.Tp

INTRODUCTION

The *Adaptive Distributed Applications in Java* (*ADAJ*) is a platform developed for efficient implementation and execution of distributed applications in Java. The *ADAJ* offers both a programming and an execution environment. To facilitate efficiency of code execution the *ADAJ* implements object observation which allows estimation of the JVM load on each of its nodes. This information, in turn, allows for utilization of load balancing mechanisms, which can dynamically adapt execution of applications to equalize loads between computers in the system.

The original *ADAJ* was implemented utilizing the JavaParty ([1, 2]) and Java/RMI platforms according to a multi-layer structure using several APIs. Here, the JavaParty provided an execution environment for running distributed applications as well as a mechanism for object migration. The main advantages of utilization of JavaParty were:

- object tracking done by the platform,
- transparency of object placement and execution,
- high efficiency of the solution.

However, it had also some serious disadvantages:

- extremely tight integration with the lower layer of RMI communication,
- utilization of proprietary protocols (KRMI),

CP1067, *Applications of Mathematics in Engineering and Economics '34—AMEE '08*, edited by M. D. Todorov
© 2008 American Institute of Physics 978-0-7354-0598-1/08/$23.00

- problems of integration with component/service oriented frameworks ([3, 4]),
- unknown future direction of development and release cycle timing.

Furthermore, the original design of *ADAJ* had the following drawbacks:

- was not service oriented and thus did not adhere to the vision of *Service Oriented Computing*,
- *ADAJ* applications had to be written utilizing the JavaParty, so the application had to be tightly coupled with the platform,
- did not provide architectural view, but multi-layer view of the platform (allowed for separation of layers but no separation of concerns).

On the basis of these premises we are in the process of re-designing of the *ADAJ* platform. The main goals for *ADAJ 2* are:

1. create *ADAJ* based on *Service Oriented Architecture* principles,
2. remove JavaParty and replace it with a more flexible brokering mechanism,
3. integrating a dynamic deployment model for the *ADAJ* service oriented applications.

In this paper, we discuss how results put forward in the *Agents in Grid* project (see [5, 6, 7, 8, 9, 10, 11, 12, 13]), and software agents in particular, can be utilized on a lower level (within the new *ADAJ*) to:

- detect load imbalance,
- facilitate migration of objects from heavily loaded machine(s) to lightly loaded ones.

To this effect we proceed as follows. In the next two sections we briefly describe the original *ADAJ* project and the *AiG* project. We follow with an outline of a solution that would combine the two project on a high level, where the *AiG* would provide user interface and resource management capability for the *ADAJ*, which would become the work handling infrastructure. Next we discuss how software agents can be introduced into the *ADAJ* to handle load observation and balancing.

ADAPTIVE DISTRIBUTED APPLICATIONS IN JAVA PROJECT OVERVIEW

Let us start from a brief overview of the original ADAJ project(see [14, 15, 16]).

General *ADAJ* Architecture

The original *Adaptive Distributed Applications in Java* (*ADAJ*) was a programming and execution system for distributed applications. The *ADAJ* provided distributed collections and the mechanism of asynchronous invocations. In addition, *ADAJ* carried out load balancing in order to improve the performance of executed applications. This

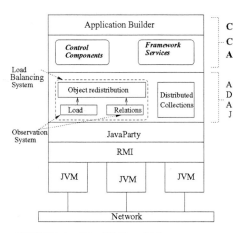

FIGURE 1. The ADAJ multi-layer structure

mechanism of load balancing was based on application of object redistribution. Its mechanisms were based on exploiting and combining information from the load observation system of the execution platform and from the observation system of dynamic relations between objects of applications running in *ADAJ*. The original *ADAJ* was implemented utilizing the JavaParty and Java/RMI platforms according to a multi-layer structure using several API's. Its main components, represented in Figure 1 were:

- Java Virtual Machine (JVM) was regarded as a homogeneous base for construction of distributed applications.
- Remote Method Invocation (RMI, [17]) allowed objects located within various JVM's to communicate between each other as if they were located on the same JVM (using a mechanism of stub/skeleton).
- JavaParty provided an environment of execution of applications distributed on a collection of workstations connected by a network. JavaParty extends the Java language to make it possible to express relatively transparent distribution as well as provides a mechanism for object migration.
- CCA component framework provided services for creating and using CCA compliant components in the platform.

Observation System in *ADAJ*

The observation system [18, 19], needed for load balancing, consisted of two mechanisms. The first is a mechanism for observation of the dynamic relations between the objects during the execution. The second is a computer load observation mechanism. Note that, knowledge of load of all the computers of the system is insufficient to ensure a correct object distribution. The introduction of a mechanism of observation of the relations (and communications) between selected objects within the JVM makes it possible

529

to enrich knowledge about actual application-related loads during the execution. This observation allows estimation of global object work by counting called method activations. In this context, the *ADAJ* recognizes two object types: *local* and *global* objects.

- **Local objects** are traditional Java objects which belong to the users. They are instantiated in the JVM. Their static part is in the same JVM and they can be used only within the JVM they were instantiated in. If these objects are needed on another JVM, they must be copied. The running state of a local object and its attributes are to be copied out in the new JVM. There is no coherence maintained between the original and the copy. A new static part is also created by copying if necessary. This copy is also a local object.
- **Global objects** are typically ADAJ objects. They can be created remotely in any JVM. They are remotely accessible and migrable.

Observation of relations between objects (through method invocation) allows for building dynamically the graph of the object application. The marking of an object consists of adding a characteristics to an object. One example is the addition of the "migrability" property to the object so that it becomes a migrable object, i.e., it can be moved from one JVM to another. To ensure transparency and facility of the object creation, we chose to use marking on the level of the class. The marking of the objects is done implicitly; the marked objects are those which inherit the class *RemoteObject* of JavaParty [2]. This inheritance makes it possible to remotely access appropriately marked objects.

The implementation of the marking process is done by post-compilation techniques. We used a tool of instrumentation of the byte code: JavaClass [20] developed at the Free University of Berlin. JavaClass allows all kinds of transformations of the bytecode: attribute and method additions, method modifications, and per instruction addition. During the post-compilation of an application within the *ADAJ*, the byte codes of all the classes is examined. Classes which inherit the *RemoteObject* are implicitly marked as global classes. The bytecode of these classes is modified by adding suitable marking information.

AGENTS IN GRID PROJECT OVERVIEW

In the *Agents in Grid* project we perceive the Grid as a distributed environment consisting of users ready to pay for usage of resources and resource owners that offer their resources for sale. Furthermore, we distinguish a *local Grid*, where some form of "organizational control" has been instantiated (e.g., Grid within a company, or Grid service put on sale by Sun Microsystems [21]). Here, there exists a specific entity, which is *responsible* for maintenance of the Grid infrastructure (and in some way assures *Quality of Service*; *QoS*). Furthermore, existence of such entity makes it reasonable to sign *Service Level Agreements* (*SLA*). On the other hand in a *global Grid* there is no entity that has a direct administrative control over resources. These resources belong to anyone; e.g., to individual owners, which makes them relatively low granularity (e.g., single PC's). In this case it is rather difficult to believe that a "business strength" *SLA* can be signed and *QoS* assured. Note that success of projects like SETI@HOME is not a counter-argument here. In this case (and in similar projects) individual results can be obtained in any order and there is no particular time limit on their completion (i.e., there is no interdependency between results). This situation is not acceptable in most cases involving business applications, where jobs have to be executed in a specific order and by a certain deadline. To remedy this situation we proposed a different approach (see, [7, 8, 9, 10, 11, 12, 13] for more details), where:

- Every *resource* as well as every *User* is represented by an agent (*LAgent*)
- Agents work in teams
- Each team has an agent-manager (*LMaster*)
- Each *LMaster* has a mirror-agent (*LMirror*); working together (sharing team-related information) they attempt at assuring long-term persistence of the team
- An agent worker (*LAgent*) joins a team that satisfies specified criteria
- Team accepts agent workers according to its own criteria
- Selecting a team to do the job involves negotiations between *LAgents* representing *Users* and *LMasters* representing teams
- The *CIC* (*Client Information Center*) agent plays the role of a central repository where information about all agent teams is stored. Specifically, it contains information about teams that look for workers (who they look for), and teams offering to execute a job (what resources they offer). Utilization of the CIC represents a "yellow page" based approach to matchmaking [22, 5].

Note that, at the current stage, the choice of the machine that will execute the *CIC* agent is out of scope of our considerations. Due to a large number of complicated questions concerning efficiency and scalability of potential solutions, we will have to address this problem in a comprehensive fashion. However, it can be assumed that in the development phase of the proposed system a solution similar to that discussed in [5] can be utilized.

As a summary, the proposed structure of the *AiG* system has been depicted in Figure 3, in the form of an AML ([23]) Social Diagram. Note that *LAgent* and *Worker* are two roles, where the role of the *LAgent* is the basic one. However, when the *LAgent* joins a team it becomes a *Worker*. Furthermore, the *LAgent* can become (start playing a role of) an *LMirror*, or an *LMaster*.

INTEGRATING *AIG* AND *ADAJ* PROJECTS

There are two levels of possible integration of software agents and the *ADAJ* project. The first one is integration of such agents directly into the *ADAJ* fabric and utilizing them for functions like load observation and balancing. We will devote to this possibility later parts of this paper. Here let us look into possibility of direct integration of the *ADAJ* and the *AiG* projects.

Observe that both projects represent different "levels of abstraction." The *ADAJ* project is focused on a relatively low-level (object-centered) design, implementation and execution of applications, and (object-level-based) load balancing of individual nodes (computers) that belong to its infrastructure. The *AiG* project, on the other hand,

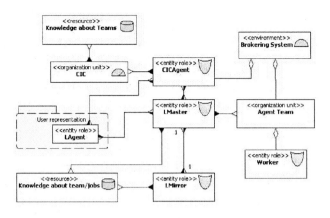

FIGURE 2. The *AiG* System; AML Social Diagram

is concerned with providing high-level infrastructure that will allow users to interact with the Grid infrastructure and contract their resources or job execution. Obviously, a resource contracted through the *AiG* infrastructure could be *ADAJ*-based. Similarly, a user-job submitted to the *AiG* could be later executed by the *ADAJ*-running collection of computers. Therefore, it should be easy to envision a situation in which the *AiG* is an agent-based infrastructure, which manages high-level functionalities (the Brain), while the *ADAJ* is (one of possible) Grid-like infrastructure(s) (the Brawn). The text in parenthesis was deliberately put there to show a direct link between the general vision advocated here and content of the seminal paper entitled "Brain Meets Brawn: Why Grid and Agents Need Each Other" ([24]).

What needs to be done to achieve integration at this level is to create an agent-based interface. Such agent, on the one hand, has to be capable of receiving messages from other *AiG* agents (e.g., the *LMaster*). On the other hand, it has to be able to "communicate" with the *ADAJ* infrastructure, e.g., to dynamically deploy a job, or to pass it in the right form to the right entity to be deployed and executed. On the "way back," it has to be able to receive the results from the *ADAJ* and pass it (wrapped in an ACL message) to the *LMaster*. An initial solution of this type has been outlined in [6].

Note also, that this approach allows the *AiG* to be integrated not only with the *ADAJ* infrastructure. Obviously, an *LAgent* can "represent" a single resource (computer), but it can also be a front-end to other Grid middleware(s). In this way the *AiG* infrastructure may be able to facilitate high-level Grid interoperability, where the responsibility for interacting with independent Grid middlewares will be left to the "gateway agents" representing them (in our case *Worker* agents).

UTILIZING SOFTWARE AGENTS WITHIN THE
ADAJ—OVERVIEW

Let us now discuss how we can utilize software agents within the *ADAJ* infrastructure. Let us recall that one of key functionalities of the original *ADAJ* system was to facilitate load monitoring and balancing and this is what we decided to utilize software agents for. Proposal presented here is somewhat similar to the one put forward by B. diMartino and collaborators in [25]. In their *MAGDA* system, stationary agents controlled workload, while a mobile agent was responsible for load balancing. One of the important problems with that approach is its heavy reliance on agent mobility, which can be costly. As a matter of fact, one of the reasons for the *AiG* system conceptualized as described in section was to overcome some limitations of the *MAGDA* system. Therefore, we propose a different approach to introduction of software agents (at this stage, primarily, as load managers) into the *ADAJ* system:

- each node (computer) has a *Local Agent* instantiated
- each *Local Agent* utilizes mechanisms developed within the initial *ADAJ* project (see above) to monitor activities within the local JVM
- *Local Agents* in specific time intervals communicate their *local workloads* to the *Central Manager* (specifically, post them to the *blackboard* under its control); this process is similar to bidding in an auction (see also [26])
- the *Central Manager* in specific time intervals calculates *average load* and sends this information as an ACL *INFORM* message to all *Local Agents*
- overloaded *Local Agents* negotiate with underloaded ones transfer of some of their load to them

Let us make a few additional comments. Process described above is completely asynchronous. *Local Agents* post their load information at certain time intervals, however this is a low priority task that should not interfere with their other work. While the *Central Manager* updates the average load at specific time intervals, this process does not depend on having all information updated by all *Local Agents*. Instead, it uses information that is currently available within the blackboard. We also assume that the *Central Manager* may be sending information about change of the average workload only if the change is larger then a certain value (e.g., 5%). This assumption is made to reduce the overall number of messages sent in the system. Finally, note also that it is possible that instead of a push-based informing about average workload (the *Central Manager* sends information to *Local Agents*) it is possible to utilize a pull-based approach in which (*Local Agents* check the average workload when they see a need for this; e.g., when their local workload changes). Finding the most efficient approach for exchanging information about local and average workloads requires further research, including experimental work, similar to that reported in [26].

LOAD BALANCING

Let us now look into more details of processes involved in utilizing software agents in load balancing within the *ADAJ* infrastructure. Since the *services* that the *ADAJ 2* will be operating on (recall that *ADAJ 2* will be fully *SOA*) consist of objects, we can apply the similar level of granularity of observed processes as in the original *ADAJ* and consider an "object" as a basic entity to be operated on. Taking this into account, in the process of load balancing we can distinguish three basic tasks:

- detecting imbalance
- determining objects to move and the target node for the migration
- migrating selected objects

In the following sections we will specify in some detail how software agents can help in completing these tasks.

Load Monitoring and Detecting Imbalance

The original *ADAJ* infrastructure incorporated a highly centralized model of observing the load of each machine and detecting the overload or underload of its nodes. Even though the information about the local load of each JVM was generated by a *local Observer module* running on every node, the whole list of objects in the JVM was transferred (through the JavaParty infrastructure) to the *global Observer module* for load analysis. This implies that the *global Observer module* needed to posses updated information not only about local workload, but also about all objects residing on all the machines in the *ADAJ* infrastructure. This solution, though satisfactory for small to medium-sized Grids (especially thanks to the highly efficient implementation of the JavaParty), was likely to cause serious scalability issues for larger systems.

The centralization of knowledge and decision making processes, as well as the amount of communication required for updating the *global Observer module* about every change in the object structure of the local JVM resulted in the solution outlined above. Note that, while it is possible to utilize only an approximate value of average load balancing, to be able to decide which objects are to be moved current information about object localization is absolutely necessary.

In the proposed approach, we will still utilize *local Observer modules* designed and implemented in the original *ADAJ*, but instead of transferring data generated by them to a central location to be analyzed, we will leave decisions to the *Local Agents*. Obviously, being over (or under) loaded depends on the state of all other nodes, while the average load value varies over time. Therefore, we have decided to introduce a *Central Manager*, that collects (within a blackboard) information about load of individual nodes. It should be noted that, even though we are introducing a global entity storing information about all the nodes (the *Central Manager*), this entity differs greatly from the *global Observer module* from the original *ADAJ*—it receives only information about the load level, not about specific objects generating this load. Furthermore, as indicated above, it periodically uses data stored within the blackboard and prepares information about the *estimated* average load. This information is sufficient for all nodes to establish if their workload is high enough to be claimed to be in a state of overload, or low enough to be considered underloaded. Regardless of a specific solution used (push or pull-based approach), after a *Local Agent* obtains/receives the current average load estimate it can compare its own load to this value and make a decision on whether it needs to act (try to off-load some of its work), or if its load can be considered "balanced." Note that a specific value of the difference between the average load and the local load that will initiate an action of the node remains to be established experimentally. Observe also that in the current design of the system only agents that are overloaded will be acting trying to reduce their loads (will be active), while underloaded nodes will be passively awaiting load increasing proposals. It is however conceivable that both sides could be actively seeking ways of balancing their load. We will consider this latter solution in the future.

Determining Which Objects to Move and Where

When the *Local Agent* determines that the node it represents is overloaded and thus decides to move some of its objects to a different one, it needs to find a machine that will accept a specific part of the load. In the original *ADAJ* this functionality was performed by the *global Load module*. Based on the fact that dynamic adaptability of workflows is one of the few areas of successful utilization of software agents in real-life applications (see, [27, 28, 29]), we have decided that this is precisely where software agents can help in increasing scalability and efficiency of the platform. Specifically, in both cases reported in literature software agents were associated with entities placed in a dynamic environment with a goal of managing local workload through utilization of negotiations. In the case of Daimler Chrysler ([27, 28]) agents representing multi-purpose machines in an assembly plant negotiated the flow of parts that were to be produced. In the transport case ([29]), agents representing trucks, load and drivers dynamically negotiated flow of goods. In both cases a performance gain of about 10-20% over static scheduling was reported.

We have, therefore, envisioned a model in which load balancing decisions in the system are made in the course of a multi-agent negotiation (see, also [30, 31]). In this solution the overloaded agent sends a *Call For Proposal* (*CFP*) message to other agents in the system, specifying objects it would like to "give away." Here we plan to use the *FIPA Contract Net Protocol* ([32]) for the negotiation process. However, negotiations may need to be modified to be able to iteratively negotiate with selected partners (which is not included in the original Contract Net, where only a single round of negotiations is expected). Selection of specific objects that the *Local Agent* would like to send to another node should be a result of (multi-criterial) analysis of: (a) object-dependencies, i.e., local and global objects, (b) cost of moving objects (or groups of objects), (c) economical constraints, etc. Agents receiving the *CFP* will utilize (multi-criterial) analysis to establish if accepting these objects "makes sense to them."

Note that we assume here that agents in the system are "benevolent." In other words, we assume that underloaded agents are "very interested" in receiving objects from overloaded agents. Obviously, this assumption reduces (if not completely eliminates) the impact of the economic considerations within the *ADAJ*. However, we would like to claim that this assumption is a reasonable one. Note that the *ADAJ* is a very tightly coupled system and can be considered an example of a "desktop Grid," where all resources belong to a single owner. In this case the assumption about benevolence of agents working together to maximize the throughput is a reasonable one. At the same time, integration of the *ADAJ* with the *AiG* will represent the moment in which the economic model will come to play with full force. Here, the agent representing the *ADAJ*-based infrastructure will try to negotiate the best conditions for selling resources it represents.

When considering the scalability of the proposed approach we need to address the overhead related to sending a *CFP* to all *Local Agents* within the *ADAJ* infrastructure. Let us assume that in the push model information about current average workload has been send to all (n) *Local Agents* in the *ADAJ* infrastructure. Next, k of them have determined that they are overloaded. As a result $k * (n - 1)$ *CFP*s are send. Obviously at least $(k - 1) * (k - 1)$ of these messages are sent completely uselessly (to agents that are overloaded to start with; while, currently, we do not allow for object swapping as a method for performance optimization). To avoid this situation, we propose a solution in which the *Manager Agent*, apart from calculating the estimated average load, also holds a list of underloaded nodes. This can be achieved in a very easy way as each node has a "placeholder" within the blackboard and thus the *Manager Agent* can immediately establish which nodes are underloaded. Therefore, when a *Local Agent* representing an overloaded machine needs to find a receiver for some of its objects, it can query the *Manager Agent* for a narrowed-down list containing only these agents that can be expected to be interested in receiving more work.

Once the agent determines (as a result of negotiations), which objects are to be sent and to which node, the object migration process can commence.

Object Migration

In the original *ADAJ* platform the task of object migration and tracking their location was handled by the JavaParty framework. However, due to reasons described above, we have decided to change this layer of the application. Thus, a need arose to develop or facilitate a solution for for transferring objects and for tracing movement of objects (to update all necessary references once an object moves). We have considered using agents in both of these challenges, but regarding the migration process itself we have found the following disadvantages of such approach:

- To be compliant with FIPA standards ([33]) and taking into account capabilities of existing agent platforms, objects would have to be transfered within ACL messages (only the *Voyager* agent platform introduced in the late 90'th and currently extinct had capabilities for transferring objects directly between agents; [34]), which would limit the scope of possible serialization methods and generate an overhead related to serialization and deserialization.
- The solution would require to be developed from scratch within the JADE ([35])

agent platform (as we are not aware of any off-the-shelf middleware frameworks taking care of such functionality)

We have therefore decided, that the task of moving an object from one machine to another can be more easily implemented using a standard approach, especially given that we have experience in this field gained when developing the original *ADAJ*.

Apart from the migration itself, it is also necessary to be able to trace movement of objects, as local objects depend on such capability of the system. This functionality, on the other hand, necessitates capability of notifying nodes holding dependent objects about migration of their dependencies. Considering this task we have come up with a simple protocol of communication between the nodes engaged in the migration process. This protocol will ensure that the migrating object is not used while it is in a transition state and that the information about its new location is delivered promptly and reliably.

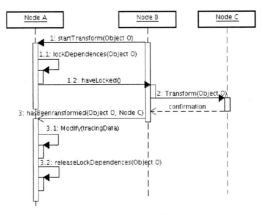

FIGURE 3.   Sequence Diagram

In the diagram we can see 3 nodes (A, B, C):

- Node A holds some objects that depend on the object O that is being transferred
- Node B is overloaded and starts the transfer of object O to node C
- Node C is underloaded and is ready to receive object O

As is shown on the diagram, the protocol specifies the interaction as follows:

- Node B sends a message to Node A telling it about the start of the transfer of object O
- Upon receiving the notification, Node A locks all remote references to object O, to prevent calling O's methods during the transferring process
- When O has been locked, Node A sends a confirmation message to B, to tell it that it is now safe to transfer the object
- Object O is transferred to Node C
- Node C sends a confirmation message to node B, to tell that transformation has been completed

537

- Node B sends a notification to Node A, telling it that the object reached its destination and providing it with the object's new location
- Node A modifies its tracing data with the new location of object A and releases locks on references of object O's

Again, we have considered the agent-based approach to this problem. The requirement for each node is to have an entity listening to notifications from other nodes and then registering the movement of the dependent objects. We have decided that this is a good place to utilize capabilities of existing software agent platforms (JADE in our case), due to the following reasons:

- JADE agents take the burden of inter-platform communication off the developer—JADE provides a uniform model of sending and receiving messages.
- JADE agents run in a separate thread, hence can be easily used as listeners to messages from the outside of the platform without interfering with the task execution
- JADE agents have built-in queuing of messages, thus enabling better scalability of the solution.

It should be noted, however, that using agents in such a scenario makes it necessary to implement an interface of integration between the non-agent (*ADAJ*) middleware infrastructure, used by the *ADAJ* services for remote method invocation and the listener agents. The possible implementation of such interaction depends mostly on the selection of the agent execution scenario.

- If the agent is to be run separately of the *ADAJ* infrastructure, then the only way for them to interact would be by some means of inter-process communication (such as pipes or shared memory). This would, of course, mean that the infrastructure would need either to constantly monitor the shared memory or check it for notifications on every remote method invocation. This solution, while separating the infrastructure from the inter-host communication method is likely to be burdened with a serious inter-process communication overhead.
- The agent can run in the same process as the *ADAJ* infrastructure. If this is the case, the agent could simply pass the reference to an appropriate module taking care of the object tracking and interact by calling the module's methods directly.

Not determining the final implementation approach (which will be determined experimentally), we can still state that the proposed solution to migration and movement notification lets us clearly separate the infrastructure responsible for object migration, tracking and remote invocation from the agent layer that takes care of providing interested parties with the information about when and to where specific objects are to be transferred.

CONCLUDING REMARKS

In this paper we have considered how software agents can be introduced into the *ADAJ* middleware. Our observations and experiences based on the *Agents in Grids* project allowed us to specify two levels of agent-*ADAJ* integration. The high level that can be used without any direct interference within the *ADAJ* and a low level that infuses software agents into *ADAJ*. We have also outlined how the latter proposal can be actually realized. We are currently investigating which infrastructure can be used to provide flexible and efficient object migration that will also be easily integrable with appropriate parts of the *ADAJ* and JADE agents. We will report on our progress in subsequent publications.

ACKNOWLEDGMENTS

Work of the Polish team was in part supported from the "Funds for Science" of the Polish Ministry for Science and Higher Education for years 2008-2011, as a research project (contract number N N516 382434). Collaboration of the Polish and Bulgarian teams is partially supported by the *Parallel and Distributed Computing Practices* grant. Collaboration of Polish and French teams is partially supported by the PICS grant *New Methods for Balancing Loads and Scheduling Jobs in the Grid and Dedicated Systems.*

REFERENCES

1. Javaparty software, University of Karlsruhe (2008), http://svn.ipd.uni-karlsruhe.de/trac/javaparty/
2. M. Phillippsen and M. Zenger, "JavaParty—Transparent Remote Objects in Java," in *ACM 1997 Workshop on Java for Science and Engineering Computation*, Las Vegas, USA, 1997.
3. I. Alshabani, *A Framework for Distributed and Parallel Software Components,* Ph.D. thesis, University of Lille, Lille, France, 2006.
4. I. Alshabani, R. Olejnik, B. Toursel, M. Tudruj, and E. Laskowski, "A Framework for Desktop Grid Applications: CCADAJ," in *ISPDC*, 2006, pp. 208–214.
5. M. Dominiak, W. Kuranowski, M. Gawinecki, M. Ganzha, and M. Paprzycki, , in *Proc. of the International Multiconference on Computer Science and Information Technology*, PTI Press, 2006, pp. 327–335.
6. R. Olejnik, B. Toursel, M. Ganzha, and M. Paprzycki, "Combining software agents and grid middleware," in *Proc. of the GPC 2007 Conference*, edited by C. Cerin, and K.-C. Li, LNCS 4459, Springer, Berlin, 2007, pp. 678–685.
7. M. Dominiak, W. Kuranowski, M. Gawinecki, M. Ganzha, and M. Paprzycki, "Utilizing agent teams in Grid resource management—preliminary considerations.," in *Proc. of the IEEE J. V. Atanasoff Conference*, IEEE CS Press, Los Alamitos, CA, 2006, pp. 46–51.
8. M. Dominiak, M. Ganzha, and M. Paprzycki, "Selecting grid-agent-team to execute user-job—initial solution," in *Proc. of the Conference on Complex, Intelligent and Software Intensive Systems*, IEEE CS Press, Los Alamitos, CA, 2007, pp. 249–256.
9. M. Ganzha, M. Paprzycki, and I. Lirkov, "Trust Management in an Agent-ased Grid Resource Brokering System–Preliminary Considerations," in *Applications of Mathematics in Engineering and Economics'33*, edited by M.D. Todorov, AIP CP946, Melville, New York, 2007, pp.35–46.
10. M. Dominiak, M. Ganzha, M. Gawinecki, W. Kuranowski, M. Paprzycki, S. Margenov, and I. Lirkov, *International Transactions on Systems Science and Applications* **3**, 296–306 (2008).
11. W. Kuranowski, M. Ganzha, M. Gawinecki, M. Paprzycki, I. Lirkov, and S. Margenov, *International Journal of Computational Intelligence Research* **4**, 9–16 (2008).
12. W. Kuranowski, M. Ganzha, M. Paprzycki, "Supervising Agent Team an Agent-based Grid Resource Brokering System—Initial Solution," in *Proc. of the Conference on Complex, Intelligent and Software Intensive Systems*, edited by F. Xhafa, and L. Barolli, IEEE CS Press, Los Alamitos, CA, 2008, pp. 321–326.

13. W. Kuranowski, M. Paprzycki, M. Ganzha, M. Gawinecki, I. Lirkov, and S. Margenov, "Agents as resource brokers in grids—forming agent teams," in *Proc. of the LSSC Meeting*, LNCS, Springer, 2007.
14. R. Olejnik, A. Bouchi, and B. Toursel, "An Object Observation for a Java Adaptative Distributed Application Platform," in *PARELEC*, 2002, pp. 171–176.
15. I. Alshabani, R. Olejnik, and B. Toursel, "Parallel Tools for a Distibuted Components Framework," in *International Conference On Information and Communication Technologies: From Theory To Applications*, Damascus, Syria, 2004.
16. R. Olejnik, V. Fiolet, I. Alshabani, and B. Toursel, "Desktop Grid Platform for Data Mining Applications," in *International Symposium on Parallel and Distributed Computing, ISPDC-06*, Timisoara, Romania, 2006.
17. Remote method invocation home, http://java.sun.com/javase/technologies/core/basic/rmi/index.jsp
18. R. Olejnik, A. Bouchi, and B. Toursel, "A Java Object Observation Policy for Load Balancing," in *PDPTA*, 2002, pp. 816–821.
19. A. Bouchi, R. Olejnik, and B. Toursel, "An Observation Mechanism of Distributed Objects in Java," in *PDP*, 2002, pp. 117–122.
20. M. Dahm, "Byte Code Engineering," in *Java-Informations-Tage*, 1999, pp. 267–277.
21. Sun utility computing, Sun Microsystems (2008), http://www.sun.com/service/sungrid/index.jsp
22. D. Trastour, C. Bartolini, and C. Preist, "Semantic web support for the business-to-business e-commerce lifecycle," in *WWW '02: Proc. of the 11th international conference on World Wide Web*, ACM Press, New York, NY, USA, 2002, pp. 89–98, ISBN 1-58113-449-5.
23. R. Cervenka and I. Trencansky, *Agent Modeling Language (AML): A Comprehensive Approach to Modeling MAS*, Whitestein Series in Software Agent Technologies and Autonomic Computing, A Birkhauser book, 2007.
24. I. Foster, N. R. Jennings, and C. Kesselman, "Brain Meets Brawn: Why Grid and Agents Need Each Other," in *AAMAS '04: Proc. of the Third International Joint Conference on Autonomous Agents and Multiagent Systems*, IEEE Computer Society, Washington, DC, USA, 2004, pp. 8–15, ISBN 1-58113-864-4.
25. R. Aversa, B. D. Martino, N. Mazzocca, and S. Venticinque, *Journal of Grid Computing* **4**, 395–412 (2006).
26. K. Wasilewska, M. Gawinecki, M. Paprzycki, M. Ganzha, and P. Kobzdej, *IADIS International Journal on WWW/INTERNET* (2008). (to appear)
27. S. Bussmann and K. Schild, "An agent-based approach to the control of flexible productionsystems," in *Proc. 8th IEEE International Conference on Emerging Technologies and Factory Automation*, 2001, vol. 2, pp. 481–488.
28. R. Schoop, R. Neubert, and B. Suessmann, "Flexible manufacturing control with PLC, CNC and software agents," in *Proc. 5th Intern. Symposium on Autonomous Decentralized Systems*, 2001, pp. 365–371.
29. M. Becker, G. Singh, B.-L. Wenning, and C. Gorg, *Intern. Journal of Services Operations and Informatics* (2007).
30. M. Tu, F. Griffel, M. Merz, and W. Lamersdorf, "A Plug-in Architecture Providing Dynamic Negotiation Capabilities for Mobile Agents," in *Proc. MA'98: Mobile Agents*, edited by K. Rothermel, and F. Hohl, Springer-Verlag, 1999, vol. 1477 of *LNCS*, pp. 222–236.
31. C. Preist, N. Jennings, and C. Bartolini, "A Software Framework for Automated Negotiation.," in *Proc. of SELMAS'2004*, LNCS 3390, Springer Verlag, 2005, pp. 213–235.
32. Fipa contract net protocol specification, http://www.fipa.org/specs/fipa00029/SC00029H.html
33. The foundation of intelligent physical agent (fipa), http://fipa.org/
34. F.-J. Wang, C.-Z. Liao, J.-W. Huang, and S.-H. Chen, "The Seventh IEEE Workshop on Future Trends of Distributed Computing Systems," 1999, chap. An Agent Platform and Related Issues, p. 219.
35. Jade—java agent development framework, TILab (2008), http://jade.tilab.com/

Security in Service-oriented Grid

R. Goranova

Faculty of Mathematics and Informatics, St.Kliment Ohridski University of Sofia, Bulgaria

Abstract. Service-oriented architecture (SOA) is the most common used approach for building architectural decisions of a system. SOA is an architectural model for developing reliable distributed systems, which functionality is provided as services. Features from this approach are services' interoperability, loose-coupling, reusability and discoverability. Grid is a new technology for using computers, network and devices. The Grid is infrastructure for coordinate use of distributed resources, shared by different institutes, computational centers and organizations. Grid is dynamic and heterogeneous system, beyond any institutional or organization boundaries. Considering this, the development and integration of Gird application in such dynamic and heterogeneous environment is a great challenge. Something more developing of Grid application is still tightly related with the underlying middleware and its security. Service-oriented Grids are one step further in integration of Grid application, because of the SOA features. These benefits from Grid and SOA however will require and new security approaches for building application. In this article, we will briefly describe some known security approaches in SOA and Grid and how they are applied in security models of two well known service-oriented grid architectures – OGSA and EGA.

Keywords: SOA security, grid security, service-oriented grid security.
PACS: 01.50.hv, 07.05.-t, 07.05.Bx

SERVICE-ORIENTED ARCHITECTURE

Service-oriented architecture [1] is an architectural model for developing reliable software solutions, distributed applications and systems. As architecture SOA provides terms for formal description of the system, defining system's functions, properties and interfaces. A main unit in SOA is the service. Most common a service can be described as software component that realizes specific system's logic and provides it to the end client by network access. The service is accomplished by the interchange of messages between three entities: a service provider, a service consumer and a service broker.

A service provider allows access to the service, creates a description of the service and publishes it to the service broker. A service broker hosts a registry of the service description and returns a list of suitable services to the consumer. A service consumer discovers and reads the service description and invokes it by message exchange.

The services have interfaces, which describes a set of public operations for each service. Interfaces are self-describing and platform independent. Only these operations of the service, which are defined as public ones by interfaces are accessible for the client.

CP1067, *Applications of Mathematics in Engineering and Economics '34—AMEE '08*, edited by M. D. Todorov
© 2008 American Institute of Physics 978-0-7354-0598-1/08/$23.00

Principles of Service-orientation

Major advantages of SOA are loose coupling, reusability, discoverability and interoperability. These features and the one defined bellow are often called principles of service-orientation. They define the contract, that systems designed in this way are service-oriented.

Principles of service-orientation:

- Loose coupling - the underlying logic of a service can be change without or with minimal impact on the other service within the same system.
- Contract - documents describing how a service can be programmatically accessed.
- Abstraction - services expose only the logic defined in service contract and hide implementation from the end user.
- Autonomy - services have control only the logic and functionality, which they capsulate.
- Reusability - services can be reused more then once and from multiple clients.
- Composition - services can be grouped in composite service, which coordinate the date exchange between composite services.
- Statelessness - services can not save the state of activity.
- Interoperability – services are platform and implementation independence.
- Discoverability – services are discoverable, there has to be a standard mechanism, which allow services to be discovered.

GRID

Theoretically the term Grid was defined for the first time in 1998 by Carl Kesselman and Ian Foster in their book "The Grid: Blueprint for a New Computing Infrastructure"[2]. According that definition "A computational grid is a hardware and software infrastructure that provides dependable, consistent, pervasive and inexpensive access to high-end computational capabilities." Few years later [3] the Grid has been defined with more clear definition. The Grid is hardware and software infrastructure, which provide coordinated resource sharing and high-end computational capabilities, based on standard open protocols, for delivering of qualities of services. So if we generalize the main idea of the Grid is to provide inexpensive and consistent access to unlimited computational and storage recourses.

Grid is layered infrastructure, often presented with "The layered grid architecture" (Figure 1) from Foster's and Kesselman's book.

Every layer (fabric, connectivity, collective, etc.) provides specific functions (high-level or low-level), protocols, services or resources.

Resources are logical units. They can be computational resources, storage systems, network resources, specific devices, sensors even clusters or pools. They are accessed through security mechanism provided by connectivity layer. This layer also defines communication and authentication protocols for data exchange. Grid security is important part from every grid infrastructure, especially when it is based on sharing.

Security mechanism defines agreements between the owner of resources and end users and authentication mechanism to verify users' and resources' identity. Such agreements often call policy of access.

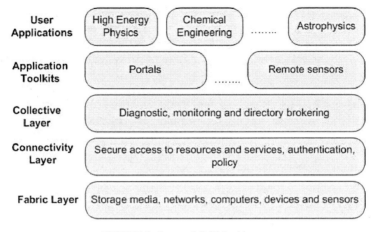

FIGURE 1. Layered Grid Architecture

Collective layer contains protocols and services for resource interaction monitoring, diagnostics, scheduling, brokering, service discovering and others. Resource brokering and scheduling are responsible for finding the best match between required resources from the user and available resources in the Grid. Information system – monitoring, accounting and diagnostic services provides statistic and accounting information for users, jobs and resources.

For scientific tasks or computation of financial, bio-medical or physical tasks, the Grid is the most suitable technology, because it provides better effectiveness decreasing the time for computation. In practice it provides unlimited access to computational power and storage resource with only one user's login (single sign on) to the system. On the other hand, the development of grid application is not an easy task, because of the dependency from the underlying grid environment. This makes grid application inflexible, dependable from the middleware and difficult for reuse.

SERVICE-ORIENTED GRID

In the recent years, it has become clear that there is considerable overlap between the goals of Grid computing and the benefits of SOA based on Web services. Service-oriented Grid is distributed environment, a grid infrastructure that follows the principles of service-orientation. Service-oriented does not mean just to provide its basic functionality as services, but also to provide and mechanism for discovering, registering and invoking these services. Since SOA is an architectural model for system design, service-oriented means designed in such a way that all principles mentioned above to be kept. Keeping principles of service-orientation in Grid will benefit grid environment with features as reusability, composability, interoperability and discoverability. Following SOA however will require new approaches, new components and new functionalities. We can say that service-oriented grids are the next step of evolution for Grid infrastructures.

Below are presented two different approaches for building service-oriented grid based on OGSA and EGA architectures.

The Open Grid Services Architecture (OGSA) describes architecture for a service-oriented grid computing environment for business and scientific use. During the years, OGSA domineer as grid architecture for building Grid systems and it is one of the less that tightly hold SOA principles.

Another direction in service-oriented grid architectures is Enterprise Grid. It specifies the use of grid computing within the context of a business or enterprise. Enterprise Grid Alliance (EGA) consortium is collaboration of well known organizations as Oracle, Novell, Sun Microsystems, HP, Cisco Systems and others with the purpose to drive the adoption of grid computing and the technologies that enable it within enterprise data centers.

SERVICE-ORIENTED AND GRID SECURITY APPROACHES

Security Requirements

When we talk about security of the system, it can be broadly defined. By security some can understand network security, other platform security. However, the purpose of this observation is application security.

Common Requirements

These aspects of security are standard for every Grid or service-oriented application. They are:
- Authentication – This is the process of verification of users' identity. The purpose of this process is to assure the system that the user belongs to it.
- Authorization – This is the process of verification of users' rights. The purpose of the process is to establish permitted user actions and access to system's resources.
- Data confidentially – Protects the confidentiality of the sensitive data.
- Data integrity – Ensures that unauthorized changes made to data may be detected by the recipient.
- Privacy – User data and action are visible only by the owner of the data and no one else can access them.

SOA Requirements

SOA requires from applications to open their capabilities and provide their functionalities for use by other application, in such a manner that it will be possible composition of services from different application in high-level services. This requires that:

- The services have to be kept as open as possible;
- The services have to be as easy to use as possible;
- The services have to be interoperable.

SOA security solutions have to deal with these requirements in such way that they have to be manageable and easy to use. The most import in this case is that interoperability as main SOA feature should not suffer because of security.

Grid Requirements

The security requirements in a grid environment can be defined according to the following categories: integration with existing system and technologies, interoperability with different hosting environment and trust relationship. The specific of them are:

- Single sign-on (logon) – This is the availability of system's user to authenticate just ones in front the system and to access system resources without need to re-authenticate for some reasonable period of time. Bearing in mind that Grid is over institutional boundaries and that Grid can give access to multiple resources which belongs to different organizations, this requirement is very essential for such kind of infrastructures;
- Delegation – This is the facility to delegate access rights from the owner (user) to services, processes, etc.;
- Local security policies of multiple trust domains cannot be overridden by the grid security policy;
- Virtual organizations (VO) - are sets of separate people, institutions and organizations, which share data or work on the same field. Virtual organizations define the requirements, the conditions and the ways how Grid to be accessed.

Security Approaches in SOA

SOA offers three security models approaches [4], to answer the requirements mentioned above: message-level security, policy-driven security and service-based security.

Services as main unit in SOA are presented by interfaces. They describe a set of public operations for each service. Operations are defined as a set of messages. Messages specify the data to be exchanged between the client and the service.

Message-level security is approach based exactly on securing data on message level. In this approach, different parts of a message can be protected differently, to make them usable only by intended parities in the message path. The protocol which implements this approach is WS-Security. This protocol contains specifications on how integrity and confidentiality can be enforced on Web services messaging. These specifications provide mechanism to secure SOAP message exchange, including signatures and encryption in SOAP messages' headers.

Other SOA approach is policy-driven security. The idea of this approach is that all security requirements should not be built in application logic, instead of this they should be declared separately as a security policy. Policy is information for requirements, capabilities and constraints for service or application. Every time when a service has to invoke another service or application, the invoking service first fetch known security policies of the invoked service (application), then compared it with its own local policies and only after that invokes the service with effective policy. The protocol which implements this approach is WS-Policy. The specification provides the basic syntax for expressing policies.

Considering the concept of SOA, the third approach service-based security approach is somehow the most intuitive. The idea of this approach is that security is realized as a service in such a way that this service has to provide abilities to authenticate, authorized, encrypt/decrypt messages, verify signatures and log messages. It is not an ordinary service but infrastructural one. This service can be invoked explicitly or not explicitly by other services. The protocols, which specify how security service can be invoked, are WS-Trust and SAML. Often, the role of security service is implemented in software infrastructure called Enterprise service bus (ESB). An ESB simplifies the integration and flexible reuse of business components within a service-oriented architecture.

Grid Security Approaches

Grid middleware [5] is software environment, which coordinates and integrates Grid resources, services and virtual organizations. The most common used approaches which Grid uses to secure the middleware are based on Grid Security Infrastructure (GSI) with Public Key Infrastructure (PKI) and Virtual Organization Management Service (VOMS) technologies.

PKI provides mechanism for certificate creation, management and verification. Certificate is a file, which combines digital signature with a Certification Authority (CA) public key and user identity.

The GSI uses PKI as the basis for its functionality. The primary motivations behind the GSI are: to secure communication between elements of the Grid, to support security across organizational boundaries, to support "single sign-on" for users of the Grid, including delegation of credentials.

GSI provides the technical framework (including protocols, services, and standards) to support grid computing with five security capabilities: user authentication, data confidentiality, data integrity, non-repudiation, and key management. GSI offers security features on transport and message level, by providing functionality to encrypt the complete communication or only the content of SOAP message. GSI also supports authentication through X.509 digital certificates, credential delegation and single sign-on.

VOMS is a grid technology which provides mechanism for user access to resources. It provides interfaces for users to apply for VO membership, for administrators to manage the users, and for other grid services to query stored information (user lists, roles, etc.).

SERVICE-ORIENTED GRID SECURITY MODELS

Requirements

Since service-oriented Grids follow the principles of service-orientation, security model of such kind of Grid has to cover additional requirements, specific for service-oriented architecture. Such requirements are:

- Interoperability – security solution has to deal with service interoperability;
- Manageability – security solution has to be eased to manage;
- Ease of development – security solution has to be easy to use.

OGSA

To standard these requirements, Global Grid Forum (GGF) – a community of users, developers, and vendors for grid computing standardization, developed OGSA.

The Open Grid Services Architecture (OGSA) [6] describes architecture for a service-oriented grid computing environment for business and scientific use. OGSA is based on Web service technologies, WSDL and SOAP. OGSA assure interoperability on heterogeneous systems so that different types of resources can communicate and share information. The goal of OGSA is to standardize practically all the services one commonly finds in a grid application (job management services, resource management services, security services, etc.) by specifying a set of standard interfaces for these services. OGSA defines a set of requirements that must be met by these standard interfaces.

OGSA represents three major logical and abstract tiers. The first tier includes the basic resources as CPUs, memory, disks, licenses and OS processes. These resources are usually locally owned and managed and they can high variability in their characteristics, quality of service availability. The second tier represents a high level of virtualization and logical abstraction. The service-oriented architecture of OGSA implies that virtualized resources are represented as services and that interaction with them can be initialized by any service from the architecture. The services from this tier need to use and manage resources from the bottom tier in order to deliver capabilities of individual service. At the third tier are the applications that use the OGSA capabilities to realize user functions and processes.

The OGSA security model is builds on Web services security with specific extensions for virtual organizations. WS-Security is used to allow service requests to provide suitable tokens, for purposes of authentication and authorization. For user authentication, delegation, and single sign-on, the OGSA uses the Grid Security Infrastructure (GSI) protocol. End-to-end message protection is provided by mechanisms such as XML encryption and digital signatures. OGSA security components supports, integrates and unify popular model, mechanism, protocols, platforms and technologies in a way that enables systems to interoperate securely.

EGA

Enterprise Grid Computing [7] is the next step in Grid evolution. It specifies the use of grid computing within the context of a business or enterprise. The accent in the Enterprise Grid Architecture is that a single organization is responsible for creating and managing a shareable networked pool of resources. Resources can be compute, network, storage and even service capabilities. Single organization is also responsible for composing higher-order components and services from individual resources and for delivering services that not only are capable of meeting a set of defined goals and requirements but also help drive value for the business.

EGA - Enterprise Grid Alliance is an open, vendor-neutral organization formed to develop enterprise grid solutions and accelerate the deployment of grid computing in enterprises. EGA promotes open, interoperable solutions, and best practices focusing exclusively on the needs of enterprise users. In addition, it addresses requirements for deploying commercial applications in a grid environment and focuses on reference models, security, and accounting.

The EGA reference model defines an Enterprise Grid as a collection of interconnected (networked) grid components under the control of a grid management entity. The Grid Management Entity (GME) is logical entity that manages the grid components, the relationships among them, and their entire life cycles. The grid component is defined as a superclass of object from which all of the components that are descended or derived. Components can be servers, network components and services. Grid components can be combined together into more sophisticated components. Components have security properties and attributes, and may define specific dependencies that can be used to support enforcement of security policies and to ensure minimal exposure.

Main security functions are covered by GME. It manages user identities and administrative roles, authentication of identities, authorization of actions taken by principals, access restrictions to the grid components, capture, storage, analysis, and reporting of audit-related events, key management, etc.

CONCLUSION

Grid is still new technology, mostly used in scientific circles, but it is also in stage of research in industrial circles. Service-oriented Grids evolve to next generation grid by establishing new standards and combining some known SOA and Grid security approaches.

REFERENCES

1. T. Erl, *Service-Oriented Architecture: Concepts, Technology, and Design*, 2005.
2. I. Foster and K. Kesselman, *The Grid: Blueprint for a New Computing Infrastructure*, 1999.
3. I. Foster and K. Kesselman, *What is the Grid? A Three Point Checklist*, 2002.
4. R. Kanneganti and P. Chodavarapu, *SOA Security*, Manning, 2008.
5. Y. Xiao, *Security in Distributed, Grid, Mobile and Pervasive Computing*, Auerbach, 2007.
6. I. Foster and K. Kesselman, *The Open Grid Service Architecture, Version 1.5*, 2006.
7. "Enterprise Grid Alliance – Reference models and use cases", online at
   `http://www.ogf.org/gf/docs/egadocs.php`

Variance Reduction MCMs with Application in Environmental Studies: Sensitivity Analysis

A. Karaivanova*, E. Atanassov*, T. Gurov*, R. Stevanovic[†] and K. Skala[†]

*IPP-BAS, Acad. G. Bonchev Str., bl. 25A, 1113 Sofia, Bulgaria
[†]CIC, RBI, 54 Bijenicka, Zagreb, Croatia

Abstract. This paper studies generator sensitivity of some variance reduction Monte Carlo methods (MCMs) with acceptance-rejection for approximate calculation of multiple integrals. This investigation is important basis for the development of the grid application Monte Carlo Sensitivity Analysis for Environmental Systems in the framework of the SEE-GRID-SCI project.

Monte Carlo are among the most widely used methods in real simulations. These methods can be considered as methods for computing an integral in the unit cube of an appropriate dimension, called the *constructive dimensionality* of the method.

Since its worst-case convergence rate of $O(N^{-1/2})$ does not depend on the dimension of the integral, Monte Carlo is sometimes the only viable method for a wide range of high-dimensional problems. Many studies show that the outcome of the simulation may be sensitive to the random generators being used, which means that obtaining unbiased estimates requires careful selection of the random generators. The random number generators based on physical events present an important option in this regard.

In this paper we study the sensitivity of several variance reduction Monte Carlo methods: importance sampling, smoothed importance sampling, weighted uniform sampling and crude Monte Carlo, to different type of generators: Quantum Random Bit Generator, pseudorandom generators and quasi-random sequences. Extensive numerical tests of several test integrals are presented.

Keywords: Monte Carlo method, variance reduction, sensitivity analysis, environmental systems, grid applications.
PACS: 05.10.Ln, 02.70.Uu, 02.60.Jh

INTRODUCTION

We consider the problem of approximate calculation of d-dimensional integrals of the form

$$\int_{[0;1]^d} f(x)dx.$$

For problems with a rather low dimensionality there are many good numerical integration rules. However, because of the curse of dimensionality, the number of points used by these quadrature formulae grows exponentially with increasing dimension. A way to overcome this problem is the use of Monte Carlo methods. In Monte Carlo integration, the integral is approximated by evaluating the function at N randomly chosen points $x_1, \ldots, x_N$ within the integration domain and then averaging over the function values. The major drawback with a Monte Carlo approach is that convergence behaves as $O(N^{-1/2})$. There are two possible ways to improve the convergence: (i) variance reduction, and (ii) change the choice of sequence.

CP1067, *Applications of Mathematics in Engineering and Economics '34—AMEE '08*, edited by M. D. Todorov
© 2008 American Institute of Physics 978-0-7354-0598-1/08/$23.00

Quasi-Monte Carlo methods (QMCMs) use quasirandom (also known as low-discrepancy) sequences instead of usual pseudorandom ones. The quasi-Monte Carlo method for integration in d-dimensions has a convergence rate of approximately $O((\log N)^d N^{-1})$.

We recall some basic concepts of QMCMs, [1]. First, for a sequence of N points $\{x_n\}$ in the d-dimensional half-open unit cube I^d define

$$R_N(J) = \frac{1}{N}\#\{x_n \in J\} - m(J)$$

where J is a rectangular set and $m(J)$ is its volume. Then define two discrepancies

$$D_N = \sup_{J \in E} |R_N(J)|, \; D_N^{\star} = \sup_{J \in E^{\star}} |R_N(J)|,$$

where E is the set of all rectangular subsets in I^d and $E^{\star}$ is the set of all rectangular subsets in I^d with one vertex at the origin.

The basis for analyzing QMC quadrature error is the Koksma-Hlawka inequality:

Theorem 1 (*Koksma-Hlawka, [1]): For any sequence $\{x_n\}$ and any function f of bounded variation (in the Hardy-Krause sense), the integration error is bounded as follows*

$$\left| \frac{1}{N} \sum_{n=1}^{N} f(x_n) - \int_{I^d} f(x)\,dx \right| \leq V(f) D_N^{\star}. \tag{1}$$

The star discrepancy of a point set of N truly random numbers in one dimension is $O(N^{-1/2}(\log \log N)^{1/2})$, while the discrepancy of N quasirandom numbers in d dimensions can be as low as $O(N^{-1}(\log N)^{d-1})$. Most notably there are the constructions of Hammersley, Halton, Soboí, Faure, and Niederreiter for producing quasirandom numbers. The theoretical properties of these point sets look promising, but are only valid asymptotically. Therefore, only when an almost infinite number of points is used, one can rely on the theoretical estimates. For smaller and more practical ranges of the number of points, there can be side- effects and the results with quasi-random points may not always be what the theory at infinity predicts.

In this paper we study the sensitivity of several variance reduction Monte Carlo methods: importance sampling, smoothed importance sampling, weighted uniform sampling and crude Monte Carlo, to different type of generators: Quantum Random Bit Generator, pseudorandom generators and quasi-random sequences. Extensive numerical tests of several test integrals are presented.

PROBLEM FORMULATION AND METHODS

Consider an integral on the unit cube $I^d = [0,1]^d$ in d dimensions,

$$I[f] = \int_{I^d} f(x)dx, \tag{2}$$

of a Lebesgue integrable function $f(x)$, and note that this integral can be expressed as the expectation of the function f,

$$I[f] = E[f(x)], \tag{3}$$

in which x is a uniformly distributed vector in the unit cube.

An empirical approximation to the expectation (3) is given with

$$I_N[f] = \frac{1}{N} \sum_{n=1}^{N} f(x_n), \tag{4}$$

if $\{x_n\}$ is a sequence sampled from uniform distribution. Equation (4) is called (crude) Monte Carlo quadrature formula. The integration error, defined as

$$\varepsilon_N[f] = |I[f] - I_N[f]|, \tag{5}$$

has a standard normal distribution, with expectation

$$E[\varepsilon_N^2] = \sqrt{\frac{\text{Var}(f)}{N}}. \tag{6}$$

Besides this, probabilistic, bound for the Monte Carlo integration error, an exact upper bound is given with Koksma-Hlawka inequality,

$$\varepsilon_N[f] \leq V[f]D_N^*, \tag{7}$$

in which V[f] in the variation of f in the Hardy-Krause sense and D_N^* is the discrepancy of the sequence. Equation (7) is valid for any bounded function and any choice of sequence.

Importance Sampling

Importance sampling is probably the most widely used Monte Carlo variance reduction method, [1]. One use of importance sampling is to emphasize rare but important events, i.e., small regions of space in which the integrand is large. One of the difficulties in this method is that sampling from the importance density is required. This difficulty can be overcome using acceptance-rejection method.

Importance sampling can increase the variance in some cases, [6]. In Hesterberg (1995, [2]) a method of defensive importance sampling is presented; when combined with suitable control variates, defensive importance sampling produces a variance that is never worse than the crude Monte Carlo variance, providing some insurance against the worst effects of importance sampling. Defensive importance sampling can however be much worse than the original importance sampling.

Owen and Zhow (1999) recommend an importance sampling from a mixture of m sampling densities with m control variates, one for each mixture component. In [5] it is shown that this method is never much worse than pure importance sampling from any single component of the mixture.

Importance Sampling with Acceptance-rejection

The problem with importance sampling is that it is very difficult to find importance function or to use it for sampling. In this case acceptance-rejection is used. On the other hand, acceptance-rejection may decrease the rate of convergence of the quasi-Monte Carlo variant of importance sampling because it breaks the smoothness. Than, smoothed acceptance rejection was introduced in [3].

If the integral (2) is rewritten as:

$$I[f] = \int_{I^d} \frac{f(x)}{h(x)} h(x) dx, \tag{8}$$

it can be interpreted as an integral with density function h. This density function should be chosen to match the behavior of f over the I^d, and yet to be either analytically, or easily numerically integrable. The variance is reduced as much as the f/h ratio is closer to a constant ($I[f]$). Now, the importance-sampled Monte Carlo estimate is written as:

$$I_N[f] = \frac{1}{N} \sum_{n=1}^{N} \frac{f(x_n)}{h(x_n)}, \ x_n \sim h(x), \tag{9}$$

where sample points are picked according to the probability density function $h(x)$.

To sample points according to h, the standard rejection method can be used. It is based on sampling from uniform distribution in $d+1$ dimensional cube. Only samples that lay below d-dimensional $h(x_{1:d})$ are accepted and variable x_{d+1} is used to make the decision, i.e., accepted points are: $x_{d+1} < \frac{h(x_{1:d})}{\gamma}$, where $\gamma \geq \sup_{x \in I^d} h(x)$.

Number of trial points is at least γ times the number of accepted points. And h has to be evaluated for all of them. However, this additional overhead is usually acceptable, since importance sampling method shall emphasize small regions of integration space and improve overall accuracy/performance of Monte Carlo integration [3].

Importance Sampling using Smoothed Rejection Method

The standard rejection method formulated above introduces discontinuities under integral (8) due to binary accept/reject decisions. Associating acceptance weights to trial points is done in smoothed rejection method. Weight function is centered at $\frac{h(x_{1:d})}{\gamma}$, and it is usually piece-wise linear, in which case it can be described with the width of transitional part, parameter δ.

Monte Carlo estimate for importance sampling with smoothed rejection method is given with:

$$I_N[f] = \frac{1}{N} \sum_{n=1}^{N^*} w_n \frac{f(x_n)}{h(x_n)}, \ (x_n, w_n) \sim h(x), \tag{10}$$

where N^* is the number of sample points required to reach total acceptance weight of N, due to acceptance points with weight less than 1. This fact also introduces some

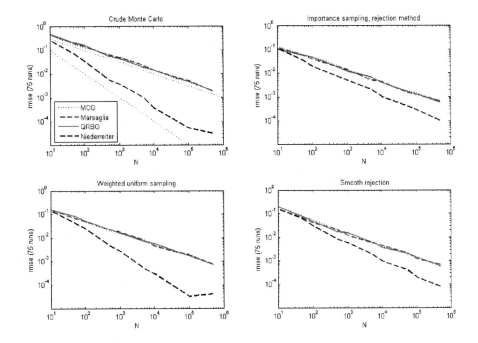

FIGURE 1. Sensitivity of Monte Carlo methods to the selection of a random number generator, experiment with f_1

additional work, which mostly can be considered small enough, compared to complexity of pure function evaluations [3].

NUMERICAL EXPERIMENTS

Two integration problems (2) were used to compare the performance of four Monte Carlo methods described above:

- Crude Monte Carlo,
- Importance sampling with acceptance-rejection method,
- Weighted uniform sampling,

  Monte Carlo weighted uniform sampling estimate is given with:

$$I_N[f] = \frac{\sum\limits_{n=1}^{N} \frac{f(x_n)}{h(x_n)} \frac{h(x_n)}{\gamma}}{\sum\limits_{n=1}^{N} \frac{h(x_n)}{\gamma}} = \frac{\sum\limits_{n=1}^{N} f(x_n)}{\sum\limits_{n=1}^{N} h(x_n)}. \tag{11}$$

The resulting estimate is biased, however it has been shown ([3]) that this bias is negligible as $N \to \infty$. Also, positive correlation between f and h reduces variation. One advantage of this method is that sample points do not need to be generated for density h, but on the other hand, a disadvantage is a long evaluation time spent in regions of low importance (for particular functions f).

- Importance sampling with smoothed rejection method

versus four random number generators (which cover random, pseudo-random and quasi-random classes of sequences):

- Multiplicative Congruential Generator (MCG),
- Marsaglia's generator,
- Quantum Random Bit Generator (QRBG), and
- Niederreiter sequence generator.

Each experiment had 75 runs. Experiments were coded and performed in MATLAB. MCG and Marsaglia generators were readily available in MATLAB, while we used Niederreiter generator from [4]. It was originally written in C++, but we wrapped it in a `mex` function to enable easier usage from MATLAB environment. QRBG sequence (of 10^9 double precision numbers) was (pre)downloaded and stored on local disk to enable faster simulations. Results are given on 1 through 4.

Test Integrals

The first experiment involved integration over $I^5 = [0,1]^5$ of the function

$$f_1(x) = \exp\left(\sum_{i=1}^{5} a_i x_i^2 \; \frac{2 + \sin\left(\sum_{j=1, j \neq i}^{5} x_j \right)}{2} \right) \tag{12}$$

using the importance function (probability distribution)

$$h_1(x) = \frac{1}{\eta} \exp\left(\sum_{i=1}^{5} a_i x_i^2 \right) \tag{13}$$

where $a = [1, \frac{1}{2}, \frac{1}{5}, \frac{1}{5}, \frac{1}{5}]$ and η is calculated so as to normalize the h_1.

The second experiment involved integration over $I^7 = [0,1]^7$ of the function

$$f_2(x) = \exp\left(1 - \sum_{i=1}^{3} \sin^2\left(\frac{\pi}{2} x_i \right) \right) \times \arcsin\left(\sin(1) + \frac{x_1 + \dots + x_2}{200} \right) \tag{14}$$

554

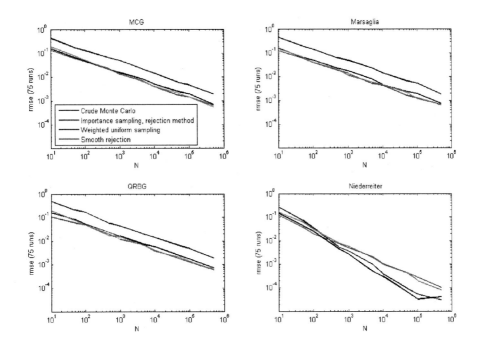

FIGURE 2. Performance comparison of Monte Carlo methods (for several random number generators), experiment with f_1

using the importance function (probability distribution)

$$h_2(x) = \frac{1}{\eta} \exp\left(1 - \sum_{i=1}^{3} \sin^2\left(\frac{\pi}{2}x_i\right)\right) \qquad (15)$$

where η is calculated so as to normalize the h_2.

The Results

The results are shown on Figures 1 through 4, using the following random generator: Multiplicative Congruential Generator (MCG) and Marsaglia's generator [8], Niederreiter sequences [4], and Quantum Random Bits Generator (QRBG) Service [7] . Even in case of acceptance rejection, quasi-Monte Carlo have better convergence than Monte Carlo with random and pseudorandom sequences. Also note that smoothed rejection managed to improve the convergence rate for both, Monte Carlo and quasi-Monte Carlo.

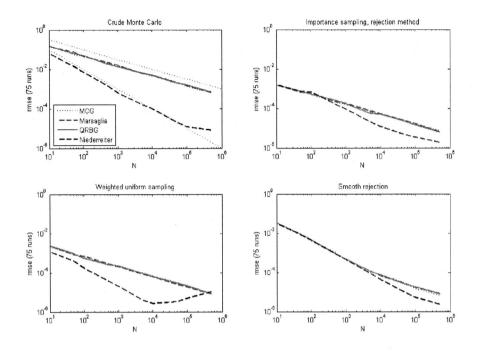

FIGURE 3. Sensitivity of Monte Carlo methods to the selection of a random number generator, experiment with f_2

CONCLUDING REMARKS AND FUTURE WORK

This study is the first part of the theoretical work for the Grid application: *Monte Carlo Sensitivity Analysis for Environmental Systems* (MCSAES). The inherent inaccuracy of the underlying physical and chemical models makes sensitivity analysis an important part of the scientific research in this field. With the Sobol sensitivity analysis [9] we obtain quantitative information about the sensitivity of the model to changes in the parameters. In our case the computational cost of each function evaluation is high because it is related to solving systems of millions of equations. This makes the comparative study of Monte Carlo and quasi-Monte Carlo methods for computing the underlying high-dimensional integrals ever more important. In any case such problems are good candidates for execution in the Grid environment, although the use of quasi-Monte Carlo or true random number generators poses certain technical challenges, related to the parallelization of the computations or the repeatability of results. This work is the base for future authors research in the field of MC sensitivity analysis for environmental systems.

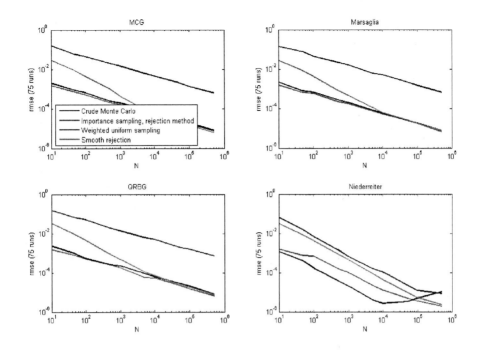

FIGURE 4. Performance comparison of Monte Carlo methods (for several random number generators), experiment with f_2

ACKNOWLEDGMENTS

This work is partially supported by the NSF of Bulgaria through grant I-1405/04 and by the European Commission under the FP7 Research Infrastructure, SEE-GRID-SCI project, contract No. 211338.

REFERENCES

1. R. E. Caflisch, "Monte Carlo and quasi-Monte Carlo methods," *Acta Numerica*, Cambridge University Press, 1998, pp. 1–49.
2. T. Hesterberg, *Technometrics* **37**(2), 185–194 (1995).
3. B. Moskowitz and R. E. Caflisch, *Math. Comp. Modeling* **23**, 37–54 (1996).
4. H. Niederreiter, *Random Number Generations and Quasi-Monte Carlo Methods*, SIAM, Philadelphia, 1992.
5. A. Owen and Y. Zhou, Safe and effective importance sampling, Technical report, Stanford University, Statistics Department, 1999.
6. E. Veach and L. J. Guibas, "Optimally combining sampling techniques for Monte Carlo rendering," in *Computer Graphics Proceedings, Annual Conference Series, ACM SIGGRAPH 95*, pp. 419–428, 1995.
7. R. Stevanović, G. Topić, K. Skala, M. Stipčević, and B. M. Rogina, "Quantum Random Bit Generator

Service for Monte Carlo and Other Stochastic Simulations," *LNCS*, Vol. 4818, Springer-Verlag (2008), pp. 508–515.

8. G. Marsaglia and A. Zaman, *Annals of Applied Probability* **3**, 562–480 (1991).

9. I.M. Sobol', *Mathematics and Computers in Simulation* **55**(1-3), 271–280 (2001).

Usage of the UNICORE Grid Technology in Scientific and Economic Domains

M. Riedel, D. Mallmann and A. Streit

*Institute for Advanced Simulation, Jülich Supercomputing Centre, Forschungszentrum Jülich,
Leo-Brandt-Str. 1, D-52425 Jülich, Germany*

Abstract. In the past years, many scientific and economic-related applications from various domains have taken advantage of Grid infrastructures that share storage or computational resources such as supercomputers or clusters across multiple organizations. Especially within Grid infrastructures driven by high-performance computing (HPC) such as the Distributed European Infrastructure for Supercomputing Applications (DEISA), the UNICORE Grid middleware has become an important tool to seamlessly access distributed resources by providing strong security and workflow capabilities. This paper highlights different usage models of UNICORE from a wide variety of scientific applications, and provides insights of approaches in economic-related scenarios with UNICORE.

Keywords: UNICORE, grid computing, supercomputing, PERMAS, NBODY, PEPC, Fluent
PACS: 01.50.hv, 04.70.-s

INTRODUCTION

The UNICORE[1] Grid technology has been developed since 1997 and provides seamless and secure access to computational Grid resources with a particular focus on supercomputers and clusters [1]. More recently, we faced new challenges since increasing complexity of Grid applications that embrace multiple physical models and consider a larger range of scales is creating a steadily growing demand for compute power. Also, more and more business use cases from economic domains seem to play more and more a crucial role in the sustainability of Grid infrastructures where UNICORE is deployed and used, for example DEISA and the German national Grid D-Grid[2].

In this work we discuss several scientific usage models of UNICORE that take advantage of recently deployed large-scale supercomputers (i.e., JUGENE at Forschungszentrum Jülich with 65536 CPUs.) in the DEISA Grid and emerging business use cases out of economic domains.

This paper is structured as follows. After a short introduction into the most recent challenges in the context of the Grid middleware UNICORE, Section 2 introduces UNICORE and its architecture. Section 3 provides a snapshot of a couple of well-known scientific domains that use UNICORE in conjunction with HPC-driven Grids. In Section 4, we shortly give insights how UNICORE is used in emerging economic domains. Finally, this paper ends with a survey of related work and concluding remarks.

---

[1] http://www.unicore.eu
[2] http://www.d-grid.org

CP1067, *Applications of Mathematics in Engineering and Economics '34—AMEE '08*, edited by M. D. Todorov
© 2008 American Institute of Physics 978-0-7354-0598-1/08/$23.00

THE UNICORE GRID TECHNOLOGY

The UNICORE Grid middleware has been developed since the late 1990s to support distributed computing applications in Grid infrastructures with a particular focus on massively parallel HPC applications. The vertically integrated design provides components on each tier of its layered architecture as shown in Figure 1. The UNICORE 5 architecture [2] mainly consists of proprietary components and protocols, while the more recently developed UNICORE 6 is based on open standards such as the Web services resource framework (WS-RF)[3]. It thus implements the concepts of Service-Oriented architectures (SOAS). UNICORE 6 conforms to the Open Grid Services Architecture (OGSA) and numerous projects such as the OMII-Europe project[4] augmented it with open standards during the last couple of years.

The loosely coupled Web services technology provides a perfect base to meet the common use case within Grid that conform to OGSA thus allowing dynamic interaction dispersed in conceptual, institutional, and geographical space. The different scientific and economic usage models presented in this paper take advantage of this service-oriented concept by using key characteristics of the UNICORE Grid middleware. These characteristics are basic job submission and data management functionalities using the UNICORE Atomic Services (UAS) [3], and the workflow capabilities of the UNICORE workflow engine and service orchestrator as shown in Figure 1.

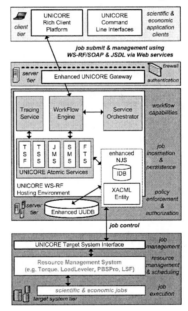

FIGURE 1. Web service based architecture of the UNICORE 6 Grid technology

[3] http://www.oasis-open.org/committees/tc_home.php?wg_abbrev=wsrf

[4] http://www.omii-europe.org

In more detail, the UAS consists of a set of core services such as the TargetSystem-Service (TSS) that represents a computational resource (i.e., supercomputers, clusters, etc.) in Grids. While the TargetSystemFactory (TSF) can be used to create an end-user specific instance of a TSS, the TSS itself is able to execute jobs defined in the Job Submission and Description Language (JSDL) [4]. Each submitted job is represented as one resource aligned with the TSS and controlled with the Job Management Service (JMS). In addition, different flavors of the FileTransferServices (FTS) are used for data transfer between remote sites and the local UNICORE client of the end-user.

While the UAS and its services (i.e., TSS, JMS, FTS, etc.) operate on the Web service level, they are supported by a strong execution backend named as the enhanced Network Job Supervisor (XNJS). The XNJS uses the Incarnation Database (IDB) to map abstract job descriptions to system-specific definitions. In terms of security, the UAS are protected via the UNICORE Gateway [5], which performs the authentication of end-users. This means it checks whether the certificate of an end-user has been signed by a trusted Certificate Authority (CA), that its still valid, and not revoked.

In terms of authorization the NJS relies on the UNICORE User DataBase (UUDB) service and checks policies based on Extensible Access Control Markup Language (XACML) [6]. The roles and general attributes of end-users as well as their group and project membership are encoded using Security Assertion Markup Language (SAML) [7] assertions, which are typically released by an Attribute Authority (AA) initially contacted from the UNICORE clients.

Typically, end-users use the UNICORE middleware to reduce their management overheads in complex scientific or business process workflows to a minimum or to automize parts of their workflow. Without Grid middleware, the workflows often start with the manual creation of a Secure Shell (SSH) connection to a remote system such as a supercomputer via username and password or keys configured before. The trend towards service-oriented Grids allows for more flexibility and orchestration of services and thus makes it easier to enable end-users with multi-step workflow capabilities. While there are many workflow-related tools in Grids (e.g., TAVERNA [8], ASKALON [9], TRIANA [10]), we shortly describe the functionalities of UNICORE that are often used in scientific and economic usage models.

To satisfy scalability requirements, the workflow capabilities of UNICORE are based on a two layer architecture. The Workflow Engine is based on the Shark open-source XPDL engine while plug-ins allow for domain-specific workflow language support. The Workflow engine is responsible to take Directed Acyclic Graph (DAG)-based workflow descriptions and map them to specific workflow steps. The Service Orchestrator on the other hand, is responsible for the job submission and monitoring of these workflow steps. While the execution site can be chosen manually, the orchestrator also supports brokering based on pluggable strategies.

All of the above described services can be seamlessly accessed using the graphical UNICORE Rich Client based on Eclipse. End-users are able to create workflows of different services that typically invoke different applications including necessary data staging elements. In this context, the Tracing Service of UNICORE allows for monitoring of each execution step of a UNICORE workflow, which can be easily followed in the client. In addition, several other clients can be used with UNICORE such as a command line client or a more recently developed portal client usable within Web browsers.

UNICORE IN SCIENTIFIC DOMAINS

N-body problems appear in many scientific areas such as plasma-physics, molecular dynamics, fluid dynamics, and astrophysics. One of these applications is the Nbody6++ program [11], which is a parallel version of the *Aarseth-type N-body code nbody6* suitable for N-body simulations on massively parallel resources like the ones accessible in DEISA. UNICORE is used to submit several sub-versions of the main code Nbody6++ that have been developed for applications in different areas of astrophysical research. These areas include dynamics of star clusters in galaxies and their centres, formation of planetary systems, and dynamical evolution of platetesimals.

Another example of N-body problem applications used with UNICORE are several routines of the Pretty Efficient Parallel Coulomb Server (PEPC) [12] in the area of plasma-physics. This massively parallel code uses the *Barnes Hut hierarchical tree algorithm* to perform potential and force summation of n charged particles in a time O(n log n), allowing mesh-free particle simulation on length- and time-scales usually possible only with particle-in-cell or hydrodynamic techniques.

The mentioned applications above use UNICORE for plain job submissions to computational Grids by submitting simple UNIX-style scripts that call the respective executables on the target systems. In addition to this, UNICORE offers a wide variety of control functionalities such as Do-N and Do-Repeat loops, Hold Job, and If-then-else that can be used in conjunction with UNICORE Grid jobs. One example that is using, for instance, the Do-N loop comes out of the field of hydrodynamics, which is related to the research of liquids in motion. In one of its sub-disciplines of fluid dynamics, UNICORE is used for hydrodynamics of active biological systems such as sperm [13].

The scientific goal is to study sperm cluster size dependence for 2D and 3D systems. In this context, experiments have revealed an interesting swarm behavior of sperm, when the sperm concentration of the system is high. But the mechanism behind the experimental phenomenon is still not clear, even after computational intensive 2D simulations. Thus the researchers use UNICORE to submit increased intensive 3D simulation jobs to HPC-driven Grids. The Do-N loop is intensively used to realize a circle of dependencies starting from the initial setup of the sperm positions, the hydrodynamics code computation, output generation, which again is the input for the next loop iteration. The hydrodynamics code in this UNICORE application is a *multi-particle-collision Dynamics (MPC) code.*

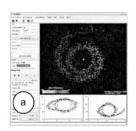 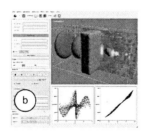 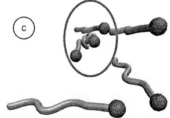

FIGURE 2. Usage of UNICORE in astro-physics (a), plasma-physics (b), and hydrodynamics (c)

UNICORE is also used in chemical domains, for instance within the Chemomentum project[5]. In this project, the strong UNICORE workflow capabilities are used in the context of the current European regulatory framework for chemical research. The goal of this so called Registration, Evaluation, Authorization and Restriction of Chemical substances (REACH) framework is to improve the protection of the human health and the environment through characterization of intrinsic properties of chemicals.

In the Chemomentum project, computational methods, such as *Quantitative Structure-Activity Relationships (QSAR) workflows* are used in the context of REACH [14]. The UNICORE QSAR workflows combine different QSAR applications in workflows, realize the access to existing databases, and finally provide results with their documentation. In more detail, the complex scientific workflow starts with a query from database to retrieve structure and toxicity data, which in turn are preprocessed in order to accomplish certain data selection and unification steps. Afterwards, a 3D and conformational space analysis is performed leading to the workflow steps descriptor calculations and thus the model building and prediction. Finally, results and statistics are available as workflow output data.

Another interesting scientific domain where UNICORE is used is the in silico drug discovery. The fundamental goal of this domain is to reduce the overall costs for drugs significantly by using computational techniques. In this context, UNICORE is used in the Wide In Silico Docking On Malaria (WISDOM) project[6], which is developing new drugs for neglected and emerging diseases with a particular focus on malaria. In the context of this project, domain scientists use UNICORE and gLite together in conjunction with the two European Infrastructures DEISA and Enabling Grid for e-Science (EGEE) [7].

In more detail, the scientists use a GridSphere portal to submit via gLite AutoDock[8] and FlexX[9] jobs to the EGEE infrastructure, which is mostly a High Throughput Computing (HTC) oriented Grid. Here, AutoDock and FlexX are used for virtual screening by docking, which predicts how small molecules bind to a receptor of known 3D structures. The output of these docking calculations is a list of best chemical compounds that might be potential drugs. The next step where UNICORE comes into play is to refine this best compound list via molecular dynamics (MD) on the HPC-driven DEISA infrastructure. In this context, the fast and highly scalable HTC calculations use the Assisted Model Building with Energy Refinement (AMBER) package[10], which is deployed on DEISA. All in all, UNICORE is used as part of the overall process to accelerate the drug discovery by using HTC resources of EGEE in conjunction with HPC resources in DEISA.

---

[5] http://www.chemomentum.org

[6] http://wisdom.healthgrid.org

[7] http://public.eu-egee.org/

[8] http://autodock.scripps.edu

[9] http://www.biosolveit.de/FlexX

[10] http://amber.scripps.edu

UNICORE IN ECONOMIC DOMAINS

Different from the approach of using UNICORE in scientific application domains are economic scenarios for example in engineering domains. In such scenarios, the usage of many applications is restricted through licenses to dedicated local hosts, and thus could not be run on a Grid with geographically dispersed resources. Therefore, the SmartLM project[11] develops a licensing architecture that provides licenses as Grid services based on service level agreements (SLAs) and integrates the solutions into the Grid middleware UNICORE. In this context, three major engineering software vendors provide their software in the Grid. First the LMS company use their software OPTIMUS, which advances automated 1D simulation processes to facilitate quick assessment of multiple design options. Second, INTES provides their general purpose finite element analysis system PERMAS. Finally, ANSYS offers their CAE tools ANSYS FEM, ANSYS CFX, and Fluent. Therefore, the economic scenarios of the SmartLM consortium demand a solution that leads to new business models.

UNICORE is also used in other economic domains such as in industrial environments. To provide one example T-Systems is using UNICORE as access method to the $C^2A^2S^2E^2$ HPC-Cluster. User communities are users from the German Space Agency DLR as well as T-Systems internal network users.

A slightly different economic domain where UNICORE is used is within the Business Experiments in Grid (BeinGrid) project, which focuses on solutions that perform a critical transition of Grid technologies from research and academic use to wider adoption by business and enterprise. Within BeinGrid UNICORE is used in business experiments related to ship building and a textile Grid Portal just to list but a few.

Finally, Figure 3 summarizes the use of UNICORE in scientific and economic domains.

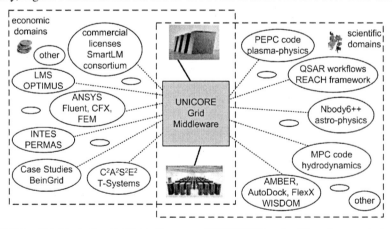

FIGURE 3. This illustration highlights the usage of UNICORE in scientific and economic domains

---

[11] http://www.smartlm.eu/

564

RELATED WORK

Related work in the field is clearly the wide variety of Grid middleware systems that are around today. Although many of them certainly reached the production quality, we only provide links here to the most known ones.

The Globus Toolkit 4 [15], for instance, is also considered to be a Grid middleware, but compared to the approach of UNICORE, it is rather a toolkit, which provides services for developers that integrate them in certain scientific and economic domain applications. UNICORE, on the other hand provides a more end-to-end solution with a wide variety of clients with many functionalities. However, in contrast to UNICORE, Globus Toolkits are able to work together with the well known Swedish Grid Accounting System (SGAS) [16], which provides a very good foundation for economic usage models. Globus is used in many scientific domains, and through the UNIVA company also more and more in economic domains.

Another well known related work in the field, is the gLite Grid middleware [17] used in the EGEE Grid. gLite differs from UNICORE mainly in his focus on HTC and brokering mechanisms. UNICORE on the other hand is clearly driven by HPC needs, which implies different usage models such as choosing in particular the right site for a job without using any brokering mechanism. Like UNICORE, gLite is used in many scientific domains, especially related to EGEE, and also provides an interesting accounting and billing system usable for economic usage models named as the Distributed Grid Accounting System (DGAS) [18].

CONCLUSIONS

In this paper we shortly described the most recent architecture of the UNICORE Grid technology, which is used in numerous scientific and economic domains. Although Grid middleware like UNICORE, as well as Globus, and gLite, has been deployed in numerous infrastructures over the last couple of years, the most use cases where always clearly driven by demands of scientific domains. More recently, business and economic models are taken more into account by Grid middleware providers since the large scientific infrastructures and technologies have to be sustained without public funding agencies moneys, which implies involvement from commercial players.

We conclude that being very efficient in scientific domains, several features of Grid middleware still lack important features in terms of production quality components required in economic domains. Pricing, billing, and accounting, for instance, are research topics that are still not broadly considered to be important although highly demanded by commercial service providers in many business cases (i.e., pay per usage payment models). Thus future work in UNICORE will be also related to improvements in this area mainly through the adoption of open standards in this field and the bridging to well-known accounting and billing systems that exist today.

Finally, we also conclude, that the first steps into the right directions are already undertaken by several projects. Most notably, the BeinGrid project is certainly very interesting for upcoming economic models and use cases that can become attractive to service providers in the commercial area.

ACKNOWLEDGEMENTS

The work described in this paper is based on work of many talented colleagues of in the UNICORE community that developed the UNICORE Grid Technology. We also acknowledge the numerous projects that contribute to the success of UNICORE 6 such as the European SmartLM Project (IST-216759), Chemomentum Project (IST-033437), and BeinGrid Project (IST-034702).

REFERENCES

1. A. Streit et al, "UNICORE - From Project Results to Production Grids.," in *Grid Computing: The New Frontiers of High Performance Processing, Advances in Par. Comp. 14*, Elsevier, 2005, pp. 357–376.
2. A. Streit, D. Erwin, T. Lippert, D. Mallmann, R. Menday, M. Rambadt, M. Riedel, M. Romberg, B. Schuller, and P. Wieder, "UNICORE - From Project Results to Production Grids.," in *Grid Computing: The New Frontiers of High Performance Processing, Advances in Parallel Computing 14*, edited by L. Grandinetti, Elsevier, 2005, pp. 357–376.
3. M. Riedel and D. Mallmann, "Standardization Processes of the UNICORE Grid System," in *Proc. of 1st Austrian Grid Symposium 2005, Schloss Hagenberg, Austria*, Austrian Computer Society, 2005, pp. 191–203.
4. A. Anjomshoaa, M. Drescher, D. Fellows, S. McGougha, D. Pulsipher, and A. Savva, *Job Submission Description Language (JSDL) - Specification Version 1.0*, Open Grid Forum Proposed Recommendation, 2006.
5. R. Menday, "The Web Services Architecture and the UNICORE Gateway," in *Proc. of the International Conference on Internet and Web Applications and Services (ICIW) 2006, Guadeloupe, French Caribbean*, 2006.
6. T. Moses et al, *eXtensible Access Control Markup Language*, OASIS Standard, 2005.
7. S. Cantor, J. Kemp, R. Philpott, and E. Maler, *Assertions and Protocols for the OASIS Security Assertion Markup Language*, OASIS Standard, 2005, http://docs.oasis-open.org/security/saml/v2.0/
8. TAVERNA Website (2008), http://taverna.sourceforge.net/
9. ASKALON Website (2008), http://www.dps.uibk.ac.at/projects/askalon/
10. TRIANA Grid Workflow Website (2008), http://www.grid.org.il
11. R. Spurzem, "Direct N-body Simulations," in *Computational Astrophysics, The Journal of Computational and Applied Mathematics (JCAM) 109*, edited by L. Grandinetti, Elsevier, 1999, pp. 407–432.
12. P. Gibbon, *Short Pulse Laser Interactions with Matter: An Introduction*, Imperial College Press/World Scientific, London/Singapore, 2005, ISBN 1-86094-135-4.
13. J. Elgeti et al, *Hydrodynamics of Active Mesoscopic Systems*, NIC Series 39, 53, 2008, ISBN 978-3-9810843-5-1.
14. Casalegno et al, *Abstracts of QSAR-related Publications: Multivariate Analysis, QSAR & Combinatorial Science*, Wiley, 2007.
15. I. Foster, "Globus Toolkit version 4: Software for Service-Oriented Science," in *Proceedings of IFIP International Conference on Network and Parallel Computing, LNCS 3779*, Springer-Verlag, 2005, pp. 213–223.
16. T. Sandholm et al, *A service-oriented approach to enforce grid resource allocation*, 2006.
17. E. Laure, et al, "Programming The Grid with gLite," in *Computational Methods in Science and Technology*, Scientific Publishers OWN, 2006, pp. 33–46.
18. R. Piro et al, "An Economy-based Accounting Infrastructure for the DataGrid," in *Proc. of the 4th Int. Workshop on Grid Comp*, 2006.

AUTHOR INDEX

A

Akbay, Ü., 483
Allison, D., 187
Alshabani, I., 527
Angelov, I., 196, 209
Angelova, I. T., 305
Ankilov, A. V., 414
Antonova, A. O., 333
Atanassov, E., 549
Avramska, S., 341

B

Baeva, S., 451
Barcia, O. E., 131
Berk, E., 483
Bijev, G., 503
Bogdanov, A. Yu., 352
Bonnaud, G., 145
Bouche, D., 145
Boyadzhiev, D., 458
Brossier, F., 271

C

Christou, M. A., 105
Christov, C. I., 3, 122

D

Dankov, D., 239
Deliiska, B., 491
Didenko, A., 187
Dimitrov, D. T., 28
Dimitrov, S., 467
Dimitrova, M. B., 361
Djulgerova, R., 279
Donev, V. I., 361
Douma, M., 196, 209
Drozdowicz, M., 527

F

Ferreira, F. A., 313, 321
Ferreira, Fl., 321
Feuillebois, F., 15

G

Ganzha, M., 527
Georgiev, G. H., 373
Gilev, B., 244
Gocheva-Ilieva, S. G., 114, 475
Goranova, R., 541
Gürler, Ü., 483
Gurov, T., 549

H

Hinov, N., 510

I

Iliev, I. P., 114, 475
Ilison, L., 155

J

Jovanovic, B., 253

K

Kandilarov, J., 253
Karaivanova, A., 549
Kazakova, Yu. A., 427
Kojouharov, H. V., 28
Koleva, M. N., 262
Komarevska, L., 451
Konstantinov, M. M., 38
Koperski, J., 279
Kovacheva, S., 467
Kraev, G., 510
Kuranowski, W., 527